Introduction to Fire Safety Management

Introduction to Fire Safety Management

Andrew Furness CFIOSH, GIFireE, Dip$_2$OSH, MIIRSM, MRSH
Martin Muckett MA, MBA, CMIOSH, MIFireE, Dip$_2$OSH

AMSTERDAM • BOSTON • HEIDELBERG • LONDON • NEW YORK • OXFORD
PARIS • SAN DIEGO • SAN FRANCISCO • SINGAPORE • SYDNEY • TOKYO
Butterworth-Heinemann is an imprint of Elsevier

Butterworth-Heinemann is an imprint of Elsevier
Linacre House, Jordan Hill, Oxford OX2 8DP
30 Corporate Drive, Suite 400, Burlington, MA 01803

First edition 2007

Notice
No responsibility is assumed by the publisher for any injury and/or damage to persons or property as a
matter of products liability, negligence or otherwise, or from any use or operation of any methods, products,
instructions or ideas contained in the material herein. Because of rapid advances in the medical sciences, in
particular, independent verification of diagnoses and drug dosages should be made

British Library Cataloguing in Publication Data
A catalogue record for this book is available from the British Library

Library of Congress Cataloging-in-Publication Data
A catalog record for this book is available from the Library of Congress

For information on all Butterworth-Heinemann publications
visit our web site at http://books.elsevier.com

ISBN: 978 0 7506 8068 4

Typeset in 9.5/12 pts Helvetica by Charon Tec Ltd (A Macmillan Company), Chennai, India
www.charontec.com
Printed and bound by MKT Print d.d., Slovenia

07 08 09 10 10 9 8 7 6 5 4 3 2 1

Contents

Contents

Contents

Preface

Introduction to Fire Safety Management has been produced for all students taking the NEBOSH Fire Safety & Risk Management Certificate, whether as part of a face to face training course or as part of a distance learning programme. It will also be of significant use for those undertaking a programme of study for Level 3 and 4 S/NVQ in Fire Safety. The book is the approved reference material for those undertaking IOSH certified Fire Risk Assessment, Principles and Practice programmes and the Fire Safety Management programmes at the Institute of Occupational Safety & Health (IOSH).

This book has been produced to provide those establishing fire safety management systems within their workplace and those undertaking fire risk assessments, on behalf of the responsible person, with an all encompassing reference book without the need to initially access the huge range of British and European Standards in relation to fire and risk management.

The material included within the text effectively covers both the National General Certificate (NGC1) material and the Fire Safety & Risk Management (FC1 & 2) materials so that those studying for the full NEBOSH programme do not have to purchase two text books. It is also useful for those who have not undertaken a course of study to have reference to general safety management principles and arrangements, as many fire safety books fail to address the management principles and detail only physical fire safety systems (fire alarm, suppression, detection, etc.).

Given that the materials included within this text book also cover previous MCI units for the Emergency Fire Service Management (EFSM) S/NVQ, the material within this book will also assist those serving in the emergency Fire & Rescue Services as part of their development programmes, particularly those starting within the Fire Safety and Community Fire Safety areas.

We hope that you find the *Introduction to Fire Safety Management* both a useful reference book for your course of study and a source of reference when undertaking fire risk assessments and establishing fire safety management systems.

Andrew Furness
Martin Muckett
July 2007

Acknowledgements

Throughout the book, definitions used by the relevant legislation, the HM Government Guides, British Standards, the Health and Safety Commission, the Health and Safety Executive and advice published in Approved Codes of Practice or various Health and Safety Commission/Executive publications have been utilised.

At the end of each chapter, there are some example examination questions, some of which have been taken from recent NEBOSH Fire Safety & Risk Management Certificate papers and a number of which have been produced by the authors as examples of possible questions covering the topic discussed. Some of the questions may include topics which are covered in more than one chapter. The answers to these questions are to be found within the preceding chapter of the book. NEBOSH publishes an examiners' report after each public examination which gives further information on each question. Most accredited NEBOSH training centres will have copies of these reports and further copies may be purchased directly from NEBOSH although as this is a relatively new course previous questions are currently limited. The authors would like to thank NEBOSH for giving them permission to use these questions.

The authors extend their gratitude for the assistance in completing their book to all the staff of Salvus Consulting. In particular to Kim and Anne who provided support throughout the project. Their assistance with gathering information and permissions for the use of the included photographs and figures was a significant contribution to the quality of the book.

We would also like to extend heartfelt thanks to Anne Black for the sterling work that she undertook in assisting in the production of Chapter 15 – The Summary of Key Legal Requirements.

About the authors

Andrew Furness is a charismatic safety professional with over 25 years' experience in both fire safety and risk management and health and safety fields. As a Fire Safety Enforcing Officer for a local Fire Authority his role was changed when he became the Health & Safety Advisor to the Fire Brigade in Buckinghamshire.

Joining IOSH in 1996 Andrew took on a number of roles within his Branch at Thames Valley before becoming the Vice Chairman of IOSH Fire Risk Management Specialist Group. In the Vice Chair's role he acted as Chairman of the Working Party that developed the NEBOSH Fire Safety & Risk Management Certificate syllabus, which was based upon the CPD fire programme produced for IOSH members.

As Managing Director of Salvus Consulting Limited he has an active involvement leading his team delivering the new NEBOSH Fire Certificate programme. Andrew is a Chartered Fellow of the Institute of Occupational Safety and Health and a Graduate Member of the Institution of Fire Engineers (IFE).

Martin Muckett has Masters Degrees in Business Administration and Local Government Management. He is been an active Member of the Institute of Fire Engineers since 1986, and was awarded the NEBOSH Diploma in 1998. He is a chartered member of the Institute of Occupational Safety and Health.

Martin has a unique experience in both fire and health and safety management. He has nearly 30 years' experience in the local authority Fire service and reached the rank of Assistant Chief Fire Officer before retiring in 2003. In 1998, Martin was appointed to Her Majesty's Fire Service Inspectorate. As Principal Health and Safety Advisor, he authored and/or edited all national fire service guidance and led the development of a suite of Home Office publications on Health and Safety for the Fire Service. He has an excellent reputation here and abroad where he lectures on health and safety risk management

He now lives and works in the Middle East for part of the year. When in the UK, he is a senior consultant for Salvus Consulting Ltd, providing health and safety management services to a variety of public and private organisations.

Illustrations credits

Front cover	Courtesy of London Fire Brigade.
Figure 1.1	Courtesy of Hereford & Worcester Fire and Rescue Service.
Figure 1.2	Courtesy of News Group International.
Figure 1.7	Cover of *Workplace Health, Safety and Welfare: Approved Code of Practice and Guidance* (HSE Books, 1996), ISBN 0717604136. © Crown Copyright material is reproduced with the permission of the Controller of HMSO and Queen's Printer for Scotland.
Figure 1.8	Cover of *Fire Safety – Risk Assessment: Factories and Warehouses* (Department for Communities and Local Government, 2006), ISBN 1851128166. © Crown Copyright material is reproduced with the permission of the Controller of HMSO and Queen's Printer for Scotland.
Figure 1.17	Adapted from the Red Guide, *Code of Practice for Fire Precautions in Factories, Offices, Shops and Properties not required to have a Fire Certificate* (Stationery Office Books, 1989), ISBN 0113409044. © Crown copyright material is reproduced with the permission of the Controllers of HMSO and Queen's Printer for Scotland.
Figure 1.19	Courtesy of EquiLift Limited.
Figure 1.25	Courtesy of Hereford & Worcester Fire and Rescue Service.
Figure 1.27	Courtesy of Hereford & Worcester Fire and Rescue Service.
Figure 1.30	Reproduced from *Guidelines on Occupational Safety and Health Management Systems,* ILO OSH 2001.
Figure 1.31	Adapted from HSG 65 *Successful Health and Safety Management* (HSE Books, 1997), ISBN 0717612767. © Crown copyright material is reproduced with the permission of the Controllers of HMSO and Queen's Printer for Scotland.
Figure 2.1	Cover of *Management of Health and Safety at Work: Management of Health and Safety at Work Regulations 1999 – Approved Code of Practice and Guidance* (HSE Books, 2000), ISBN 0717624889. © Crown Copyright material is reproduced with the permission of the Controller of HMSO and Queen's Printer for Scotland.
Figure 3.9	Courtesy of Draper.
Figure 4.4	Courtesy of Stocksigns.
Figure 4.5	Source HSE. © Crown Copyright material is reproduced with the permission of the Controller of HMSO and Queen's Printer for Scotland.
Figure 4.6	Redrawn from HSG 65 *Successful Health and Safety Management* (HSE Books, 1997), ISBN 0717612767. © Crown copyright material is reproduced with the permission of the Controllers of HMSO and Queen's Printer for Scotland.
Figure 4.15	Redrawn from HSG 48 *Reducing Error and Influencing Behaviour* (HSE Books, 1999), ISBN 0717624528. © Crown Copyright material is reproduced with the permission of the Controller of HMSO and Queen's Printer for Scotland.

Figure 8.29   Redrawn from *Fire Safety – An Employer's Guide* (The Stationery Office, 1999), ISBN 0113412290. © Crown Copyright material is reproduced with the permission of the Controller of HMSO and Queen's Printer for Scotland.

Figure 8.30   © pavingexpert.com

Figure 9.2   Cover of *Building Regulations Approved Document B – Fire Safety, Volume 2 – Buildings other than Dwellinghouses, 2006 Edition (The Stationery Office, 2005)*. © Crown Copyright material is reproduced with the permission of the Controller of HMSO and Queen's Printer of Scotland.

Figure 9.6   Redrawn from *Fire Safety – Risk Assessment: Sleeping Accommodation* (Department for Communities and Local Government, 2006), ISBN 1851128174. © Crown Copyright material is reproduced with the permission of the Controller of HMSO and Queen's Printer for Scotland.

Figure 9.8   Courtesy of Bodycoat Warrington Fire.

Figure 9.16   Courtesy of Hart Door Systems Ltd.

Figure 9.17   Courtesy of Environmental Seals Ltd.

Figure 9.37   Courtesy of Ingersoll Rand.

Figure 9.45   Redrawn from *Fire Safety – Risk Assessment: Offices and Shops* (Department for Communities and Local Government, 2006), ISBN 1851128158. © Crown Copyright material is reproduced with the permission of the Controller of HMSO and Queen's Printer for Scotland.

Figure 9.47   Redrawn from *Fire Safety – Risk Assessment: Large Places of Assembly* (Department for Communities and Local Government, 2006) ISBN 1851128212. © Crown Copyright material is reproduced with the permission of the Controller of HMSO and Queen's Printer for Scotland.

Figure 9.48   Courtesy of Stocksigns.

Figure 9.56   Courtesy of Stocksigns.

Figure 9.59a   Courtesy of Stocksigns.

Figure 9.59b   Courtesy of Stocksigns.

Figure 9.59c   Courtesy of Stocksigns.

Figure 9.59d   Courtesy of Stocksigns.

Figure 9.61   Adapted from *Building Regulations Approved Document B – Fire Safety, 2006 Edition (The Stationery Office, 2005)*. © Crown Copyright material is reproduced with the permission of the Controller of HMSO and Queen's Printer for Scotland.

Figure 9.62   Courtesy of Viking Group Incorporated.

Figure 9.67   Courtesy of Marioff Corporation.

Figure 9.71   Image reproduced by courtesy of Chubb Fire Ltd.

Figure 9.75   Courtesy of Apollo Fire Detectors Limited.

Figure 9.76   Courtesy of Apollo Fire Detectors Limited.

Figure 9.77   Courtesy of Apollo Fire Detectors Limited.

Figure 9.78   Courtesy of Apollo Fire Detectors Limited.

Figure 9.84   Courtesy of Wiltshire Fire and Rescue Service.

Figure 9.86   Redrawn from *Building Regulations Approved Document B – Fire Safety, 2006 Edition (The Stationery Office, 2005)*. © Crown Copyright material is reproduced with the permission of the Controller of HMSO and Queen's Printer for Scotland.

Figure 9.91	Redrawn from *Building Regulations Approved Document B – Fire Safety, 2006 Edition (The Stationery Office, 2005)*. © Crown Copyright material is reproduced with the permission of the Controller of HMSO and Queen's Printer for Scotland.
Figure 10.2	Reprinted from *Introduction to Health and Safety at Work Second edition,* Hughes and Ferrett, page 49, fig 4.5c, 2005, with permission from Elsevier.
Figure 10.8	Courtesy of Evac+Chair International Ltd.
Figure 10.12	Courtesy of Greater Manchester Fire and Rescue Service.
Figure 10.15	Courtesy of Apollo Fire Detectors Limited.
Figure 11.1	Adapted from HSG 65 *Successful Health and Safety Management* (HSE Books, 1997), ISBN 0717612767. © Crown copyright material is reproduced with the permission of the Controllers of HMSO and Queen's Printer for Scotland.
Figure 11.2	Courtesy of Hereford & Worcester Fire and Rescue Service.
Figure 12.1	Courtesy of Hereford & Worcester Fire and Rescue Service.
Figure 12.6	Cover and form from *Accident Book* (The Stationery Office, 2003), ISBN 011703164X. © Crown Copyright material is reproduced with the permission of the Controller of HMSO and Queen's Printer for Scotland.
Figure 12.7	Cover of *A Guide to the Reporting of Injuries, Diseases and Dangerous Occurrences Regulations 1995* (HSE books, 1999), ISBN 0717624315. © Crown Copyright material is reproduced with the permission of the Controller of HMSO and Queen's Printer for Scotland.
Figure 12.8	Courtesy of Hereford & Worcester Fire and Rescue Service.
Figure 12.9	Courtesy of Hereford & Worcester Fire and Rescue Service.
Figure 12.13	From FDR1 (94) Fire incident reporting form. © Crown Copyright material is reproduced with the permission of the Controller of HMSO and Queen's Printer for Scotland.
Figure 12.14	Courtesy of Hereford & Worcester Fire and Rescue Service.
Figure 12.15	Courtesy of Wiltshire Fire and Rescue Service.
Figure 12.17	Courtesy of Wiltshire Fire and Rescue Service.
Figure 12.18	Courtesy of Hereford & Worcester Fire and Rescue Service.
Figure 13.8	Courtesy of Oil Technics.
Appendix 12.1	From *A Guide to the Reporting of Injuries, Diseases and Dangerous Occurrences Regulations 1995* (HSE books, 1999), ISBN 0717624315. © Crown Copyright material is reproduced with the permission of the Controller of HMSO and Queen's Printer for Scotland.
Appendix 12.2	FDR1 (94) Fire incident reporting form. © Crown Copyright material is reproduced with the permission of the Controller of HMSO and Queen's Printer for Scotland.
Appendix 12.3	Adapted from BS 5839-1:2002, ISBN 0580403769.
Figure 13.1	Courtesy of Hertfordshire Constabulary.
Figure 13.2	Courtesy of Hereford & Worcester Fire and Rescue Service.
Figure 13.5	Courtesy of Wiltshire Fire and Rescue Service.
Figure 13.6	Courtesy of Hereford & Worcester Fire and Rescue Service.
Figure 13.7	Courtesy of Forbes Technologies Ltd.
Figure 13.8	Courtesy of Oil Technics.
Figure 13.9	Courtesy of Darcy Products Ltd.

Figure 13.10   Redrawn, courtesy of Biffa.

Figure 13.11   Courtesy of Wiltshire Fire and Rescue Service.

Figure 14.1   Redrawn from *Fire Safety – Risk Assessment: Offices and Shops* (Department for Communities and Local Government, 2006), ISBN 1851128158. © Crown Copyright material is reproduced with the permission of the Controller of HMSO and Queen's Printer for Scotland.

Figure 14.6   Cover of *Fire Safety – Risk Assessment: Means of Escape for Disabled People Supplementary Guide* (Department for Communities and Local Government, 2007), ISBN 1851128743. © Crown Copyright material is reproduced with the permission of the Controller of HMSO and Queen's Printer for Scotland.

Figure 15.1   Redrawn from INDG 350 *The Idiot's Guide to CHIP: Chemicals (Hazard Information and Packaging for Supply) Regulations 2002* (HSE Books, 2002) ISBN 0717623335. © Crown Copyright material is reproduced with the permission of the Controller of HMSO and Queen's Printer for Scotland.

Chapter 15
Page 395   Courtesy of Stocksigns.

Chapter 15
Page 396   Courtesy of Stocksigns.

Chapter 15
Page 396   Courtesy of Stocksigns.

Chapter 15
Page 396   Courtesy of Stocksigns.

Chapter 15
Page 396   Courtesy of Stocksigns.

Fire safety foundations

To enable successful management of both fire and health and safety it is vital to develop a solid base of understanding and the key elements that will provide a foundation upon which to build. For students and safety professionals alike the information presented in this book outlines the legal requirements and management considerations that will assist the reader to successfully minimise the risk of harm from fire in the workplace.

1.1 Definitions

The terms relating to the management of safety in this chapter are defined by a variety of publications. To clarify the meaning of the text, it is important to establish a common understanding of the following, frequently used basic terminology:

This chapter discusses the following key elements:

➤ The scope and nature of both fire and occupational health and safety
➤ The moral, legal and financial reasons for promoting good standards of safety within an organisation
➤ The legal framework for the regulation of fire and health and safety
➤ The legal and financial consequences of failure to manage safety
➤ The nature and significance of key sources of fire and health and safety information
➤ The basis of a system for managing safety.

Occupational health and safety – factors and conditions that can affect the well-being of persons within the workplace, i.e. employees, contractors, temporary workers and visitors.

Safety – the freedom from unacceptable risk from harm.

Fire/combustion – a chemical reaction or series of reactions involving the process of oxidisation, producing heat, light and smoke. There are two classes of fire: conflagration (where combustion occurs relatively slowly) and detonation (where combustion occurs instantaneously).

Ill health – the term ill health includes acute and chronic physical or mental illness which can be caused or made worse by physical, chemical or biological agents, work activity or environment.

Accident – an undesired event resulting in death, ill health, injury, damage, environmental loss or other loss.

Incident – an undesired event that does not result in any harm or loss. Incidents are often referred to as near misses; some organisations refer more accurately to 'incidents' as 'near hits'.

False alarm – an unwanted fire signal resulting from a deliberate operation of a fire safety system, the unintentional electrical actuation of a fire safety system, or the actuation of a fire safety system with good intent (believing there to be a fire).

Environmental protection – management arrangements to cover the protection of the environment, including mitigating the effects from fire fighting and other emergency operations from pollution, caused by workplace operations.

Hazard – a source or situation with the potential to cause harm (death, injury, ill health, damage to property or environment).

Risk – the combination of the likelihood and severity (consequences) of a hazard causing harm.

Further definitions will be provided throughout the book.

1.2 Scope and nature

In today's complex world effective safety management is the cornerstone of managing an economically viable business. The requirement to manage safety effectively extends to all private and public business sectors. Legal responsibilities for safety performance extend throughout all organisations from the management board to the student on work experience.

Every operation within any organisation has an impact on the safety not only of those undertaking and managing the work but also of others who may be affected by their work activities. Any product or service provided to any body must be designed or delivered in such a way as to reduce the risks to the end users to an acceptable level. Therefore it can be seen that safety is inextricably linked with all facets of work.

The failure to manage safety adequately all too often results in death or injury, chronic ill health and damage to property and/or the environment. Such results have a significant impact on the physical and economic well-being of society.

In the Health and Safety Commission's (HSC) revitalising health and safety strategy statement the cost of health and safety failures to society as a whole was estimated as being as high as £18 billion annually. In

Figure 1.1 Aftermath of fire – the cost of failing to manage fire safety

terms of the cost of fire alone in the UK, the direct costs were estimated to be in the region of £8 billion for 2003, which is equivalent to about 1% of the gross domestic profit (GDP) of the English and Welsh economy.

In Europe, the guiding philosophy of legislation since the early 1990s has been for those who work with hazards and risks in relation to fire and health and safety to effectively control them. This requires organisations and individuals to assess the potential risks associated with their work activities and to introduce effective measures to control such risks.

High profile prosecutions in the UK have reinforced the message that the responsibility for effective safety management rests not only with the body corporate but also with individuals within an organisation.

1.3 The moral, legal and financial reasons for promoting good standards of safety within an organisation

1.3.1 Moral (humane) reasons

There are a number of convincing arguments for the promotion of good safety standards. The human consequences of fires, accidents and incidents are widespread and affect a number of different people in different ways.

The most obvious result of a fire or an accident at work is that the persons directly involved are likely to suffer. The impact on these individuals ranges from death through to relatively minor injuries. In addition to the physical impact on a person, it is often the case that individuals involved in any form of safety event suffer some form of physiological ill health.

Less obvious, but no less real, are the effects upon the families and dependants of those who suffer injury or ill health caused at work. The impact on these groups can be significant and wide ranging, e.g.:

➤ The emotional stress of seeing a family member suffer
➤ Financial hardship due to loss of earnings
➤ The loss of social amenity
➤ The potential requirement for the provision of long-term care.

In addition to those directly affected, work colleagues and other witnesses of any serious work-related injury are proven to be susceptible to a number of related physiological disorders such as post-traumatic stress disorder (PTSD).

There can also be significant emotional and physiological consequences for those who may consider themselves to some degree directly responsible for killing or injuring a work colleague, member of the public or a product/service user by either failing to manage safety effectively or through simply making a mistake.

Every working day in Great Britain, at least one person is killed and over 6000 are injured as a result of work activities and about one million people take time off because of what they regard as work-related illness. In total, accidents and work-related ill health result in about 30 million lost work-days.

1.3.2 Financial reasons (economic costs)

It is widely accepted and understood that safety events (accidents, incidents, fires, environmental damage, etc.) cost money. The financial costs to an organisation following a fire are substantial. There is a perception that the majority of such costs are insurable; however, as can be seen from the research undertaken on behalf of the HSE, *The Cost of Accidents at Work*, there are numerous areas which are not covered by insurance.

Financial surveys undertaken on behalf of the insurance organisations identify clearly the cost of fires to the British economy, details of which are shown in the graph in Figure 1.3.

As can be seen from the graph the estimated financial losses attributed to fire are based upon those

Figure 1.2   The human cost of fire

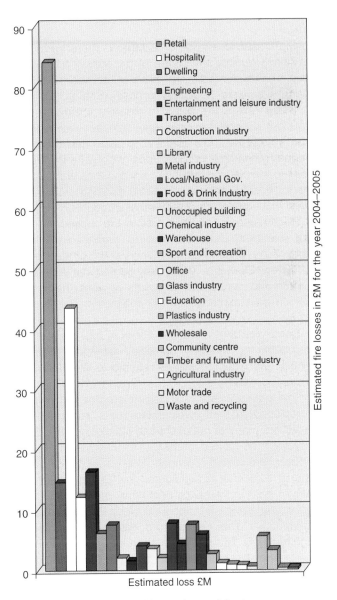

Figure 1.3   Graph detailing estimated fire losses

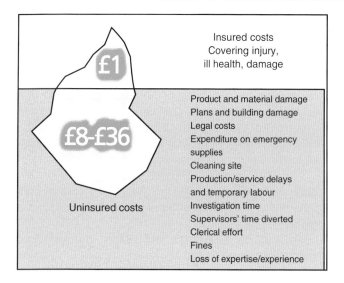

Figure 1.4 Uninsured costs

causing losses in excess of £250 000. There are many fires falling outside the scope of the survey that result in an organisation failing to recover, particularly those relating to small to medium enterprises (SMEs).

These reflect the costs to insurers (claims settlement) but do not, however, take into account a wide range of non-insurable costs. It is also often the case in industry now that organisations underwrite their own losses, particularly in relation to fire and thus are responsible for finding the financial sums to cover claims and losses which are often considerable amounts.

Regardless of whether people are injured or not, there will be a financial cost to organisations. The Accident Prevention Advisory Unit (APAU) of the Health and Safety Executive (HSE) has carried out extensive research into the cost of accidents at work, the results of which are summarised in the publication *The Cost of Accidents at Work* (HS(G)96).

Some accident costs are obvious, e.g. compensation payments, property damage, damaged product, sick pay, etc. These costs are referred to in HS(G)96 as the direct costs.

The indirect costs of accidents are not so obvious, e.g. replacement staff, investigation costs, poor publicity. In addition, many of the direct and indirect costs are not recoverable as insured losses.

The relationship between insured and uninsured costs of accidents is highlighted in HS(G)96 where for every £1 paid in insurance premiums, the average non-recoverable costs were about 10 times the amount paid in premiums. The losses from day-to-day accidents range from 8 to 36 times the amount paid in premiums.

For most organisations, the cost of insurance premiums can be compared to the tip of an iceberg with the majority of the costs (uninsured and non-recoverable) lurking beneath the water line.

1.3.3 Legal reasons

The United Kingdom has, over time, developed a set of rules and standards. These rules and standards are reflected in civil and criminal laws, which regulate, among other things, our work activities.

In civil law, it has been established that employers must take reasonable care of their employees. Failure to meet these obligations can result in a claim for compensation by the individual/s who have suffered a loss.

The criminal law places statutory duties on employers, responsible persons and others to ensure the health and safety of employees and other persons who may be affected by the work activities.

Legislative control over fire safety matters in the UK was rationalised in 2006 with the introduction of the Regulatory Reform (Fire Safety) Order 2005. The Order sets out in detail the roles and responsibilities of those charged with managing fire safety within organisations (the 'responsible person') and that an assessment of fire risk has been undertaken. The order is enforced by local fire authorities (see section 1.4 below).

The Health and Safety at Work etc. Act 1974 (HSWA) together with the Management of Health and Safety at Work Regulations 1999 require employers to demonstrate that they have assessed and are managing their risks to their employees and other persons who could be affected by the work activity.

Failure to comply with any of the general safety or fire specific legislation can result in significant fines for companies and their managers, custodial sentences and enforcement action by the enforcement authorities.

1.3.4 The business case for managing fire safety

The moral, economic and legal consequences of a failure in any safety system can have a significant impact upon a business. A serious fire in a workplace that results from inadequate management of fire safety matters can begin a spiral of events that may result in total business failure.

1.4 The legal framework for the regulation of fire and health and safety

There are two main branches of law of interest to the safety professional, civil and criminal.

Each has a bearing on the conduct of both employers and employees while carrying out their work activities. Table 1.1 provides a comparison of some significant aspects of both branches of law and the following paragraphs discuss the key aspects in more detail.

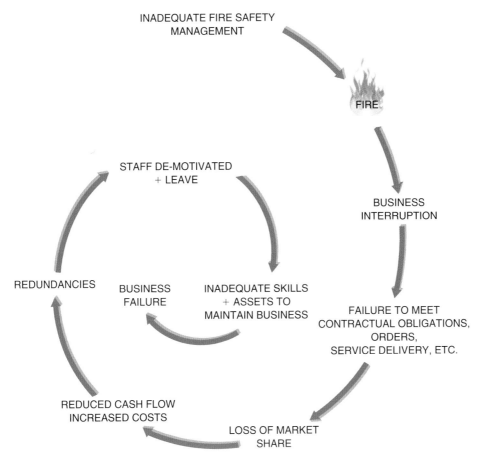

Figure 1.5 Business case for managing fire safety

Table 1.1 Two main branches of law

Aspect	Criminal law	Civil law
Primary source	Statutes, Orders, regulations issued by Parliament	Legal precedent – the accumulation of decisions made in various cases
Purpose	To regulate society by punishing the offender and discouraging others	To compensate those who have suffered loss or harm as a result of others' action or omissions
Judged by	Criminal court system	Civil court system
Burden of proof	In general the defendant is presumed innocent until proved guilty. *However, for prosecutions under section 40 of the HSWA it is the defendant who must prove he has in fact acted 'reasonably'*	If there is sufficient evidence to infer a breach has occurred the burden of proof moves to the defendant who must prove he has in fact acted 'reasonably'
Standard of proof	Must be proved beyond reasonable doubt. *However, for prosecutions under section 40 of the HSWA the defendant need only prove his case on the balance of probabilities*	Need only be proved on the balance of probability
Outcomes	Fines, imprisonment and/or official orders	Orders to pay compensation
Insurance	Cannot be insured against	Can be insured against and in the case of employer liability, *must be insured against* (see below)

1.4.1 Criminal law

The criminal branch of law deals with offences against the state. The purpose of criminal law is to deter people from breaking the law and to punish them accordingly when they do, rather than to merely compensate the wronged party. Many types of criminal law exist for many different purposes and the most important of these in relation to fire and health and safety are as follows:

Acts of Parliament

Acts are sometimes referred to as statutes. The primary Act relating to health and safety in the UK is the Health and Safety at Work etc. Act 1974 (HSWA). This Act, among other statutes, imposes a number of legal duties on an employer and failure to comply with these duties may give rise to criminal liability, resulting in fines and/or imprisonment.

Regulations and Orders

Many Acts of Parliament confer power on a Secretary of State or Minister of the Crown to make Regulations and Orders, these are also known as Statutory Instruments.

The specific section of the HSWA that relates to this power is section 15. Statutory Instruments specify the more detailed rules of the parent Act. Regulations and Orders are referred to as 'subordinate' or 'delegated' legislation because the power to make them is delegated by an Act of Parliament. Parliament does not debate regulations and Orders and so they are able to be made and implemented quickly. Although Parliament does not debate Regulations they are legally binding and enforceable in the same way as statutes.

Regulations and Orders are written using the same legal terms as the statutes that enable them. To define the regulations and Orders in language that is easily understood, Approved Codes of Practice and Guidance notes are issued.

Approved Codes of Practice (ACoPs)

These Codes of Practice are issued by the Health and Safety Commission (HSC) and approved by the Secretary of State. They provide practical guidance on the requirements which are set out in the legislation. Although Approved Codes of Practice are not legally binding in themselves they are used as a minimum standard in a court of law.

They have a quasi legal status in that they give practical advice on how to comply with the law. If the advice in an ACoP is followed, those following it will be doing enough to comply with the law in respect of those specific matters on which it gives advice. Alternative methods to those set out in an ACoP in order to comply with the law may be used.

Figure 1.6 Law courts

Figure 1.7 Approved Codes of Practice

If a prosecution is brought for a breach of health and safety law and it is proved that you did not follow the relevant ACoP you will need to show that your alternative method has enabled compliance with the law in some other way.

As an example, the Provision and Use of Work Equipment Regulations 1998 require that all persons who use work equipment have received adequate training for the purposes of health and safety. This is then interpreted by the ACoP, which states that in the case of chainsaw users this training would be supported by a certificate of competence or national competence award unless they are undergoing training and are adequately supervised. An ACoP helps the reader to understand the requirements of the law to which it applies. Following ACoPs will ensure the law is complied with.

Guidance notes

Guidance notes are issued by governmental bodies such as the HSC HSE or HM Government as opinions on good practice. An example of a fire safety guidance note would be Fire Safety Risk Assessment. This guide explains in plain language what the reader must do to comply with the law. The guides are not legally binding unlike ACoP but they may be referred to in court as establishing a minimum standard.

Figure 1.8 Guidance notes

However, following guidance will normally ensure that the relevant law is being complied with.

European Union/British Standard (EU/BS) and Industry Guidance

EU/BS standards contain detailed information on the specific standards for complying with health and safety and fire safety requirements, e.g. BS 5839 Part 1 – Fire Detection and Fire Alarm Systems for Buildings. Although compliance with EU/BS standards should assist to ensure legal compliance, this cannot be relied on as a defence in a court of law.

Industry guidance, such as *SG4:05 – Preventing Falls in Scaffolding in False Work* produced by the National Access and Scaffolding Federation, is simply regarded as best practice in that industry and has no formal legal status.

The relationship between Acts, regulations, Orders, ACoPs and guidance notes is illustrated in Figure 1.9.

Legal standards

In law some requirements placed upon employers and employees are more stringent than others. It is necessary to distinguish between:

➤ Absolute duties
➤ The duty to do what is practicable
➤ The duty to take steps that are reasonably practicable.

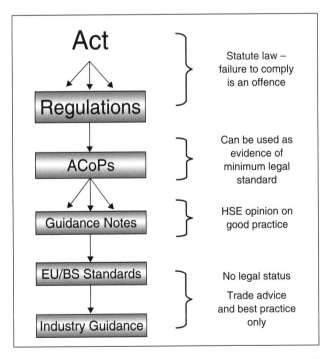

Figure 1.9 Acts, Regulations, Orders, ACoPs and guidance notes

Absolute duties

These are requirements laid down in law which usually state that something 'shall' or 'shall not' or 'must' or 'must not' be done. There is no effective legal defence against a breach of an absolute standard, including ignorance. An example of an absolute standard would be the requirement laid down in Article 9(1) of the Regulatory Reform (Fire Safety) Order that a responsible person must make a suitable and sufficient assessment of the risks to which relevant persons are exposed for the purpose of identifying the general fire precautions he needs to take.

Regulation 3 of the Management of Health and Safety at Work Regulations 1999 requires every employer to carry out suitable and sufficient assessments of risk. Article 23(1) of the RRFSO or section 7 of HSWA states that employees must take care of themselves or others (relevant persons) who may be affected by their acts or omissions.

Practicable duties

These require steps to be taken in light of what is actually possible using current knowledge and technology, e.g. it is technically possible. A good example would be the requirement under regulation 11 of the Provision and Use of Work Equipment Regulations 1998 for all dangerous parts of machines to be guarded so far as it is practicable to do so. Obviously it would be impossible to guard every part of a grinding wheel, for example, so only those parts which it is *practicable* (technically possible) to guard need to be covered.

However, unlike reasonably practicable below there is no quantum relating to the cost of provision.

Reasonably practicable duties

These require the employer to assess the risks associated with a particular work activity and then take appropriate measures to counteract those risks, taking into account the costs of the proposed controls. The controls may be measured in time, effort or money, and there will be an optimum balance point at which further risk reduction

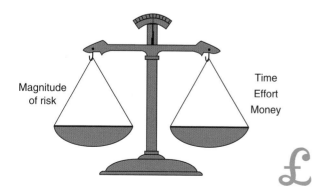

Figure 1.10 Balancing risk against cost

would not be cost effective. The requirement is that the employer must reduce the risks using controls commensurate with those risks; a balance needs to be struck.

1.4.2 Specific fire and health and safety legislation

There are three key pieces of legislation relating specifically to fire and health and safety in England and Wales:

➤ The Health and Safety at Work etc. Act 1974
➤ The Management of Health and Safety at Work Regulations 1999
➤ The Regulatory Reform (Fire Safety) Order 2005.

As it is the basis upon which our current safety legislation is founded we will look at the Health and Safety at Work etc. Act first.

The Health and Safety at Work etc. Act 1974

This Act came into force as a result of work undertaken by the Rubens Institute, in 1972. In essence the Act confers duties on a number of key parties in relation to health and safety. The primary responsibilities are held by:

➤ Employers
➤ Occupiers of premises
➤ Designers, manufacturers, suppliers, importers, installers, etc.
➤ Employees
➤ Personal liabilities
➤ HSC and HSE.

Employers
The general duty of employers under the Act is to ensure, so far as is reasonably practicable, the health safety and welfare at work of all his employees. This general duty is extended to include the following specific requirements:

➤ The provision of safe plant and systems of work
➤ The safe storage, handling, use and transportation of articles and substances used at work
➤ The adequate provision of information, instruction and training with supporting supervision
➤ A safe place in which to work with adequate means of access and egress
➤ A safe working environment with appropriate provision of welfare facilities.

The Act also places a duty upon an employer to produce a health and safety policy which if there are five or more employees should be written down.

There is also a general duty for an employer to consult with duly appointed trade union safety representatives and to form safety committees given certain criteria.

Figure 1.11 Employers are responsible for all persons affected by their work

Employers are also responsible for ensuring the safety of other persons who may be affected by their work activities and thus the law implies that assessment of risk, in relation to such persons, should be undertaken.

Such persons may be:

➤ Contractors undertaking works
➤ Visitors
➤ Members of the public
➤ Emergency service personnel (undertaking their duties)
➤ Enforcement agency staff.

The employer is also required to make provision for other items in relation to safety for which they are not able to charge, e.g. personal protective equipment (PPE) for protecting a person's eyes.

Occupiers (persons in control of premises)
Occupiers having overall control of premises also have duties to ensure the safety of persons while on the premises, e.g. a council allowing organisations/persons to utilise council land (parks) for which they have overall control are responsible for all persons coming onto them, so far as is reasonably practicable.

Occupiers therefore need to ensure, so far as is reasonably practicable:

➤ The safe access and egress of persons to and from the premises they have control over

➤ That plant or substances that are made available are safe and without risk.

Occupiers also have a duty under the Occupiers Liability Acts 1957 and 1984. The original 1957 Act places a duty upon those in control of premises to ensure that any visitor is reasonably safe, having been invited or permitted by the occupier to be there. This duty includes children for whom there is a higher duty of care. The Act does also require that any person on site also acts in a reasonable manner.

The 1984 Act extends the duty to other persons and takes into account trespassers. In these circumstances the occupier must take reasonable care to ensure that anyone on the premises, invited or uninvited, will not be harmed by a condition or activity and it is therefore the occupier's responsibility to know of the dangers that people may face and if trespassers operate in the vicinity.

Designers, manufacturers, suppliers, importers, installers, etc.
The HSWA places duties on persons who design, manufacturer, import, supply or install any article or substance used at work.

The duty that the Act imposes on these persons is, so far as is reasonably practicable, to:

➤ Ensure goods and substances used at work are safe and without risks to health when properly used

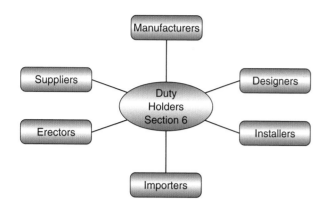

Figure 1.12 HSWA section 6 duty holders

➤ Conduct such tests and examinations as are necessary
➤ Provide adequate, up-to-date safety information
➤ Conduct research to identify, eliminate or minimise any risks to health and safety
➤ Ensure that nothing about the way that the article is installed or erected makes it unsafe.

This section is also very important when considering fire risk, particularly that which relates to the fire retardant nature of products or installing fire safety systems.

Personal liabilities
All employees The Act places three key duties on employees:

➤ To take reasonable care of their own health and safety and of others who may be affected by their acts or omissions at work
➤ To cooperate with their employer and others in the discharge of their legal obligations
➤ Not to interfere or misuse anything that is provided for their safety (although this refers to all persons not just employees).

Senior managers/directors In addition to their own personal liabilities as 'employees', section 37 of the Act enables the enforcement authorities to prosecute senior manager/directors as well as the corporate body, where the individual (holding a senior position and can be seen as 'the controlling mind') has been deemed to have consented, connived or neglected with a duty by an act or omission in breach of any statutory legislation.

HSC & HSE
The Act established both the health and safety commission and the health and safety executive identifying their roles and responsibilities. The Act also lays down the mechanisms by which the Act will be enforced, which will be dependent upon the type of business activity being undertaken.

Table 1.2 indicates which enforcing body is responsible for various business sectors.

Table 1.2 HSE enforcing body table

HSE	Local enforcers
Construction	Offices
Off shore	Shops
Factories	Restaurants
Petro-chemical works	Hotels

The Management of Health and Safety at Work Regulations 1999 (MHSW)

The Management of Health and Safety at Work Regulations 1999 (MHSW) originally arrived on UK statute books in 1992 as part of the requirements to implement the European Framework Directive of 1989.

The regulations are detailed and accompanied by both an ACoP and guidance, and provide a cornerstone in the overall management of health and safety within the UK. They also form the basis from which the Regulatory Reform (Fire Safety) Order 2005 was produced.

Risk assessment
Employers have an absolute duty to make suitable and sufficient assessment of health and safety risks, including risks from fire. They must take into account employees and others who may be affected by their work activities. The purpose of the risk assessment is for the employer to be able to ascertain what they have to do to comply with their legal obligations. Particular attention has to be given to assess risks where young persons (that is, people under 18 years of age) may be at risk. This regulation also requires employers to record the results of risk assessments and to review risk assessments.

Principles of prevention to be applied
The MHSW requires employers who are implementing risk control measures (referred to as preventive and protective measures) to follow the principles set out in the regulations. These principles are a hierarchy of risk control measures of the type described in Chapter 5. The hierarchy begins with 'avoiding risks' and ends with 'giving appropriate instructions to employees'.

Health and safety arrangements

Employers are required to make, give effect to and in certain cases record appropriate health and safety arrangements. These arrangements should cover planning, organisation, control, monitoring and review of preventive and protective measures.

Health surveillance

It is a requirement of the regulations that employers ensure that, where appropriate, adequate health surveillance is provided for employees at risk of exposure to substances and activities that have the potential to cause ill health.

Health and safety assistance

Employers are required to appoint one or more competent persons to assist them to comply with their obligations under safety legislation. Should more than one person be appointed, there must be adequate arrangements for ensuring cooperation between them. Preference should be given to appoint 'in company' where competent persons are available.

Procedures for serious and imminent danger

Employers must establish and implement procedures to be followed in the event of serious and imminent danger to persons working in their respective undertakings. Common procedures are likely to be established for fire, bomb and environmental release. More specific procedures will also be required for danger areas such as exposure to asbestos, or rescues from confined spaces or from activated fall arrest devices (harnesses).

Contacts with external services

In further support of the arrangements for serious and imminent danger employers are required to ensure that any necessary contacts with external services are arranged, particularly as regards first aid, emergency medical care and rescue work.

Information for employees

Employers are duty bound to provide information to their employees on, for example, the arrangements for emergencies, competent persons and the risks to their health and safety identified by assessments. Information should also be provided on the preventive and protective measures required to prevent harm occurring. Where a child is employed (under school leaving age) by an employer, such information that would normally be provided to an employee will also be required to be provided to a parent.

Cooperation and coordination

Every employer and self-employed person who shares a workplace with any other employer or self-employed person is required to cooperate with that other person so far as is necessary to enable him to comply with his statutory safety obligations.

In addition there is also a requirement to coordinate the measures taken in compliance with statutory health and safety obligations with measures by other persons and to provide those other persons with specified health and safety information particularly in relation to the findings of risk assessments.

Figure 1.13 Cooperation and coordination with others

Persons working in host employer's or self-employed person's undertakings

Host employers must ensure that the employers of people working in the host employer's undertaking are given comprehensible information on risks to the employees' health and safety and any control measures taken by the host employer to minimise the risks. Host employers also have to provide employees of other organisations with comprehensible information on the risks to their health and safety such as emergency response procedures.

Capabilities and training

This part of the regulations requires employers to consider their employees' capabilities prior to assigning tasks and also ensure that in specified circumstances their employees are provided with adequate health and safety training. In addition employers should also establish a system to enable them to provide refresher training where appropriate and to adapt training to take account of new or changed risks to health and safety. Such health and safety training must be conducted during working hours.

Employees' duties

Employees' duties are twofold in that they should use all work items provided by their employer in accordance with the training and instructions they have received. Employees are also duty bound to inform their employer (and other persons who could be at risk) of any work situation which they consider represents a serious and imminent danger, and any shortcoming in the employer's protection arrangements.

Temporary workers

Temporary workers, together with those workers working on a site on behalf of another company (e.g. contract cleaners) and self-employed persons, must also be provided with specified health and safety information before they commence their work for the employer.

New or expectant mothers

The regulations relate to employers of women of child-bearing age to include in their risk assessments risks to new or expectant mothers. Where there are such risks,

Figure 1.14 New and expectant mothers

and they can be avoided by altering working conditions or hours of work, this should be done. If it is not reasonable to alter working conditions, or hours of work, or this would not avoid the risk, the employee should be suspended from work.

New or expectant mothers who work at night should be suspended from work if they have a certificate from a medical practitioner or midwife showing that this is necessary for their health or safety.

Employers are not duty bound to avoid risk unless they have been notified in writing by an employee that she is pregnant (certificate from a medical practitioner or midwife), given birth within the previous six months, or is breast feeding.

Protection of young persons

Every employer must ensure that young persons employed by them are protected from the additional risks which they are exposed to as a consequence of their lack of experience, low awareness of risks, and lack of physical and mental maturity. Young persons must not be employed for some specified tasks including tasks which are beyond their physical or psychological capacity, for example tasks:

➤ Which involve harmful exposure to agents which chronically affect human health
➤ Involving harmful exposure to radiation
➤ Which pose a risk from extreme heat or cold, noise, or vibration.

It is permitted for young persons to carry out such tasks in the workplace provided that adequate training and supervision are given.

An employer will therefore be required to complete a full assessment of risks prior to young persons undertaking work.

The Regulatory Reform (Fire Safety) Order 2005 (RRFSO)

The RRFSO 2005 came into effect on 1 October 2006, at which time over 100 separate pieces of fire-related legislation have been revoked or amended. The RRFSO covers, with very few exceptions, all non-domestic premises and stands as the primary legislation for fire safety in England and Wales, with Scotland and Northern Ireland being responsible for their own fire safety legislation. Scotland's fire safety legislation is enacted by the Fire (Scotland) Act 2005.

In general terms the RRFSO reflects the duties and approach contained within the MHSW and employees' duties under the HSWA. However, there are significant fire specific issues that the RRFSO covers in some detail, these include:

➤ The duty for employers ('responsible persons') to take 'general fire precautions'
➤ The need to conduct a fire risk assessment
➤ The notion of a 'responsible person'
➤ The control of risks from dangerous substances
➤ Fire fighting and detection
➤ Emergency routes and exits
➤ Powers of inspectors and enforcement actions (differing from general health and safety).

General fire precautions

The RRFSO identifies a number of general fire precautions that all 'responsible persons' have an absolute duty to provide for all non-domestic premises. These general fire precautions are the measures that are taken:

➤ To reduce the risk of fire and fire spread
➤ In relation to the means of escape (MoE) from the premises
➤ For ensuring that the MoE can be safely and effectively used at all material times
➤ In relation to fire fighting on the premises
➤ In relation to detecting and giving warning in case of fire
➤ In relation to emergency action to be taken in the event of fire, including training and mitigating the effects of fire.

The need to conduct a risk assessment

The RRFSO has as its basis for ensuring safety from fire in all non-domestic buildings, an absolute requirement for responsible persons to conduct risk assessments.

Figure 1.15   Means of escape

As a result of the risk assessments, employers as responsible persons have a duty to do all that is reasonably practicable to safeguard 'relevant persons'* who may be employees and people who are not employees but who may be exposed to risk in the event of fire (see Chapter 14).

Responsible person

The legislation refers to a 'responsible person', who is defined as a specified individual who is responsible for fire safety.

> The meaning of the term 'responsible person' is defined by the RRFSO as being twofold:
>
> ➤ In relation to a workplace, the employer, if the workplace is to any extent under his control;
> ➤ In relation to any premises not falling within the above
>   ➤ the person who has control of the premises (as occupier or otherwise) in connection with the carrying on by him of a trade, business or other undertaking (for profit or not); or
>   ➤ the owner, where the person in control of the premises does not have control in connection with the carrying on by that person of a trade, business or other undertaking.

The responsible person is the main duty holder for fire safety and has, as a result, overall responsibility for:

➤ Undertaking the fire risk assessment
➤ Putting precautions in place to safeguard employees and non-employees
➤ Ensuring that testing and maintenance are carried out for such aspects as
  ➤ Fire detection and alarm systems
  ➤ Fire fighting equipment
  ➤ Emergency exit routes and fire exits
  ➤ Fire evacuation drills and assembly points.

---

* Note relevant persons
The RRFSO specifies 'relevant persons' as being any person who is or may be lawfully on the premises and also includes any persons in the immediate vicinity who may be at risk from a fire. Significantly, the RRFSO specifically excludes fire fighters who are carrying out emergency actions as there is no expectation that the responsible person will know how fire fighters will go about their duties.

Moreover, the responsible person is accountable for appropriate training, provision of information and a variety of other duties relating to the management of fire safety, such as the protection of young people, managing risks from explosive atmospheres, consulting with employees and other relevant persons, etc.

The control of risks from dangerous substances

If there are dangerous substances present in or near a work premise, the responsible person must ensure that the risk to all persons, including those on and off site, is either eliminated or reduced so far as is reasonably practical. Where the substance cannot be eliminated, the judgement relating to what is reasonably practical must be informed by the risk assessment and measures must be introduced that:

➤ Control the risk at source
➤ Mitigate the effects of a fire
➤ Ensure the safe handling, storage, transportation of dangerous substances
➤ Maintain the necessary measures.

Additional emergency arrangements also need to be made to reduce the effects of an accident, incident or emergency relating to dangerous substances.

Fire fighting and detection

Where the risk assessment deems it necessary, the responsible person must ensure that the premises are provided with appropriate:

➤ Easily accessible fire fighting equipment
➤ Fire detectors and alarms.

In addition to the provision of fire fighting equipment, detection and alarms, the responsible person must, in the light of the findings of the risk assessment, also:

➤ Take the appropriate measures in the premises for fire fighting
➤ Nominate competent persons to implement those measures and ensure such persons have adequate training and equipment
➤ Arrange contacts with any external emergency services with particular regard to:
  ➤ Fire fighting
  ➤ Rescue work
  ➤ First aid, and
  ➤ Emergency medical care.

Figure 1.16   Highly flammable substances

Emergency routes and exits

The responsible person must ensure that the routes to emergency exits and the exits themselves are kept clear at all times. In addition the order requires that:

➤ Emergency routes and exits must lead as directly as possible to a place of safety
➤ It must be possible to evacuate the premises quickly and safely
➤ The number, size and distribution of exits must be adequate for the maximum numbers of persons who may be present
➤ Emergency doors must open in the direction of escape
➤ Sliding/revolving doors must not be used as emergency exits
➤ Exit doors must be easily and immediately opened by any person who may need to use them in an emergency
➤ Escape routes must be indicated by signs
➤ Emergency routes and exits must be adequately illuminated in the event of a failure of the normal lighting.

Procedures for serious and imminent danger and for danger areas

The responsible person also has a duty to establish emergency procedures and test that those procedures work. In the light of this they are also responsible for nominating sufficient numbers of competent persons to implement the evacuation procedures and ensuring that employees and others have been provided with sufficient information regarding hazards, and their controls, together with appropriate fire safety and other related safety training.

Fire door
Keep shut

Door closer

Figure 1.17 Emergency routes and exits

Maintenance of facilities, equipment and devices

The premises and any facilities, equipment and devices provided to minimise the fire risk must be subject to a suitable system of maintenance and necessary testing to ensure that they are maintained in an efficient state, an efficient working order and in good repair.

Establishing a maintenance scheme can fall on a number of responsible persons either individually or together. These include the employer or occupier of the premise or any other premises forming part of the building, or dependent upon contract terms of the owner or management company.

Safety assistance (competent person or persons)

The responsible person is required to appoint one or more competent persons to assist in undertaking the preventive and protective measures. The need for appointing more than one person will be dependent upon the size and the distribution of the risks throughout the premises and activities together with the time and means available to fulfil their functions.

It may be that appointments are made from outside the organisation, e.g. a consultant, if this is the case adequate information must be provided by the responsible person to enable the competent person/s to undertake their role.

Under the RRFSO a person is to be regarded as competent where they have sufficient 'training and experience or knowledge and other qualities' for them

to provide assistance in undertaking the preventive and protective measures.

Information, training and consultation

Information must be provided by the responsible person as to the following:

➤ The risks identified by the risk assessment and any preventive and protective measures required to minimise the risks
➤ The emergency procedures
➤ The identities of nominated competent persons.

This information should be provided to employees and other relevant persons including contractors and others who may be affected by fire in the workplace.

In addition similar information regarding dangerous substances should also be provided (materials safety data sheets, etc.). Where children may be affected by the responsible person's operations the information should be provided to their parent/guardian, etc.

The RRFSO requires that 'adequate safety training be provided', to ensure persons have a knowledge of fire safety measures and their role in securing them. Such a programme of training should start at induction and be refreshed regularly taking into account changes in risk, job, technology etc.

As in the case of health and safety legislation the RRFSO requires that under Safety Representatives and Safety Committees Regulations 1977 and the Health and Safety (Consultation with Employees) Regulations 1996 the responsible person has the duty to consult the employees on matters relating to safety.

Legal enforcement of the RRFSO

Enforcement of this Order is generally the responsibility of the local fire and rescue authority for the area in which the premises are situated.

The HSE still retains responsibility for the enforcement of some specified areas, these are:

➤ Construction industry
➤ Ship building/repair
➤ Nuclear installations
➤ Defence bases
➤ Crown owned premises.

The relevant local authority also retains enforcement for premises which consist of:

➤ Sports ground designated as requiring a safety certificate under the Safety of Sports Grounds Act 1975
➤ A regulated stand under the Fire Safety and Safety of Places of Sport Act 1987.

Inspectors have the authority to enter and inspect premises if they have reason to believe such a visit is appropriate. In addition to inspecting the premises themselves, such a visit could involve interviewing responsible persons, inspecting records, taking samples and making whatever other enquiries are deemed necessary to determine whether or not the responsible persons are complying with the provisions of the Order.

1.4.3 Other supporting legislation

There are numerous pieces of legislation that support the HSWA, MHSW and RRFSO. These are dealt with later in the book.

In terms of the principal legislation that has an impact on the management of fire safety the list includes:

- Electricity at Work Regulations 1989
  - The management of electrical supplies, systems and equipment has a direct bearing on fire safety due to the high incidence of fires started by electrical sources – the regulations set a minimum standard for the management of electrical supplies, systems, etc.
- Workplace (Health Safety and Welfare) Regulations 1992
  - A range of issues are covered by the regulations and include the maintenance of workplace equipment, ventilation and lighting (including emergency lighting) doors and windows.
- Construction (Design and Management) Regulations 2007
  - All construction work comes under the CDM Regulations. These regulations require the production of a 'Construction Phase Plan' for activities taken during construction and a 'Health and Safety File' to be produced following completion of the project. The regulations allocate role and responsibilities to key parties from client or development company through to contractors undertaking the actual works.
  - Also covered in these regulations is the requirement to plan for emergencies, including fire during construction, alterations and maintenance operations.
- Supply of Machinery (Safety) Regulations 1992 (as amended)
  - Require all UK manufacturers and suppliers of new machinery to make sure that it is safe and fit for purpose. The regulations also apply to refurbished secondhand machinery. Manufacturers have duties which include the provision of safety information to those who may use, maintain, etc. the equipment (fires can result from bearings running dry, friction, etc.)

Figure 1.18 Emergency plans required under CDM

- Provision and Use of Work Equipment Regulations 1998
  - As in the case of the Supply of Machinery (Safety) Regulations 1992 (as amended) the duty is to ensure that work equipment is safe and fit for the purpose; however, in PUWER the responsibilities are directed toward the employer. Training and the requirement to establish an ongoing maintenance regime are also included the regulations.
- Environmental Protection Act 1990
  - This key piece of legislation identifies roles and responsibilities for key parties in the management of pollution from emissions and discharges into the environment. It also covers the management of waste and establishes the requirement for the management of controlled waste including that as a result from fire.
- Dangerous Substances and Explosive Atmospheres Regulations 2002
  - The Dangerous Substances and Explosive Atmospheres Regulations 2002 (DSEAR) are concerned with the preventive and protective requirements against the risks (fire or explosion) from dangerous substances that are used, stored, etc. in the workplace. These regulations interact strongly with the RRFSO with many similar facets.
- Control of Substances Hazardous to Health Regulations 2002 (COSHH) (as amended)
  - COSHH is the key set of regulations that are concerned with the management of hazardous materials within the workplace. A detailed piece of legislation that places duties upon employers, employees and the self-employed to manage hazardous chemicals safely, from avoidance through to personal protective equipment (PPE).

Figure 1.19 Stair lift in situation

➤ Chemicals (Hazard Information and Packaging for Supply) Regulations 2002 (CHIP)
  ➤ CHIP requires that chemical products provided from an approved supplier are required to be provided with a Material Safety Data Sheet (MSDS) which provides information under 16 headings. The MSDS provides information among other things on flammability, storage arrangements, boiling points, etc. all of which are extremely useful when determining fire risks and control measures.
➤ Fire and Rescue Services Act 2004
  ➤ This piece of legislation places prevention at the heart of the role of the fire and rescue service, but also introduces powers for fire fighters to gain entry in the event of fire, and undertake fire investigation.
➤ Disability Discrimination Act 1995
  ➤ The Act principally identifies who has duties and responsibilities, the need to ensure that 'reasonable adjustments' have or will be made so as to ensure that those employing or providing services to people who are disabled have adequate arrangements in place which would include arrangements for effective safe evacuation in the event of an emergency, e.g. a fire.
➤ Building Regulations 2000
  ➤ The Building Regulations are designed to secure the health and safety, welfare and convenience of people in or about buildings and of others who may be affected by buildings or matters connected with buildings. In addition they develop and control the conservation of fuel and power and prevent waste, undue consumption, misuse or contamination of water.
➤ Building Regulations approval is required under the following circumstances:
  ➤ The erection or extension of a building (including loft conversions)
  ➤ The installation or extension of a service or fitting (e.g. washing and sanitary facilities, hot water cylinders, foul and rainwater drainage, replacement windows and fuel burning appliances)
  ➤ Alteration works which may have an effect on the building in terms of structure, fire safety and access and facilities for all (e.g. openings in walls, removal of fire doors, changes to accessibilities)
  ➤ The insertion of insulation into a cavity wall
  ➤ The underpinning of a building's foundation
  ➤ The change of use of all or part of a building (e.g. conversion of a shop into a dwelling, conversion of a house into flats and conversion of a garage into a room)
  ➤ Replacement windows
  ➤ Electrical installations in dwellings.
➤ Building Regulations are enforced by local building control authorities and in relation to fire they consult with the local fire and rescue authorities.

1.4.4 Fire (Scotland) Act 2005 (FSA)

Identified within Part 3 Chapter 1 and Schedule 2 of the FSA are the duties of employers to employees and duties in relation to relevant premises.

In essence, the principles of the FSA cover the same ethos as the RRFSO, while not directly following the MHSW Regulations. The FSA identifies the overall duty of an employer to 'so far as reasonably practicable' ensure that the employer's employees are not put at risk from fire in the workplace. The Act goes on to require that a fire risk assessment should be completed and control measures be put in place.

As in the case of the RRFSO and the MHSW Regulations the risk assessment should be subject to review and fire safety measures should be put in place to comply with the law. The latter are contained in Schedule 2 and are outlined below:

➤ measures to reduce the risk of:
  ➤ fire in relevant premises; and
  ➤ the risk of the spread of fire there;

- measures in relation to the means of escape from relevant premises;
- measures for securing that, at all material times, the means of escape from relevant premises can be safely and effectively used;
- measures in relation to the means of fighting fires in relevant premises;
- measures in relation to the means of:
  - detecting fires in relevant premises; and
  - giving warning in the event of fire, or suspected fire, in relevant premises;
- measures in relation to the arrangements for action to be taken in the event of fire in relevant premises (including, in particular, measures for the instruction and training of employees and for mitigation of the effects of fire).

Those who have 'to any extent' control of premises are also duty bound to ensure that the above measures (employer's duties) are carried out, as per the RRFSO.

Employees are also given responsibilities that fall under very similar scope to those that are contained within the RRFSO, MHSW Regulations and the HSWA.

1.4.5 Civil law

Civil law has its roots in ancient laws from the 11th century and beyond and is most likely to be encountered relative to workplace conduct as 'common law'. It deals with the manner in which individuals should conduct their affairs in modern society.

It is not laid down by statute but rather is found as an accumulation of decisions made by judges in individual cases. This process is referred to as a precedent whereby as each case is decided in court principles of law are established.

A fundamental principle that has been established by common law in the UK is that people have a 'duty of care' towards others who may be affected by what they do. In the case of health and safety, this duty has been expanded by a judicial precedent to be a duty of 'reasonable care'. When determining what is meant by reasonable care the courts will take into account the qualifications, experience, age, locality, intelligence, seniority and skills of the individuals concerned.

For example, it is likely that the courts will find that reasonable care has been exercised by a company when making arrangements for the safe evacuation of a building in the case of fire if it has ensured sufficient arrangements for the management of the young, infirm, disabled and sensory impaired persons.

Negligence

The term 'tort' means a civil wrong committed by one party against another. In the case of occupational health and safety, the tort of negligence is of particular interest.

The tort of negligence involves more than simply careless conduct causing a loss to an individual. For a civil law claim of negligence to succeed, the claimant must prove three elements:

1. That a duty of 'reasonable care' was owed to the claimant by the defendant
2. That the duty of 'reasonable care' was breached
3. That the claimant suffered a loss as a result of the breach of duty of reasonable care.

Duty of 'reasonable care' owed

When considering whether or not a duty of reasonable care was owed courts will always seek to identify an established relationship between the claimant and the duty holder. Relationships that have been established by judicial precedent include employer/employee, doctor/patient and teacher/pupil. For claims relating to the duty of reasonable care the landmark case that established the 'neighbour' principle was the case of *Donoghue* v. *Stevenson* (1932).

In this case, the claimant, Miss Donoghue, and a friend went into a café and her friend bought two bottles of ginger beer. The bartender served the beer in their original bottles that were dark green and opaque.

The claimant drank part of the contents of one bottle and on refilling her glass, she discovered the part decomposed body of a snail in the remaining beer. As a result of drinking the beer she became ill.

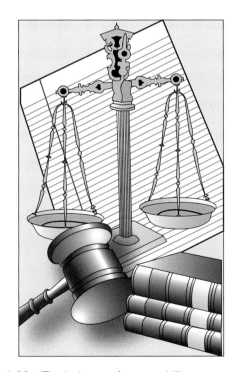

Figure 1.20 The balance of reasonability

So, whom would she sue for her loss?

Her options were to sue the friend for supplying the beer, the bartender for selling the beer, or the original manufacturer and bottler of the beer. As was proven in the case the party most responsible for her loss and in best position to compensate Miss Donoghue was the manufacturer.

During the case, Lord Aitken made the following statement regarding the manufacturer, which is now known as the 'neighbour principle':

> *You must take reasonable care to avoid acts or omissions that you can reasonably foresee would be likely to injure your neighbour. Who then in law is my neighbour? They are persons who are so closely or directly affected by my acts or omissions that I ought to have them in mind.*

Ultimately the manufacturer does not only owe duty to those purchasing, but also those consuming the ginger beer. This principle is used in negligence claims under safety grounds. Therefore using this neighbour principle, the following groups of people could be regarded as the 'neighbours' of employers:

➤ Employees
➤ Agency staff
➤ Contractors
➤ Members of the public
➤ Emergency services
➤ Visitors.

Duty of 'reasonable care' breached

The second test of negligence is that there is a breach of the duty of reasonable care. In order to demonstrate that there is a breach, a claimant must prove that the defendant negligently did something or omitted to do something that a reasonable man would do. This appears to be a somewhat circular argument; however, the courts use the case of *Blythe* v. *Birmingham Waterworks Co.* (1856) as the test of whether the duty of reasonable care has in fact been breached. In that case the judge set the precedent that:

> *Negligence is the omission to do something which a reasonable man, guided upon those considerations which obviously regulate the conduct of human affairs, would do or something which the prudent and reasonable man would not do.*

This definition itself raises two further questions which the court must decide, i.e.:

1. What is a reasonable man?
2. What are those considerations?

What is a reasonable man?

A reasonable man is defined in the case *Glasgow Corporation* v. *Muir* as:

> *An imaginary being who is neither imprudent nor over cautious . . . he is in effect the man on the Clapham omnibus.*

It is therefore the behaviour of a hypothetical (average) person that courts have to consider when deciding whether or not the standard of 'reasonable care' has been achieved.

An important factor is that the standard of reasonable care owed varies dependent upon the skill, experience and competence of the person so that the standard of care owed by the skilled, experienced, competent person is greater than that expected from the unskilled, inexperienced, less competent person. The same can be said for an organisation, namely the employer who owes a duty of reasonable care to employees and others.

For example, the level of care taken by a young person working as a trainee chef in a busy kitchen may not be the same as the chef. The courts would expect the chef to be able and motivated to take more care, therefore the standard for the chef is higher than for the trainee.

What are those considerations?

The considerations that guide the reasonable man relate to the degree of risk associated with the activity and the cost incurred in averting the risk.

Risk is a combination of the likelihood of injury or harm and the severity of the injury or harm risked:

$$Risk = Likelihood \times Severity$$

The cost incurred in averting the risk is not simply financial but should include the time and effort required in implementing the precautionary measures. The result is a scale of risk v. cost as illustrated in Figure 1.21.

The judgment of risk v. cost must be made *before* not after the injury, damage or loss is suffered.

Figure 1.21 The balance of risk against cost

Figure 1.22 Employers are vicariously liable for the actions of their employees

Loss sustained as a result of the breach

The third and final test of the tort of negligence is for the claimant to prove that injury, damage or loss was sustained as a result of the defendant's failure to take reasonable care. It is important that the loss sustained is directly linked to the breach. Cases have been lost where the loss that was undoubtedly sustained had not been directly caused by the specific breach of duty.

The employer's liability

Due to the nature of the common law relating to negligence in the UK it is obvious that an employer is liable to be sued for compensation for any loss or damages suffered by his employees. It is for this reason that employers in the UK are obliged by law to hold compulsory 'employers liability insurance'. This insurance covers the employers for claims from their employees for up to £5 million. Employers will also take out public liability insurance to cover themselves from claims made by third parties who are seeking compensation for a loss, although this is not compulsory.

Vicarious liability

An important principle in negligence cases is that of vicarious liability. In essence this renders the employer directly liable for the actions of his employees.

The reason why this principle is applied is to allow the courts to order compensation from the employer who is generally in a much better position to be able to pay large amounts of compensation to the claimant.

To avoid being held vicariously liable, an employer must be able to demonstrate that the employee who acted negligently was doing so on his own volition.

The courts refer to the independent action by an employee as his being on a 'frolic of his own' and in these cases the employer cannot be held vicariously responsible.

Defences against claims for compensation due to negligence

In the cases of civil actions for compensation for negligence, the defendant has available the following defences:

➤ **There was no duty owed to the claimant** – the defendant may claim the claimant was not a neighbour
➤ **There was no breach of the duty of reasonable care** – the defendant may claim that all that could reasonably be done was done
➤ **The loss was not caused by the breach** – the defendant may suggest that the loss suffered by the claimant was not connected with the breach
➤ **_Volenti non fit injuria_** – the defendant argues that the loss was caused after the claimant had accepted the risk voluntarily.

Once the court has established negligence on behalf of the defendant, it may then be argued that the claimant displayed **contributory negligence**. In other words that the loss was wholly or partly as a result of the claimant's own unreasonable behaviour.

Limitations Act 1980 applies
This Act of Parliament gives a specific time period during which claims can be made in the civil courts. For personal injury cases, this period equates to three years from the date of the unreasonable acts which caused them, or for industrial diseases three years from the date of diagnosis. Any claims submitted outside this deadline are not accepted by the civil courts.

Figure 1.23 Law and fire and safety management

1.5 The legal and financial consequences of failure to manage fire and health and safety

1.5.1 Legal

The legal consequences of failing to manage safety effectively are divided into two key areas, namely, civil and criminal breaches, which we will now discuss.

Civil court system
The civil court system comprises the County Court and High Court. Given that civil law is to provide compensation for loss, the county court deals with minor compensation claims of up to £50 000. This system is also supported by the small claims court for claims for compensation less than £5000.

For claims in excess of £50 000 the High Court is used; however, should appeals be heard in relation to awards then these will be heard by the Court of Appeal. There is also the potential that an appeal can be routed as far as the European Courts (European Court of Justice and European Court of Human Rights) via the House of Lords.

Criminal court system
The majority of cases in relation to breaches of safety legislation will be heard initially in the magistrates' courts. The magistrates' court can only fine up to £20 000 (summary conviction) for employers who breach the law and £5000 for employees who breach the law. It is also possible for the magistrate to sentence persons for a term of up to six months in jail.

Where death has occurred and for serious health and safety cases the magistrates' court will defer to the

Crown Court where sentences are less restrictive. As can be seen from Table 1.3 the indictable cases (heard within the Crown Court) can lead to up to two years' imprisonment.

Appeals in relation to criminal law take the same route as those for civil law via Courts of Appeal, the House of Lords and European Courts.

In addition to the criminal and civil courts there is a further court system dealing with employment law. However, the role of this system has been extended to incorporate appeals against safety enforcement notices and disputes between safety representatives and employers.

The employment tribunal is supported by an appeals system which is ultimately routed through the same channels as both criminal and civil appeals.

Enforcement arrangements
The HSE and local authorities (county, district and unitary councils) are the enforcing authorities for health and safety standards. Building Regulations are enforced by local authorities. Local fire and rescue authorities enforce virtually all matters in relation to fire safety in the UK.

The Environment Agency (Scottish Environmental Protection Agency) are responsible for both authorising and regulating emissions for industry as a whole and are also responsible for enforcing and providing guidance in relation to any form of environmental pollution.

Enforcers options
Regardless of the enforcement authority, e.g. HSE, local authority or fire authority, enforcers have a variety of

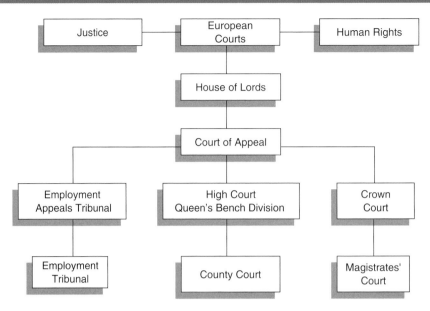

Figure 1.24 Diagram of court structure

Table 1.3 Table itemising maximum penalties under the HSWA

Offence	Summary conviction	On indictment
Breach of sections 2–6 HSWA	£20 000 per breach	£ Unlimited
Breach of sections 7–8 HSWA	£5000 per breach	£ Unlimited
Breach of regulations or Orders	£5000 per breach	£ Unlimited
Contravening section 20 items	£5000	Not indictable
Preventing a person from appearing before an inspector	£5000	£ Unlimited
Making a false statement	£5000	£ Unlimited
Contravening an improvement or prohibition notice	£20 000 and/or 6 months' imprisonment	£ Unlimited and/or 2 years' imprisonment

options dependent upon the seriousness of the breach. They may:

➤ Provide verbal advice and direction
➤ Provide the guidance and direction by way of a letter
➤ Serve an enforcement notice
➤ Prosecute the offending organisation or person should there be grounds for so doing.

Enforcement of the legislation

Dependent upon the enforcing authority there are a variety of types of enforcement notices that can be served.

Health and safety improvement notice – this notice may be served by a health and safety enforcing officer when, in their opinion, there has been a breach of statutory duty or that there is likely to be a breach of statutory duty. For example, if risk assessment records have not been recorded or specified training has not been completed.

Figure 1.25 Powers of enforcement authorities

Fire safety alterations notices – the enforcing authority may serve an alterations notice on a responsible person under two circumstances:

1. When the authority is of the opinion that the premises constitute a 'serious risk' to 'relevant persons' or
2. The premises may constitute a serious risk if a change is made to them or to the use to which they are put.

An alterations notice must detail those issues which constitute a serious risk and state the reasons why the enforcing authority has formed its opinion. On both occasions when an alterations notice has been served the responsible person must notify the enforcing authority of any proposed changes to those premises.

It should be noted that an alterations notice is merely a notice that provides information to a responsible person relating to the magnitude of the perceived fire risk. If the enforcing authority is of the opinion that works need to be carried out to reduce the risk, they may issue an enforcement notice.

Fire safety enforcement notice – this is very similar to the health and safety improvement notice and is served when, in the opinion of the enforcer, the responsible person or other person has failed to comply with any aspect of the RRO.

Fire fighters' switches for luminous discharge tubes – luminous tube signs designed to work at a voltage exceeding a prescribed voltage (1000 volts AC or 1500 volts DC if measured between any two conductors; or 600 volts AC or 900 volts DC if measured between a conductor and earth) must have a cut-off switch which is placed, coloured and marked to satisfy such reasonable requirements of the fire and rescue authority to ensure that it is readily recognisable by and

accessible to fire fighters. Where this is not the case the fire authority may issue a notice.

Health and safety and fire safety prohibition notices – this type of notice, whether for health and safety or fire safety, is served when the enforcer is of the opinion that the activity or premises involved is so serious that the activity should be stopped or the use of the premises be prohibited or restricted. Examples of these would be operating a machine with no guard or chaining up a final fire exit door.

Powers of inspectors/enforcers

In order to carry out their duties inspectors/enforcers are given a variety of powers. These can be wide ranging; their key powers are listed below:

➤ Enter premises at any reasonable time taking with them a police officer (to maintain the peace) if required
➤ Request and inspect documentation and records and if required take a copy
➤ Require any person to provide assistance to take samples of any articles or substances
➤ To examine, conduct investigations and where appropriate require the premises or items to be left undisturbed
➤ Remove, render harmless or destroy articles or substances
➤ Issue verbal advice, written advice or serve notices.

Local Building Control enforcement officers have additional powers in respect of breaches of Building Regulations. These additional powers include the ability to refuse permission to build or make alterations to buildings and take action to render structures unsafe at the expense of the owner/occupier.

Appeals

As previously discussed, an appeal system exists to enable those being prosecuted and those who have been served with enforcement notices to take their case to appeal. Appeals can be made against any enforcement notices by lodging an appeal with an employment tribunal. Table 1.4 details the consequences of appealing within 21 days against the notices served.

1.5.2 Prosecutions

In addition to issuing notices, enforcement authorities may also seek prosecution for breaches of legislation. Enforcing authorities may prosecute any persons who hold a legal duty for fire, health and safety, e.g. responsible persons, employers and individuals. It is now documented that on occasions the enforcement authorities

Table 1.4 Consequences of appealing within 21 days against notices

Actions/options	HSE/LA/FA improvement/enforcement	FA alterations	HSE/LA/FA prohibition
Action of appeal	Suspends the operation of the notice until hearing	Suspends the operation of the notice until hearing	Does not suspend the operation of the notice until hearing
Court options	Cancel, affirm, affirm with modifications	Cancel, affirm, affirm with modifications	Cancel, affirm, affirm with modifications

Note: although there is provision to appeal against a notice relating to switches for luminous discharge tubes, the RRFSO fails to mention whether the notice remains in force during the appeal.

Table 1.5 HSE prosecution of employers

Year ending	Number of convictions	Total fines imposed	Average fine per prosecution
2001	495	£5 573 525	£41 676
2002	530	£8 225 466	£58 167
2003	451	£4 710 051	£39 888
2004	512	£8 928 600	£45 882
2005	396	£6 857 200	£40 895

from both fire and health and safety seek joint prosecutions for breaches that result in deaths as a result of fires and explosions.

While there appears to be little information regarding the number of prosecutions taken in relation to fire safety, the HSE produces summaries of prosecutions which are made available on their website. Table 1.5 shows the number and nature of prosecutions taken by the HSE over a five year period. The figures shown are merely the fines imposed on employers under the HSWA. They do not include the indirect costs incurred by organisations who are prosecuted, in terms of bad publicity and legal fees.

Where a safety breach is committed by a body corporate and is proven to have been committed with the:

➤ consent or
➤ connivance of
➤ or to be attributable to any neglect

on the part of any director, manager, secretary or other similar officer of the body corporate (or any person purporting to act in any such capacity) they, as well as the body corporate, are guilty of that offence, and are liable to be prosecuted and punished accordingly.

1.5.3 Financial

In addition to the fines imposed as a result of successful prosecutions, the financial ramifications of failing to manage safety effectively, as discussed previously, can impact considerably upon an organisation. It is clear, given a variety of surveys undertaken over past years, that many organisations do not fully appreciate the true consequences of an incident and believe that the insurance premiums that they pay cover the costs.

Over recent years insurance companies have begun to influence the management of safety by rewarding organisations that have an effective management safety record (low claims history) by maintaining premiums and excess; alternatively they have raised the poorer safety performers' premiums and excess and on some occasions refused to cover them at all.

Insurance influences in relation to fire, given the financial impact that even a small fire can have, are even more prominent, with specific codes being produced such as 'Fire Prevention on Construction Sites', the *joint code of practice on the protection from fire on construction sites and buildings undergoing renovation*. Published by the Construction Confederation and the Fire Prevention Association with the support of the Association of British Insurers, the Chief and Assistant Chief Fire Officers Association (now CFOA) and the London Fire Brigade, the joint code provides clear guidance on minimum standards for the construction industry relating to fire safety management.

Compliance with the code for insurance purposes is often included as part of contract terms to secure insurance cover. Therefore if the code is not followed it may result in a breach of contract and thus could result in insurance ceasing to be available or in certain circumstances it may result in a breach of legislation requiring the provision of insurance.

The organisation SIESO (Sharing Information and Experience for Safer Operations) has identified that over 50% of businesses that have been involved with a major incident have ceased trading within 12 months of the

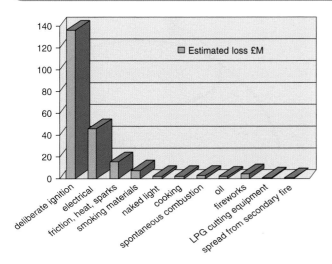

Figure 1.26 The cost of fire per initiation source for 2004/2005

incident. This is in part due to the supply chain continuing to require commodities and that if one organisation (having suffered an incident) cannot deliver then an alternative will need to be utilised.

Following a number of recent large scale disasters (Buncefield Oil Depot, Festival Fireworks Storage facility/factory) statistics have unfortunately been proven correct for those operating in close proximity to the disaster areas. Many smaller organisations have failed to recover and have closed their operations for good.

Financial consequences to industry (Fire Protection Association figures)

Figure 1.26 identifies the results for statistical analysis gathered from the Fire Protection Association who have close ties with the British insurance market. The statistics indicate the estimated losses as a result of fires (£250 000 and above) for the UK in 2004–2005. It should be noted that these estimates show only the direct financial costs and exclude the losses that will result from loss of business and brand image.

1.6 The nature and sources of safety information

Gathering, using and giving information relating to safety is critical to successful safety management. In this section we will review the nature and sources of information available, which will not only assist in providing information for this programme but will also provide information and a direction for future reference.

1.6.1 Internal information

The majority of organisations have a variety of sources of information at their fingertips. These will range from reactive data such as the information gathered from accident reports/investigations, incident reports, fire alarm/detection actuations, damage reports and breakdown maintenance reports and records to active (proactive) information gathered from safety surveys, staff questionnaires, inspection reports, planned preventive maintenance reports, health and safety reviews and audits.

The nature of information will be discussed throughout this book and the above paragraph identifies only an outline.

Risk assessment records will provide a rich source of information from which an organisation can draw. In addition, an organisation's health and safety policy, its written procedures and safe systems of work will also provide a valuable source of information, together with any posters and information sheets produced internally.

Internally it is anticipated that much information will be drawn from safety advisers who have a key role in gathering and passing on safety information. Further information is likely to be available from the facilities management team, human resources/personnel department, occupational health teams, etc. Each company will differ in the roles that its departments or teams undertake and the size of the organisation.

1.6.2 External information

As can be seen from the preceding section all organisations are required to provide safety information in relation to the products and services that they manufacture, produce or provide. Information can therefore be gleaned from manufacturers' documentation and records, e.g. sound levels, chemical hazards, weights of items and flammability ranges.

An extremely valuable source of information is HSE Books who produce free publications, guidance documentation and report books.

The HSE and the Department of Communities and Local Government (DCLG) provide information by way of their respective websites which can include reactive data such as health, safety and fire statistics, together with a wealth of information relating to safety campaigns.

The Office of Public Sector Information (OPSI) provides information relating to legislation and actual statutes and can again be accessed via the worldwide web.

The legislation itself, together with any ACoP or Guidance produced by the HSE, is an equally valuable source, as are the Fire Safety Guides produced by DCLG. The British Standards Institute (BSI) can also provide a valuable source of information regarding items such as the requirements for safety management systems, together with physical standards such as those appertaining to emergency lighting, etc.

Figure 1.27 The provision of fire safety information

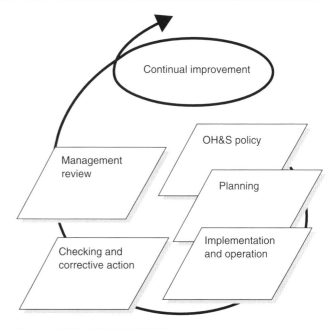

Figure 1.28 OHSAS 18001

The Institution of Occupational Safety and Health (IOSH) and the Institution of Fire Engineers (IFE) produce monthly journals, as does the British Safety Council, all of which provide a valuable insight into specific safety areas.

In relation to fire the Passive Fire Protection Federation (PFPF) and the Fire Protection Association (FPA) also produce a range of industry information and guidance on both practical and managerial aspects of fire safety.

1.7 The basis of a system for managing safety

Over a number of years a variety of management systems have been produced, any of which can be utilised to assist in establishing a safety management system.

Many organisations are familiar with BSEN ISO 9001, a quality management system which can be accredited by an external organisation. The same can be said for BSEN ISO 14001, the environmental management system.

The production of an integrated management system such as OHSAS 18001 (Occupational Health and Safety Assessment Series) has enabled mapping to be achieved between the systems mentioned above and a health and safety management system.

While OHSAS 18001 has not been designated as a British Standard it has been produced by the British Standards Institution (BSI) to enable external validation of an organisation's safety management system to be achieved.

Originally when first published in 1994 the British Standard BS 8800, which is a guide to Occupational Health and Safety Management Systems, only concentrated on a choice between HSG65 and ISO 14001. In

its review and update in 2004 it also shares common management system principles with ISO 9000 series, ISO 14000 series and is consistent with ILO-OSH 2001 (International Labour Organisation – Occupational Safety and Health) and OHSAS 18001.

A schematic view of each of the available systems has been included for reference within this book, although the system known as HSG65 produced by the HSE will be discussed in more depth.

The basic elements of each of the above systems include, in one definition or other, the same phases which are:

➤ Planning
➤ Performance
➤ Performance assessment
➤ Performance improvement.

The HSE's guidance simplifies these phases into 'Plan, Do, Check, Act' as detailed below:

Plan – this involves the setting of standards for safety management that reflect legal requirements and the risks (risk assessment findings) to an organisation.
Do – involving putting into place or implementing the plans to achieve overall aims and objectives set during the planning phase.
Check – the measurement of progress against the plans and legal standards to confirm compliance.
Act – consider and review the status against the plans and standards and taking of action when appropriate.

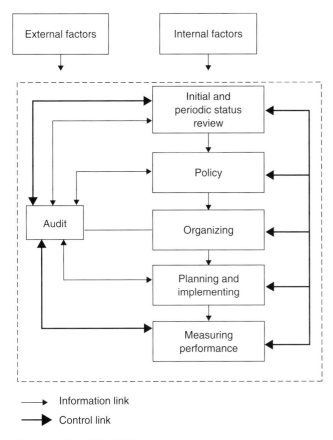

Information link

Control link

Figure 1.29 BS 8800

Figure 1.30 ILO OSH

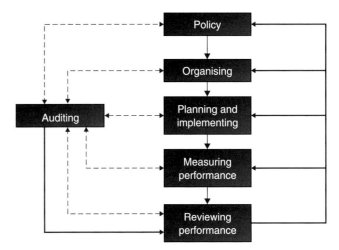

Figure 1.31 HSG 65 approach

1.7.1 HSE's Guide – Successful Health and Safety Management (HSG65)

HSG65 is based upon a framework comprising six core elements, each of these elements interacts with the others enabling a continuous improvement cycle to be obtained.

A brief outline of each element is described below.

Policy

The Policy sets a clear direction for the organisation in its aim to achieve successful and continuous improvement of its safety management system. It also clearly identifies the standards to be achieved in promoting and demonstrating its drive towards improved safety performance. As will be discussed in the following chapter it is evident that senior management are taking their responsibilities for the safe management of people and the environment seriously.

Organising

Organising for safety is more than allocating roles and responsibilities; it must also take into account how activities that influence an organisation's safety and its culture, such as the use of contractors, procurement of materials and the management of information imparted from the organisation, will affect its safety management.

Establishing both organisational and individual competencies and the involvement of staff at each level are key elements in effectively organising a company to manage safety effectively.

Planning and implementing

Establishing a clear health and safety plan based upon the analysis of risks (risk assessment) is an essential element of safety management. The risk assessment process will enable an organisation to prioritise and set objectives for short-, medium- and long-term strategies for controlling risk. Implementing the preventive and protective measures identified by the risk assessment process, setting measurable performance standards against which to measure the implementation programme must be seen as an essential element in producing an effective management system.

Measuring performance

Performance measurement will include both active and reactive monitoring programmes; ultimately this monitoring will enable an organisation to see how effectively it is managing health and safety from investigating an accident which seeks to identify how systems have failed through to actively monitoring safe systems of work, providing a benchmark from which to analyse the organisation's ability to meet its short-, medium- and long-term goals.

Reviewing performance

The results from internal reviews which seek to benchmark performance, both internally and externally, will enable the organisation to ensure that it is achieving the minimum legal standard in terms of compliance, or if it has set them, performance targets that meet its own standards, which may be higher than those required by legislation.

The review process is required by the MHSW Regulations RRFSO and thus to comply with the law must be undertaken. The results of an annual review should be made available to all stakeholders.

Audit

Each of the previously mentioned elements in the management system must be audited to ensure that the performance of any one element does not have a detrimental effect upon another. In order to achieve unbiased results the HSE believes that an audit should be conducted independently from those who can influence the safety management system, e.g. it would be virtually impossible for an internal safety team to audit its own management systems as in all likelihood they would have established many of the key elements.

Note: Students may wish to note that the requirement to manage under both the MHSW and the RRFSO are the same. However, they both differ from the guidance contained in HSG65 in that the regulations require that employers and responsible persons need only plan, organise, control, monitor and review. The need to establish a policy and audit the elements of the management system is not explicitly mentioned in either the

MHSW or the RRFSO; however, as will be discussed later in this book they are critical elements required for successful safety management.

1.8 Case study

Prosecution of a fabrication company by Kent and Medway Fire and Rescue Authority (KMFRA) and the Health and Safety Executive (HSE).

A fire in a fabrication company on 9 June 1999 resulted in the death of an employee. In Maidstone Magistrates' Court on 3 July 2000 the company pleaded guilty to three breaches of fire legislation. In tandem with the Kent and Medway Fire and Rescue Authority investigation and prosecution, the Health and Safety Executive also prosecuted under section 2 of the Health and Safety at Work etc. Act 1974. The investigation and subsequent prosecution was a classic example of two prosecuting agencies (HSE and KMFRA) working together.

The extended length of time between the fire and the court case was due to the delay for an inquest into the fatality and the Crown Prosecution Service's consideration of securing a conviction for 'manslaughter'.

The charges, and subsequent guilty pleas, were based on the following:

Kent and Medway Fire and Rescue Authority

One charge each on the basis that the company was found to be reckless in their failure to provide:

➤ Suitable fire fighting and fire detection
➤ Suitable emergency routes and exits
➤ Suitable maintenance arrangements.

Health and Safety Executive (HSE)

One charge based on the general duty of care under section 2 of the HSWA in that the company was negligent in their duty.

The resultant fines awarded were as follows:

➤ Failure to comply with the Fire Regulations – a total of £6000 with a contribution towards the costs of the fire authority of £2500
➤ Failure to comply with the duties imposed by the HSWA – £10 000 for the one offence with a contribution of £1500 towards the costs of the authority.

A prominent feature of this case was the emphasis by the magistrates on the failing of a business to satisfy the legal and moral obligations to employees. The fact that the company had fully complied with the requirements of the fire authority and the HSE within six weeks was put as mitigation by the solicitor for the defence. Even with a plea of 'poverty' on behalf of the company did not deter the bench from imposing substantial and significant fines.

1.9 Example NEBOSH questions for Chapter 1

1. **Outline** the main responsible person's duties under the Regulatory Reform (Fire Safety) Order 2005 (8)
2. **Describe** the options open to a responsible person should they feel an enforcement notice issued by the enforcing authority is unjustified? (8)
3. **Explain** the meaning, status and roles of:
   (a) Health and safety regulations (3)
   (b) HSC Approved Codes of Practice (3)
   (c) HSE guidance. (2)
4. (a) **Explain**, giving an example in EACH case, the circumstances under which a fire safety enforcer may serve:
   (i) A prohibition notice (2)
   (ii) An enforcement notice. (2)
   (b) **State** the effect on EACH type of enforcement notice of appealing against it. (4)
5. (a) **Outline** the three standard conditions that must be met for an injured employee to prove a case of negligence against their employer
   (b) following an accident at work.
   (c) **State** the circumstances in which an employer may be held (6)
   (d) vicariously liable for the negligence of an employee. (2)
6. **Outline** reasons for maintaining good standards of health and safety within an organisation. (8)

2

Safety policy

Regardless of the type of organisation, its activities and the specific management issues that it faces, clear unequivocal policies relating to general health and safety matters as well as fire safety matters are needed in order to establish effective organisational control of its activities. Management policies that organisations implement are likely to include those for: quality, environmental, safety, equality and fairness. Many of these policies are required to enable an organisation to comply with the law.

A policy is the basis of an organisation's management strategy; providing direction, enabling it to organise, plan, set targets and implement its organisational objectives. However, a policy in itself cannot be effective unless the words are turned into actions.

When managing fire safety (in common with managing all health and safety issues) it is vital that a clear and effective management system is developed and implemented. The model provided by the HSE in the guidance document HSG65 – Effective Health and Safety Management – recommends that the first step to effective management is to have effective policy, a policy that sets the aspirations and direction for an organisation.

This chapter discusses the following key elements:

➤ The importance of setting policy in safety
➤ The key features and appropriate content of an organisation's safety policy
➤ Specific fire safety-related arrangements within a policy.

2.1 The importance of setting policy

2.1.1 Satisfying the law

As has been explained in the previous notes a key feature of any safety management system will be the preparation and implementation of a policy if an organisation is to effectively manage its safety obligations.

The law establishes a minimum standard for the requirements of any health and safety policy. Section 2(3) of the HSWA requires that every employer prepare and revise a written statement of his general policy for the management of health and safety. The requirement extends to ensuring that employees are provided with information in relation not only to the policy statement but also the organisation and arrangements for managing and carrying out the policy.

The requirements for recording the policy inevitably revolve around the size of an organisation. Such a policy should therefore be recorded when five or more employees (or an aggregate of that number) are employed.

In addition to the HSWA, the RRFSO and MHSW Regulations also place a duty upon the responsible person and/or employer to 'make and give effect to such arrangements as appropriate to the nature of his activities and the size of his undertaking, for the effective planning, organisation, control, monitoring and review of the preventive and protective measures'.

The Approved Code of Practice to the MHSW Regulations go on to state that employers should set up an effective health and safety management system to implement their health and safety policy which, as discussed in Chapter 1, is as good (quasi legal) as requiring a policy in law.

Figure 2.1 ACoP of Management of health and safety at work

Figure 2.2 Maximising staff involvement

2.1.2 Maximising staff involvement

Those organisations that involve their staff in planning and managing safety issues in the workplace not only are complying with the requirements under the RRFSO to consult with staff, but will also reduce the fire risks associated with their undertaking. Well-organised companies manage all their functions effectively and recognise the key role that effective health and safety management has in terms of business development and survival.

A clear, effectively communicated, policy that has involved staff and those who may interact with it (contractors, co-employers, etc.) will make an organisation more efficient in the day-to-day decision-making process that reflects and supports the management system in which they are made.

When a line manager who is responsible for a high risk process in a petrochemical plant is empowered to make safety critical decisions and the same manager is aware of the company's safety policy and the ethos behind it, the decisions made for undertaking the process are more than likely to reflect the safety of the site personnel, visiting contractors and the effects on members of the public in addition to the production schedules that need to be met.

The best safety policies are integrated with human resource management, acknowledging that people are a key resource in the management of both safety and production. Human resource management policies can be undermined by poorly written, poorly structured safety policies.

However, organisations that experience higher output, higher quality of service delivery and an enhanced, motivated workforce do the same things, they:

➤ Recognise the benefits of a competent, committed, enthusiastic and fit workforce
➤ Establish arrangements for the promotion of accident-free work practices
➤ Positively promote ill-health management systems.

An effective safety management system and the policy contained within have a direct bearing upon an organisation's safety culture and therefore how a policy is written will have a significant effect.

When conducting safety audits and reviews, which take an in-depth look at an organisation's safety management system, the initial documentation that is likely to be requested will be a copy of the safety policy. The policy and the statement of intent itself are generally seen as being an underpinning requirement of any successful safety management system. Therefore a well-written, well-presented policy is also likely to attract business opportunities, particularly when work is being contracted out, as in the case of the construction industry sector.

2.1.3 Avoiding loss

A clear, effective safety policy will aid the prevention of human suffering and financial loss as a result of the work activities of any organisation.

The safety culture of an organisation will reflect the safety policy. Thus a well-written policy will enable the organisation and those within it to identify hazards and risks before they cause injuries, ill health or other loss outcomes. As has been discussed earlier in this chapter, avoiding financial losses attributed to safety failings is one of the many factors of business success.

For example, in the event of fire in the workplace, following the response procedures contained within a safety policy will mitigate the potential effects of the emergency situation, e.g. shutting down a process safely, preventing injury to persons or damage to the equipment or machinery.

2.2 Key features of a safety policy

In order to satisfy the legal requirements of section 2(3) of HSWA and guidance contained in HSE, DCLG and BS 5588 Part 12 (managing fire safety), to ensure that the safety policy is effective it must include the following:

➤ Policy statement – statement of intent
➤ Organisation
➤ Arrangements.

2.2.1 Statement of intent

The statement of intent can be seen as an organisation's 'mission statement' for effectively managing safety. The policy statement will set the direction that the organisation will take; it will establish standards upon which to measure performance. In addition the policy will set objectives and targets on which to measure the success of the management of safety and the system itself.

In HSG65 the HSE consider that written statements of policy should, at the very least, set the direction of the organisation by demonstrating senior management commitment, placing safety in the context with other business objectives and make a continuous improvement in safety performance.

The statement is likely, therefore, to include naming the director or senior manager with overall responsibility for both the formulation and the implementation of the policy and that the document is duly signed and dated by the director or chief executive. The policy statement briefly explains the responsibility of all persons from board level through to staff members, which also recognises and encourages the involvement of all employees and safety representatives.

Included in this statement will be an outline of the basis for effective communications and how adequate resources will be allocated. It will also commit the organisation's senior management and leaders to plan, review and develop the policy. The need to ensure that all staff are competent and where necessary that external competent advice will be sought, should also assist the

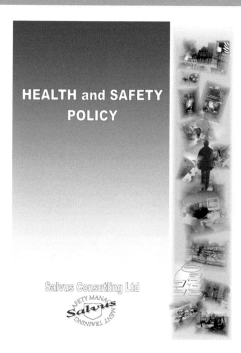

Figure 2.3 Health and safety policy document

organisation in developing a framework for managing safety.

The policy statement will need to be made available to a wide variety of groups that will include staff members and their representatives, contractors and other persons who may be affected by the organisation's activities or undertakings. It often forms part of vetting procedures when considering whether to engage an organisation in tendering processes.

A large proportion of organisations therefore ensure that the policy statement of intent is displayed on safety notice boards within the organisation. The policy can also be included in team briefings and toolbox talks, during induction or refresher training courses. Some companies issue individual copies of the policy in booklets and include reference to it in newsletters and e-mails. The policy is also discussed as an agenda item in a variety of safety meetings including those of a safety, health and welfare committee.

The policy itself can be enhanced by setting clearly defined Specific, Measurable, Achievable, Realistic and Time oriented (SMART) safety objectives.

For an example of a policy statement please refer to Appendix 2.1.

2.2.2 Organisation

This section of the policy is quite often confused with the 'organising for safety' element of a safety management system. This is due to a large number of similarities between the two.

The organisation element of a safety policy should clearly identify and define the roles and responsibilities for all levels of management and operatives within an organisation. It may be the fact that different sections of the policy, specifically fire, may have differing organisational structures due to the need to outsource some of the competent person duties due to a lack of 'in-house' expertise.

It is generally acknowledged that the role of safety adviser, whether fire or health and safety, is to provide support to managers and staff alike. Those to whom they are likely to provide support, advice and guidance are:

➤ Management board
➤ Safety director
➤ Human resources/personnel
➤ Managers and supervisors
➤ Fellow competent persons, i.e. first aiders, fire wardens
➤ Safety committee and employee representatives
➤ Employees.

Each of the above groups should have clearly identifiable roles and responsibilities allocated to them as part of the policy and should be reflected in job descriptions appropriate to the role.

Additional roles and responsibilities will also need to be assigned for other key areas such as:

➤ Fire wardens/marshals
➤ First aiders/other emergency response team members
➤ Safety event (accident/incident) investigators
➤ Occupational health departments
➤ Chairperson of the safety committee
➤ Procurement/purchasing department.

With clearly allocated roles and responsibilities those employees and groups with specific roles and responsibilities will be able to fulfil their functions effectively. Additionally allocating specific roles to individuals facilitates the monitoring of performance of those individuals against the expectations of their role.

2.2.3 Arrangements section

Supporting both the statement of intent and the organisation for safety is the arrangements section of the policy. This section details the guidelines and procedures for the implementation of the safety management system.

Dependent upon the nature of the activities of the organisation, the arrangements section can be either one, that is, included with both the statement and the organisation, section (as a complete document), or a framework section identifying where further guidance for implementation may be found.

Regardless of the mechanism that is used, i.e. integrated or framework, many core elements remain the same. The following bullet pointed list is indicative of the areas included in an arrangements section:

➤ Access and egress	➤ Housekeeping
➤ Accidents/incidents	➤ Lone working
➤ Alcohol/drugs	➤ Manual handling
➤ Asbestos	➤ Noise at work
➤ Consultation with employees	➤ Personal protective equipment
➤ Control of contractors	➤ Permit to work
➤ Display screen health and safety	➤ Plant and machinery
	➤ Office equipment
➤ Electrical safety	➤ Ladders and other access devices
➤ Employment of young persons	➤ Risk assessment
➤ Environmental safety	➤ Security and staff safety
➤ Hazardous substances (COSHH)	➤ Signage
	➤ Smoking
➤ Health surveillance	➤ Stress
➤ Fire and emergency arrangements	➤ Temporary and casual staff
	➤ Training/information
➤ First aid	➤ Utilities and services
➤ Food hygiene	➤ Visitors
➤ Safety inspections and safety tours	➤ Waste disposal
	➤ Working at height
➤ Hours of work	➤ Working environment

Some or all of the above arrangements in larger or more complex organisations may have their own individual policy, organisation and arrangements sections.

Regardless of whether each element of the arrangements section is included within the overall health and safety policy or whether it is an independent policy and/or set of procedures, a critical issue is that those persons nominated in the organisation section of the policy are aware of the arrangements for which they have responsibility.

Fire safety policy

The fire safety policy of an organisation should be structured along the same lines as the general health and safety policy, i.e. it should have a statement of commitment, roles and responsibilities of persons charged with implementing the policy and detail arrangements.

The arrangements detailed below will invariably be included in any fire arrangements section; however, in order to fully appreciate whether all areas are covered the completion of a fire safety review or fire risk assessment will help to identify any shortcomings in the arrangements section, e.g. if contractors are seen conducting hot work operations without a hot work permit this will identify either that no such arrangement exists or that the arrangements are inadequate.

Specific elements that may be included in the fire safety arrangements section therefore may be as indicated in Table 2.1.

Table 2.1   Elements of a fire safety arrangements section

Actions in the event of a fire	Fire safety and electrical equipment
Catering fire safety management	Fire safety furnishings and fabrics
Contingency planning	Fire safety inspections, reviews and audits
Emergency lighting systems	Fire safety training and instruction
Emergency planning	Fire service liaison
Evacuation exercises	Fixed fire fighting equipment/installations
Highly flammable liquids – transportation storage and use	Management of contractors and hot work permits
Fire alarms and detection systems	Means of escape (fire doors and routes)
Fire investigation and reporting (false alarms and incidents)	Personal emergency evacuation plans (disabled persons)
Fire plans (the production of)	Portable fire fighting equipment
Fire risk assessment	Security against arson

While it will be necessary for all staff members to be aware of key elements within the arrangements section, namely actions in the event of fire, evacuation exercises and means of escape, specific personnel with allocated roles and responsibilities will need to have a far greater and in-depth knowledge of specific arrangements, e.g. for those involved with assisting the escape of disabled persons or those involved in conducting and reviewing fire risk assessments.

2.2.4   Review and revision

To ensure that the safety policy remains up to date and reflects the organisation's safety management systems and the risks associated with the organisation's activities, the policy should be subject to regular review and, where necessary, revision. To ensure that the policy

remains valid regular reviews are seen as 'best practice' in line with each of the previously mentioned standards (BS 8800, OHSAS 18001, HSG(65)) and in 'Health and Safety in Annual Reports: Guidance from the Health and Safety Commission (HSC)' available on the HSE's website at www.hse.gov.uk

Changes in circumstance that may affect the validity of a safety policy are numerous and therefore policies could be reviewed under the following headings:

➤ People – may include a change in management structure, director, or the reporting lines within an organisation; may also include change in persons employed, or visiting the premise, such as young people and those with disabilities
➤ Place – may include changes to the types of premises being occupied, or the number of premises being operated out of, or operating in shared accommodation
➤ Plant – may include changes in the type, numbers, or risks attributed to the plant and machinery used by the company, such as moving from air powered tools to battery operated units
➤ Processes – may include changing the process activities of a company relating to the risk level such as moving from construction operations into facilities management activities, or using less harmful products such as water-based paint instead of solvent-based paint
➤ Enforcement action – may include reacting to enforcers' guidance, serving of notices, or prosecution
➤ Legislation and standards – may include changes to legislation, approved codes of practice, or guidance issued by the DCLG, HSC or HSE; may also include changes to British or European Standards
➤ Audit/review results – may include reacting to findings of both internal and external audits which may identify areas of weakness
➤ Consultation – following discussions during safety committee meetings with representative bodies or employees direct
➤ Adverse safety events – may include fire, false alarms, personal injury accidents, ill-health occurrences, damage-only incidents or near miss incidents.

These are but a sample of such changes and there may well be many more that will affect an organisation's policy validity.

2.3   Case study

Following a fire risk assessment, an office-based organisation whose headquarters operated out of a three-storey detached building, which included offices, meeting rooms, a plant room and roof-mounted equipment, identified a number of areas in which improvements could be made.

The fire action plan produced from the risk assessment identified that no one in the organisation had been allocated the responsibility for coordinating a fire incident procedure (gathering information from fire wardens) or providing information to oncoming fire crews. This was addressed by allocating a role and responsibilities for an incident controller, nominating staff members and the introduction of formulating plans and information sheets to be provided to oncoming fire crews.

During the risk assessment process it was also identified that the installation of IT cabling had breached a number of the fire compartment walls (fire resisting) as no one had been allocated the responsibility for reviewing the works undertaken by contractors. This was addressed by reviewing the roles of the facilities management team (FMT) to incorporate this aspect and providing the FMT members with appropriate training.

(b) **Outline** the circumstances that may give rise to the need for a health and safety policy to be revised. (6)

2. **Outline** the issues that are typically included in the arrangements section of a health and safety document. (8)

3. (a) **Explain** the purpose of the 'statement of intent' section of a health and safety policy. (2)

 (b) **Outline** the circumstances that would require a health and safety policy to be reviewed. (6)

4. **Outline** the issues that are typically included in the fire safety arrangements section of a health and safety policy document. (8)

5. (a) **State** the legal requirements whereby employers must prepare a written statement of their health and safety policy. (2)

 (b) **Outline** the various methods for communicating the contents of a safety policy to the workforce. (6)

2.4 Example NEBOSH questions for Chapter 2

1. (a) **Outline** the legal requirements whereby employers must prepare a written statement of their health and safety policy. (2)

Appendix 2.1

Policy statement

General statement

ACME Corporation acknowledges their legal responsibilities as an employer under current national and European legislation and will, as far as is reasonably practicable, provide a safe and healthy environment for staff and visitors to its premises.

The standard set by the legal requirements is seen as the absolute minimum standard by which ACME Corporation will operate, we will strive to meet the highest standards of occupational safety and health.

We will take steps to ensure that our statutory duties are met at all times, including the provision of sufficient funds and facilities to meet the requirement of this policy.

The maintenance of a safe and healthy environment in which to work is seen as an equally key objective as the generation of income-related business.

Each employee will be given such information, instruction and training as is necessary to enable the safe performance of all their work activities.

We also fully accept our responsibility for other persons who may be affected by our activities and will seek to provide those others with appropriate information to ensure their safety.

Bill Bloggs has been appointed the director responsible for health and safety to whom issues related to health, safely and welfare management should be addressed.

It is the duty of management to ensure that all processes and systems of work are designed to take account of health and safety and are properly supervised at all times.

Adequate facilities and arrangements will be maintained to enable employees and their representatives to raise issues of health and safety.

Competent persons will be appointed to assist us in meeting our statutory duties including, where appropriate, external specialists.

Fundamental to the success of this policy is the responsibility and cooperation of all employees for health and safety at work.

Each individual has a legal obligation to take reasonable care for their own health and safety, and for the safety of other people who may be affected by their acts or omissions. Full details of the organisation and arrangements for health and safety will be set out in Chapters 2 and 3 of this document.

Affiliated health and safety policies are listed at Annex A.

Key health and safety objectives

The key objectives contained within this policy will assist us to ensure:

➤ The health and safety of all employees, visitors and others who may be affected by our undertaking;
➤ That all employees have a clear understanding of their individual and collective responsibilities regarding health and safety;
➤ The adequate and appropriate training of all employees;
➤ Effective monitoring of policies/procedures by inspection/audit;
➤ Employees are actively involved via the localised consultation process;
➤ All health and safety policies and procedures are regularly reviewed; and
➤ That standards are set with regard to health and safety management within the organisation.

Chapter 3 of this policy document outlines ACME Corporation's arrangements for health and safety for staff, visitors and others.

This policy will be regularly updated to reflect operational and legislative changes and any 'lessons from experience'. A copy of this policy is available to all ACME Corporation's members of staff and consultants, contractors, other building users and their employees.

We request that all our visitors and those who may come onto our premises respect this policy, a copy of which can be obtained on demand.

Signed:
(Director)
Date:

Organising for safety

3

3.1 Introduction

Key to effective fire and health and safety management is an understanding of the roles and responsibilities that individuals and organisations have for ensuring safe workplaces, safe activities and safe products, etc. This chapter covers a number of important individual and organisation roles and outlines the legal framework that shapes their contribution to safety.

This chapter discusses the following key aspects:

➤ Safety management roles and responsibilities
➤ Joint occupation of premises
➤ Consultation with employees
➤ The supply chain
➤ The supply and use of work equipment
➤ The supply of hazardous substances
➤ Construction safety
➤ Consumer products
➤ Contractor management.

3.2 Safety management roles and responsibilities

3.2.1 The responsible person (RP)

The RRFSO defines a key role for the management of fire safety in organisations as a 'responsible person'.

The Order details the duties placed upon this person to ensure the effective management of fire safety.

The responsible person is required to establish general fire precautions arrangements to safeguard any of his employees and any relevant persons (any person who may lawfully be on the premises and any person who may be in the immediate vicinity of the premises who is at risk from a fire) who are not his employees.

The above paragraph reflects the duties so far as is reasonably practicable and is in line with current health and safety legislation.

The responsibility for completing suitable and sufficient fire risk assessments and implementing the findings of such assessments also falls upon the responsible person. As previously discussed, fire safety arrangements including the provision of fire fighting and detection, emergency routes and exits, procedures for serious and imminent danger, danger areas and emergency measures in respect of dangerous substances also fall under the remit of the 'responsible person'.

Further details relating to the role of the responsible person will be covered later. However, it is worthy of note that the responsible person will invariably also be an employer.

3.2.2 Employers

As discussed in Chapter 1, all employers have duties imposed by both criminal and civil law. In criminal law any employer has a general duty to ensure, so far as is reasonably practicable, the health and safety of his employees while they are at work. In addition, employers have a similar statutory duty to ensure, again so far as is reasonably practicable, the health and safety of any other persons who may be affected by his work undertaking.

In civil law all employers have a duty to take reasonable care of the safety of anyone who may be affected by his work undertaking. This includes, for example, employees, customers, end users of products and other members of the public.

The way in which employers discharge these duties will vary with the size and nature of an organisation. However, in general terms it will be the employer's role to ensure that effective policies are in place, sufficient resources are allocated to ensure that all work can be carried out safely and that an effective safety culture (see Chapter 4) is established and maintained.

It is often the case in larger organisations that the employer may himself be remote from the workplace. For many public service organisations, e.g. NHS trusts, police authorities and social services, the employer is a board of elected representatives of the public. In these circumstances it can be seen that it is even more critical for the employers to set clear policies in order that the directors and senior managers of the organisation can fulfil their roles to assist in discharging the legal duties of their employers.

As with all persons who have safety responsibilities within any organisation the exact roles and responsibilities of the employers to achieve effective health and safety management will be detailed in the health and safety policy required by section 2(3) of the HSWA.

3.2.3 Directors and senior managers

The Health and Safety Commission has identified health and safety as a boardroom issue. The chairman and/or chief executive are seen as having a critical role to play in ensuring risks are properly managed. They further argue that those at the top of any organisation have a key role to play. In order to allow directors and senior managers to contribute to the HSC it is recommended that a director is given specific responsibilities for health and safety. These again will be detailed in the organisation section of the company's health and safety policy.

Current authoritative guidance suggests that the 'boards of companies' take the following five actions:

1. **The board needs to accept formally and publicly its collective role in providing safety leadership in its organisation**. Strong leadership is seen as being vital in delivering effective risk control. Everyone should understand that the most senior management is committed to continuous improvement in safety performance.
2. **All members of the board should accept their individual role in providing safety leadership for their organisation**. Board members are encouraged to ensure that their actions and decisions always reinforce the messages in the board's safety

Figure 3.1 The responsibility for safety management starts and stops in the board room

policy statement. Any mismatch between board members' individual attitudes, behaviour or decisions and the organisation's safety policy will undermine employees' belief in the intentions of the board and will undermine good fire and health and safety practice.

3. **The board needs to ensure that all its decisions reflect the organisation's safety intentions, as stated in the policy statement**. Many business decisions will have both fire and health and safety implications. It is particularly important that the safety ramifications of investment in new plant, premises, processes or products are taken into account as the decisions are made.
4. **The board must recognise its role in engaging the active participation of workers in improving safety and risk management**. Effective safety risk management requires the active participation of employees. Many successful organisations actively promote and support employee involvement and consultation. Employees at all levels should become actively involved in all aspects of a safety management system.
5. **Current best practice is that boards appoint one of their number to be the 'safety director'**. By appointing a 'safety director' there will be a board member who can ensure that fire and health and safety risk management issues are properly addressed throughout the organisation.

It is important that the role of the safety director should not detract either from the responsibilities of other directors for specific areas of fire and health and safety risk management or from the safety responsibilities of the board as a whole.

Directors and senior managers often carry out the function of the top-level executive. In order to achieve this, directors and senior managers will need to establish the policy and put in place and monitor the detailed organisation and arrangements section of the safety policy.

As the top level of management, directors and senior managers are responsible for putting in place arrangements to support a positive safety culture including:

➤ Establishing the methods of management **control** throughout the organisation
➤ Securing effective **cooperation** between individuals, safety representatives and groups
➤ Ensuring effective **communication** throughout the organisation is maintained
➤ Facilitating and monitoring the necessary individual and organisational **competencies**.

BS 5588 Part 12 (fire safety management) confirms current 'best practice' when it argues that fire safety management is best achieved through the appointment of a single individual who is made responsible for all fire safety matters within the organisation. Experience shows that where the responsibility for fire safety management is spread throughout an organisation, break down in communication and control results in the ineffectual management of fire safety. BS 5588 confirms the view that if the role of the manager responsible for fire safety is ill-defined the standard of management is likely to be poor.

As previously discussed senior managers and directors have specific personal liabilities placed upon them by the HSWA and the RRFSO. Section 37 of the HSWA and Article 32(8) RRFSO allow that, in addition to the liability of a corporate body to be prosecuted for a breach of statutory duty, an individual manager may also be prosecuted for the same offence.

In its own guidance the HSE advise that the health and safety duties of the board are to:

➤ Review the health and safety performance of the organisation regularly (at least annually)
➤ Ensure that the health and safety policy statement reflects current board priorities
➤ Ensure that the management systems provide for effective monitoring and reporting of the organisation's health and safety performance
➤ Be kept informed about any significant health and safety failures, and of the outcome of the investigations into their causes
➤ Ensure that the board addresses the health and safety implications of **all** the board decisions
➤ Ensure that health and safety risk management systems are in place and remain effective. Periodic

audits can provide information on their operation and effectiveness.

The above issues can easily and succinctly be adapted to meet fire safety needs of an organisation.

3.2.4 Middle managers and supervisors

The role of middle managers and supervisors will normally be centred on implementing the detailed arrangements for all functions of the organisation, including health and safety.

Although used in the HSWA and supporting ACoPs and guidance, the term supervisor is becoming increasingly redundant. The HSE, recognising that the term 'supervisor' can give a negative impression of an overly autocratic role, acknowledges the alternative role of 'team leader'.

Whichever term is applied, middle managers and team leaders have a key role to play implementing health and safety policies. Their roles are likely to include:

➤ Providing information, training, instruction and supervision for those staff they have responsibility for
➤ Providing technical input to the formulation of polices and work practices
➤ Providing feedback to senior management on the effectiveness of health and safety policies and their implementation.

3.2.5 The competent safety adviser (safety assistance)

Both Article 18 of the RRFSO and Regulation 7 of the MHSW Regulations require that the responsible person and/or the employer appoint one or more competent persons to assist him undertaking the measures he needs to take to comply with the requirements imposed upon him under the relevant statutory provisions.

This is an absolute legal duty to appoint one or more persons who have adequate knowledge, training, experience or other qualities to enable them to assist the responsible person and/or the employer discharge their legal duties.

In deciding who and how many persons to appoint as competent advisers the employer/RP must take into account:

➤ Nature and scope of work activities and the size of the undertaking/premises
➤ The work involved
➤ The principles of risk assessment and prevention
➤ Any current legislation and standards

➤ The capability of the individual to apply their competence to the task to which they are assigned
➤ The time available for them to fulfil their functions and the means at their disposal.

Once again, the role of the competent safety adviser will be detailed within the organisation section of the safety policy. Dependent upon the size and nature of the organisation the functions of the adviser will include:

➤ Assisting in the development of the safety policy
➤ Facilitating and assisting in the production of risk assessments
➤ Advising the responsible person and/or the employer and the management board on hazards and risk control measures
➤ Monitoring safety management systems
➤ Active monitoring – tours and inspections, health surveillance or air quality sampling
➤ Reactive monitoring – investigating safety events, reports, etc.

Figure 3.2 The responsible person assisted by a competent adviser should manage fire safety in the workplace

➤ Supporting the works safety committee by the provision of:
  ➤ Statistics
  ➤ Technical information
  ➤ Details of any relevant statutory provision
➤ Liaising with external bodies, e.g.:
  ➤ Enforcers
  ➤ Contractors
  ➤ Scientific specialists
  ➤ Insurance companies
  ➤ Other employers
  ➤ Trade bodies
➤ Providing health and safety advice to all managers and staff as required
➤ Communicating changes in current safety legislation and best practice to the responsible person and/or employer, safety representatives, employees and the management board.

The competent advice and safety assistance in relation to fire includes the fire safety manager and those who provide assistance to them (fire warden, fire incident controller, alarm verifier – see Chapter 10). The law also allows for this competent person or persons to be appointed from outside an organisation such as the use of consultants.

The RRFSO requires that any person appointed by the responsible person is given appropriate information about the factors that could affect the safety of all persons and about any person working within the organisation who may be more at risk (disabled persons, sensory impaired persons, etc.).

3.2.6 Fire safety manager

There is no specific terminology for a variety of roles and responsibilities for those managing fire safety. British Standard 5588-12:2004 Fire Precautions in the Design, Construction and Use of Buildings – Part 12: Managing Fire Safety provides an outline of the responsibilities of the fire safety manager which include:

➤ Conducting fire risk assessments to assist in identifying and reducing the likelihood of fire occurring
➤ The development and implementation of a fire strategy
➤ Undertaking training of staff, including evacuation exercises and the maintenance of training and evacuation exercise records
➤ Conducting inspections, arranging for maintenance and testing of potential fire hazards
➤ Monitoring and maintaining the means of escape, fire evacuation procedures and the behaviour of occupants and reviewing fire plans accordingly

➤ Ensuring access and egress and other appropriate provisions for disabled people
➤ Arranging maintenance, testing and, where appropriate, inspection of:
  ➤ Fire safety equipment
  ➤ Systems and procedures
  ➤ Emergency communication systems
➤ Monitoring any general maintenance/building works that may affect fire safety arrangements, including the supervision, monitoring and instruction of contractors
➤ Management of non-routine high fire risk activities including the issue of hot work permits
➤ Notification to the authorities (fire authority, building control) of any changes that may affect fire precautions, e.g. changes to internal layout, extensions or structural alterations and where appropriate if highly flammable or explosive mixture levels alter
➤ Ensuring that relevant British Standards and other appropriate standards are complied with.

The above list of responsibilities assigned to the fire safety manager appears quite extensive and it is likely that additional competent persons will need to be appointed to assist in the management of fire safety. Therefore, additional responsibilities of a fire safety manager in a large, complex organisation may also include the following:

➤ Appointment of fire marshals/fire wardens, fire alarm verifiers, fire incident controllers and members of first responder/fire teams
➤ Undertaking a training needs analysis and development of a training policy to ensure staff members have necessary competencies
➤ Organising/conducting fire safety reviews and audits

Figure 3.3 Fire fighting equipment must be appropriate and adequately maintained

➤ Ensuring the continued effectiveness of any automated fire safety system, e.g. fire alarm and detection systems
➤ Pre-entry fire safety checks of the premises prior to entry by members of the public
➤ Monitoring and reviewing the fire safety manual (log book) including false alarms and near miss fire events
➤ Recording of changes to the building for inclusion in the operations and maintenance manuals or health and safety file under CDM
➤ Contingency planning:
  ➤ Abnormal occupancy levels – failure or planned preventive maintenance of fire safety critical systems
  ➤ Abnormal inclement weather conditions
  ➤ Collaborating with local authorities involved in disaster planning and the assessment of potential environmental impact of fire (see Chapter 13).

The level of responsibility and accountability of the fire safety manager will need to reflect the organisation; however, any safety management structure should provide for clear lines of responsibility, authority, accountability and resources, in particular in relation to common areas within multiple occupied buildings/premises.

The above management responsibilities may well be assigned to a number of persons within an organisation. Training records, for instance, may be the responsibility of the human resources department and the maintenance of fire safety systems records may rest with the facilities management or property services department.

3.2.7 Fire safety coordinator

In some cases the fire safety manager will wish to delegate some of his duties to an individual to coordinate the management of fire safety. This is particularly relevant in the construction industry for larger projects, particularly those that fall under CDM, where it is recommended that each site has a fire safety coordinator whose role is to:

➤ Ensure all preventive and protective measures on site are maintained
➤ Assist in the development of the construction site fire risk assessment
➤ Nominate and manage fire marshals
➤ Ensure contractors operate in a way that minimises the risk of fire
➤ Manage the permit to work system for hot works
➤ Ensure effective arrangements for liaising with the fire service and other emergency service prior to and at the time of an incident.

Figure 3.4 The reactive role of a fire warden during evacuation

3.2.8 Fire wardens/marshals

The fire safety manager and/or fire safety coordinator will, in most cases, require the assistance of competent fire wardens or fire marshals (these terms are interchangeable). Fire wardens/marshals have two distinct roles; first, they will assist in the overall management of fire safety in a workplace (active management) and second, they will undertake specific duties in the event of a fire emergency to ensure the full and safe evacuation of all people in the premises and to ensure, as far as is possible, the risk of fire spread is minimised (see Chapter 10).

3.2.9 Fire alarm verifier

As part of the evacuation procedure of a premise fitted with an automatic fire alarm the role of an alarm verifier should be considered. The fire risk assessment will indicate at what stage an emergency evacuation may be initiated and, depending upon the risk, this will either occur prior to a verification of the alarm, at the same time as the alarm is being verified or as a result of the verification of the alarm.

The specific duties of the alarm verifier include:

➤ Attending the fire alarm indicator panel at the time the alarm sounds
➤ Identifying the reported location of a fire
➤ Conducting a physical check of the location of the fire to ascertain if there is a fire or a false alarm
➤ Communicating the situation to the fire safety manager or fire incident controller.

3.2.10 Fire incident controller

The role of fire incident controller is given to a person who will be responsible for the proactive and reactive roles of the fire wardens/marshals and alarm verifier. In addition they are the person responsible for coordinating any emergency evacuation including liaison with the fire and other emergency services.

Further details relating to the above fire specific roles will be included later in the book (Chapter 10).

3.2.11 Employees and self-employed persons

The safety role and responsibilities of individual employees will vary depending upon the size and nature of the organisation and the job description of the individual. For example, some employees will hold designated roles, e.g. fire wardens whose duties will include the proactive monitoring of certain fire risk control measures; some employees may hold designated roles relating to the purchase of plant or equipment that satisfies certain statutory requirements; some individuals, as discussed above, may be directors or managers of companies.

Irrespective of the specific job role of any individual employee, the law imposes certain statutory duties on them as employees. Table 3.1 summarises those individual duties. It will be remembered that in addition to these duties, senior managers and directors are liable to be prosecuted as well as the body corporate.

In recent years in the UK there continues to be a growth in self-employed persons. These range across all industry sectors from home-based and peripatetic business consultants and advisers to the more traditional workforce of the construction or vehicle maintenance industries.

It is generally the case that the bureaucratic burden on such people is considered overwhelming. For a self-employed person business continuity and cash flow are often priorities which leave little resources for researching and implementing well-founded safety practices.

In the higher risk industries, for example, such as construction, the self-employed, although working as individuals, represent a significant risk to others who may be affected by their work activities. Significant statutory control of the risks associated with the self-employed is imposed by, among other legislation, the HSWA and CDM.

THE RRFSO does not specifically place duties upon self-employed persons, in so far as the significant duty holder is the responsible person although it may be the case that the responsible person is, incidentally, self-employed for the purposes of the management of fire.

The health and safety duties imposed directly on the self-employed are limited and contained primarily within the HSWA and the MHSW Regulations.

Table 3.1 A summary of individuals' duties of care

The law	Section/ Regulation/ Article	The duty
The Health and Safety at Work etc. Act 1974	7	To take reasonable care of their own health and safety and of others who may be affected by their acts or omissions at work
	7	To cooperate with their employer and others in the discharge of their legal obligations
	8	Not to interfere or misuse anything that is provided for their safety
The Management of Health and Safety at Work Regulations 1999	14	To inform his employer of any work situation which he would reasonably consider being of immediate and serious danger to health and safety
	14	To inform his employer of any matter which he would reasonably consider as representing a shortcoming in the employer's protection arrangements for health and safety
The Regulatory Reform (Fire Safety) Order 2005	23(1)a	To take reasonable care of their own health and safety and of *relevant persons* who may be affected by their acts or omissions at work
	23(1)b	To cooperate with their employer and others in the discharge of their legal obligations
	23(1)c(i)	To inform his employer of any work situation which he would reasonably consider represents a serious and immediate danger to safety
	23(1)c(ii)	To inform his employer of any matter which he would reasonably consider as representing a shortcoming in the employer's protection arrangements for safety

Under the HSWA, self-employed persons have general duties to ensure, as far as is reasonably practicable:

➤ The safety of themselves while at work
➤ That others who may be affected by their work activities are not exposed to risks to their health and safety.

The general duties imposed by the HSWA are extended by the MHSW Regulations in that the self-employed are required to:

➤ Conduct a risk assessment – there is no legal requirement to record the findings
➤ Cooperate with other people who work in the same premises and, where appropriate, cooperate in the appointment of a health and safety coordinator
➤ Provide clear understandable information to other people's employees working in their undertaking.

3.2.12 Persons in control of premises

The term premises includes buildings, outdoor areas such as farms or railway lines but specifically excludes domestic premises. Persons in control of non-domestic premises include such persons as:

➤ Owners
➤ Leaseholders
➤ Tenants
➤ The major employer occupying the premises.

Under section 4 of the HSWA, persons in control of premises have a duty to ensure, as far as is reasonably practicable, the safety of:

➤ The premises
➤ The means of access and egress and
➤ Equipment or substance provided for use in the premises.

It may be the case that the controller of the premises has to provide information, instruction and/or training to persons having access to the premises. It follows that unless the premises are intrinsically safe for all cases of people who may gain access, including trespassers, the controller of the premises will need to exercise some form of entry control.

In the example of railway premises, access to the hazardous areas, i.e. the railway line itself, is prevented by the provision of security fencing.

In the example of a tenanted, multi-storey office block the safety of people using/maintaining the lift is a statutory duty of the controller of the premise.

The RRFSO places additional legal duties for fire safety management on any person who is in 'control' of a workplace. The RRFSO identifies the person in control of a premises as either:

➤ The employer, if the workplace is to any extent under his control or
➤ If the workplace is not to any extent under the control of an employer:
  ➤ The person who has control of the premises whether for profit or not or
  ➤ The owner.

Having identified the person who has control over a premise the RRFSO defines them as the 'responsible person'. The responsible person (see Chapter 1) then becomes the main duty holder for fire safety and has, as a result, overall responsibility for:

➤ Undertaking the fire risk assessment
➤ Putting precautions in place to safeguard employees and non-employees

➤ Ensuring that testing and maintenance is carried out for such aspects as:
  ➤ Fire detection and alarm systems
  ➤ Fire fighting equipment
  ➤ Emergency exit routes and fire exits
  ➤ Fire evacuation drills and assembly points.

3.3 Joint occupation of premises

Where two or more responsible persons and/or employers share or have duties in respect of a workplace whether or not it is a permanent or temporary arrangement, each party has absolute legal duties to:

➤ Cooperate with all other responsible persons and/or employers in order that they may all secure adequate fire precautionary measures for the premises
➤ Take all reasonable steps to coordinate the general fire precautions for the premises
➤ Inform all the other responsible persons and/or employers of the risks to safety arising from their work activities
➤ Where two or more responsible persons share the premises and the potential for an explosive atmosphere may occur it is the responsibility for the person who has overall charge of the premises to coordinate the risk controls to manage any potential explosive atmospheres.

These duties extend to the self-employed, which means that employers have a duty to cooperate, coordinate and inform self-employed persons as well as other employers sharing the workplace. In addition, of course, the self-employed have identical duties to cooperate, coordinate

Figure 3.5 Common circulation areas are the responsibility of those in control of the premises

Figure 3.6 Light industrial units jointly occupied

and inform employers and other self-employed persons with whom they share a workplace.

In order to ensure that all the fire risk control measures are taken for the whole workplace it is absolutely essential for people sharing a workplace to work together. For example, they will all need to conduct a fire risk assessment (see Chapter 5). It will only be possible to ensure the assessment is suitable and sufficient and the resultant control measures are effective if the premises as a whole are being considered.

> It may be the case that one of the employers sharing a premises has control over the premises. Such would be the situation when a small portion of a premises is sublet. In this case the person in control of the premises should take the lead role in ensuring adequate fire safety arrangements and exchange of health and safety information is managed adequately. When there is such a situation the subletters will need to assist the person in control of the premises.
>
> Where there is no controlling employer the employers and self-employed people sharing a particular workplace will need to agree joint arrangements for discharging their duties under common statutory provisions. For example, they will all need access to competent safety advice and may achieve this communally by appointing a safety coordinator.

Similarly the responsible person (e.g. a landlord) although they do not have employees on site will hold duties under section 4 of the HSWA for ensuring safe egress and access to common areas of the premise and those that fall under 'landlord domain'. This will extend to ensure that effective emergency arrangements for all those on the premises are secured.

In terms of information relating to fire safety matters, the responsible person, in addition to providing information to employees, also has a duty under the RRFSO to provide relevant and complete information relating to both the risks and preventive and protective measures including the emergency arrangements in the event of a fire to:

➤ Persons working in their undertaking who are not employees and
➤ Employers and employees who share premises or may be affected by their activities.

3.4 Consultation with employees

The vast majority of safety legislation requires the provision of information to employees. The effectiveness of a safety culture relies heavily upon the quality and standard of the information provided.

If employees are to comply with systems of work to keep themselves and others safe specific information should be provided such as:

➤ The details of any risk assessment carried out, particularly the risks inherent with the facility or task by which they may be affected
➤ The control measures such as those that prevent staff and others being harmed and any appropriate protective measures
➤ The emergency procedures, whether for fire, bomb threat, chemical spillage, asbestos release, etc.
➤ The names, roles and responsibilities of the competent persons, e.g. safety adviser, fire wardens, first aiders, etc.

The provision of information is generally seen as being a one-way mechanism by which the responsible person and/or employer provides the details as above. However, to be successful in safety management and to comply with the law a prudent responsible person and/or employer will establish a consultation process with employees and others.

Employee involvement supports a positive safety culture (see Chapter 4) where safety is everyone's responsibility and they feel ownership for safety matters. The best form of participation is a partnership, where workers and their representatives are involved in early consultation

Figure 3.7 Consultation with employees

of changes and in the solving of any potential problems, rather than being merely informed after decisions have already been taken.

One of the most effective methods of establishing a safety culture is to consult with employees over safety-related matters, and be seen to act on their suggestions, where reasonably practicable. This is the main difference between consulting with employees and merely informing them as required by current legislation. This statutory consultation raises the profile of safety among the workforce and ensures that the management team is seen to be committed to the partnership approach to safety.

Current safety legislation provides for the appointment of recognised trade union safety representatives (this has to be completed in writing).

The employer has to consult with the safety representative to ensure the adequacy of arrangements to safeguard health and safety at work.

In addition, the law allows for the formation of safety committees to ensure that consultation can be managed effectively and which can monitor and review the measures taken to ensure health and safety at work.

3.4.1 Specific legislation that requires consultation (other than HSWA)

The Safety Representatives and Safety Committees Regulations 1977 (SRSC), which deal with the appointment of representatives in unionised workplaces, and the Health and Safety (Consultation with Employees) Regulations 1996, which applies where the workplace in question is not covered by a recognised trade union.

The Schedule to the MHSW Regulations had the effect of updating the SRSC Regulations by placing an obligation on employers to consult safety representatives in good time with respect to:

➤ Workplace measures which may affect the health and safety of employees
➤ The arrangements for appointing competent persons
➤ Any statutory obligations to provide health and safety information to employees
➤ Any statutory obligations to plan and organise the health and safety training of employees and
➤ The health and safety consequences of the introduction of new technology.

In addition to these duties, the employer must also provide any such facilities and assistance as time off and training that safety representatives may reasonably require in order to carry out their full range of functions.

Safety Representatives and Safety Committees Regulations 1977

Under these regulations, the appointment of representatives is undertaken in writing by a recognised trade union who will then inform the employer of the name of the representative and the group of employees they will represent.

A safety representative does not require any specific qualifications to fulfil the role, but the minimum requirement should be at least two years' service with the present employer or two years in similar employment. Once appointed, a safety representative remains in the position until they resign the appointment, leave the employer or the nomination by the appointing trade union is withdrawn.

There is no statutory requirement on the number of representatives required for a particular size of workforce. The guidance notes offer the advice that when determining the numbers appointed, consideration should be given to:

➤ The total numbers employed
➤ The variety of the different occupations involved
➤ The size of the workplace and the locations involved
➤ Particulars of shift systems and
➤ The inherent dangers of the work activities.

Functions of the representatives

It is important to note that representatives are given functions only, as opposed to duties. A function is defined as an activity that a safety representative is permitted to carry out by legislation but does not have a duty to perform. It is, therefore, treated as an advisory action. A safety representative cannot be held accountable for failing to carry out a function or for the standard of the advice given when performing a function, and the HSE has issued statements to the effect that no action will be taken against them in respect of these regulations.

This does not mean that representatives have no duties at all. They are still required to act reasonably and to cooperate with the employer, but there are no *additional* duties under the SRSC Regulations.

The reason for this specification was that if a representative were to be held liable for advice he gave to management, or for not carrying out safety inspections, for example, there would be a severe shortage of those willing to volunteer for the role and safety culture would suffer accordingly.

The main functions are as follows:

1. To take reasonably practical steps to keep himself informed of:
   (a) The legal requirements relating to the health and safety of people at work, particularly the group of people they directly represent

Figure 3.8 Safety representatives have a right to conduct inspections of the workplace quarterly.

(b) The particular workplace hazards and the measures necessary to eliminate or minimise the risks deriving from these hazards

(c) The employer's health and safety policy, as he is included in it somewhere

2. To encourage cooperation between the employer and his employees in promoting and developing essential measures to ensure the health and safety of the workforce and in checking the effectiveness of these measures

3. To investigate:
(a) Hazards and accidents in the workplace, specifically those which are reportable to the enforcing authority and
(b) Employee complaints relating to health, safety and welfare

4. To carry out inspections in the workplace, with a right to do so every three months

5. To alert the employer, in writing, to any unsafe or unhealthy working practices or conditions, or unsatisfactory arrangements for welfare at work

6. To represent the employees to whom he has been appointed in consultation with HSE inspectors and any other enforcing authority. This may be as a result of an accident, an employee complaint or simply a routine visit by the inspector

7. To receive information from HSE inspectors regarding site visits, findings of inspections or investigations, and any future action to be taken.

8. To request the formation of a safety committee (two or more representatives must make the request).

9. To attend safety committee meetings in connection with the previous items.

A recurring theme throughout these regulations is that the employee representatives have the right to perform certain functions, and the employers are then required to provide them with such facilities and assistance as they may reasonably require them to carry out.

Specifically, an employer must:

➤ Provide time off with full pay to enable the representative to carry out his functions and to undergo such safety training as may be required. There is no statutory requirement determining the type or level of training required, but guidance suggests it should be appropriate to the level and type of risks encountered within the workplace in question

➤ Provide information which the representative may need to fulfil their functions. The guidance notes to the regulations suggest that this should include:
(a) Information about the plans and performance of the workplace, with particular regard to any proposed changes which may have health and safety implications
(b) Technical information about hazards in the workplace and the precautions necessary to overcome them. This might include safety manuals, materials safety data sheets, manufacturers' instructions, etc.
(c) Records of any accidents and diseases and statistics relating to these
(d) Any other information which would be relevant, for example results of inspections, air monitoring, risk assessments, noise surveys, etc.

In general, the functions laid down for union appointed safety representatives under the SRSC are taken as a recommendation for the conduct of non-union appointed representatives. To properly deal with representatives from these other workplaces it is necessary to look at the regulations which were introduced to include them in the consultation process.

The Health and Safety (Consultation with Employees) Regulations 1996 (HSCER)
The HSCER covers consultation with employees who are not in groups covered by trade union elected safety representatives. These employees can be consulted directly or through their own elected representatives (the consultation method should be one suitable to all the parties involved).

Employers' duties
Where there are employees who are not represented by safety representatives under SRSC, the employer has to consult those employees or their elected representatives (representatives of employee safety) on matters relating

to their health and safety at work, in particular with regard to:

(a) The introduction of any measure which may affect employee health and safety
(b) The arrangements for appointing health and safety assistance and procedures for serious and imminent danger and for danger areas
(c) Any health and safety information which may be required under any legislation
(d) Planning and organisation of any training in connection with health and safety
(e) The health and safety consequence of any new legislation.

Employer's duty to provide information
Employees or representatives should be given enough information to allow them to take a full and effective part in consultation.

Employers do not have to provide information if it:

➤ Is contrary to national security or unlawful
➤ Concerns individuals who have not consented to information being divulged
➤ Would harm the business, unless this coincides with effects on health and safety or
➤ Involves information connected with legal proceedings.

Functions of Representatives of Employee Safety (ROES)
ROES have the following functions, similarly to their union counterparts:

➤ To make representations to the employer on hazards, risks and dangerous occurrences in the workplace which affect, or could affect, the group of employees they represent
➤ To make representations to the employer on general matters affecting the group of employees they represent and, in particular, any such matters the employer consults them about
➤ To represent the group of employees in any discussion with health and safety inspectors.

3.5 Safety committees

When two or more union appointed safety representatives request, in writing, the formation of a safety committee, the employer must establish such a committee within three months.

The membership and structure of the safety committee is determined by consultation between the management and the representatives concerned. This should be aimed at keeping the total size as compact as possible but with adequate representation of the interests of management and employees.

To this end it may be necessary in large workplaces to appoint subcommittees who will discuss areas of common concern and then nominate one spokesperson to represent them all at the main committee level. As a general rule a large safety committee not only has longer meetings but it is less effective than a smaller committee.

For a safety committee to operate effectively it is necessary to determine its objectives and functions. To succeed, both management and employees have to demonstrate a commitment and positive approach to a programme of accident prevention and the establishment of a safe and healthy environment and safe systems of work.

3.5.1 Objectives

The primary objectives of a health and safety committee should be:

➤ The promotion of health, safety and welfare at work by the provision of a forum for discussion and
➤ The promotion and support of normal employee/employer systems for the reporting and control of safety issues.

Typically terms of reference for a **safety committee** may include:

➤ Examining the safety adviser's report and other reports on the agenda with a view to recommending actions and assessing priorities
➤ Monitoring the effectiveness of health, safety, fire and welfare communication and publicity
➤ To develop, progress and assist in the implementation of health at work initiatives
➤ Monitor, evaluate and assist in the development of the organization's safety-related policies and operational procedures
➤ To review accidents, incidents, collect statistics, analyse trends on occupational health and study reports on health, safety and fire safety issues
➤ To review new, existing and recurring issues revealed by safety audits
➤ Report to individual senior managers or an organisation's senior management team on matters concerning safety at work
➤ To monitor safety training programmes and standards achieved
➤ To act as a consultative forum on workplace measures which may affect health, safety and welfare of employees

➤ To consider enforcing authority reports and information releases

➤ To assist in the development of safety rules and safe systems of work

➤ To review health, safety and welfare aspects of future development and changes in procedure

➤ To review safety aspects of purchasing equipment and materials and in contract procurement

➤ To review renewal/maintenance programmes.

3.5.2 Composition of the committee

To function effectively it is necessary for the committee to be composed of a representative sample of the workforce. This usually requires a balanced approach consisting of equal representation from management and workforce.

The suggested composition is:

➤ **A senior manager** who must have **adequate authority** to give proper consideration to views and recommendations and ensure that budgets are allocated to meet them. If this manager is the person who signed the safety policy this helps to affirm the management's commitment to health and safety issues.

➤ **Supervisors, or line managers**, who have the **local knowledge and expertise** necessary to provide accurate information on company policy, production needs and technical plant, machinery and equipment.

➤ **Specifically competent people** may need to be present if a **particular topic of concern** is to be discussed which exceeds the level of knowledge of those normally present. It is suggested that a bank of experts is available from which relevant persons may be selected as necessary, rather than having large numbers of extraneous personnel present at each meeting.

➤ **A safety adviser and/or the fire safety manager** to provide **guidance** on statutory requirements, regulations, codes of practice and safe systems of work, capable of interpreting and suggesting methods of compliance and implementation.

➤ **Safety representatives** need no specific qualifications for committee membership, but due consideration should be given to **experience**. There is no guidance on the length of service of representatives but a minimum term of two years is acceptable with half the committee retiring every year. This has the effect of livening up the committee every year and allows the safety message to be spread among more participating members.

The number of representatives varies depending on the nature and activities of the business, as previously discussed. Special consideration should be given to minority groups within the company, such as apprentices, religious minorities or the disabled, if these have specific safety concerns.

3.6 The supply chain

3.6.1 General

The term supply chain refers to all those parties involved in the design, manufacturing, importation, erection and installation of any article or substance used either at work or by an end user of any articles and substances used at work.

The supply chain concept is useful when considering the dangers associated with any particular article or substance used at work in that it allows consideration of a number of sources of hazard and risk. For example, purchasing non-flame retardant curtains and furniture coverings for an area in which people are allowed to smoke, such as a work's social club.

It is extremely useful to note that the supply chain involves those materials that are not only purchased but are hired, borrowed or donated and also include the procurement of services (contractors, etc.) and for this reason it is helpful to use the term procurement.

Among the problems associated with procuring articles, substances and services for work are the provision and use of:

➤ Substandard equipment

➤ Poorly equipped, poorly trained contractors

➤ High risk chemicals and substances in inappropriate working environments

➤ Work equipment outside its design parameters (using it for what it is not designed to do)

➤ Inadequate information relating to:

   ➤ Hazards associated with chemicals and substances

   ➤ How contractors undertake their activities

   ➤ Safe means of maintaining and testing machinery.

3.6.2 The law relating to the supply chain

Section 6 of the HSWA places duties upon *all* those who design, manufacture, import, or supply any article for use at work. The duty that is placed upon them is threefold:

1. To ensure so far as is reasonably practicable, that any article used at work in the UK is designed and constructed as to be safe and without risks to health when properly used

49

2. To carry out or arrange for the carrying out any tests and examinations as may be necessary to ensure that work articles are safe and without health risks when used properly
3. Take steps to ensure that there will be information available relating to the design use and limitations of any article used at work. The information should include any instructions on how to use the article safely. It should also detail where applicable how to install, handle, erect and dispose of the items to ensure they remain safe and fit for purpose and any residual risks inherent with the items have been reduced to the lowest level reasonably practicable. This information must be provided to all persons those who may require it.

In addition to the duties placed on all those in the supply chain to reduce risks and provide information, specific polices and practices are required to ensure the safe supply, provision and use of:

➤ Work equipment
➤ Hazardous substances
➤ Buildings.

The following sections discuss the management of these subjects.

Figure 3.9 PUWER covers all work equipment

3.7 Work equipment

In addition to the duties placed on designers, manufactures, importers and suppliers of articles used at work, there are further, specific duties placed on those who supply, provide and use *work equipment.*

The term work equipment covers a vast range of tools, vehicles and machinery. Those articles used at work that present a low risk are always covered by the general duties in section 6 of the HSWA; however, for work equipment and machinery there are additional requirements imposed by the Provision and Use of Work Equipment Regulations 1998 (PUWER).

3.7.1 The Provision and Use of Work Equipment Regulations 1998 (PUWER)

In these regulations work equipment is defined as:

> *any machinery, appliance, apparatus or tool for use at work*

This includes hand tools, single machines, lifting equipment, ladders and other access equipment, vehicles, office equipment, etc.

The term use applies to not only the actual use for which the equipment has been designed and manufactured but importantly when the equipment is:

➤ Installed
➤ Programmed and set
➤ Started and stopped
➤ Maintained and serviced
➤ Cleaned
➤ Dismantled and
➤ Transported.

Suitability of work equipment

The regulations require that when an employer is considering the provision of a piece of work equipment he first must consider the suitability of the proposed equipment for the work. When considering the suitability of a piece of equipment an employer must take into account:

➤ The person/s who are going to use the equipment, in terms of their size, physical and mental abilities, levels of competence and motivation
➤ The task/s to be performed
➤ The frequency of use for the equipment
➤ The working conditions and environment in which it will be used.

If a piece of work equipment that is not intrinsically safe (i.e. spark proof) is installed within a potentially flammable or explosive atmosphere it could cause a fire or explosion.

Employers must ensure that work equipment conforms to current EU standards and requirements. Most new work equipment particularly any machinery must display the European CE mark. The CE mark is merely a claim by the manufacturer that the equipment meets all relevant standards and is safe.

In addition to researching and controlling the hazards associated with work equipment the manufacturers are also obliged to keep a technical file relating to the findings and outcome of their research and issue a 'Declaration of Conformity', which simply states that the equipment complies with the relevant functional health and safety requirements within the EU.

Inspection and maintenance

All work equipment must also be adequately and suitably maintained in an efficient state to prevent breakdown and/or risks of fire, health or safety. A programme of planned preventive maintenance (PPM) must be implemented to ensure that work equipment is kept in efficient working order and in good repair and the fire risks are reduced to a reasonable level. The frequency of maintenance should be determined by:

➤ The frequency of usage
➤ The environment in which the equipment is used

In relation to fire safety it is particularly important to ensure that moving parts of machinery are subject to planned preventive maintenance to prevent the equipment being a source of ignition.

If a rotating or moving part is not suitably lubricated in line the manufacturers' guidance the machine may be liable to dry out (running dry) causing excessive friction. This, in turn, creates a source of heat.

This source of heat, if added to the potential for it to come into contact with contaminated lubricant, can increase the potential for a fire, as was the case at Kings Cross Underground station where a serious fire started in the escalator.

➤ The variety of operations the equipment is used for
➤ The criticality of failure.

If a logbook is provided with the equipment, it should be kept up to date and available. Where a logbook is not provided with the equipment, the employer does not need to provide one although they should develop a system for recording maintenance work that has been carried out.

Where work equipment involves a specific risk (e.g. risk of fire) the employer should restrict the use and maintenance of the equipment to those given the task to use it. An example of which would be an electrician using electrical testing equipment.

In all cases the inspection and testing must be carried out by suitably qualified and competent staff. It is therefore important that prior to providing any piece of work equipment an employer or responsible person must consider, plan and take account of the detailed arrangements for the safety inspection and planned preventive maintenance of any such equipment.

Information, instruction and training

In order to reduce the risks associated with work equipment, all persons, including operatives, supervisors, service engineers and cleaners, must be provided with adequate information and instruction covering the safe use of the work equipment.

The exact level and nature of the information, instruction and training provided by an employer will obviously be dependent upon the complexity of the equipment and the magnitude of the associated risks. There are various ways in which the training can be delivered, for example:

➤ From the manufacturer/supplier – suppliers or manufacturers of work equipment should in all cases provide some basic safety information as required by section 6 of HSWA. In addition they may also provide specialist training either on or off site for more complex equipment or tasks, e.g. the servicing of pressure vessels
➤ From the user – it is sometimes the case that larger companies will have a training department that will ensure that professional and competent trainers provide initial and routine information, instruction and training to all those who use work equipment
➤ On the job – often training relating to equipment will be given on the shop floor as part of the supervisory process. This may take the form of a competent operative talking through the job and directly supervising the actions of those receiving the instruction and training.

51

Protection against specified hazards

Under PUWER, employers are required to implement suitable controls for specific risks including the risk of fire or overheating of the work equipment, e.g. excessive heat building up as a result of insufficient lubrication of a moving part.

> The ACoP relating to PUWER specifically identifies examples of hazards that the regulations cover. The ACoP provides an example of a risk of overheating or fire due to: 'Friction (bearing running hot, conveyor belt on jammed roller), electric motors burning out, thermostat failure, cooling system failure. '
>
> A fire risk assessment should therefore consider each of these areas as part of the primary hazard category in the risk assessment process.

Emergency controls

These regulations require that work equipment is fitted with suitable controls for starting, stopping and emergency stopping, as well as any other controls that may be necessary.

Where appropriate, work equipment must be provided with a suitable means to isolate it from all its sources of energy. The method of isolation will depend on the type of energy being isolated and should:

➤ Be clearly identified
➤ Be readily accessible
➤ Ensure complete and positive isolation

Figure 3.10 Isolation switch for gas-fired kitchen equipment

➤ Protect all persons at risk
➤ Prevent accidental or unintentional reconnection.

In order to isolate work equipment that is involved in a fire the control must be capable of being isolated outside the potential fire danger zone. For example, the emergency isolation for kitchen equipment, such as deep fat fryers, ovens, etc., should be located in such a position as to enable safe isolation during evacuation.

A permit to work system should be used to support isolation procedures for high-risk work equipment (see Chapter 6).

3.8 Hazardous substances

Managing hazardous substances in the workplace is an important aspect of achieving a reasonable level of fire safety as overprovision or a lack of management of flammable or reactive chemicals will impact upon the level of fire risk. The supply and use of hazardous substances used at work are covered by a number of regulations. The most fundamental in terms of the supply chain is the Chemicals (Hazard Information and Packaging for Supply) Regulations 2002 (CHIP). These regulations require the suppliers of any substances to identify the hazards associated with the substances they are supplying and classify them by hazard type. They must then label the package containing the substance (Appendix 3.1 shows an example of a label required by CHIP) and provide information relating to the hazards to those that they supply the chemicals to and to package them safely including:

1. The name of the substance or preparation
2. Its chemical composition as regards hazards contents
3. Hazard classifications it holds, e.g. toxic, corrosive, etc.

Figure 3.11 Hazardous substances in the workplace

4. First aid information
5. Fire fighting information
6. Procedures for dealing with accidental spillage
7. Handling precautions and storage requirements
8. Controls necessary during use, and recommended personal protection
9. Physical properties and appearance
10. Stability of the substance or preparation, including shelf life
11. Toxicological data, if applicable
12. Known effects on the environment
13. Methods or precautions for the disposal of the chemical or by-products
14. Transport information for road and rail systems
15. Legislation which also applies to the substance or preparation
16. Other relevant information, e.g. supplier's contact address or number.

It is the duty of the person supplying the substance or preparation to supply the relevant data sheet, no later than dispatch of the chemical. The data sheet should be dated and significant changes in its contents should be brought to the attention of the customer up to one year from the date of supply.

Once the suppliers of harmful chemicals have classified and packaged them, all employers have duties to ensure that the hazards associated with all substances harmful to health are adequately controlled. The legislation requiring employers to control these hazards is the Control of Substances Hazardous to Health Regulations 2002 (COSHH) as amended.

Substances that represent a danger of fire or explosion are described as 'Dangerous' substances. The storage, handling and transportation of dangerous substances in the workplace is controlled by the Dangerous Substances and Explosive Atmosphere Regulations 2002 (DSEAR) (see Chapter 8).

Comprehensive details of the requirements of CHIP, COSHH and DSEAR appear in the summary of legal requirements at the back of the book. The following paragraphs summarise the key elements of both pieces of legislation.

In the case where chemicals present a specific fire or explosion-related risk they must be assessed by those who intend to use them. The duty to assess the risk and explosion risks in respect of dangerous substances is placed upon the 'responsible person' as part of their duties under the RRSFO (see Chapter 14).

3.8.1 The Control of Substances Hazardous to Health Regulations 2002 (COSHH) as amended

CHIP specifically covers the aspects relating to the supply of dangerous substances. Once the substances have been supplied the responsibility rests with the employer, and in the case of the RRFSO to the responsible person. COSHH is the principal legislation that covers the ongoing use, transportation, storage and disposal of substances hazardous to health.

The COSHH Regulations apply to the following classifications of substances:

➤ Substances classified by CHIP as very toxic, toxic, harmful, corrosive or irritant
➤ Substances assigned a WEL according to EH 40
➤ Other substances that create a hazard to human health
➤ Biological agents
➤ Dusts in concentrations great enough to be hazardous.

Figure 3.12 Controlling hazardous substances in the workplace

However, it is important to note the COSHH Regulations do not apply to the following substances:

➤ Those that only pose physical hazards, e.g. where the only hazard is due to flammable or explosive properties (where DSEAR applies, see Chapter 7)
➤ Those that are covered by other specific legislation, e.g. lead and asbestos.

Under COSHH, an employer is required to carry out a suitable and sufficient assessment of the risks to the health of their employees created by work involving hazardous substances. The assessment must be kept under review and any necessary changes made following the review. In addition, control measures to reduce the risks posed by the substances must also be subject to maintenance, e.g. local exhaust ventilation systems (LEV) for the removal of flammable mixtures from the atmosphere.

Emergency arrangement have to be considered as part of the control measures relating to hazardous materials, under the regulations, particularly in relation to fire and explosion. These are detailed elsewhere in this book.

Extract from Dangerous Substances and Explosive Atmospheres ACoP:

Certain gases (hydrogen, methane, propane etc) are extremely flammable and come within the scope of DSEAR. However the gases themselves can also act as asphyxiants, reducing the quantity of oxygen in the workplace to the extent that life can be put at risk. As a result, they will also satisfy the definition of a substance hazardous to health for the purposes of COSHH. In these circumstances, employers will have duties to control the risks from those substances under both sets of regulations.

3.9 The provision of buildings

As part of their responsibilities employers and responsible persons are required to ensure that the workplaces under their control are safe and fit for purpose so far as is reasonably practicable.

Whether the building or work facility is owned, leased, or used as part of a shared building (e.g. where more than one employer occupies a building) those in control are obliged to establish formal arrangements for the management of all aspects of safety including fire.

Much depends upon the responsible person and/or employer undertaking an assessment of, not only their work activities but also the suitability of the building or facility to allow them to undertake their operations safely. It is true to say that a large proportion of organisations do not undertake sufficiently rigorous examination of these areas prior to procuring their facilities.

When considering undertaking furniture restoration in a small section of a multi-occupied warehouse the fire safety impact of introducing highly combustible materials and highly flammable chemicals into a warehouse situation must be assessed.

This assessment will need to draw information from a range of sources. Perhaps the most significant sources of information relating to the safety of any building including the suitability and limitation of its use are:

➤ Operations and maintenance manuals which should be provided by those who manage the facilities detailing how equipment operates and its maintenance schedule
➤ Any previously issued fire certificate or conditions of licence issued by an enforcing authority
➤ The Health and Safety File as required under the Construction (Design and Management) Regulations 2007 (CDM).

Following the introduction of CDM in 1995 and subsequent changes that came into effect in 2007, the requirement was established to complete a health and safety file when undertaking any new building or refurbishment. The contents of this file are now probably the single most important source of information relating to newer buildings for those involved in the ongoing management of fire safety.

The main intention of the regulations is aimed at the design and management aspects of the project and will have the greatest effect at the planning stages. To that end the regulations identify certain 'key parties' involved in a construction project and impose duties upon all of them for the safe completion of the project.

These key parties are:

➤ The Client
➤ The Designer
➤ The CDM Coordinator
➤ The Principal Contractor
➤ Other Contractors.

3.9.1 Duties of designers

Designers play a key role in construction projects. Contractors have to manage risks on the site, but designers can often eliminate or reduce them at source.

Any designer of a structure or part of a structure must:

➤ Advise clients of their duties
➤ Take positive account of health and safety hazards during design considerations
➤ Apply principles of prevention during the design phase to eliminate, reduce or control hazards
➤ Consider measures that will protect all workers if either avoidance or reduction to a safe level is possible
➤ Make health and safety information available regarding risks which cannot be designed out
➤ Cooperate with the coordinator and any other designers involved
➤ Design the building and its mechanical and engineering services to fully comply with:
  ➤ The functional requirements of the Building Regulations which includes the provision of adequate fire safety

Figure 3.13 Designers have a responsibility for the safety of all others

➤ Any relevant British/European standard, e.g. BS 5839 Part 1/fire detection and fire alarm systems in buildings.

3.9.2 Construction Phase plan

A Construction Phase plan is essentially a collection of information about the significant health, safety, environmental and fire risks of the project. The plan will initially be based upon information provided by the client, the designer and the coordinator's own assessment of the risks inherent in the project, and will provide the health and safety management plan during the construction phase.

The initial plan should include:

➤ A general description of the nature of the project and intended completion times
➤ Information regarding the existing structure and environment. This will include fire hazard information from the client
➤ Details of the organisation for fire safety management and liaison during the project

➤ Details of significant risks which cannot be avoided at the design stage and any fire risks created by the design or specification of the structure
➤ Information regarding the overlap with the client's undertaking including the arrangements for fire emergency and any other site rules laid down by the client, e.g. smoking policy
➤ Other safety information, e.g. means of escape in case of fire and fire fighting facilities, signage rating to the arrangements for highly flammable liquids and liquefied petroleum gases.

This will amount to a list of hazards for which the principal contractor must develop suitable control measures.

Health and safety file

The CDM coordinator has the legal duty under CDM to ensure that the health and safety file is compiled for the completed building and mechanical and electrical services, etc. within. The file is then handed to the client on completion of the project. The client must then ensure that the file is kept safe and is available for any interested party in the future.

The purpose of the health and safety file is to provide all of the relevant details regarding the structure and services, etc. within to anyone who in the future may be carrying out further construction work, repairs, maintenance or refurbishment of the building or facility.

The precise content of the file will depend upon the nature and complexity of the structure. However, the information contained in the file will be provided in part by the designer and in part by the principal contractor and, in reality, it may well be that the bulk of the work in compiling the file will be done by the principal contractor as they will be in first capture of the information.

3.10 Case study

In 1996 the owners of a large chemical plant were prosecuted under section 33 of the Health and Safety at Work etc. Act 1974 as a result of their failure to comply with section 3(1) of HSWA when an employee of the specialist contractors, who had been hired to carry out a repair to the lining of a tank within the plant, was seriously injured as a result of inadequate fire safety management.

During the case it was heard that acetone was ignited by a damaged light bulb.

The plant was shut down for its annual maintenance and a small firm of specialist contractors was repairing the lining of a tank. An employee of the contractors was working in the tank by the light of an electric light bulb attached to a lead.

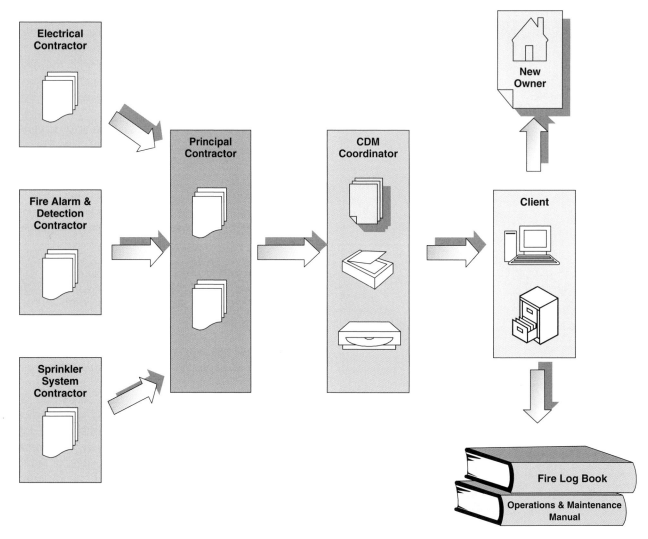

Figure 3.14 Development and flow of health and safety file

After grinding the damaged area of the lining, the employee had to clean it down with acetone before applying a fibreglass matting patch with resin. Acetone is a volatile liquid that gives off highly flammable vapour. The acetone was an old paint bucket which he had found in a refuse bin. While he was applying the acetone with a brush, the light bulb broke. Some of the liquid had probably dripped onto it. As the acetone was in an open bucket there was a significant amount of vapour in the tank.

The broken bulb caused a flash fire as a result of which the contractor was badly burned.

The chemical company argued that the specialist contractor was in effect responsible in law for the injury to their own employee as the work was outside their own 'undertaking' and that the chemical company had in fact undertaken an appropriate level of selection and management of contractors and therefore could not be held liable.

This was countered by the HSE who argued that an employer is responsible for the safety of contractors if the contractor is engaged in work as part of the employer's 'undertaking'.

On appeal, the court established the precedent that, if a company, in the course of its undertaking (which it was found in this case), creates a risk to contractors or members of the public, it has a duty to reduce those risks to the lowest level reasonably practicable. In this case, the court found that the chemical company had failed to discharge this duty and that they, as well as the specialist contractor, were in breach of their duty of care.

The company was fined a substantial sum, there was also significant and ongoing damage to its corporate reputation from such a high profile case.

In addition, the contractor's employee was seriously injured. This might have been avoided if the company had accepted their full responsibility and looked beyond its own staff in terms of the application of health and safety procedures.

3.11 Example NEBOSH questions for Chapter 3

1. (a) **Outline** the benefits to an organisation of having a health and safety committee. (4)
   (b) **Outline** the reasons why a health and safety committee may prove to be ineffective in practice. (8)
   (c) **Identify** a range of methods that an employer can use to provide health and safety information directly to individual employees. (8)
2. (a) **Identify** the particular requirements of regulation 3 of the Management of Health and Safety at Work Regulations 1999 in relation to an employer's duty to carry out risk assessments. (3)
   (b) **Outline** the factors that should be considered when selecting individuals to assist in carrying out risk assessments in the workplace. (5)

3. **List** the factors that might be considered when assessing the health and safety competence of a contractor. (8)
4. Following a significant increase in accidents, a health and safety campaign is to be launched within an organisation to encourage safer working by employees.
   (a) **Outline** how the organisation might ensure that the nature of the campaign is effectively communicated to, and understood by, employees. (8)
   (b) Other than poor communication, **describe** the organisational factors that could limit the effectiveness of the campaign. (12)
5. With reference to the Health and Safety (Consultation with Employees) Regulations 1996
   (a) **Identify** the particular health and safety matters on which employers must consult their employees. (4)
   (b) **Outline** the entitlements of representatives of employee safety who have been elected under the regulations. (4)

Appendix 3.1

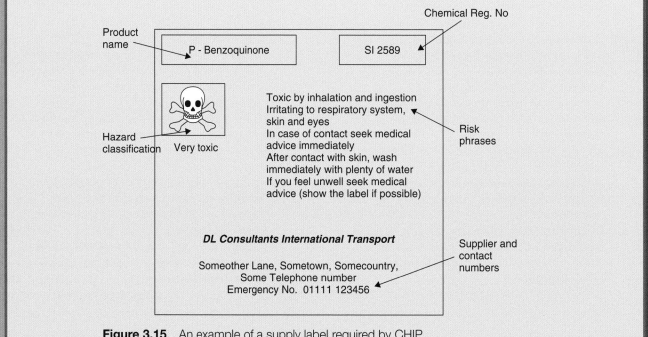

Figure 3.15 An example of a supply label required by CHIP

The size of the label is dictated by the regulations to prevent the suppliers from having tiny, invisible labels on large barrels full of toxic substances. On the other hand, a large label would be impracticable on a small bottle of oven cleaner, for example.

The requirements are as follows:

Less than 3 litres : label at least 52 × 74 mm
3–50 litres : label at least 74 × 105 mm
50–500 litres : label at least 105 × 148 mm
More than 500 litres : label at least 148 × 210 mm

Safety culture

4.1 Introduction

The culture of an organisation is said to have a significant impact upon performance. Organisational culture is manifested in behaviours and attitudes of both workforce and management. Often difficult to define and measure it is widely accepted that the prevailing culture within an organisation is an important factor in order to achieve successful management of any safety issues.

The 'safety culture' of an organisation is a concept that describes the attitudes and beliefs of an organisation in terms of its safety performance. It must be recognised that the so-called safety culture within which a company operates is driven by the pervading culture within the wider organisation. The safety culture will impact upon the effectiveness of all safety functions including the management of fire safety matters.

This chapter discusses the following key elements:

> The concept of safety culture and its various components
> How to assist in the development of a positive safety culture within an organisation
> Factors promoting a negative culture
> External influences on safety standards
> Internal influences on safety standards
> Human factors
> Human error
> Affecting cultural change.

This chapter explores safety culture and examines key factors that affect the culture, both positive and negative, in order to understand how better to manage both general safety and fire safety within the workplace.

4.2 The concept of safety culture and its various components

4.2.1 Defining safety culture

In terms of safety culture, there is a range of definitions cited by public enquires and research bodies. These definitions invariably cite poor management control as a key factor leading directly to serious accidents or disasters. For example, the absence of a safety culture is said to have played a major part in the nuclear reactor disaster at Chernobyl in 1986.

The team in control of the reactor, being influenced by the need to complete an unusual test quickly, removed layer after layer of the safety controls – introduced to keep them safe – in order to carry out a test. This resulted in the reactor being operated under conditions which gave rise to serious instability in the reactor, resulting in the disaster.

The subsequent enquiry found that 'the control team operated in a managerial culture that failed to discourage the taking of risks where other priorities intervened, e.g. the need to complete the test quickly'.

In order to be truly effective in the management of fire or health and safety, the organisation must develop what has become known as a positive safety culture based on proactive management of safety issues.

A number of studies carried out by the HSE have shown that significant numbers of major injury and fatal accidents could have been prevented by positive action by management.

These studies established that in the order of 70% of fatal accidents in construction and maintenance activities could have been prevented by direct, positive, management action.

The studies emphasise the critical role of the organisation in establishing a robust health and safety culture and a safety management system aimed at preventing human error, as well as establishing the 'hardware' controls for health and safety.

In the field of safety management, the most widely used definition of culture is that suggested by the Advisory Committee on the Safety of Nuclear Installations. This definition was used to shape the guidance given by the HSE in its guidance HSG65 'Successful health and safety management':

> The safety culture of an organisation is the product of individual and group values, attitudes, perceptions, competencies and patterns of behaviour that determine the commitment to, and the style and proficiency of, an organisation's health and safety management.
>
> Organisations with a positive safety culture are characterised by communications founded on mutual trust, by shared perceptions of the importance of safety and by confidence in the efficacy of preventive measures.

The term safety climate is used to refer to psychological characteristics of employees, in other words the way that people feel about the safety culture within an organisation. An investigation into safety culture by the HSE into the two major rail crashes in 2000 and 2001 concluded that the safety climate within an organisation is an expression of the values, attitudes and perceptions of employees with regard to safety within an organisation.

So it can be seen that the safety climate of an organisation is an important influence on its overall safety culture. If employees' own values, attitudes and perceptions do not motivate them to support and/or comply with safety rules the safety culture will be a negative one.

The investigation into safety climate argued that there were five drivers for a positive safety climate:

Leadership

Whereby senior managers provide a demonstration of safety as having a high status in the organisation by providing:

➤ An adequate budget and resources for managing safety (including safety specialists)
➤ Opportunities for effective safety communication
➤ Training and support to personnel
➤ High visibility of management's commitment to safety
➤ Effective safety management systems led by a strategic level safety management team.

Communication

Whereby there are effective:

➤ Channels for top-down, bottom-up and horizontal communication on safety matters
➤ Systems for reporting safety issues
➤ Transfer of information between individuals, departments and teams.

Involvement of staff

Effective employee participation should be supported with good systems and training, and allow employees to be responsible personally for areas of safety.

The existence of a learning culture

Whereby systems are in place that allow:

➤ Employees to contribute ideas for improvement in procedures
➤ Effective analysis of incidents, and good communications of the outcomes.

The existence of a just culture

Organisations with a blame culture overemphasise individual blame for human error, at the expense of correcting defective systems. Blame allocates fault and responsibility to the individual making the error, rather than to the system, organisation or management process.

To reduce the impact of a blame culture organisations should:

➤ Promote accountability
➤ Understand the mechanism of human error
➤ Demonstrate care and concern towards employees
➤ Maintain confidentiality
➤ Enable employees to feel that they are able to report problems without fear of reprisal.

There are various 'safety climate' assessment tools available which allow employers and responsible persons to assess the state of their safety culture. These normally take the form of staff questionnaires which can be used along with tangible indicators of safety culture (see 4.3 below).

4.2.2   The benefits of a positive safety culture

The importance of creating a positive safety culture cannot be overstated. In extreme cases a positive safety culture can make the difference between the success or failure of an organisation's survival.

In his report on the Ladbroke Grove rail enquiry Lord Cullen stated: 'A key factor in the industry is the prevailing culture, of which safety culture is an integral part. There is a clear link between good safety and good business'. He then went on to quote in submission of the HSE: '. . . the need for a positive safety culture is the most fundamental brought before the Inquiry'.

If an organisation can create and sustain a positive safety culture each and every member of staff will be competent and committed to work safely, and business and

Table 4.1   The benefits of a positive safety culture

Benefits to the organisation	Less time lost through accidents Reduced risk of civil claims for compensation Reduced risk of enforcement action Enhanced company image Greater efficiency Less production downtime Minimised insurance premiums Having a competent and committed workforce Making better quality decisions as a result of involving the workforce
Benefits to the individual	Less risk of injury Less risk to work-related ill health Less risk of work-related stress Increased chances of continued employment Working for a competent and committed team with competent and committed management Being clear and confident about what is expected from management Increased job satisfaction from being empowered to contribute to safety management

individual losses will be minimised. Table 4.1 summarises some of the more obvious benefits of a positive safety culture, for both the organisation as a whole and the individuals working within it.

4.3   Tangible indicators of safety culture

It is never simple to discern the state of the safety culture within organisations. Some say that one can get a 'feel' for a positive or negative safety culture by visiting the workplace or in some cases reviewing the products or documentation of an organisation. These subjective assessments are of limited value and rarely allow the situation at any one time to be quantified, e.g. is the safety culture worse now than it was last year?

Without quantifiable measures safety culture (as is the case with all aspects of an organisation) cannot be effectively managed. It is therefore vital for managers to have some idea of the culture within their organisations. Moreover, safety culture is critical to the success or failure of risk management within an organisation; it is important to attempt to understand whether the safety culture is working for or against the aims of management. Quantifiable data is available to assist managers assess the safety culture, from the following indicators:

➣   Accident/incident occurrence and reporting rates
➣   Sickness and absenteeism
➣   Staff turnover
➣   Compliance with safety rules
➣   Complaints
➣   Output quality
➣   Staff involvement.

When attempting to determine the nature of an organisation's safety culture, is it vital not to draw too firm a conclusion from one single indicator. Astute managers will recognise that there are a number of variable factors that may influence any one of the indicators. More confidence can be placed in conclusions that are drawn from a cluster of indicators, tending to suggest a similar picture.

4.3.1   Accident/incident occurrence and reporting rates

The most obvious implication to draw from accident/incident rates or the number of fires or false alarms is that if they are going up period on period it is likely that there is a negative culture. However, it is important to consider, not only the numbers of these safety events, but also the seriousness of the outcomes. If it is the case that there are many reports of personal injury accidents but the injuries sustained are negligible, it may

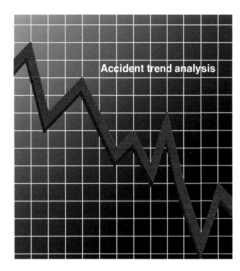

Figure 4.1 Downward trend of accidents at work

well indicate a positive culture in that staff are motivated to report minor incidents because they believe management will take some action to improve safety. Conversely minor accidents may be reported in order to flood management with paperwork or frustrate, or express disapproval of, the introduction of new equipment or work practices.

As with all other indicators, caution must be used to draw too firm a conclusion from this important but single indicator.

Managers will normally take comfort if in general the accident/incident rates are in ratios that would normally be expected, there are few serious injuries and the rate of injuries maintains a downward trend – as would reports from employees highlighting fire safety issues, making a positive contribution to the management of fire risk in the workplace.

4.3.2   Sickness and absenteeism rates

Sickness and absenteeism rates are often used in an attempt to draw conclusions about workplace culture for two reasons. First, they are easy to quantify and second, there appears to be a logical co-relationship between an absence from and a lack of desire to work.

In addition to the numbers of days lost through sickness and absenteeism, it is equally pertinent to examine the nature of the sickness or absence in terms of the duration of absences, the timing of absences through the calendar or work cycle and the reasons given for absence.

Absenteeism rates can also be affected by contractual conditions of work, for example it is a recognised phenomenon that public service workers have significantly higher absenteeism rates than their private sector counterparts.

Nevertheless sickness and absenteeism rates are seen as a reliable indictor of the levels of motivation and commitment to work. Job roles that provide little satisfaction for the employee will often experience high levels of sickness.

4.3.3   Staff turnover

Again, staff turnover is often quoted as a good indicator of organisational culture, but again caution must be exercised when drawing inferences about the culture in a particular organisation. In general it is thought that the more positive a safety culture in an organisation the greater the feelings of loyalty among the staff and therefore the less the turnover. There are, however, a number of factors that affect staff turnover that have little relationship with safety culture, including:

➤ Industry norms
➤ Remuneration packages of competitors
➤ Life/work balance issues for the individual
➤ Personal career aspirations.

4.3.4   Compliance with safety rules

An examination of the degree to which safety rules are applied or breached at work gives a good indication of the effectiveness of the safety culture. This is because almost all factors that affect the implementation of safety rules are in the direct control of the management. Therefore it is possible to draw reasonable inferences from an analysis of breaches of safety rules.

With a positive safety culture at work, individuals will be aware of the rules, have sufficient physical resources, time and competence to apply them and be motivated to do so.

Active monitoring systems (see Chapter 6) will enable instances of breaches in safety rules. Particularly effective formal active monitoring systems include safety tours, inspections and workplace audits. Less formal but none the less relevant are those occasions when observing behaviour in the workplace occurs as a result of visiting the workplace for other reasons, for example quality control or welfare.

Fire doors wedged open, bad housekeeping and breaches of security arrangements that are identified during active monitoring will provide a good indication that routine fire safety management is poor.

4.3.5   Complaints

The level of complaints in a workplace can be indicative of a dissatisfied workforce. Dissatisfied workers are not motivated to comply with any management systems. Again it is a fairly good indicator of the state of the culture because if there is dissatisfaction within the workplace it is likely that it is as a result of the actions or omissions of management.

However, a complete absence of complaints from the workforce may indicate a atmosphere of fear or uncertainty. A workforce that is competent and confident of management's ability to respond positively to issues raised will result from a positive culture.

Managers need to analyse the levels and nature of complaints with some caution in order to get a feel as to whether they are a positive or negative indicator.

4.3.6 Output quality

In organisations that produce either goods or services, quality of the output is fundamental to business success. Poor output quality can be indicative of a poor safety quality. If poor output quality demonstrates low levels of management control, and if quality control management is poor, it is likely that management is poor across the organisation thus having a direct bearing upon safety. Poor management will adversely affect employees' motivation. Employees that are not motivated are likely to take less care of the outputs from their work and so the cycle continues.

4.3.7 Staff involvement

The degree to which staff are willing to become involved in non-core or social aspects of the work may provide management with a useful insight into the state of workplace culture. Among the activities for which quantitative data can be made available are:

➣ Suggestion schemes
➣ Work committees
➣ Social activities
➣ Response to attitude surveys.

4.4 How to assist in the development of a positive safety culture within an organisation

While the guidance contained in current fire safety documentation is a little sparse on fire safety culture, the HSE describe in their guidance document HSG65 – 'Effective health and safety management' that there are four building blocks to an effective safety culture. The blocks are often classified as the 'four Cs' of control, cooperation, communication and competence. The following sections discuss these four Cs:

➣ Control – the methods by which an organisation controls its safety performance

➣ Cooperation – the means that an organisation will secure the cooperation between individuals, safety representatives and groups
➣ Communication – the methods by which the organisation will communicate in, through and out of, the organisation
➣ Competence – the means by which the organisation manages the competency levels of individuals and teams.

4.4.1 Control

Establishing and maintaining control is fundamental to all management activities.

Control over safety management starts by allocating clear and unequivocal roles and responsibilities throughout an organisation. The roles and responsibilities will be formalised in the safety policy (see Chapter 2) and will enable all those with responsibilities to be clear as to what is expected of them, together with the level of resources at their disposal and the degree of authority they have to act and/or make decisions.

In addition to allocating clear roles and responsibilities, it is equally important to ensure that individuals and teams are made accountable for their performance. This is not to say that there need be an inappropriate level of oppressive monitoring or supervision, but rather a system whereby the individuals are aware that they will be required to account for the way in which they have discharged their responsibilities.

Without ensuring individuals are accountable for their action the exercise of allocating roles and responsibilities is merely academic. Management systems that are used to ensure individual and team safety accountabilities include:

➣ Written job description that contains reference to safety responsibilities and objectives
➣ Job appraisal and performance review systems that measure and reward good safety performance
➣ Systems that deal with failures and that identify a range of actions that can be taken to rectify the failures. (This is often achieved through the normal discipline arrangements of the organisation.)

Once roles, responsibilities and accountabilities have been established, it is then necessary to set some key safety objectives both for the organisation as a whole, and where appropriate for individual members of staff. For example, a company may wish to adopt a measured reduction of unwanted fire alarm actuations and may do so by linking the reduction of false alarms to a maintenance engineer's bonus pay scheme.

Safety objectives need to be 'SMART' and supported by plans that will identify both key milestones towards achieving the targets and the resources necessary to achieve the objectives.

It is important to set such objectives for individuals and teams because it helps to provide the organisational impetus necessary to continue to improve workplace safety. In management terms it is often observed that 'what gets measured gets done', so for an organisation striving to achieve improvements in safety, clear safety objectives must be established and monitored.

Finally management must retain effective control by monitoring and supporting the implementation of the plans and ensuring that the objectives are achieved, or when circumstances dictate, reviewed and revised. The credibility of the whole system of management control will be undermined if unrealistic or no longer relevant safety objectives remain current at a time when they should have been modified.

An important aspect of maintaining effective management control of safety is dealing with non-compliance of safety rules. For example, if an individual infers with something provided for safety or a line manager fails to correct an unsafe act, such as wedging open a fire door or blocking a fire escape route, the organisation *must* respond in a way that secures the future compliance by the individual. A positive management response will send a message throughout the organisation which reinforces the safety culture.

4.4.2 Cooperation

A positive safety culture can only be built in an atmosphere of true cooperation where management and staff work together in a partnership to establish and pursue safety objectives. In order to support a positive safety culture, it is crucial that employees are involved in decisions that affect safety performance. Staff will not be fully

Figure 4.2   Cooperation – the key to success

committed to safety targets or safe procedures if they are presented with them as a 'fait accompli'. It is therefore necessary to ensure that staff are involved as early as possible in the planning process. It is often observed that there is no cooperation without consultation.

Employers have a legal duty to consult with employees (see Chapter 1), for example employers must consult their safety representatives when making arrangements to secure competent health and safety advice. However, successful organisations are prepared to go much further than meeting the minimum statutory obligations. They will actively encourage employees to become involved in setting targets and assisting the organisation in problem solving. In the best cases, safety representatives are trained alongside management, which enables the development of a shared understanding of the issues, which in turn provides common ground on which to continually improve the safety performance of the organisation.

Once trained employees' representatives and other members of staff will be able to, and should be encouraged to, become involved in a range of safety critical issues including, for example:

➢ Conducting and reviewing risk assessments
➢ Reviewing the procurement of new work equipment
➢ Developing and introducing safe systems of work or other procedures
➢ Carrying out workplace safely inspections and audits.

It is often the case that when an organisation first begins to actively seek the cooperation of their staff there is a potential for conflict. This arises, in part, due to managers feeling challenged by staff who will be able to identify failings in management systems. For example, an employee who has received training relating to conducting a fire risk assessment will undoubtedly be able to highlight where risk control systems should be improved. The potential for conflict will soon reduce as mutual trust is developed between all parties. The benefits of the resultant cooperation will not only have a significant impact upon the safety performance of the organisation but also assist in enhancing the quality of management decision making.

4.4.3 Communication

Successful organisations have an effective communication strategy, which in turn enhances and improves the pervading safety culture. In its guidance HSG65 (successful health and safety management) the HSE propose that effective communication within an organisation relies on the management of information; coming in, flowing within and leaving the organisation.

Communication is defined as:

The imparting, conveying, sharing and exchange of ideas, knowledge and information.

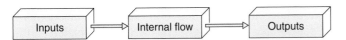

Figure 4.3 Diagram of effective organisational communications

Inputs

The first step of effective safety management is to formulate policy. It is vital for good sources of information to be readily available and to be used when developing safety policies and practices. This can only be achieved if an organisation monitors the changes of health and safety legislation together with any associated guidance; developments in technology that are relevant to risk control systems; and the application of good management practices that have been demonstrated to enhance safety performance.

Internal flow of information

The type of information that has to be communicated internally includes:

- The health and safety policy and practices
- The level of commitment by senior management to the policy and its implementation
- What the safety objectives are and how they are to be achieved
- How performance is to be measured and the outcome of that measurement
- The outcome of the investigation of any safety events including fire incidents and personal injury accidents or near misses and false alarms.

There is a wide range of methods for communicating safety information within an organisation. Successful managers will adopt sufficient methods to ensure that safety information is widely known and understood. Key to effective internal communication is the visible behaviour of senior managers; the quality of written information and face to face discussion.

Managers will need to lead by example, for instance if there is a requirement for fire doors to be closed in specified locations on site, senior managers will always be seen closing them. Similarly if the policy of the organisation is to encourage the active participation of employees the senior managers' behaviour will reflect that policy in their dealings with staff.

Senior managers will also take an active interest in safety matters and be directly involved in, for example,

taking part in fire safety tours of the workplace, chairing safety committee meetings, being actively involved in safety event investigations and responding positively to suggestions made by staff.

As important as the visible behaviour of senior managers is the information that is communicated in writing. In many cases there is a legal duty to produce written documentation, for example there is a requirement to have a written health and safety policy and written records of the significant findings of fire risk assessments (see Chapters 1 and 2). What is absolutely crucial is that written communication is clear, concise and understood by those it is provided for.

Examples of the range of written communication:

- Policy documents
- Notices
- Posters
- Newsletters
- Electronic information via e-mails or intranet
- Handouts at training programmes.

In addition to formal written policies, successful organisations will use a range of other media to communicate the safety information. The information may be supported with photographs, diagrams and cartoons. It will also be provided at strategic locations in the workplace that is convenient to view and be provided in a language that will be understood in the workplace. The language used in some cases will be other than English, particularly when the workforce is composed of differing nationalities. Written communication may also need to use the relevant vernacular to gain acceptance and understanding of the target work group.

A key component, for instance, with regard to communicating fire safety arrangements is the need to provide safety signage to direct people along an emergency escape route. In this case the written communication takes the form of internationally recognised pictograms.

The provision of some very specific information for employees is required by current UK legislation (see 'Summary of key legal requirements' at the rear of the book). The main requirement is for the employer to display prominently an approved poster or to issue each employee with a leaflet.

The poster and the leaflet contain the relevant information relating to the HSWA, and providing the name and contact address of the relevant enforcing

Figure 4.4 Typical example of a fire alarm call point notice

Figure 4.5 Health and safety poster

only the technical content, but also the more subtle messages relating to the importance of the information and the people it is being communicated to. The opportunities to reinforce safety messages in this way are extremely valuable, as they will often allow for the flow of information in a number of directions, e.g. from shop floor to stores, and from sales to management. Face-to-face communication will also reinforce the successful organisation's ethos of employee involvement at all levels which directly affects the safety culture.

Examples of opportunities to achieve all the benefits of effective face-to-face communication of safety information include:

➤ Safety tours and inspections
➤ Team briefings
➤ Management meetings
➤ Tool box talks by supervisors
➤ Problem-solving workshops
➤ Appraisal interviews.

A common problem with face-to-face communication results from the communicator simply passing the message without stopping to check that it has been received and understood. This can result in confusion and inevitable conflict where the communicator thinks that they have passed the message, the receiver thinks that they have received the message but the message has been confused in the processing stage. To avoid the harmful effects of misinformation it is useful to build in a feedback loop, where the person receiving the information explains what he/she has understood.

Outputs

It is necessary for safety information to be communicated to external bodies. For example, certain specified work-related injuries, diseases and dangerous occurrences must be notified to the enforcers as soon as practicable after they occur (see Chapter 12). Other information relating to the safety performance of an organisation will need to be provided to insurance companies and potential customers. In the public services, information has to be provided to governing bodies or other government agencies responsible for monitoring public sector performance.

It is a requirement of current UK legislation that employers ensure that they make arrangements with external services, in particular with regard to first aid,

authority and the local office of the Employment Medical Advisory Service (EMAS).

In addition to written material, the flow of information within an organisation is supported by face-to-face communication. Face-to-face communication is said to be a 'rich' source of information, in that it conveys not emergency medical care and rescue work. It is therefore necessary to communicate with those external bodies that will provide assistance in the case of emergencies including the police, fire and ambulance services.

Further examples of information that should be communicated outside an organisation include:

- Details contained within the health and safety file produced upon completion of construction operations under CDM
- Hazards associated with a particular site, e.g. chemical or biological risks
- Environmental management systems including facilities for hazardous waste disposal
- Emergency procedure for visitors and contractors
- Hazard information to prospective customers as part of the supply chain management.

Again, a variety of media is used when communicating information to external parties. Increasingly the use of electronic communication technology is simplifying and accelerating the flow of information. An obvious disadvantage of the level of use of electronic media is the possibility of information overload, where safety critical issues can be lost in a plethora of trivial information.

4.4.4 Competence

Competence is the fundamental requirement to allow any task to be completed safely. A number of defining safety and employment bodies have attempted to provide a succinct definition of the term competence.

Safety competence can therefore be described as being a combination of knowledge, skills and experience that ensures roles are fulfilled and tasks completed with due regard to the hazards involved and the risk control measures necessary to achieve the required levels of safety. The RRFSO also adds the term 'other qualities' without giving an indication of what they are referring to.

Examples of definitions of competence:

The ability to perform the activities within an occupation or function to the standards expected in employment (Manpower Services Commission 1991).

The ability to use knowledge, understanding, practical and thinking skills to perform effectively to the national standards expected in employment (DfEE 1998).

Competence gives confidence to those who have it and ensures that policies will be developed and implemented effectively by all those having responsibility for safety in the workplace.

It follows that the competence of the most senior manager is as relevant to shop floor safety as the competence of the contract cleaner. Each has a different but nonetheless vital role to play in the control of risk.

Each needs to have an adequate level of knowledge, skill and experience to enable them to do what is necessary to make the workplace safe. The senior buyer may be responsible to ensure, for example, the provision of safe work equipment, whereas the contract cleaner is responsible for ensuring the waste by-products of the equipment do not present a fire hazard. Each must be able to recognise what needs to be done and each must have the necessary skills to play their part effectively.

Individual competence may be affected by the following factors:

- The ability to acquire knowledge or develop skills
- Mental approach (e.g. powers of concentration, maturity, motivation)
- The ability to apply existing knowledge and 'common sense'
- Skills acquired through experience, instruction, or training
- The physical ability to develop work skills.

Organisations that effectively manage safety need, as a minimum, to monitor levels of competence at various times and will normally have formal arrangements for measuring competence on the following occasions:

- On recruitment of employees
- On promoting and internal transfer of staff
- Prior to engaging contractors
- The introduction of new:
  - Work equipment
  - Procedures
  - Working routine
- To assess training needs
- To assess training effectiveness
- When investigating incidents and accidents.

The Department of Employment define training as:

The systematic development of attitude, knowledge and skill patterns required by an individual to perform adequately a given task or job.

The use of the term 'systematic training' means the training must be designed and planned so as to ensure the employee has the necessary information and skills to deal with workplace risks. The training should be structured and delivered by people competent to do so.

Safety training may be incorporated into routine skills training, or may be delivered as a separate training programme; the main objectives should be:

➤ To provide necessary safety skills and information to the workforce
➤ To support the safety culture within the organisation
➤ To ensure the success of any safety programmes and support the safety management system
➤ To ensure compliance with risk control strategies, e.g. fire wardens training to assist in the safe evacuation of a building.

All health and safety training (including fire safety training) must be provided for all employees relevant to their duties and responsibilities, and the training programme must include temporary or short-term employees as well as permanent staff. The training should take place during working hours. Where this is outside of normal working hours it must be treated as an extension of working hours, i.e. it should be considered as overtime.

Training is not a 'one-off' activity; to be effective it is conducted systematically throughout an organisation and is cyclical in nature.

Training needs analysis

The specific requirement for training will vary between organisations and between departments or sections

The main benefits gained through systematic training cycle include:

➤ Trained employees will understand the risks they face and the necessary risk control procedures, reducing accidents and increasing efficiency
➤ Legal obligations will be met, reducing the likelihood of claims or enforcement action
➤ Employee morale and teamwork should improve, increasing job satisfaction
➤ Attitudes to safety should improve, supporting the safety culture
➤ Management time spent dealing with non-compliance issues and investigating accidents/incidents will be reduced
➤ The workforce will be more flexible and responsive to safety initiatives.

within the organisation. A detailed analysis of 'training needs' is required in order to determine:

➤ **What** training is necessary to meet the needs of the task and the individual, e.g. safety procedures, skills training, supervisory/management training, etc.
➤ **When** the training will be needed – short-, medium- and long-term training requirements
➤ **Where** the training is best conducted, e.g. on or off the job
➤ **How** the training will be delivered – the precise format of the training
➤ **Who** will require the training – taking note of any specific individual requirements
➤ The **standard** of performance required on completion of the training.

The training needs analysis forms part of a systematic approach to the management of training.

Appendix 4.1 provides a diagrammatic representation of a systems approach to training which illustrates the role that the training needs analysis plays in the overall management of competence in an organisation.

Getting the message across

Whenever a training programme is developed, it should be borne in mind that people undergoing training will remember the information at a different rate depending upon their personal learning preferences. The way the information is presented also has a significant bearing

Figure 4.6   Training for safety – a typical training cycle

on how much people will remember. As a guide people will generally remember at the following rates:

- 10% of what they READ – notes, handouts, etc.
- 20% of what they HEAR – audio presentations
- 30% of what they SEE – pictures/diagrams
- 50% of what they SEE and HEAR – audio-visual presentations
- 70% of what they SAY – case studies and feedback
- 90% of what they SAY and DO – role play and simulation.

Safety training

As stated above, the specific needs for training within an organisation will vary greatly. However, the common types of training having relevance to safety are:

- Induction training
- Job or skills training
- Refresher or continuation training
- Remedial or corrective training, e.g. on identification of non-compliance with procedures.

Induction training

Statistics show that significant numbers of accidents occur to people within a relatively short time after entering new workplaces, either on initial induction or on job transfer. Induction training is, therefore, vital to enable a person to quickly and efficiently fit into the work environment. It should provide the employee with the information they need in order to act safely during the first few days on the job and to carry out their tasks without creating risks to themselves or their colleagues. The training will also assist in the employer and employee integration process and reduce damage to equipment or premises due to ignorance.

Figure 4.7 Safety training is conducted on a variety of occasions

The common areas covered during induction training may include:

- The safety rules and procedures as defined in the safety policy
- The responsibilities for health and safety in the organisation, including their own responsibilities
- The reporting procedures for hazards, accidents, near-miss situations, etc.
- The major hazards on site which may affect their safety
- Safety monitoring procedures in operation
- The access, egress and safe travel within the work areas
- Areas they should not enter, or where specific additional safety controls or training are required
- Who they report to and who will oversee their initial training and introduction to the workplace
- The availability and location of facilities, e.g. toilets, hygiene facilities, first aid, etc.
- The personal and occupational hygiene requirements
- The personal protective equipment available and how it should be used and maintained
- The emergency procedures, such as fire, evacuation and rescue, including the location and operation of emergency alarms and refuges
- The person who will take control of emergency situations in their work area, e.g. local fire wardens
- The terminology used in the workplace, especially any verbal 'shorthand' used.

It is likely that a new employee, especially a young person, entering the workplace will be unable to assimilate all of the relevant information on day one or in a single training session.

The training programme should be planned, where necessary, to take account of this and may require the training to be delivered in stages, with reinforcement and feedback sessions to confirm the learning.

Job or skills training

The relevant safety topics should be included in any job or skills training required and the training provided should be based on task analysis. Inclusion of safety issues at this stage reinforces the importance and commitment attached to safety by the organisation.

Again, the training content will vary greatly; however, the common topics may include:

- Legal responsibilities
- Site-wide safety rules and practices
- Specific practices for both on and off the job safety
- Current workplace procedures and codes of practice relevant to the task

- Housekeeping requirements and how to achieve acceptable standards
- The meaning of safety signs and workplace notices
- Relevant operating manuals, checklists, forms and necessary records
- Manual handling techniques
- The procedure for the supply, use, maintenance and replacement of personal protective equipment
- Correct operation of machinery, tools and equipment
- Their role during emergency situations, including the use of fire fighting and other emergency equipment
- First aid procedures and skills
- Accident/incident and workplace hazard reporting procedures
- The objectives of the accident investigation procedure
- How to contribute to safety committee meetings.

Specific fire safety-related training for key post holders such as the 'responsible person', fire wardens and fire incident controllers is also required (see Chapter 10).

Refresher or continuation training

The organisation may need to develop a programme of refresher training to reinforce the original training message and to introduce specific training, or information, regarding any changes which may have occurred since the original training was delivered.

The subject areas covered will, again, vary depending on circumstances but will include the relevant aspects of the original training, plus any relevant new topics necessary.

Refresher and continuation training will also need to be provided on those occasions when there have been significant changes to:

- Individual responsibilities
- Work equipment
- Technologies employed
- Safe systems of work.

Remedial or corrective training

Where specific non-compliance issues are identified in the workplace, e.g. the unauthorised discharging of fire extinguishers, the employer should introduce measures to rectify the situation. This may include the need for training to reinforce the original training message or to introduce new procedures.

Such training may be a general requirement across the whole workforce or may only concern specific groups or individuals. The training must be handled very carefully in order to avoid, as much as possible, alienating individuals or causing them undue embarrassment.

The form and content of the training will depend on the desired training objective, but will contain the relevant elements as described above.

Training records and certificates of training achievement

In some circumstances the maintenance of training records and the provision of training certificates is a requirement, for instance for fire wardens or forklift truck operators. Where this is not the case the employer will need to develop records and suitable certification in support of the training. The benefits of providing some form of certification for training that has been received include:

- The organisation and the individual are able to demonstrate they have set and achieved specific levels of performance
- The organisation is better able to demonstrate competence to outside agencies, such as the fire authority and HSE or client organisations
- The award of a certificate often provides a degree of status and achievement (and therefore motivation) to the employees
- Management can exclude employees not in possession of certification from certain activities, thereby enhancing operational control
- They allow for periodic training reviews and the structuring of refresher training.

Figure 4.8 Recognising training achievements

Fire safety training

Like all training the scope and depth of fire safety training will depend upon the nature and role of the individual receiving the training, the nature of the fire hazards and risks, the type of work involved and the nature of the workplace.

Individual factors affecting the scope and nature of the fire safety training they will receive include their age, physical capability, existing levels of competence and the criticality of their fire safety role. For example, the fire safety training given to a young person on work experience will differ greatly from that given to an employer or contractor conducting hot work in an area of high fire risk.

Those individuals who are required to undertake specific roles relating to fire safety management such as fire wardens and fire incident coordinators will require additional specialised training (see Chapter 10).

When considering the factors relating to the nature of the fire hazards and risks, it will always be the case that far more extensive fire safety training will be required for those fire hazards that involve high fire risks, e.g. the training required for electrical engineers on service station forecourts will be more than for electrical engineers maintaining standard office equipment.

In all cases the structure of the fire safety training given to individuals will need to cover as a minimum the following general topics:

➢ Fire hazards in the workplace, e.g. arson, faulty electrical equipment, hot work
➢ Risks associated with fire, e.g. smoke inhalation, business disruption
➢ Key risk control measures, e.g. security, user checks, permit to work system
➢ The emergency procedure for the workplace, e.g. the sound of the fire alarm, the location of the assembly point
➢ Actions that should be taken in the event of fire, i.e. what to do on:
  ➢ Discovering a fire
  ➢ Hearing the alarm.

4.5 Factors promoting a negative culture

It can be safely assumed that the absence of all the factors discussed above (see section 4.3) will promote a negative safety culture. If an organisation fails to provide a working environment that nurtures a positive culture it will have a direct impact upon the organisation and the employees.

The factors that promote a negative safety culture include:

➢ Management behaviour and decision making
➢ Staff feeling undervalued
➢ Job demands
➢ Role ambiguity.

People respond to a negative work culture in a number of ways, some will become cynical and ambivalent towards work, others will seek to want to deliberately sabotage the organisation's plans. A common outcome of a negative culture is individual work-related stress. If stress is intense and goes on for some time it can lead to mental and physical ill health, i.e. depression, nervous breakdown, heart disease.

4.5.1 Management behaviour and decision making

The behaviour of managers at work has a massive impact on their subordinates. The impact is far greater than many managers may realise and it sends strong messages to the staff as to how they ought to behave. Examples of management behaviour that have a negative impact on the safety culture of an organisation are:

➢ Failure to follow or deal with non-compliance of safety rules:
  ➢ Blocking or obstructing escape routes
  ➢ Not wearing PPE
  ➢ Not using safety guards
  ➢ Moving extinguishers
  ➢ Not evacuating during an excercise.

In addition to the physical behaviour of managers, the way that staff feel about safety will be adversely influenced by management decisions that demonstrate that safety is not a high priority. This is even more damaging if the organisation has a good safety policy in place as it not only undermines the safety culture but it indicates that management do not consider their own policies important.

4.5.2 Staff feeling undervalued

Problems that can lead to stress include the lack of adequate and meaningful communication and consultation on safety matters that will affect the individual employees. Management will also demonstrate that they undervalue employees if they operate a so-called 'blame culture' that results in the blame for problems always identified as a failure of an individual rather than accept that there has been a failure of a management system.

As well as finding fault with individuals and/or teams organisations often promote a negative safety culture by failing to recognise good work, for example when safety objectives are met or when accidents are avoided by the personal intervention of a conscientious member of staff.

4.5.3 Job demands

An organisation promoting a negative culture will impose unrealistic and unnecessary job demands on the workforce, in particular poor organisations will fail to:

- Prioritise tasks
- Cut out unnecessary work
- Try to give warning of urgent or important jobs
- Ensure that individuals are matched to jobs
- Provide training for those who need it
- Increase the scope of jobs for those who are overtrained
- Ensure workplace hazards, such as fire, noise, harmful substances and the threat of violence, are properly controlled
- Allow staff to control any aspects of their own work or make decisions about how that work should be completed and how problems should be tackled.

4.5.4 Role ambiguity

Role ambiguity is experience by individuals who are uncertain of their work roles and responsibilities. People will often find themselves in the situation where they do not know what they are supposed to be doing and feel ill equipped to do what they think may be expected. Organisations will create role ambiguity by ill-defined job descriptions in order to cover all possible eventualities. This has the benefit, for the negative organisation, of being able to identify and blame individuals for a broad range of failures.

Role ambiguity has been proven to be a significant cause of work-related stress with all of the serious consequences of chronic mental ill health for the sufferers. Specific factors that lead to role ambiguity among the workforce are the lack of:

- Clearly written and communicated policies, supported by clear and realistic job descriptions
- Proper supervisory support
- Adequate training and supervision for the job holder
- Adequate performance review.

4.6 External influences on safety culture

No organisation operates in isolation. Despite all the good intents and actions of management, the safety culture of organisations is influenced significantly by external forces which are, in the main, outside the control of management.

It is true to say that safety culture cannot be separated from the wider culture of an effectively managed business or operation, and that a reliable view of it should not focus on safety alone but rather on the delivery of the business objectives as a whole, including those for quality and service delivery.

Key among the external influences impacting on the safety culture are:

- Legal
- Economical
- Stakeholder expectations
- Technical.

4.6.1 Legal

The legal framework within which organisations manage safety issues should have a significant positive impact

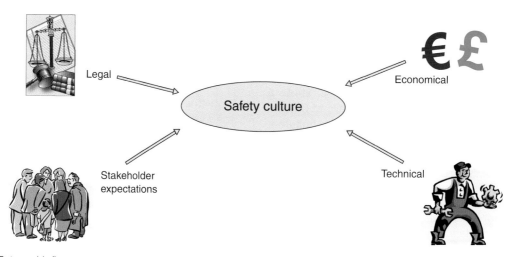

Figure 4.9 External influences on safety culture

on the culture of an organisation. For example, the law relating to how employers consult with their employees should if complied with result in early and effective engagement of the workforce in discussions relating to any matter that may affect their safety.

The mere act of employers consulting effectively with their employees as a result of complying with the law will result in improved working relationships and more effective risk management systems.

It is not only the law in terms of the acts and regulations that shape an organisation's culture, but also enforcement action taken by the Health and Safety Executive and other enforcement bodies. For example, if the Health and Safety Executive were to issue an improvement notice on, for instance, a local outlet of a national retail organisation, the impact of that local improvement notice will be felt across the entire organisation.

4.6.2 Economical

The external economic pressures that are applied to organisations can vary in their effect upon the safety culture. Those organisations who understand the relationship between reduced risk and reduced cost will want to put in place effective risk control measures; this in turn will require an excellent safety culture.

Conversely those organisations feeling themselves under acute financial pressure often pay less attention to key health and safety matters such as maintenance of equipment and training of personnel. This in turn will lead to a feeling throughout the organisation that safety is not as important as survival. In these circumstances both managers and workforces will be motivated to cut corners and reduce standards. It is often seen that the safety culture within an organisation that has restricted cash flows or business opportunities will be negative.

4.6.3 Stakeholder expectations

Every organisation will have individuals and groups who feel they hold a 'stake' in the success or failure of the organisation. Internal stakeholders include managers and employees. However, there are a number of significant stakeholders who may be outside of the undertaking including, for example:

- National unions
- Trade organisations
- Public pressure groups
- Local and national politicians
- Enforcement authorities
- Business competitors

- Society
- Insurance companies.

These groups exert influence on an organisation's safety management which can be negative or positive. Insurance companies have a stake in reducing the risks within the organisations they insure whereas the pressure from unions may be to resist the introduction of new technologies and from society.

4.6.4 Technical

One of the key principles of prevention, contained within both the RRFSO and the MHSW Regulations, is that responsible persons and employers should where reasonably practicable take advantage of developments in technology to control risk. Therefore if a specific technological development is made the management of an organisation may come under pressure to adopt it.

On many occasions it is likely that the development in technology is initially expensive and will require additional costs such as training for those who will need to operate with it.

The development of new technology therefore poses both a threat and an opportunity to the safety culture of an organisation in that, at the time that the new technology is embraced, the workforce is liable to feel positive about the benefits (providing that it doesn't result in a worsening of their employment conditions). Conversely if new technology is available and not adopted by the organisation employees are likely to feel that their needs and safety are not a priority.

4.7 Internal influences on safety culture

In addition to those forces that are outside an organisation there are obviously huge influences within an organisation that do have a significant impact upon the development and maintenance of the safety culture.

4.7.1 Management commitment

Perhaps the most singular and most important influence on an organisation's safety culture is the visible commitment of all managers and in particular senior management. A visibly committed management has two significant impacts. First, a genuine commitment to safety management will result in effective risk assessment and risk control systems. Second, when management routinely demonstrate their commitment by their behaviour at work, they will induce similar behaviour among their employees.

The culture of an organisation is sometimes described as 'the way we do things around here'.

In general people at work will want to conform to 'the way things are done'. When management reinforce the importance of risk control systems by, for example, always wearing personal protective equipment when necessary, employees will understand that conforming to safety rules is 'how things are done'. A management that fails to visibly demonstrate their commitment to safety will send mixed and confused messages if they attempt to introduce safety measures which they themselves do not comply with.

Management can demonstrate their commitment to safety in the workplace through a variety of means including:

> Providing clear and workable policies
> Allocating sufficient resources including time for safety matters
> Complying with all the safety rules in the workplace
> Attending safety committee meetings
> Responding positively to safety events and the safety concerns of the workforce
> Taking part in safety tours of the workplace
> Dealing effectively with non-compliance to safety rules.

The need for commitment at senior levels is ongoing. All too often safety procedures fall into disuse because of management neglect or due to other business pressures taking precedence such as, for example, setting unreasonable production or financial targets.

4.7.2 Production demands

An important way that management can signal their commitment to safety is to ensure it remains a priority. There are some occasions where this becomes more difficult. Pressure on production can lead to managers taking short cuts in safety. There are many instances of serious accidents that have been directly caused or exacerbated by pressures on production. The case of Piper Alpha, the North Sea oil production platform that suffered a series of explosions and fire which resulted in the tragic loss of life, was caused in part by the pressure on engineers on the platform to maintain production and other production platforms in the area continuing to pump product to the stricken Piper Alpha for fear of losing vital output from the oil field.

Of course the consequences in less hazardous workplaces of the commercial pressure to maintain production are fortunately nowhere near as dramatic; however, lives are lost every year by managers and the workforce coming under pressure to sacrifice safety for production.

Figure 4.10 Pressure of work has led to failures in safety management systems

4.7.3 Communications

It counts for little if the management of an organisation feels a genuine commitment to safety and maintains safety as a priority in the light of changing demands of the market, if their intentions, policies and practices are not communicated clearly and consistently to the workforce.

The tools available for effective communications in organisations are discussed later in this chapter, but whatever methods are used to communicate safety throughout the workplace effective communication will have the minimum key components:

> Messages will be clear and unambiguous
> Messages will be consistent through a range of formal and informal media
> It will be made in a medium that is understood by the target audience
> There will be informal and formal feedback opportunities
> Communication will be effective horizontally across an organisation and vertically both up and down lines of responsibility.

4.7.4 Employee representation

The degree and quality of both the formal and informal representation that employees have in the workforce can greatly influence safety standards. At one end of the scale, a workforce that reflects management's positive and consistent commitment to safety standards will be motivated and empowered to influence day-to-day decisions that affect safety. They will also be involved in longer-term decision making regarding the development of the business in terms of product and market innovations and improvements to production and safety.

Figure 4.11 Consultation with employees increases commitment to safety systems

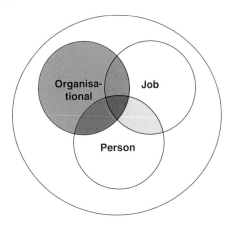

Figure 4.12 The interrelated job, person and organisational factors that affect the safety culture of an organisation

Employees who believe themselves to be an integral part of the decision-making processes at work feel motivated to safety standards and contribute in often quite unexpected ways that have a positive impact on safety standards. In contrast a workforce that feels disconnected with management and develops a feeling that their views are of little value can very easily adopt a cynical approach to safety which results in a massively negative impact on safety standards.

This is a rather simplified view of the cause and effect of human behaviour on the safety standards of organisations. The full picture is somewhat more complex, and the next section discusses human behaviour in terms of what it is, how it can affect safety in the workplace and how knowledge of how humans behave at work can be used to improve safety standards.

4.8   Human behaviour

In attempting to understand how individuals may behave in the workplace, it is important to consider what is termed 'human factors'. An understanding of human factors will enable organisations to understand and manage the effects that humans have upon risk control systems.

The safety of the employees will always depend, to a greater or lesser degree, on their own skill and ability to work 'safely', based on their training, knowledge and experience. Under normal conditions, the competence of individuals makes an essential contribution to workplace

safety. The knowledge, experience and training are often of even greater importance if events take an unexpected turn.

The acceptance of safety issues by people at work, and therefore their contribution to them, depends on the importance placed on safety by the organisation and all of the people within it.

A number of factors affect and impact upon human behaviour in the workplace. The most important of these factors relate to the organisation in which the individual works, the job being done and the person undertaking the work.

The relationship between the individual, the job they perform and the organisation in which they work is both complex and interrelated. An effective safety culture is one that recognises and manages these interdependent spheres of influence and manages the interfaces between work and:

➢ The organisational characteristics which have an influence on safety-related behaviour at work
➢ The influence of equipment and system design on human performance
➢ The perceptual, mental and physical abilities of people and the interaction between them and their job and working environment.

4.8.1   The organisation

Where management fails to take positive action on non-compliance with safety procedures, or worse still actively promote such breaches, the individuals within the organisation will perceive that such actions are condoned.

As individuals, our behaviour is influenced by the various organisations, or groups, to which we belong. Where the influences are complementary our behaviour, good or bad, is encouraged and reinforced. However, where the influences are in conflict human behaviour will generally follow the strongest influence.

75

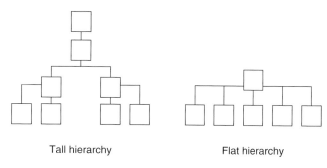

Tall hierarchy   Flat hierarchy

Figure 4.13 Examples of tall and flat hierarchical structures

Organisations can be structured or designed in a number of different ways. It is often considered that the form a particular organisation takes is a reflection of the function of the organisation. For example, the organisation of an academic institution will normally be a loose confederation of autonomous departments that are coordinated by a figurehead. In contrast, an organisation providing a service directly to the end user will tend to be far more hierarchical. Some structures may be referred to as 'tall' hierarchies and others as 'flat', with a greater divestment of authority and responsibility at lower levels.

Whichever type of organisational structure is in place it is vital that there are, at all levels of the organisation, clearly defined responsibilities for safety.

Within the work organisation will be a number of subgroups, each of which will in their turn exert an influence on the behaviour of the individuals within the organisation. Some of these groups will be formal groups and under direct control of the organisation, such as the various departments and sections, while others will be informal groups which simply form within the work organisation, such as groups based on rest room relationships or internal club relationships.

Again, the organisation must ensure that the influences exerted by the various groups are complementary and supportive of the safety culture. Where this is not the case the pressure brought by a subgroup upon its members can disrupt even the most strict control mechanisms within the formal organisation. Where subgroups do not have the same acceptance of safety goals as the employer, the members of the groups will tend to disregard the organisational safety procedures. This is often based on differing attitudes to safety within the groups, e.g. there can be a 'macho' attitude often displayed in the construction industry, leading to safety measures being sidelined.

Conflict can exist between the various groups due to a number of reasons, including:

➤ Differing priorities and goals
➤ Differing motivations and acceptance of safety issues

➤ Differing cultures and objectives
➤ Misunderstanding of individual roles
➤ Poor communication between groups.

The nature of the formal and informal organisational structures in a workplace will affect how individuals feel about safety. Some individuals may feel more comfortable working in a flat organisation whereas some will prefer the certainty of a rigid hierarchy where the individual's role is tightly defined and controlled.

Organisational structures themselves have an impact on how safety is managed with a hierarchical structure often providing an effective basis for action but a sometimes more difficult environment to achieved employee involvement in decision making. A flat structure, on the other hand, may be better for allowing individuals to contribute but does not provide the most effective framework for driving forward change.

Common problems associated if ineffective or inappropriate organisational structures include:

➤ Communication failures
➤ They are hierarchical and rigid in nature and therefore impersonal
➤ They ignore the emotional impact of organisational decisions or procedures
➤ They are often seen as uncaring and lacking commitment, especially to safety issues.

4.8.2 Job factors

All work tasks should be designed to take account of the limitations of human performance, both in normal operating circumstances and during any foreseeable emergency conditions. Matching the job to the person will ensure they are not overloaded and that they make the most effective contribution to the organisation.

This requires the employer to consider the physical and mental match between the task and the worker. Mismatches between job requirements and the capabilities of the individual increase the likelihood of human error.

Physical match

Employers need to consider the ergonomic design of the whole workplace, working environment and the work equipment within it to ensure the impact on the workers is reduced as much as possible.

Mental match

Employers should also consider the various job demands made on individuals in respect of decision making, the receipt and understanding of information and perception of the requirements of the task – particularly if additional responsibilities are added such as an evacuation coordinator or first responder.

Job design

The design of the job and working environment should be based on task analysis of the activities required of the worker to ensure the correct match between the demands of the job and the abilities of the worker exists. The main considerations during job design include:

» Identification and comprehensive analysis of the tasks expected of the individuals, including any foreseeable errors they may make
» Evaluation of the decision-making requirement placed on the individual
» The optimum balance between human and automatic safety actions
» Ergonomic design of all man/machine interfaces, including control devices, panel layouts and information and warning systems
» Design and format of operational instructions or procedures
» Provision of correct tools and equipment
» Design of work and shift patterns
» Arrangements for emergency operations and procedures
» Short- and long-term communication procedures.

Job safety analysis

Job safety analysis involves the identification of all the accident prevention measures appropriate to a particular job or area of work activity and the behavioural factors which most significantly influence whether or not these measures are taken.

Derived from task analysis, job safety analysis examines the:

» Operations incorporated in the job
» Hazards which could arise, e.g. fire explosion
» Skills required by operators in terms of knowledge and behaviour
» External influences on behaviour:
  » Nature of the influence, e.g. noise from a boiler alarm
  » Source of the influence, e.g. machinery, and activities involved – machine loading, procedure
» Learning method.

This technique is often used for higher risk or complex tasks enabling the production of safe operating systems or safe systems of work.

4.8.3 Personal factors

Personality – individuals bring to their job their own personal habits, attitudes, skills, personality and desires, all of which can affect their individual behaviour, and which may be either a strength or a weakness in relation to the task in hand. Individual characteristics influence behaviour in complex and significant ways.

Some individual characteristics, such as personality, are developed throughout the person's life and are not generally amenable to change.

Others, such as skills and attitudes, can be modified or enhanced by correct motivation, training and by the influence of the organisational culture. It is vital that the employer carefully selects individuals to ensure, as much as possible, that they match the requirements of the job.

Each human can be said to be a unique individual with a unique blend of attributes that characterise him or her. These 'individual differences' are often described as an individual's personality. People's personalities develop into adulthood and by the time they enter the world of work their personality is often fixed and tends to become more so as they continue to mature.

Attitude is an important facet of personality that has a significant impact on managing safety is an individual attitude. There have been many definitions of attitude, including the following:

» A predisposition towards a particular response in relation to people, objects and situations – not evacuating when the alarm sounds
» A predetermined set of responses built up as a result of experience of similar situations – no fire last time
» A shorthand way of responding to a particular situation
» A preconception of the way a situation is going to develop – no fire – loss of productivity = less pay.

The factors that affect the development of attitudes at work include:

» Self image and levels of confidence
» The degree of perceived control over a particular situation
» The influence of peer groups and group norms
» Superstitious and inaccurate perceptions, e.g. 'accidents are bound to happen – there is nothing we can do to stop them'.

Attitudes at work can be ingrained and built up in an entire workforce over a period of years. Changing attitudes can be a difficult and lengthy process and is, most successfully, undertaken sensitively.

In order to change attitudes it is first necessary to convince people that there is a valid reason to change, demonstrating to the individual that their attitude is based on misperceptions – generating behaviour that is disadvantageous to both themselves and the organisation can do this. Once the case for change has been accepted it is important to identify the benefits of the new behaviour that is being suggested. Provided that individuals still feel that the decision to change is theirs they will be motivated to adopt new behaviours at work and more likely to view any changes with a positive attitude. Finally the new behaviour needs to be reinforced by recognising those displaying it and dealing with non-compliance.

A useful example of changing attitudes of the public relates to the introduction of the mandatory wearing of seat belts in cars.

When compulsory use of belts was first introduced many drivers had never used them. Initially it was found that drivers did not use them, stating that:

➢ They found them uncomfortable
➢ They could not see the need
➢ They cost lives by trapping you in the car after a crash
➢ They had never been hurt in a crash so why did they need a belt.

Most of these objections were based on the general public's attitude to the imposition of seat belts on them. Over time people have changed their attitudes and, therefore, their behaviour has changed. This has been achieved through various means, such as advertising, law enforcement, provision of information, etc.

Perception – the way in which people interpret or make use of information. For instance, the way people identify risk is dictated by a range of factors, such as their age, individual attitudes, skills, training, experience, personality, memory and their ability to process sensory information.

As a result of this, if there is a mismatch between a person's ability to perceive the risk accurately, and the real life extent of the risk, the person can be misled into under- or overestimating the level of the risk.

For example:

➢ Most construction workers would perceive offices as 'safe' work environments based on their experience in a high risk industry and would, therefore, probably feel perfectly comfortable and safe in even the most unsafe office
➢ An office worker, on the other hand, taken to even a well-run construction site would feel threatened by

the apparent chaos. Again, their perception of the risk would be inaccurate based on their experience of office environments
➢ The perception of a child to the risks associated with playing with matches alone in his bedroom while the family are asleep will differ from the parents' perception of the risks associated with that activity.

How people perceive the risks associated with fire is discussed further in Chapter 10.

Motivation in its simplest form is considered to be the reaction of humans to stimulus or perceived need. It has been defined as 'a willingness to exert effort to achieve a desired outcome which satisfies a need', which implies that motivation is 'need satisfaction'. A motivator is therefore said to be something that provides the drive to produce certain behaviour or to change behaviour.

Important factors in motivating people to achieve better safety performance at work have been shown to include:

➢ Involvement in the safety management process through consultation and active participation in planning work organisation
➢ Active involvement in working parties and committees, assisting in defining health and safety objectives
➢ Clear demonstration of commitment by management to safety issues
➢ The attitudes of management and other workers towards safety
➢ Active participation in day-to-day management and monitoring of safety performance
➢ Effective communication of information to and from management
➢ The system for communication within the organisation
➢ The quality of leadership at all levels: management, trade unions, government.

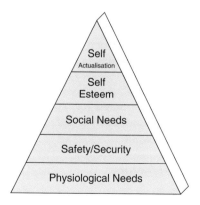

Figure 4.14 Maslow's hierarchy of needs

4.9 Human failure

Human beings are prone to fail due to limitations in perception of risk and their ability to interpret and act on information. There is a variety of ways in which human failure may manifest itself in the workplace. On some occasions people may omit owning up to something they should have done, or taking an inappropriate course of action. These failures occur very often for complex reasons that involve individual perceptions, motivation, job design and organisational factors.

There are two broad types of human failure – 'errors', where the failure to follow safety rules are made unconsciously, and 'violations', where there is a deliberate failure to comply.

4.9.1 Human errors

Human errors are occasioned by lapses of attention or mistakes

➤ **Lapses of attention** – although the intentions and objectives of the individual are correct they make an error in performing a task, perhaps due to competing demands for their attention. People at work, particularly in jobs that are routine and/or pressured, may either forget to follow a particular safety rule or even forget that there is a rule to follow

➤ **Mistaken acts/omissions** – there are two distinct types of mistakes that are made in the workplace. The first type is where the individual knows the rules but applies the wrong rule to a situation. These types of errors are caused when a person incorrectly interprets information based upon their expectation of what should happen. The other type of mistake can be made as a result of a lack of knowledge, skills or experience of the individual which leads to the individual either not taking an action they should take or taking an inappropriate course of action.

Minimising human error

Organisations can only reduce the incidence of human error by addressing the entire relevant job, organisational and individual factors. This will require effective systems for consultation, training, supervision and active monitoring.

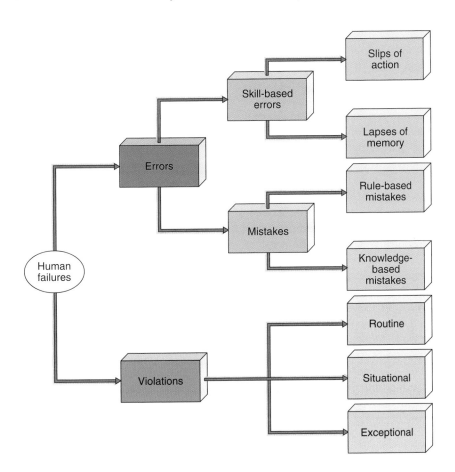

Figure 4.15   Types of human failure

4.9.2 Violations

In contrast to human errors that are made unconsciously there are many occasions when humans will deliberately breach safety rules. On some occasions the breach will be of a rule that is not physically possible to follow or may just be inconvenient, for instance the removal of machinery guards to speed production and achieve enhanced productivity payments. On other occasions the deliberate breaching of safety rules will be as a result of personal antipathy for the work or the organisation.

Violations will only occur when the individual is motivated to ignore them. This decision will be influenced by factors relating to the job, the organisation or the individual. In general deliberate breaches of safety rules occur in three distinct circumstances; routine, situational and exceptional.

➢ **Routine** – routine violations of safety rules are made when the job becomes part of day-to-day practice in a workplace. Examples include when fire doors are routinely wedged open with fire extinguishers or when PPE is routinely not worn. It will only continue if managers fail to challenge the non-compliance early. The longer a safety rule is routinely breached the more difficult it is for individuals to be persuaded of the need to comply

➢ **Situational** – situational violations occur when the design of the job make it difficult to comply with safety rules. For example, there may be occasions when an individual will need to access high level storage but is not provided with convenient access equipment such as a step ladder. In this situation access may be gained by standing on chairs or other office equipment. In an organisation with a positive health and safety culture the occasions when there is inadequate provision of work equipment will be bought to the attention of a commitment management who would resolve the situation and positively recognise those individuals who bought the issue to their attention

➢ **Exceptional** – exceptional violations rarely occur in part because they are only made when a new or unusual task is required to be undertaken or an emergency is required to be responded to. This type of violation is more likely to occur in organisations that have a high tolerance of routine and situational violations in the place. Typically with exceptional violations, only achieving the desired outcome of the task is considered and any other issues are excluded, such as cost or safety.

Minimising violations

Violations can only be reduced to a minimum by creating a positive safety culture in the workplace, i.e. a culture

Figure 4.16 Routine violations of basic safety systems indicate a poor safety culture – wedged fire door

that involves and empowers employees to be actively involved in decisions that affect their own safety.

A culture where the most senior management are visibly committed to high safety standards and a culture that encourages standards of safety to be maintained despite internal problems and external pressures will also assist in minimising employee violations.

4.10 Effecting cultural change

In order to effect cultural change with an organisation, it is necessary for the most senior levels of management to understand the implication of a negative culture. Experience has shown that organisations with a poor culture are liable to suffer serious accidents with subsequent loss of life, business assets and reputation. In addition they will be liable to lengthy legal action by enforcing authorities and the families of those who have been killed or injured.

Without this understanding of the risks that are being run, if safety is not seen as a central element of managing a business, managers are not likely to be committed to safety in a way that takes their whole organisation with them.

Once management are fully committed to improving the safety culture they need to establish and promote high standards of safety in the workplace. This must be reinforced by clear leadership including acting in a way that provides a clear and positive example for all the individuals in the organisation.

Management then need to ensure that individuals understand their role at work and are competent to undertake that role. Everyone in the workplace should have sufficient relevant knowledge skills and experience to operate confidently and safely.

People at work cannot guess what they are supposed to be doing. This is particularly true if what they are expected to be doing is new. Therefore adequate, focused training which, where possible, is tailored to the individual training needs will play a major part in effecting any cultural change. Opportunities will arise for training and thereby improving the safety culture when there is, for example, a change in the law, or the introduction of new equipment or procedures or when individuals are recruited or change roles.

Management need to constantly reinforce their commitment with clear, consistent, understandable communications. The safety ethos of an organisation should be reflected across a wide range of communication media. Care must be taken to ensure arrangements are in place for effective communications both vertically, up and down a hierarchy, and horizontally, across an organisation.

Neither cultural change nor changes in any behaviour at work can ever be achieved without convincing employees of the need to change. Once the need to change is accepted it can only be effectively implemented if the workforce feels that they have a serious and important contribution to make. Therefore among the tools available for managers to achieve cultural change are the formal and informal methods of consultation with employees.

4.11 Case study

On the evening of 6 March 1987 a British car ferry, the *Herald of Free Enterprise*, left the dockside at Zeebrugge, Belgium, for a routine crossing of the North Sea. The ship was of a design called 'roll on roll off' (RORO) whereby vehicles drove through large doors at one end of the ship when loading and drove out through similar doors at the other end to disembark. RORO vessels are constructed with large, unrestricted car decks for maximum capacity to allow them to load and unload quickly.

Shortly after leaving the port, while many of her 500 plus passengers were in the restaurant or buying duty-free goods, the ship suddenly began to list to port.

Within 90 seconds, she had settled on her side on the seabed and, despite rescue craft being on the scene in as little as 15 minutes, a total of 193 passengers and crew died. It was the worst British peacetime accident since the *Titanic*.

The subsequent public inquiry found that the bow doors through which cars and lorries were loaded had not been closed before she left her berth. As a result water began entering the car deck and very quickly affected her stability, even though the sea was calm.

The accident involved a phenomenon called the 'free water effect', which can cause catastrophic instability in vessels when even a few centimetres of water enter a hold or deck and, moving when the vessel rolls or turns, destroy its stability.

It was the policy of the company at the time that the ship did not sail with the bow doors open. However, the routine practice had evolved to leave port with the doors open in order to allow the fumes which had built up in the hold during loading to dissipate. Members of the crew were very well aware of the risks associated with this routine violation of a fundamental safety rule and attempted to bring it to the attention of senior managers.

The inquiry into the disaster, conducted by Sir Richard Sheen, found that workers had in fact raised their concerns about the risk of leaving the bow doors open on five separate occasions, but the message got lost in middle management. The inquiry concluded that the operating company, Townsend Thoresen, was negligent at every level and 'From the top to the bottom, the corporate body was infected with the disease of sloppiness.'

A coroner's inquest into the capsizing of the *Herald of Free Enterprise* returned a verdict of unlawful killing. Many of the victims' families made it clear they wished to see the Townsend Thoresen company directors (now part of P&O) face prosecution but due to the existing legal framework it was not possible in this particular case.

Charges of manslaughter were bought against the company on the basis that the company could be held criminally liable for manslaughter, that is, the unlawful killing by a corporate body of a person. However, the prosecution of P&O for corporate manslaughter ultimately failed, since it was ruled that a prosecution can only succeed if within the corporate body a person who could be described as 'the controlling mind of the company' could be identified as responsible, and that the identified person was guilty of gross criminal negligence.

4.12 Example NEBOSH questions for Chapter 4

1. **Outline** the factors that might cause the safety culture within an organisation to decline.   (8)

2. **Outline** the factors that will determine the level of supervision that a new employee should receive during their initial period of employment within an organisation. (8)

3. **Give** reasons why a verbal instruction may not be clearly understood by an employee. (8)

4. **Outline** ways in which the health and safety culture of an organisation might be improved. (8)

5. (a) **Explain** the meaning of the term 'motivation'. (2)

   (b) Other than lack of motivation, **outline SIX** reasons why employees may fail to comply with safety procedures at work. (6)

Appendix 4.1

Shown below is the diagrammatic representation of a systems approach to training.

THE ACTIVITY
Describe
Analyse tasks:
1. Competencies
2. Knowledge
3. Skills
4. Attitude

THE TRAINEE
Level of:
• Competence
• Knowledge
• Skills
• Attitudes

CONFIRMED LEARNING

IDENTIFYING TRAINING NEEDS
• Who and for What
• Overall Aim

DEFINE OBJECTIVES
• Performance
• Conditions
• Standards
CRITERION TEST

SYLLABUS

DESIGN TRAINING PROGRAMME
Available:
• Methods
• Materials
• Place
• Time etc.

CHANGE

DELIVER

EVALUATE

FEEDBACK

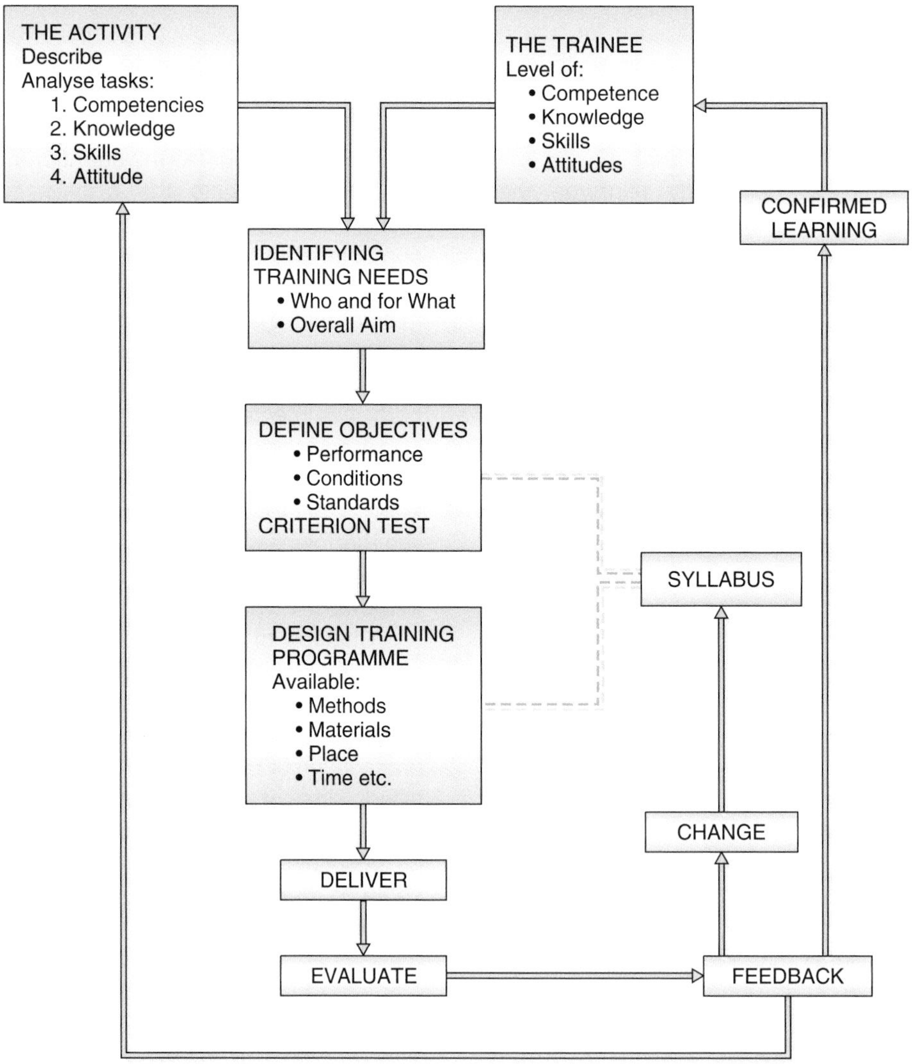

Figure 4.17 A systems approach to training

Principles of risk assessment

5

5.1 Introduction

The purpose of risk assessment is to assist an employer and/or a 'responsible person' to identify the preventive and protective measures required to comply with the law and in doing so, ensure, as far as reasonably practical, the safety of their workforce, premises and those around them who could be affected by their activities.

This chapter aims to provide the basic principles which are used during activity-based assessments and the more general risk assessment techniques. Chapter 14 will discuss the fire risk assessment process which is based upon many of the issues discussed in this chapter. An effective risk management strategy is a critical element in assisting an organisation to successfully plan their management of safety. Risk assessments conducted by an organisation will help to identify and quantify areas of strengths and weaknesses, prioritise actions for controlling fire risks and provide a basis for developing a positive safety culture.

As discussed in Chapter 1, the legal argument for managing fire and health and safety is only one of the three reasons that an organisation should manage safety effectively.

The same can be said for risk assessment, in so far as while assessments are required by law, the moral argument for reducing personal injuries and incidents of ill health by the use of risk control measures should also not be overlooked. The financial and economical

This chapter discusses the following key elements:

➤ Definitions relating to risk assessment
➤ Risk assessment and the law
➤ Competencies required for risk assessment
➤ The process of risk assessment
➤ Risk assessment recording and reviewing procedures.

Figure 5.1 In order to assess risks from fire it is necessary to understand the general principles and practice of conducting risk assessments

implications of failing to identify risks generally, particularly those related to fire safety, can also have serious implications on an organisation's ability to fulfil its contractual obligations, maintain its position in the market, protect its reputation and potentially, and secure its survival.

The risk assessment process will vary depending upon an organisation's activities. It may be that the assessment will be a highly technical and complex scientific analysis, such as in the case of COMAH site activities (a site defined under the Control of Major Accident Hazards Regulations 1999). At the other end of the scale, the assessment may simply be a fairly succinct analysis of the hazards, risks and control measures relating to the work activities conducted in a small office environment.

5.2 Definitions relating to risk assessment

The first chapter of this book included basic definitions relating to the management of safety. There is a wide range and some diversity of terminology contained within British and European standards and HSE and industry guidance. The following definitions, drawn from the standards and guidance, are used in this book.

5.2.1 Hazard

Something with the potential to cause harm. A source or situation that could cause harm such as chemicals, electricity, working at height, hot work processes and in the case of an emergency an inability to respond and escape to a place of safety.

5.2.2 Harm

Harm includes the effects relating to human injury and ill health, damage to the environment or loss to an organisation.

5.2.3 Risk

A combination of the likelihood (chance or probability) of a specified event occurring and should it do so, the severity (or consequences) of the outcome.

5.2.4 Risk assessment

The process of identifying hazards and evaluating the level of risk (including to whom and how many are affected) arising from the hazards, taking into account any existing risk control measures.

It is critical to distinguish between the two elements contained in risk so that a judgement of risk magnitude can be identified. For example, if assessing the risks associated with a contractor starting a fire during hot work processes it is necessary to differentiate between the likelihood of a fire being initiated and the severity or consequences of the outcome. As if hot works are being undertaken outside in the open air the consequences of a fire occurring are potentially less disastrous than if the hot work was being undertaken inside a building, where smoke levels may prevent people escaping.

5.2.5 Risk controls

Workplace precautions, for example a guard on a dangerous part of machinery, sprinkler systems within a building, safe systems of work (procedures), personal protective equipment (PPE), safety signs.

Figure 5.2 Inspecting contractors arrangements for working at height

5.2.6 Risk control systems (RCS)

These are arrangements that ensure that risk controls (workplace precautions) are implemented and maintained. For example: the provision for ensuring that an adequate level of supervision is maintained during work processes; a system for planned preventive maintenance for work equipment and specific safety systems; and establishing a programme of inspections and audits for buildings, sites and workplaces.

5.3 Risk assessment and the law

The HSWA requires employers to understand 'the risks inherent with their work' to ensure that they keep their employees and others who may be affected by their work activities safe, so far as is reasonably practicable.

The RRFSO and the MHSW Regulations enhance and indeed make more specific the requirements and duties placed upon 'responsible persons' and/or employers for risk assessment.

In relation to fire, the responsible person, in addition to other duties, must make a suitable and sufficient assessment of the risks to which relevant persons are exposed in order to identify the general fire precautions that are required to comply with the requirements and prohibitions imposed by the Order.

The Order also identifies requirements relating to assessing the risks arising from the presence of dangerous substances (see Chapter 14) and risks in relation to young persons.

Figure 5.3 The responsible person and/or the employers must conduct an assessment of the risks in the workplace

Under the MHSW Regulations every employer has an 'absolute' legal duty to 'make a suitable and sufficient assessment of the risks' to the health and safety of:

➤ His employees to which they are exposed while they are at work
➤ To persons not in his employment, arising out of or in connection with his activities.

The requirement to conduct suitable and sufficient risk assessments placed upon employers is extended to self-employed persons, who have an absolute legal duty to assess the risks to their own health and other persons who may be affected by their work activities.

The MHSW ACoP indicates that, in practice, the term 'suitable and sufficient' requires employers, the self-employed and responsible persons to:

➤ Identify the risks arising or connected with their work clearly differentiating between the significant risks and the insignificant risks (trivial)
➤ Prioritise the necessary control measures to comply with the law
➤ Take reasonable steps to assist themselves in identifying risks, including those that they could reasonably be expected to know or foresee
➤ Ensure the assessment is appropriate to the nature of their work
➤ Identify a period of time for which the assessment is likely to remain valid.

> The HSE provide guidance for employers and the self-employed on how to conduct risk assessment in which they identify the following five key steps to risk assessment:
>
> 1. Identify the hazards associated with a work activity
> 2. Identify persons who may be at risk
> 3. Evaluate the risks and existing control measures
> 4. Record the findings
> 5. Review when necessary.

While the MHSW Regulations are the umbrella under which risk assessments are required by law, many other pieces of legislation also require the completion of risk assessments. Detailed below is a list of such regulations.

➤ The Noise at Work Regulations 2005
➤ Manual Handling Operations Regulations 1992*

➤ Health and Safety (Display Screen Equipment) Regulations 1992*
➤ Control of Lead at Work Regulations 1998
➤ Ionising Radiation Regulations 1999
➤ Dangerous Substances and Explosive Atmospheres Regulations 2002
➤ Control of Substances Hazardous to Health Regulations 2002
➤ Control of Vibration at Work Regulations 2005
➤ The Regulatory Reform (Fire Safety) Order 2005.

*As amended by the Miscellaneous Amendments Regulations 2002.

5.4 Competency to conduct risk assessments

Employers and responsible persons need to determine who should be part of the risk assessment team. A team approach is often the most effective way to ensure that all the appropriate risks have been identified, this is because:

Perception – individuals' perception of risk will be different. Depending upon age, experience, attitude, knowledge of the area, etc., one person may have a completely different view of what constitutes a tolerable (acceptable) or intolerable (unacceptable) risk than another person. Some people are happy to accept risks which another person would not tolerate, and so a different assessment of risk in the same area would be made

by different people. To reduce this element of subjectivity inherent in any risk assessment, it is recommended that the team approach is adopted and the majority decision is accepted in areas where disagreements occur.

Limits of knowledge – one person does not have the overall knowledge required of each process, person, activity, machine, area, etc., to be able to adequately identify sufficient hazards. More than one pair of eyes is necessary to ensure that nothing is missed.

Familiarity – a person carrying out a risk assessment in their own area is likely to miss a number of the present hazards due to familiarity or complacency. They may also be likely to accept certain risks as tolerable either because that it is the way they have always been or because they know they will be the person responsible for implementing any necessary controls.

Teams – a team approach may be used to involve the people who actually carry out the task or work in the area and thereby gain their input into any likely hazards that could be identified together with any current risk control measures. If the workforce is able to suggest control measures which are later implemented, this not only makes the controls more likely to be complied with but also has positive effects on the whole organisation's safety culture.

An effective risk assessment team could therefore involve any or all of the following:

➤ Health and safety/fire safety advisers
➤ Department managers
➤ Supervisors
➤ Workforce
➤ Competent risk assessors

Figure 5.4 Training to support risk assessment

➤ Safety representatives
➤ Designers
➤ Process engineers.

Once the team has been assembled a list of all activities within the chosen area, or list of areas, would need to be compiled as part of the initial inventory preparation. (This key issue is discussed in Chapter 14.)

It will normally always be necessary to support the assessor and/or assessment team with appropriate training to fulfil their role. The exact nature and duration of such training will be dependent upon their role, existing levels of knowledge and the work activities being assessed.

Any training designed to support a risk assessment process is likely to include:

➤ Organisational policy on risk assessment
➤ Legal requirements for risk assessment and the interpretation of legal standards
➤ How to identify hazards using sources of information (HM Government guides, HSC ACoPs, safety event reports and inspection reports)
➤ Evaluating risks using qualitative and quantitative mechanisms
➤ The identification and selection of appropriate control measures (taking into account those that are reasonably practicable)
➤ Recording the assessment (forms, reports and recording skills)
➤ Communication and dissemination of the outcomes of the assessment.

While the above list is not exhaustive and any training programme will not make a person or persons 'competent', a basic programme will provide underpinning knowledge from which an assessor can become competent.

5.5 The risk assessment process

Because of the fundamental role risk assessments play as a starting point for developing safety management systems, they must be conducted systematically. A systematic approach will help satisfy the law and ensure that nothing which could present a risk is inadvertently omitted. What appears to many to be the daunting task of conducting all the necessary risk assessments for any given work undertaking can be relatively easily achieved by a straightforward progression through a number of logical steps.

There are a number of different methodologies that are currently used throughout industry and commerce to achieve a systematic approach to risk assessment. In its guidance *Five Steps to Risk Assessment* the HSE has suggested the following stepped approach to assessing risks to health and safety in the workplace:

1. Identify the hazards
2. Decide who might be harmed and how
3. Evaluate the risks (in terms of likelihood and severity) and decide whether the existing precautions are adequate or whether more should be done
4. Record the significant findings
5. Review the assessment and revise if necessary.

Figure 5.5 HSE's *Five Steps to Risk Assessment* guide

In its own fire safety guidance documents HM Government has adopted a very similar approach in its guidance on how an assessment of fire risks can be achieved (Chapter 14).

The above steps identify the basic process of risk assessment and are discussed in more detail later in this chapter.

In order for organisations to conduct suitable and sufficient risk assessment and ensure that all risks arising from work activities are identified, evaluated and effectively controlled it is necessary to adopt a systematic approach to conducting risk assessments. It is likely that any such approach which require organisations to perform the following stages:

➤ Preparing an activity inventory that clearly identifies the types of workplace/s, any activities and processes that are to be assessed

➤ Identifying the significant hazards involved (something with the potential to cause harm, etc.) for each area, or task included in the inventory

➤ Identifying all persons who may be harmed by identifying various groups of people and considering the numbers of individuals from each group who may be harmed

➤ Evaluating the levels of residual risk taking into account the risk control measures that are already in place

➤ Applying (where necessary) additional control measures that may be required to reduce the risk to the lowest level reasonably practicable

➤ Recording the significant findings of the assessment, which is a legal requirement when five or more persons are employed

➤ Reviewing and where necessary revising the risk assessment

➤ Communicating the findings to all appropriate persons.

5.5.1 Preparing an inventory of activities

The first stage in producing a risk assessment programme is to identify all the buildings and areas (including the activities and tasks undertaken) that will need to be assessed; this is often termed inventory preparation. This process provides an initial overview of each building or area, task or activity that will need to undergo a suitable and sufficient assessment. This initial analysis of the sources and distribution of risk arising from a work activity also enables organisations to prioritise the premises, tasks and activities that will need to undergo a more detailed assessment.

When faced with an entire workplace or a number of buildings on one site to assess, the first step would be to break the site down into smaller, more manageable

Figure 5.6 Preparing inventory of activities to assess

sections, then go on to list the specific areas or premises to be assessed. The usual method is to select geographical areas which would contain similar hazards or sites of similar nature or construction. In addition this approach provides a register of locations that can help employers/responsible persons to identify when changes take place that may require the assessments to be reviewed.

> The HSE has produced useful guidance on inventory preparation in HSG65. The guidance suggests that an inventory should include:
>
> ➤ Risks brought into an organisation
> ➤ Risks associated with its activities and
> ➤ Risks associated with outputs and by-products.

The only downside to producing an inventory is that, for it to be an effective part of the risk assessment and risk management process, it must be kept up to date and therefore requires management resources.

5.5.2 Identifying the significant hazards

Once the inventories have been created and a programme of assessment has been established the next stage is to identify all the hazards that are involved with the premises, tasks and activities.

Possibly the most valuable source of information relating to the hazards, risks and effectiveness of existing risk control systems in the workplace are the employees and their representatives. Often those undertaking a specific task or working within a particular facility are only too aware of the hazards and potential risks associated with the work and in many cases will have views on how improvements to safety management could be made.

In addition to consultation with the workforce, there are a number of sources for information relating to hazards that assessment teams will want to consider; the most important of these include:

➤ Reviewing records:
  ➤ Safety event (accident/incident, fire/false alarm) records, hazard information records and COSHH assessment records
➤ Reviewing documents:
  ➤ HM Government and HSE Guidance, industry guidance and manufacturers' data sheets. Company safety policies, method statements, emergency procedures, etc.

➤ Location inspection:
  ➤ Site safety tours and site safety inspections
➤ Activity observation:
  ➤ Job safety analysis/task safety analysis (such as hot work processes).

As part of the identification of hazards, it is useful to use HSE's RIDDOR 'accident categories' as a means of identifying the hazards by considering the causes of injury which may arise. These categories cover slips, trips and fall, falls from height and falling objects, collision with objects or being trapped or crushed beneath or between objects, manual handling, contact with machinery or hand tools, electricity, transport, contact with hazardous chemicals, asphyxiation/drowning, contact with animals, and violence, not to forgetting of course fire and explosion.

These categories have been adopted when trying to identify hazards and risks in relation to general health and safety; however, they may equally be applied when considering the types of risk in relation to fire which may include:

➤ Slips, trips and falls (including those from height) while evacuating
➤ Handling, lifting or carrying portable fire fighting equipment
➤ Being trapped by a wall collapsing
➤ Being asphyxiated from the inhalation of smoke
➤ Coming into contact with the release of harmful substances
➤ Being exposed to fire or explosion while undertaking fire fighting action
➤ Coming into contact with live electrical equipment due to degradation of wiring during a fire
➤ Coming into contact with moving machinery while trying to shut down in the event of an emergency

➤ Being struck by a moving vehicle while evacuating
➤ Being assaulted by a person panicking in the event of a fire.

It is likely that the above list will not necessarily be included in a fire risk assessment (see later in the module); however, each should be considered as part of a general risk assessment.

In the same way, categories of health risk, i.e. chemical, biological, physical, physiological, must also be considered as part of a general risk assessment process. Chemicals can obviously present a risk when coming into contact with humans, whether or not it is as a result of fire, and the release of asbestos fibres may also present a significant risk (see Chapter 7).

Many of the physical and potentially psychological health risks can also be linked to a fire scenario and thus must be taken into account during the risk assessment process.

It is also prudent when assessing hazard and risk to make the distinction between acute (single instant contact) and chronic (prolonged/repeated exposure) ill health. It may be that the inhalation of smoke containing toxic chemicals will cause death rapidly or that the inhalation or exposure to less hazardous chemicals during the fire process may worsen conditions of a person who has already been exposed over a period of time.

5.5.3 Identifying who is at risk

The third stage of the risk assessment process involves identifying who could be harmed (i.e. who is at risk) and how. When considering who is at risk, it is important to consider all those who may be affected by the activity or be on the premises, as well as those who are directly involved. Groups of people who could be at risk may include the following:

➤ Those directly involved with work within a facility, or undertaking an activity – skilled operatives, trainees, new workers
➤ Contractors – new contractors or regular maintenance contractors
➤ Visitors – clients, business representatives, consultants, regular visitors, first time visitors, those who only use facilities for meetings
➤ Members of the public and passers-by

Figure 5.7 It is important to understand the nature of the risks from fire from hazards such as LPG cylinders

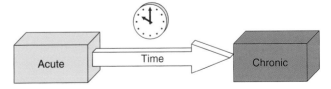

Figure 5.8 Diagram illustrating 'acute' and 'chronic'

➤ Adults, children (under school leaving age), including other people on whose site the activity is taking place, etc.

➤ Young persons – under the age of 18 (due to age, have limited knowledge, awareness, strength, etc.)

➤ Those who share the building or facilities – other employers, staff members and the self-employed

➤ Other persons – those whose building or facilities are close enough to be affected by the organisation's activities or the result of an emergency within

Figure 5.9 Building contractors have the potential to increase fire risk

➤ Lone workers – those persons who work on their own either in a remote location within a workplace or those whose work is peripatetic moving from site to site.

In addition, within each group, individual people may be at greater risk due to their age, inexperience, competence, physical condition, etc.:

➤ Disabled/sensory impaired persons

➤ Those with physical impairments or sensory impairment that may prevent awareness or response

➤ Expectant/nursing mothers, who may be more susceptible to physical/mental stresses.

The assessment should also take into account how many people are involved, anyone who is particularly at risk and why.

5.5.4 Evaluating residual risk

The fourth stage in the risk assessment process is to evaluate the level of residual risk, i.e. taking into account any current control measures (workplace precautions and risk control systems) that are already in place. Some risks may also be adequately controlled leaving only a low residual risk; these will require no further action other than a review of the validity of the assessment on a periodic basis (see below).

An example is a contractor who has been engaged to install IT cabling within an existing workplace. The work location will include the main access corridor to the main staircase within the building. Those that could be put at risk from the operation would be the contractor, the occupier of the premises, visitors to the premises, other occupiers (if the building is occupied by more than one) and other contractors.

Risks could include the more obvious falling from height, falling materials, slips and trips. In addition the less obvious, but nevertheless safety critical, is the potential obstruction of escape routes and breaches of fire compartment walls. Many people could be affected during the works and potentially many more over the life of the building if fire compartmentation is breached by running cables or trunking through fire breaks.

Hazard presents a **risk** which causes an **accident/incident** which in turn causes **injury, ill health, damage, or loss**.

As an example, oil can be considered as a hazard. When it is in the container it has potential but no likelihood of achieving its potential, therefore there is no risk. When it is spilt on a table there is a small possibility of causing minor harm (ill health perhaps), therefore the risk exists but is low. When the oil is spilt on the floor there is a better chance of it causing greater harm (slip), so the risk could be termed medium. When the oil is spilt on a heater unit or close to other potential ignition sources there is an even greater chance of it causing serious harm or even death (starting a fire), so the risk could be termed high or very high. It can be seen then that a hazard is an inherent quality of something, whereas risk is dependent entirely upon the circumstances surrounding the hazard.

There are a number of methods for evaluating risk. The method applied for any particular risk will depend on a number of factors, such as the complexity of the activities carried out and the type and nature of the workplace. For many of the day-to-day risks that people in the workplace are exposed to, including fire, a simple qualitative assessment will suffice, for more complex risks a quantitative or semi-quantitative assessment may be needed.

> **Qualitative analysis – describes the quality of risk using words.**
>
> **Quantitative analysis – quantifies the risk with numerical data.**
>
> **Semi-quantitative analysis – uses numbers to quantify qualitative data.**

Qualitative analysis

Qualitative analysis describes a quality of the risk. Typical of the qualities most often described is that of quantum, i.e. size or magnitude. For example, when assessing the means of escape in the event of a fire, fire risk has historically been rated as high, normal or low.

Qualitative analysis of risk is a subjective measure, based upon the risk assessor's judgement. As with all methods of risk evaluation, a qualitative assessment will need to allow consideration of the two aspects of risk, i.e. the likelihood of a particular occurrence and the severity of the consequences.

The HSE, in its guidance document HSG65, have suggested a basic estimator as appropriate for a simple evaluation of risk. The HSE estimator, shown in Table 5.1, uses purely subjective measures of likelihood and uses RIDDOR events to ascertain the qualitative descriptors of the seriousness of the outcome.

As no numerical scales have been introduced with the above simple estimator it is difficult for a precise indication of the level of risk to be determined. Neither does this approach provide an easy mechanism to confirm if the risk has been reduced to the lowest level reasonably practicable.

However, by introducing the above subjective estimations of the two elements of risk into a simple risk matrix, a qualitative assessment of risk can be made. Figure 5.10 is an example of a simple risk matrix.

When undertaking risk assessments for a number of work activities the application of a consistent method of qualitative evaluation of risk will allow for actions to be prioritised.

For example, a systematic qualitative evaluation of an organisation's risk will result in some risks that are evaluated as 'high', some that are 'low' and some that are 'insignificant', management decisions can then be made of a basis of risk vs cost. This qualitative method provides a basic evaluation of risk and will allow an organisation to consider what may or may not be considered a reasonably practical level of safety.

	Major injuries may occur	Serious injuries may occur	Slight injuries may occur
High chance of an event	High risk	Medium risk	Low risk
Medium chance of an event	Medium risk	Medium risk	Low risk
Low chance of an event	Low risk	Low risk	Insignificant risk

Likelihood of an accident/incident occurring
Outcome of the potential accident/incident

Figure 5.10 Example of a simple risk matrix using the two aspects of risk to determine the magnitude of risk, expressed in qualitative terms

Table 5.1 The HSE estimator

Likelihood	Severity		
High	Where it is certain or near certain that harm will occur	Major	Death or major injury (as defined by RIDDOR) or illness causing long-term disability
Medium	Where harm will often occur	Serious	Injuries or ill health causing short-term disability
Low	Where harm will seldom occur	Slight	All other injuries or illness

Quantitative analysis

In this method risk is evaluated, not by subjective judgement, but by numerical data. Quantitative evaluation of the risk is therefore more demanding than a qualitative approach but provides a more rigorous evaluation.

A quantitative approach, when used to recalculate the effectiveness of controls (after implementation of risk control measures) also provides clear evidence that confirms that risks have in fact been reduced.

It should be noted that specific raw data is required when using this technique to analyse the magnitude of risk. Data that will be required to allow a quantitative evaluation of risk will be found in the local and national records of:

➤ Hazard reports
➤ Injury accidents and incidents
➤ Ill health and sickness
➤ Health monitoring systems
➤ Environmental monitoring systems
➤ Fire-related incidents.

Details relating to the types of safety events that occur together with the likelihood of occurrence gathered from such records are an essential tool when evaluating risk as they provide statistics that confirm the severity (or potential severity if a near miss) and how frequently the event occurs.

For most organisations gathering statistics in relation to fire to provide a quantitative evaluation of fire risk can be difficult to achieve. However, data from the insurance industry via the statistics produced by the Fire Protection Association (FPA) and from the fire services via DCLG is available. Generally the statistics produced by the insurance industry focus upon estimated financial losses, whereas those produced by DCLG relate to fire deaths and injuries.

Both sets of statistics provide information on the causes of fire and the numbers of fires occurring in given areas or sectors. As the severity rating or potential outcome of a fire is death or multiple deaths, reducing the likelihood of a fire occurring and managing the secondary hazards associated with fire (not being able to escape, etc.) must be seen as being a high priority.

Gathering meaningful data that gives a clear indication of frequency and severity is very often difficult (unless the industry, sector, or organisation is large enough to have sufficient statistics available).

While personal injury accident data and ill-health data are generally readily available due to the frequency of events, the statistics relating to the numbers of fires and false alarms reflect the relatively low numbers of incidents; however, the outcomes are quite often more severe, so pure quantitative analysis can be very often difficult to achieve.

Pure quantitative analysis is generally only needed in a small select group of high risk industries, such as nuclear and offshore. The final and probably most widely used evaluation of risk is a combination of the qualitative and quantitative approaches and is referred to as 'semi-quantitative' risk.

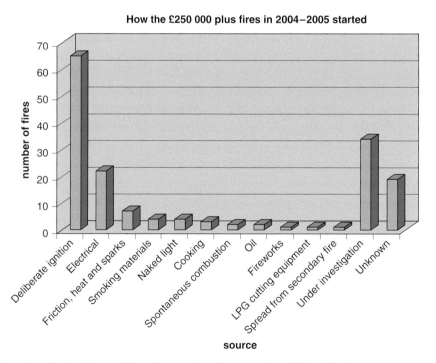

How the £250 000 plus fires in 2004–2005 started

Figure 5.11 Graph showing the initiators of fires that resulted in more than £250 000 damage

Semi-quantitative analysis

The use of semi-quantitative techniques for risk assessment is widespread and it is often referred to as a quantitative method; however, it is easily seen that risk is expressed as a numerical value, the estimation of the magnitude of the risk is in fact subjective and therefore qualitative. A semi-quantitative evaluation of risk allows numerical values to be assigned to both severity and likelihood in the absence of data.

Like the qualitative approach, the semi-quantitative technique is essentially subjective and is based upon the risk assessor's personal interpretation of the level of risk, but with the added assistance of a risk matrix so that a risk factor can be calculated numerically.

For example, the likelihood of a particular event occurring is assigned numerical values as shown in Table 5.2.

Table 5.2 Example of numeric values given to subjective, qualitative descriptions of risk, of a typical semi-quantitative assessment

The qualitative description of the likelihood of a particular event occurring	Numerical value that may be assigned
No evidence of occurrence	1
Foreseeable but remote	2
Has occurred but only infrequently	3
Has occurred fairly frequently	4
Has occurred regularly and will occur again	5

While it is easy to differentiate between the two opposing definitions at the end of each scale, there is often confusion with the words used to separate out the mid range.

There is no laid down criteria for either the words used to describe the value or the numerical values within the scale. A 5 by 5 scale is arbitrarily used here, though there are a wide variety of scales used from 3 through to 8 and sometimes even 10.

Just as the likelihood rating is assigned numerical values, so too is the severity rating. Again using a scale of 1 to 5 and adapting the HSE's simple qualitative estimations of severity, an example of how numerical values may be assigned is shown in Table 5.3.

In addition to the severity of the outcome of an event for an individual, the total numbers of people that may be affected must also be considered, as in the case of calculating the likelihood of the rise in fires started by discarded smokers' materials in places of public entertainment (if a smoking policy exists). The resultant effect upon persons being able to escape safely may also revolve around the numbers of persons present.

Having assigned numeric values to each element of risk, a risk matrix can then be completed to provide a calculation of an overall 'risk factor'.

The resultant risk matrix can be used to provide and develop an action plan which may also be assigned numbers so that priorities can be identified. Risk matrices are often colour coded to provide a visual concept of whether or not the residual risk is tolerable or acceptable.

An example of a semi-quantitative risk matrix, which incorporates the numerical scales discussed above, is shown in Figure 5.12.

Table 5.3 An example of how numerical values may be assigned

The qualitative description of the severity of a particular event	Numerical value that may be assigned
First aid injury or illness requiring minimal attention (plaster, etc.)	1
Minor injury or illness. Includes those where a person could spend *up to* 3 days away from work	2
'3 day' injury or illness. Those where the person would be off work more than 3 days (as per RIDDOR)	3
'Major' injury or illness (as defined per RIDDOR)	4
Fatality or disabling injury or illness preventing return to work	5

Note: When assigning numerical scales to severity the assessor must take into account chronic or long-term effects of any particular event and/or the effects of the long-term exposure to a perceived 'low' hazard.

Figure 5.12 Example of a semi-quantitative risk matrix

Once individual risks have been evaluated using a semi-quantitative approach, it is possible to establish formalised action plans based upon the risk grading – a sample is shown in Table 5.4.

The type of risk analysis or evaluation method used will vary from organisation to organisation; whether a numerical matrix or high, medium, low matrix is used will very much be dependent upon the risk levels of an organisation's activities and workplaces.

5.5.5 Applying additional risk control measures

Having analysed the risks taking into account any existing control measures, the next stage in the risk assessment process is to consider the effectiveness of the controls and additional control measures that will be required to reduce the risk as low as is reasonably practicable.

A wide range of risk control measures are available, many of which can be identified through reading HM Government, HSE and industry guidance documentation, or seeking advice from specialist advisers and companies.

In assessing the suitability of controls not only must guidance be sought, but reference to the MHSW Regulations and RRFSO hierarchy of controls should be considered. These are detailed below:

1. Avoiding risks
2. Evaluating the risks which cannot be avoided
3. Combating the risks at source
4. Adapting to technical progress
5. Adapting the work to the individual, especially as regards the design of workplace, the choice of work equipment and the choice of working and production methods, with a view, in particular, to alleviating monotonous work and work at a predetermined work rate and to reducing their effect on health
6. Replacing the dangerous by the non-dangerous or less dangerous
7. Developing a coherent overall prevention policy which covers technology, organisation of work and the influence of factors relating to the working environment
8. Giving collective protective measures priority over individual protective measures and
9. Giving appropriate instructions to employees.

The above principles of control are used in the MHSW Regulations. The same principles are used in the RRFSO and the Fire (Scotland) Act, which omits point 5 and slightly rewords the paragraphs to focus upon fire.

Chapter 6 will detail in more depth with the range of risk control measures available in a practical approach, which while not exactly mirroring the above serves to provide the basis from which to consider the management of any residual risks.

To ensure that a risk assessment can be demonstrated as being 'suitable and sufficient', one of the key factors to be considered is that of prioritisation of

Table 5.4 Example of recommended actions against residual risk values

Risk level	Guidance on necessary actions and timescales
High risk 15–25	These risks are unacceptable; significant improvements in risk control are required. The work activity or use of the workplace should be halted with immediate effect until risk controls are implemented that reduce the risk so that it is no longer high
Medium risk 6–12	Efforts should be made to reduce the risks associated with the activity or workplace. Medium- and longer-term risk reduction measures should be introduced within a specified time frame following the introduction of a series of short-term risk control measures
Low risk 2–4	Minimal control measures are required to be implemented to satisfy the level of risk. Any control measures required to further reduce the risks are of a low priority. Arrangements should be made to maintain current arrangements for risk control
Insignificant risk 1	These risks are considered acceptable; no further action is required other than to secure arrangements to maintain current risk control

the risk controls, particularly with regard to an implementation programme.

It is therefore essential that a mechanism for prioritisation be considered. Any action plan emanating from a risk assessment must be SMART:

➤ Specific
➤ Measurable
➤ Achievable
➤ Realistic
➤ Time bound.

In addition any action plan arising from a risk assessment must seek to identify the appropriate resources required to complete the action. This must include time and people required. It is also important to ensure that the person in the organisation who is responsible for its implementation is identified to enable follow-up of the implementation plan.

Figure 5.13 A fire door retaining device – an example of the application of technical progress

Following a fire risk assessment in a small office facility it was identified that no early warning system by way of detection was installed, particularly in storerooms and other unoccupied out of the way areas.

While the action plan identified that a detection system complying with BS 5839 Part 1 was required, the action plan also identified that domestic smoke alarms could be installed as an interim measure, while awaiting the design, installation and certification of the new system.

5.5.6 Recording of risk assessment findings

Where five or more persons are employed by an employer, the risk assessment findings (the significant ones) must be recorded. It is good practice that all findings of risk assessments are recorded, not only if they are legally required, but also so that they can be used for providing staff and management with information and may help defend claims for negligence.

A wide variety of differing forms are used for recording risk assessments and there is no universal layout. In general forms reflect the requirement of the law and guidance issued by the enforcing authorities. The *Five Steps to Risk Assessment* document produced by the HSE suggests the minimum standards of what should be recorded on a risk assessment form.

Regardless of the approach adopted to record the findings, key elements should be recorded and are likely to include:

➤ The building, task, or location being assessed
➤ The name of the assessor/s
➤ The date of assessment
➤ The scheduled date of review of the assessment (arbitrary see section 5.4.7)
➤ A breakdown of all the identified hazards (if a task, hazards at each stage of the task)
➤ Persons at risk and why
➤ Current control measures
➤ An evaluation of risk (with existing controls in place)
➤ Identification of controls required to reduce the risk (prioritised in an action plan)
➤ An evaluation of any residual risk (with the additional controls in place).

Better assessments identify named persons and target dates for implementing findings as part of the action plan. An example of a form to record a task-based risk assessment is provided at Appendix 5.1.

5.5.7 Reviewing and revising the assessment

The assessment must be kept under review to ensure that it remains valid for the activity or premises to which it relates. If circumstances change, the assessment must be reviewed and where new hazards are introduced or the overall risk changes, the assessment must be revised, amended or a new assessment completed to take account of the changes that have invalidated the original assessment.

It should also be noted that when a prioritised system for implementing controls is used the assessment should be reviewed at each stage to ensure that the required level of control is maintained.

Circumstances that may require a review of the any risk assessment may include:

➤ Changes of:
  ➤ The premises – internal or external layout
  ➤ The people affected – numbers, ages, disabilities
  ➤ The plant or tools – power supplies, heating systems
  ➤ The procedures in place – systems shutdown, emergency arrangements
➤ The introduction of new work processes – hot work, confined spaces mechanical handling
➤ Changes in legislation
➤ As a result of hazard reports
➤ Following a safety event (fire/false alarm or accident/incident)
➤ Results of occupational health assessment or monitoring
➤ As a result of enforcement action
➤ Following consultation and discussion in a safety committee meeting.

Having completed the initial assessment, it is important to review the whole risk assessment process to determine:

➤ The accuracy of the initial assessment
➤ The effectiveness, applicability and practicability of the selected precautions
➤ Whether the risk assessment team operated effectively.

Risk assessment should not be seen as a 'one-off process' but as an evolving exercise which continues to improve the overall management of safety within the workplace.

5.6 Communicating the findings of a risk assessment

If risk assessments are going to be more than just words in a document then the significant findings need to be acted upon. Managers and other staff need to understand the risks in the workplace, what needs to be done to control them and what role they have to play in the process. The only way for the findings of the risk assessment to be effectively implemented is for them to be effectively communicated.

Besides being a requirement of current legislation, it is good business sense for the findings of any risk assessment to be communicated to all those who may be affected. Groups of people who should be provided with information should include, but not necessarily be limited to:

➤ Employees
➤ Members of the management team (who are likely to be involved with implementing the action plan)
➤ Visitors and contractors
➤ Other employers/employees sharing the workplace
➤ Landlord and 'responsible persons' (particularly in the case of fire).

There are a variety of different mechanisms by which the significant findings can be communicated. It may be that providing written copies of the assessment may be effective, or the information is included in staff meetings, tool box talks or other formalised programmes. The information is also often included in pre-planned training sessions as part of both induction and ongoing training programmes.

The key issues that should be included in the information provided from the fire risk assessment are

Figure 5.14 Reviewing and revising risk assessments

Figure 5.15 Communicating the significant findings of a risk assessment

the hazards that arise from the activity or workplace, the groups of people at risk, the level of risk and the control measures that are required to reduce the risk. It may also be appropriate to emphasise what to do if hazards still exist or control measures are not working effectively.

5.7 Case study

In November 2000 an 11 year old child was playing, as was the custom, with a group of friends in a multi-storey car park when he became trapped in the roll-up machinery of an automatic roller shutter door.

The company who owned and managed the car park had, at the time of the accident, not conducted an assessment of the risks to children who may have gained unauthorised access to the car park.

The child was seriously hurt and died as a result of the injuries he suffered.

Following an investigation of the circumstances of the accident the HSE prosecuted the company involved for failure to ensure the health and safety of persons not in their employment in that over a significant period of time, local children used the car park as a play area thereby exposing themselves to risks to their health and safety.

The company had also breached regulation 3 of the Management of Health and Safety at Work Regulations 1999 by failing to conduct a suitable and sufficient risk assessment. In fact the company concerned had conducted a health and safety risk assessment of the roller shutter door prior to the fatal accident but it was deemed to be not suitable and sufficient.

The company was fined a total of £50 000 and were ordered to pay an additional £50 000 in costs to the HSE.

5.8 Example questions for Chapter 5

1. (a) **Identify** the particular requirements of regulation 3 of the Management of Health and Safety at Work Regulations 1999 in relation to an employer's duty to carry out risk assessments. (3)
   (b) **Outline** the factors that should be considered when selecting individuals to assist in carrying out risk assessments in the workplace. (5)

2. (a) **Explain**, using an example, the meaning of the term 'risk'. (2)
   (b) **Outline** the key stages of a general risk assessment. (6)

3. An organisation has introduced a new work process for which a risk assessment is required under regulation 3 of the Management of Health and Safety at Work Regulations 1999.
   (a) **Outline** the steps that should be used in carrying out the risk assessment, identifying the issues that would need to be considered at EACH stage. (8)
   (b) **Explain** the criteria that must be met for the assessment to be deemed 'suitable and sufficient'. (4)
   (c) **Identify** the various circumstances that might require a review of the risk assessment. (8)

4. (a) **Explain**, using an example, the meaning of the term 'risk'. (3)
   (b) **Outline** the content of a training course for staff who are required to assist in carrying out risk assessments. (5)

5. (a) **Outline** the factors that should be considered when selecting individuals to assist in carrying out risk assessments in the workplace. (5)
   (b) **Describe** the key stages of a risk assessment. (5)
   (c) **Outline** a hierarchy of measures for controlling exposures to hazardous substances. (10)

Appendix 5.1   Example of a general activity risk assessment record and action plan

RISK ASSESSMENT RECORD	Acme Company Ltd		Number	
Risk assessment for	Assessor's name/s		Date of assessment	Review date
Tasks involved				

No.	Hazards	Persons at risk	Risk rating (H,M,L)	Control measures in place (Control measures required)	Action yes/no

No.	Hazards	Persons at risk	Risk rating (H,M,L)	Control measures in place (*Control measures required*)	Action yes/no

Risk matrix:

	Major	Serious	Slight
High	High risk	Medium risk	Low risk
Medium	Medium risk	Medium risk	Low risk
Low	Low risk	Low risk	Insignificant risk

Control hierarchy

| Eliminate |
| Reduce/substitute |
| Isolate |
| Control – engineering/SSOW |
| PPE |
| Discipline/signage |

RISK ASSESSMENT ACTION PLAN		Acme Company Ltd		Number	
Tasks involved					

No.	Control required	Priority/Target date	Person responsible	Re-assessment by (Name)	Revised risk rating

1	Immediate action required within 24 hrs
2	Short term action required within 1 week
3	Undertake action within 1 month
4	Action within 3 months or agree plan within 6 months
5	Review as part of Business Plan

General principles of control

6

6.1 Introduction

The key to effective safety management, once the risks have been identified, is to establish and implement a control strategy. The control measures that are implemented to secure the safety of all those at work or who may be affected by the work or work processes should reflect the legal requirements, as a minimum standard, and any technological advances that have been made.

The term 'prevention' when used in relation to fire has often been interpreted to mean preventing a fire starting in the first place, which clearly is the most effective way to deal with fire risk management. In current legislation and guidance, however, the term prevention relates to the need to prevent persons being harmed or loss being sustained.

The management of fire risks both in relation to primary fire hazards (ignition, fuel, oxygen) and secondary fire hazards (those preventing people escaping safely in the event of a fire) will be discussed in greater detail later within the book allowing this chapter to focus on strategies of control as they relate to safety management as a whole.

This chapter discusses the following key elements:

- Principles of prevention
- The hierarchy of risk control measures
- Supervision and monitoring
- Safe systems of work
- Emergency procedures.

6.2 Principles of prevention

The fundamental 'principles of prevention' that are included in both the RRFSO and the MHSW are supported by a range of additional legislation, approved codes of practice and guidance each providing assistance when considering the control measures that may be required to reduce the risks to the lowest level reasonably practicable.

A responsible person (under the RRFSO) and/or employer (under the MHSW) are legally required to apply the 'principles of prevention'. These are principles that have been adopted throughout the European Community and are contained in both the RRFSO and the MHSW. The principles state that:

- Where possible risk should be avoided
- Risks which cannot be avoided should be evaluated (i.e. assessed)
- Risks should be combated at source
- Control measures should adapt to technical progress
- Dangerous substances/articles should be replaced by the non-dangerous or less dangerous
- A coherent overall prevention policy should be developed, which covers technology, organisation of work and the influence of factors relating to the working environment
- Priority should be given to collective protective measures (i.e. those which protect the most people) over individual protective measures and
- Employees should be given appropriate instruction.
- An additional control included in the MHSW is that of adapting the work to the individual, especially as regards the design of workplaces, the choice of work equipment and the choice of working and production methods, with a view, in particular, to alleviating

monotonous work and work at a predetermined work rate and to reducing their effect on health. In addition the overall prevention policy also includes working conditions and social relationships.

This additional 'principle of prevention' contained within the MHSW does not directly relate to fire; however, it can be seen that individual human errors that result in a fire in the workplace may well reflect a monotonous work pattern and/or poor working relationships.

> The principles of prevention are particularly useful when considering controlling the risks from fire. For example, when a 'responsible person' is developing a fire risk control strategy, they should attempt to completely eliminate the risk of a fire occurring. This may be done by having effective security, and ensuring that the workplace contains no electrical equipment, dangerous substances or potential arsonists. In most cases this will not be a practical solution. Therefore the responsible person will need to evaluate the fire risk, in other words make an assessment of the nature and magnitude of the risks from fire in the workplace. This will include assessing the risk of a fire breaking out *and* the resultant risks to people, the building and the business assets.
>
> Once the nature and magnitude of the fire risk is known the responsible person must consider applying the other principles of prevention.

There follows some more examples of how the principles of prevention may be understood and applied:

Avoiding risks – if it is possible avoid a risk altogether perhaps by undertaking the work a different way, for instance using compression fittings when joining pipe work together rather than using a naked flame and LPG.

Evaluating the risks which cannot be avoided – this is undertaken by completing a risk assessment or series of risk assessments, for instance completing a risk assessment for handling a reactive flammable chemical and/or completing overall fire risk assessment for a premises.

Combating the risks at source – this means taking steps to reduce the risks at source such as protecting the external metal fire escape staircase from the vagaries of the weather rather than putting up warning signage e.g. slippery when wet.

Adapting to technical progress – where new technology exists there is a requirement to adapt the control measure to take any progress into account, such as fitting intumescent strips to fire doors to enable them to present a more effective fire stop rather than relying on a larger door rebate.

Replacing the dangerous by the non-dangerous or less dangerous – for instance, replacing a low flash point chemical with either a non-flammable chemical or a chemical with a higher flash point (Chapter 7).

Developing a coherent overall prevention policy which covers technology, organisation of work and the influence of factors relating to the working environment – the preparation of safety policy documentation including organisational and arrangements sections and the development of a positive safety culture.

Giving collective protective measures priority over individual protective measures – give priority to the measures that protect a number of employees rather than those of individual employees such as providing effective fire compartmentation within a building which prevents rapid fire spread. So that all occupants can safely escape in case of fire should be given a higher priority than developing a personal emergency evacuation plan (PEEP) for a disabled employee.

Giving appropriate instructions to employees – ensure that all relevant persons (employees, self-employed persons, other employers) understand their responsibilities and what they should do, for instance emergency procedures or when undertaking potentially hazardous hot work processes.

Figure 6.1 Practical fire evacuation exercises are an essential way of providing information.

Consultation and the arrangements for gathering information from workers and their representatives are particularly important if adaptation to the individual is to be addressed as part of the principles of control to prevent human error.

6.3 The hierarchy of risk control measures

The principles of control outlined above provide a basis from which to consider the adequacy of existing control measures, both those already in place and those that are likely to be required. The principles themselves do not provide a hierarchy from which to assist confirmation that the risks have been reduced so far as is reasonably practicable, neither do they establish a hierarchy of which control or series of controls to use.

The following is a summary of the preferred hierarchy of risk control principles as outlined in the HSE publication HSG65 as they relate to general health and safety.

Eliminate risk by substituting the dangerous for the less dangerous, e.g.:

➤ Use less hazardous substances
➤ Substitute a type of machine that has a better guarding system to make the same product
➤ Avoid the use of certain processes entirely perhaps by contracting out.

Combat risks at source by engineering controls and giving collective protective measures priority, e.g.:

➤ Separate the operator from the risk of exposure to a hazardous substance by fully enclosing the process
➤ Protect the dangerous parts of a machine by guarding
➤ Design process machinery and work activities to minimise the release, or to suppress or contain, airborne hazards (LEV)
➤ Design machinery to be operated remotely and to which materials are fed automatically, thus separating the operator from danger areas such as moving blades, etc.

Minimise risk by:

➤ Designing suitable systems of working
➤ Using personal protective clothing and equipment; this should only be used as a last resort.

HSG65's hierarchy reflects that risk elimination and the use of physical engineering controls and safeguards to control risk can be more reliably managed rather than those that rely solely on people, particularly when they are prone to error.

The hierarchy shown above relates to the management of occupational safety and health issues, the hierarchy is very often extended slightly following a very similar layout enabling the effective management of both the primary and secondary fire hazards to be considered.

The hierarchy is shown below:

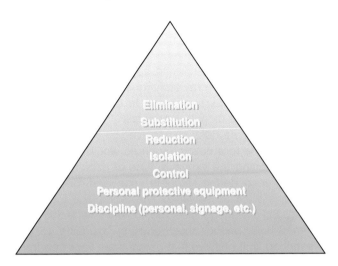

Figure 6.2 The hierarchy of risk control measures

6.3.1 Elimination

The most effective way of managing risk – removing an entire process, task or activity eliminates the risk entirely. Perhaps changing a process that requires heat via a naked flame to a cold process may be possible given the technological advancement made over recent years. With regard to fire risk management elimination can only effectively be used to manage primary fire hazards although if it can be used effectively it will have an impact on managing the secondary fire hazards. On many occasions it is not possible to stop undertaking a process and thus an alternative control measure or selection of control measures from within the hierarchy must be considered.

6.3.2 Substitution

The replacement of a high risk hazard, such as a solvent-based paint, with that of a risk-free or low risk alternative such as a water-based paint. A further example would be the use of battery powered hand tools as a substitute for 220/240 volt electrical tools which possess greater fire risks not to mention slip/trip risks from the cables.

6.3.3 Reduction

Reducing can take a number of forms such as the reduction in the numbers of persons exposed to the hazard. For example, changing the work pattern by undertaking the refurbishment of a main staircase in a busy office building concourse could be undertaken out of normal working hours thereby reducing the numbers of people exposed to the work or who may be affected by the work (means of escape partially obstructed). In relation to the management of dangerous substances in the event of a fire, a mitigating measure would be the reduction of the number of persons potentially exposed by undertaking the work in a remote or segregated area.

When considering a method of work it may be possible to apply a different technique to reduce the risks such as using a roller to apply a coating material rather that using a spray technique which may present an additional explosion risk due to the atomised spray and the presence of ignition sources.

Reduction may also be achieved by the reduction of exposure time to a hazard, such as would be the case when managing the exposure time of persons involved in the transportation, storage, use and disposal of chemicals such as solvents. The Workplace Exposure Limit (WEL) assigned to acetone, for example, is 500 ppm in an 8-hour period, thus reducing exposure to the prescribed limit ensures that persons are kept free from ill health arising from the hazardous chemical. Acetone is, however, a chemical with a low flashpoint so even a small amount could present a fire and explosion risk and thus reducing the volume is also a fire control measure.

The single biggest reduction measure in relation to dangerous substances is to reduce the quantity or amount of the substance transported, stored, handled, etc. to the absolute minimum. This could be achieved by establishing a 'call-off' system whereby minimum stock levels are supported by regular deliveries keeping the stock levels down to the absolute minimum.

6.3.4 Isolation

Isolating the hazard from people by enclosing the hazard should also be considered when elimination, substitution or reduction cannot be achieved. Isolation can be achieved, when handling flammable and reactive materials, by the provision of a glovebox enclosure to prevent ill-health effects from inhaling the vapours, in addition to preventing the flammable vapours finding an ignition source causing a fire or explosion.

Isolation of electrical supplies throughout a building (service risers) can be achieved by ensuring that the risers are fire resistant enclosures, thus should a fire start the secondary hazard (such as the production of smoke) is contained within the enclosure and will not have an adverse effect upon those escaping.

It may also be possible to undertake hot work within an enclosure which is fire retardant and/or fire resistant,

thus isolating the risk of fire progressing outside the enclosure by containing it within.

6.3.5 Control

Control can be achieved in two separate ways, although like many of the controls discussed in this chapter they may well be used together in an overall control strategy.

Engineering controls – the use of a guard or an interlocking system is an effective engineering control for preventing access to dangerous moving parts on machinery. Engineering controls are widely used in relation to fire safety and risk management solutions, for example smoke extract systems manage smoke levels to allow greater time to escape from large buildings and premises such as shopping centres. Sprinkler and other types of fixed fire fighting installations are installed to manage and suppress a fire. Detection and alarm systems are engineering controls that provide early warning and a communications system in the event of a fire (Chapter 9).

Safe systems of work (SSOW) – these are formal written procedures that describe how a process or activity is to be undertaken. The HSWA requires the production of SSOWs and that all employees know them. There is a wide range of terminology used across industry to describe an SSOW. In the construction industry, for example, the term 'method statement' is used, in many production-based organisations the SSOWs are referred to as safe operating or standard operating procedures. Regardless of the terminology used and the way they are recorded, SSOWs tend to include very similar items, such as a description of the work, its sequence, the risks and controls, etc.; further details on SSOWs are discussed later in this chapter.

6.3.6 Personal protective equipment (PPE)

Personal protective equipment should only be issued as a last resort, where risks cannot be controlled by more positive means, i.e. with controls further up the hierarchy. Where it is issued the employer is not allowed to make any charge for its provision. The equipment must be readily available and employees must have clear information on how to obtain it, use it, maintain it and store it. In order to ensure the proper level of protection and comfort the PPE must be available in a range of sizes to suit the entire workforce. PPE must conform to the relevant standards and carry the CE mark.

Any PPE supplied must be suitable for the operating conditions and it must protect against the specific risk present. This will require the employer or responsible person to carry out an assessment of the precise need for protection and specify PPE which will meet those needs. This assessment must also take into consideration the needs of the wearer in terms of fit and usability.

Any PPE issued must have a supporting system to ensure it is properly maintained or replaced as necessary. The level of maintenance support required will depend on the nature of the equipment and may range from simple cleaning to a planned inspection and maintenance programme.

While PPE is seen as a last resort, there are a number of benefits to its use these are:

> It supports and enhances other control measures such as SSOWs and permit to work systems (confined space entry)
> It provides a solution to risk reduction where other controls are impracticable such as wearing bump caps in low head height plant room areas
> It can be used as an immediate interim measure to allow work to continue prior to establishing further controls such as guarding, enclosure, etc.
> It will be required as part of the emergency response arrangements for effectively rescuing a person from areas such as confined spaces or enabling isolation of power sources, service supplies, etc. in the event of an emergency.

There are a number of limitations attached to the use of PPE, the majority of which revolve around the fact that if the PPE is not worn, is damaged, does not fit or is not maintained it will not only offer limited or no protection, it can also lead the wearer to believe that they are protected even if they are not.

6.3.7 Discipline

The use of signage markings and personal discipline is the final control measure covered by ERICPD hierarchy. The reliance upon individuals following the sign's direction is immense; it is also well known that an overprovision of signage actually results in a person becoming 'sign blind' where they cannot identify what the specific signage denotes and will therefore not take the appropriate course of action – this will be discussed further in Chapter 10. All signs must display a pictogram to identify their meaning which may, if necessary, be supported by suitable wording.

Mandatory signs
Signs prescribing specific behaviour
Circular sign
Blue ground (50% of area of sign)
White border and pictogram.

E.g. fire door keep shut.

Figure 6.3 Typical fire door keep shut sign

Prohibition signs
Signs prohibiting behaviour likely to cause or increase danger.

Circular sign
Red edging and cross stripe (35% of area)
White ground
Black pictogram.

E.g. No smoking.

Figure 6.4 A typical no smoking sign

Warning signs
Signs giving warning of a hazard or danger.

Triangular sign
Yellow ground (50% of area)
Black edging and pictogram.

E.g. Explosive atmosphere.

Emergency escape or first aid signs
Fire safety signs (also known as safe condition signs). Signs giving information on emergency exits, first aid, rescue facilities, etc.

Rectangular or square sign
Green ground (50% of area)
White pictogram.

E.g. Emergency exit.

Figure 6.5 A typical explosive atmosphere warning sign

Figure 6.6 A typical emergency escape sign

Fire signs

Signs showing the direction to, or location of, fire fighting equipment.

Rectangular or square sign
Red ground (50% of area)
White pictogram.

E.g. fire extinguisher.

Figure 6.7 A typical fire extinguisher location sign

Clear, well-positioned and well-maintained signage will enhance the control measures identified earlier in this chapter. As will be seen in Chapter 10 the use of escape signage is a critical control in assisting persons to leave the building safely.

Behavioural controls in the form of personal discipline is required to follow not only the safety signage but also the conditions laid down in SSOWs and for an individual to utilise engineering controls as part of an overall risk management strategy. The safety culture pervading in an organisation will directly affect each of the elements of the hierarchy of controls if elimination has not been achieved. Therefore discipline in the shape of supervision and monitoring will also form part of the control process.

It is likely that rather that placing reliance upon any one individual control measure covered in ERICPD a combination of controls outlined in the hierarchy will often be required and then supported by an appropriate level of supervision and monitoring.

6.4 Supervision and monitoring

With the exception of elimination, the remainder of the hierarchy of controls will require an adequate level of supervision and monitoring. This will ensure that the controls are implemented and maintained, particularly those that place a heavy reliance upon humans.

For example, effective management of both primary and secondary fire hazards is essential and the role of effective supervision and monitoring must not be underestimated.

A well-trained and competent person will be able to undertake a level of 'self supervision' reducing the need for more formal management supervision. The level of self supervision will also need to take into account the level of risk.

When considering a programme of hot works the levels of supervision will be dependent upon both the competence of the individual undertaking the task and the risk of fire starting and spreading. Therefore for this higher risk activity effective management systems with appropriate levels of formal supervision are essential to control the risks. The levels of supervision and monitoring required will need to take into account the following factors:

➤ Individual factors, e.g.:
  ➤ Competence
  ➤ Capability
  ➤ Age
➤ Task factors, e.g.:
  ➤ Complexity
  ➤ Frequency of operations
  ➤ Environment
➤ Risk factors, e.g.:
  ➤ To the individual
  ➤ To other groups or persons
  ➤ To the business
  ➤ To the environment.

The type and level of safety monitoring will depend upon the workplace precautions or controls in place. Monitoring of engineering controls such as fire detection and alarm systems will be based on formal systems such as those

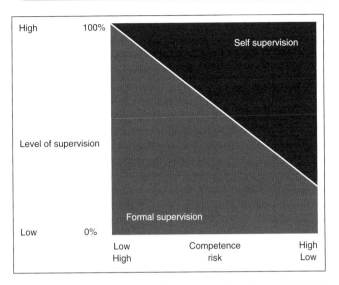

High   100%

Self supervision

Level of supervision

Formal supervision

Low   0%

| | Low | Competence | High |
| | High | risk | Low |

Figure 6.8   Diagram of levels of supervision from HS(G)65

found in BS 5839 Part 1, whereas the mechanisms for ensuring that persons can safely evacuate in the event of an emergency (means of escape) will depend upon active monitoring such as inspections and safety tours.

The monitoring and management of contractors is also required at a high level particularly when works involve high risk processes such as the use of flammable substances or where works are undertaken on access and egress routes.

Consideration must also be given to how an organisation monitors emergency exercises and its training and competency levels, as the protective measures implemented by it to keep people safe will rely upon them following procedures for which they have been trained.

6.5   Safe systems of work

6.5.1   What is a safe system of work (SSOW)?

A critical element in controlling primary fire and other workplace hazards is the use of a safe system of work (SSOW). This SSOW, in simple terms, defines a way of undertaking a task in a safe manner, for instance it would define the safe way of refuelling a forklift truck

Definition of a safe system of work:

A step-by-step procedure based upon the integration of people, equipment, materials, and the environment, to produce an acceptable level of safety.

powered by LPG, or the safe way of undertaking hot work processes.

Safe systems of work appear in a variety of forms including:

> Standard operating procedures
> Safe operating procedures
> Method statements.

Such a system of work would on the majority of occasions be formally recorded and documented, taking into account all foreseeable risks associated with the operation and detail the procedure for minimisation of the risk, or where reasonably practicable the complete elimination of the risk.

As can be seen from the above definition the safe system of work should provide a step-by-step approach which integrates all the facets that will secure the safety of not only those undertaking the task or job, but also those who may be affected by it.

As previously discussed, it is the risk assessment that determines the need for a safe system of work to be introduced, as part of the mechanisms by which the hazard and associated risks are controlled. The safe system of work relies heavily upon human behaviour and discipline and therefore selecting a safe system of work above other control strategies, such as elimination, should not be considered.

The procedural controls developed in the creation of the safe system of work will also have to take into account technical control measures including design features such as the use of guards on cutting equipment, noise reduction systems and workplace modifications. Each of these will be supported by behavioural controls which address the individual in terms of selection and training, which are included in the 'safe person approach'.

The actual control measures should be selected on the basis of their effectiveness, applicability, practicability and, finally, cost. Some controls may be dictated by legal requirements, e.g. equipment guards. In most cases, however, it is usual to apply a hierarchy of measures such as that given in the publication *Successful Health and Safety Management* (HS(G)65).

It is often the case that complex or unusual tasks are required to have a formally documented SSOW. Over recent years the formal system has been adopted for the majority of routine tasks as their production assists in defining the safety standards required of an operation or process and importantly provides a basis for the provision of information, instruction and training programmes.

A number of pieces of current legislation refer to the requirement for the production of safe systems of work, together with the provision of information, instruction and training. Therefore in addition to introducing measurable standards, the introduction of safe systems of work assists an employer or responsible person to comply with the law.

In order for an organisation to effectively use safe systems of work as part of its risk control strategy it will be necessary to consider the follow:

➤ When a safe system of work is required
➤ How safe system of work will be developed
➤ How the safe system of work will be documented
➤ How the system will be communicated and persons trained
➤ How the system will be monitored, reviewed and when necessary revised.

6.5.2 When is a safe system of work required?

As discussed above a safe system of work may well be recorded and documented for the majority of tasks or operations undertaken. In general the risk assessment of an operation will determine the need to document the safe system or consider that an informal approach is appropriate. To assist in determining the level of formality that a safe system of work takes the following key factors may be considered:

➤ The level of risk identified in the risk assessment
➤ The legal requirements, i.e. (Construction Design Management Regulations ACoP)
➤ Guidance from trade bodies in relation to 'best practice' (Fire Protection Association)
➤ Complexity of the task or operations
➤ Previous experience (safety events, accidents, incidents)
➤ Level of resources required to implement and monitor the SSOW:
  ➤ Documentation
  ➤ Training
  ➤ Supervision.

Safe systems of work do not always require documenting, for example the changing of a light bulb on a desk lamp. If, however, there is no formal safe system of work the mechanisms by which people undertake tasks and operations may vary and there is the potential that people may be harmed as a result. A formalised system enables standards to be communicated, implemented, monitored and reviewed which will assist in managing the risk.

6.5.3 Developing a safe system of work

The development of a safe system of work for a work activity requires an adequate level of resources to be committed. The level of resources required will reflect the complexity of the operation, those that will be involved and the resources available.

One key area in the development of a safe system of work is to gather sufficient information from a variety of sources. Table 6.1 identifies such sources of information and provides a brief overview of the information that may be provided.

Once the information has been gathered the development of the system of work will need to be coordinated and recorded. The responsibility for this task rests with the employer or responsible person. Assistance of a competent person will on many occasions be required, so that the coordination of the team producing the SSOW and the development of the system reflect the preventive and protective measures required by law.

The competent person is likely to have knowledge of a variety of techniques for analysing the operation including:

➤ Job safety analysis (JSA)
➤ Hazard and operability study (HAZOP)
➤ Fault tree analysis (FTA)
➤ Failure modes and effects analysis (FMEA).

The last three may assist in preparing the safe system of work, particularly when the operation may have a high loss potential.

The JSA technique is frequently used to assist in the development of a system for a variety of operations and includes the following steps:

Select	the job to be analysed
Record	the component parts of the job in chronological order
Examine	each component part to determine the risk of harm
Develop	control measures for each step to reduce the risk of harm
Record	the job safety instruction
Communicate	the information in the instruction to operators and supervisors
Maintain	the safe system to ensure it remains effective

Regardless of the technique adopted to develop the system of work, there are four essential factors that must be considered when developing it: people, equipment, materials and the environment.

Table 6.1 Sources of information available when developing a safe system of work

Source of information	Comment
Employees that undertake the operation or task	Vital source of information on how the work is actually undertaken, or how it is likely to be undertaken. Previous knowledge of similar activities, the hazards and risks. History of faults, failures and success Very valuable asset as part of the legally required consultation process
Legislation, ACoP and guidance issued by enforcing authorities	Provide minimum standards required by law. For example, Confined Space Regulations require a system of work for confined spaces and the ACoP goes on to identify the main elements to consider when designing the safe system including the use of permit to work
Enforcement bodies themselves	Verbal or written information provided while in face-to-face contact during an inspection
Guidance from trade bodies	Written information from these bodies can take the shape of guidance notes such as in the case of *Fire Prevention on Construction Sites* published by the Construction Confederation and the Fire Protection Association. This guide provides information on, among others, arrangements for storing flammable liquids and LPG
British, European and International standards	These documents detail the minimum industry standards that should be applied and are often cross-referenced by other guidance such as in the case of Building Regulations Guidance Approved Document B that cross-references BS 5839 Part 1 Code of practice for system design, installation, commissioning and maintenance of fire detection and alarm systems for buildings
Manufacturer's information and guidance	The range of information available is vast, e.g. how to test and maintain systems, chemical hazards and personal protective equipment that should be used to protect against residual risks
Risk assessments and job safety analysis	The information from previously completed risk assessment and JSA records provide a backbone on which to formalise a safe system of work
Organisation's policy, procedures and standards	Information from existing policies and procedures such as those relating to the management of vehicles may provide a basis from which to determine the safe systems for reversing or refuelling vehicles
Safety event investigations	The findings and recommendations contained in an investigation report may assist in reviewing the safe system of work
Results of health surveillance	The results are likely to be able to confirm that the effective use of PPE has reduced exposure or not and that the system is effective or not

People – a safe system of work should be designed to:

➤ Ensure the person doing the work has the right mental and physical capabilities
➤ Ensure adequate training is provided
➤ Promote safe behaviour
➤ Ensure employees are properly motivated towards safe working
➤ Ensure the correct level of supervision is exercised to ensure compliance with the system.

Equipment – a safe system of work should ensure that:

➤ All equipment, including any personal protective equipment, is suitable and fit for its intended use
➤ Safety specifications are considered such as ergonomics, noise, etc.

➤ Any specific risks presented by the equipment are controlled
➤ There are adequate maintenance procedures to ensure equipment remains in an efficient state
➤ Emergency shutdown procedures have been established
➤ Any training required for safe operation of the equipment is provided.

Materials – materials must be safe during use or processing:

➤ Any materials used during the work process must be considered, including any by-products created during the process
➤ Appropriate waste disposal measures should be employed

➤ End products must conform to the required quality and safety standards.

Environment – a safe system must consider (where necessary):

➤ Control of temperature, lighting and ventilation
➤ Appropriate controls for dust, fumes, vapours, radiation, chemical and biological hazards
➤ Safe access and egress
➤ Provision of adequate welfare facilities
➤ Noise and vibration
➤ Variations:
   ➤ In climatic conditions
   ➤ Due to the time of day or year
   ➤ Due to other persons in the work environment
➤ Evacuation in the event of an emergency.

Any safe system of work should be designed to combine these four elements to produce an integrated method of working which will ensure that tasks are carried out in the safest way that can be achieved under the circumstances.

To illustrate how an effective safe system of work might be developed, it is useful to consider the requirements for a 'lone worker' who by the very term works by themselves without close or direct supervision.

There are a range of work situations that may involve lone working and include:

➤ Cleaners (out of hours)
➤ Security staff
➤ Delivery drivers
➤ Installation and maintenance engineers
➤ Warehouse persons
➤ Police officers
➤ Social workers.

In addition to those listed above, any member of staff may work out of normal work hours or be working in a remote location and as such may be deemed to be lone working.

Using the people, equipment, material and environment approach detailed above the procedure adopted would take into account as a minimum:

People – ensuring that the selection process takes into account the psychological capabilities required of the role and that adequate training assists in attaining the required level of competence and confidence. Regular periodic visits from their line manager to ensure that adequate monitoring of operations is maintained.

Equipment – the provision of safe equipment for the tasks being carried out taking into account only one person is available. Other equipment including communications equipment (mobile phone, pager, radio)

Figure 6.9 Consideration of the people, equipment, materials and environment is required when conducting a job safety analysis

to enable regular contact with the line manager and others in the team must also be considered. Mobile first aid kit and automatic warning devices in the event of an emergency should also be made available.

Materials – each work situation would dictate consideration of a variety of chemicals (cleaners, delivery drivers, installation and maintenance engineers, warehouse workers) that persons could come into contact with or in the case of installation and maintenance engineers the products conform to the required quality and safety standards.

Environment – safe access and egress arrangements which would include arrangements in the event of an emergency. Any specific arrangements for the provision of adequate welfare facilities, perhaps through liaison with a third party, must be considered. The safety of those working in the hours of darkness must be taken into account and work patterns considered accordingly.

This list is not exhaustive but is used to indicate how a safe system of work may be considered and then produced.

6.5.4 Documenting a safe system of work

Having assessed when a safe system of work is needed, and a competent person has been engaged to assist in its development, the next stage of the process is to document the system. As discussed previously it should be noted that not all safe systems of work will need to be documented; this decision is likely to be made by the competent person.

How the system is to be documented should be considered and reflect among other issues; the nature of the operation, the level of control required, who will be operating it.

The documentation should be written in such a way as to be easily intelligible whosoever will refer to it. It should be user friendly and avoid the use of technical jargon as far as possible. Many safe operating procedures within industry are reproduced as summary sheets, which are on many occasions posted adjacent to or on the machinery to which they relate.

Documentary control is an essential part of managing a successful system and in particular when introducing a safe system of work. An effective document management system will ensure that the most up-to-date system of work is being operated to, particularly when amendments have been made to reduce the overall risk.

To assist an organisation to record and document its safe systems of work many will adopt a standard format such as the use of a template. This is particularly prevalent in the construction industry sector where templates for the production of method statements exist. Whichever format is developed it is likely that it will need to include as a minimum:

> The name of the task or operation
> The address or location
> A brief description of the operation
> The sequence of work or operations that are to be carried out
> The plant, tools or machinery involved
> Any chemicals or hazardous substances that may be involved or produced
> Foreseeable hazards associated with the operation (cross-reference to the risk assessments, COSHH assessments, manual handling assessments, etc.)
> Workplace precautions to be used to minimise the risks to all persons who may be affected, for higher risk operations a permit to work may be required
> Competency or specific training requirements of the persons undertaking the work or operations
> Supervisory management levels
> Emergency procedures
> Monitoring arrangements
> Revision number and associated documentary control measures.

6.5.5 Permits to work

In circumstances where it is necessary to adopt a highly formalised safe system of work it is often necessary to introduce a 'permit to work' system. A permit to work system documents details relating to the work to be done, the hazards involved, the precautions necessary and the persons responsible. Typical examples of work that may require a permit to work system include:

> Working with pressurised systems
> Working adjacent to overhead crane tracks
> Working with asbestos-based materials
> Work involving high voltage electrical equipment
> Work involving underground services
> Any work in confined spaces
> Hot work.

The permit to work system is more fully described in Chapter 8 specifically as it relates to the control of hot work.

6.5.6 Safe system of work – communication

Having documented the safe system of work and prior to its implementation a critical factor that will affect the success of this procedural control measure will be ensuring that the information contained in the safe system of work is communicated to all those involved with the operation, including the operators, supervisors and managers.

How the information is communicated will vary from a briefing sheet to a full training programme dependent upon the complexity, familiarity, competence, etc. of those needing to receive the information.

It may also be necessary to provide information to other persons who may directly or indirectly be affected by the operation, such as the facilities management team (if isolations are required as part of a permit to work system).

The content of a typical training programme for those involved with the operation of a safe system of work is detailed below:

> What a safe system of work is
> Why the safe system of work is required
> What the work involves
> Hazards associated with the work and the findings of the risk assessment
> The control measures that will be adopted including (if appropriate):
>> Permit to work
>> Guarding systems
>> Isolations
>> Personal protective equipment
>> Monitoring (air sampling, health, etc.)
> Emergency procedures
> How the work will be supervised.

6.5.7 Monitoring, reviewing and revising

To ensure that any safe system of work remains effective an organisation must establish a management programme for monitoring, reviewing and where necessary revising the system of work.

Monitoring of the system is often undertaken by first line management as part of their job function. It is the role of supervisors to identify the effectiveness of the system in controlling the hazards and risks relating to the operation. Where there appears to be a shortcoming in the arrangements this should be reported to the senior management so that a review of the system can be undertaken.

In many organisations formal systems exist to undertake periodic reviews of safe systems of work regardless of whether shortcomings have been identified. This 'active' or 'proactive' approach has some distinct benefits such as:

➤ Re-enforcing the organisation's safety culture
➤ Identifying where technological advances can be introduced to the safe system of work, to reduce the risk
➤ Assist the organisation to fulfil its obligation to consult with its workforce
➤ Prevent conditions worsening that could result in injury, damage or loss.

Part of the monitoring and review programme should also take into account emergency procedures, which will not necessarily be tested on a regular basis, such as a gearing mechanism breaking down on an industrial machine or a tower crane operator suffering an ill-health effect that prevents descent via the access ladder. Emergency procedures and arrangements are discussed later in this chapter.

Systems of work that are out of date, ineffective or present an obstacle to production or service delivery will be unlikely to be fully implemented. It is therefore essential that an organisation ensures that it has a robust system for monitoring, reviewing and revising its safe systems of work.

6.6 Emergency procedures

To ensure that his legal duties of care are discharged, an employer will need to consider events that represent a serious and imminent danger which will trigger emergency actions such as:

➤ Fire or explosion (see Chapter 10)
➤ Personal injury
➤ Acute ill health
➤ Serious process failure which may cause fire
➤ Spillage or flood of a corrosive agent which may make contact with skin
➤ Failure to contain biological or carcinogenic agents
➤ Process failure leading to a sudden release of chemicals

➤ A threatened significant exposure over a hazardous substance
➤ Environmental release (see Chapter 13).

When considering emergency procedures, regardless of the type of emergency, a management system should be established. The emergency planning arrangements should include establishing a policy, preparing for an emergency both in the provision of equipment and training, etc. and making arrangements for those who will have to deploy in the event of an emergency, such as those dealing with spillages, fighting fires or rendering first aid.

Specific information relating to fire emergency actions is covered in Chapter 10.

Initial actions that are likely to be taken in the event of an emergency will depend upon the type; however, generally the following steps will be taken:

1. Make the scene safe
2. Render first aid (where necessary)
3. Prevent conditions worsening (spill control, fire, etc.)
4. Consider evacuation (where appropriate)
5. Contact emergency services
6. Report to management team (internal)
7. Report to enforcers (external where appropriate)
8. Initiate investigation procedures
9. When safe to do so restart of business operations
10. Report findings of investigation.

6.6.1 First aid

In order to save life and minimise the result of any injury it is important that persons suffering acute illness or are injured at work receive attention straight away to stabilise

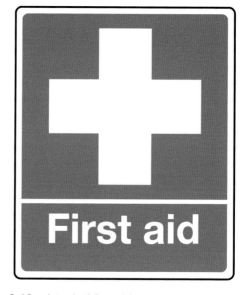

Figure 6.10 A typical first aid at work sign

their condition. Once these often life saving first aid actions are taken the patient can be transferred to the care of medical professionals to continue their recovery.

The Health and Safety (First-Aid) Regulations 1981 require the employer or responsible person to provide adequate and appropriate equipment, facilities and personnel to enable first aid to be given to employees if they are injured or become ill at work. Employers should make an assessment of their first aid needs in terms of the equipment and staff required and arrangements for liaising with emergency services. The assessment should include consideration of the following factors:

- Any specific risks, e.g.:
  - Hazardous substances
  - Dangerous tools
  - Dangerous machinery
  - Dangerous loads or animals
- The disposition of different risks, e.g. offices and process areas
- The number and types of previous injuries
- The numbers of persons on site
- Any employees with disabilities or special health problems
- Hours of work, e.g. 24/7 or normal office hours
- Arrangements for any lone workers.

6.6.2   Equipment

There is no standard list of items that should be contained within a first aid box. The contents of the first aid box will be dictated by the outcome of the assessment. However, where there is no special risk in the workplace, a minimum stock of first aid items is likely to include:

- A leaflet giving general guidance on first aid, e.g. HSE leaflet 'Basic Advice on First Aid at Work' (see 'Where can I get further information?')

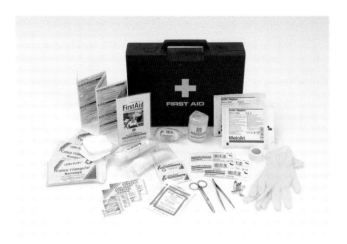

Figure 6.11   First aid box

- 20 individually wrapped sterile adhesive dressings (assorted sizes)
- Two sterile eye pads
- Four individually wrapped triangular bandages (preferably sterile)
- Six safety pins
- Six medium sized (approximately 12 cm × 12 cm) individually wrapped sterile unmedicated wound dressings
- Two large (approximately 18 cm × 18 cm) sterile individually wrapped unmedicated wound dressings
- One pair of disposable gloves.

There should be no medicines stored in any first aid kit.

6.6.3   Staff

The Health and Safety Executive suggest that there are two classes of competent person that should be in a workplace; appointed persons and first aiders.

The role of an appointed person is to:

- Take charge when someone is injured or falls ill, including calling an ambulance if required
- Look after the first aid equipment, e.g. restocking the first aid box.

Appointed persons should not attempt to give first aid for which they have not been trained, though short emergency first aid training courses are available.

The role of a first aider is to be trained in administering first aid at work; they should hold a current first aid at work certificate. The training provider has to be approved by the HSE. A first aider can undertake the duties of an appointed person.

Assessing the level of provision

HSE provide guidance on how many appointed persons and/or first aiders may be required in any specific circumstance. The details in Table 6.2 are suggestions only – they are not definitive nor are they a legal requirement. It is for employers to assess their own first aid needs in the light of their particular circumstances.

It should be borne in mind that both appointed persons and first aiders should be available at all times.

Liaising with emergency and rescue services

The assessment of the need for first aid arrangements will identify if any special procedures may be required to deal with accidents or illnesses at work. In most cases the local National Health Trust emergency service will suffice. However, in some cases, for example in remote areas or where there is a likelihood of an unusual rescue,

Table 6.2 HSE suggested provision of appointed persons and first aiders

Category of risk	Numbers employed at any location	Suggested number of first aid personnel
Lower risk e.g. shops and offices, libraries	Fewer than 50 50–100 More than 100	At least one appointed person At least one first aider One additional first aider for every 100 employed
Medium risk e.g. light engineering and assembly work, food processing, warehousing	Fewer than 20 20–100 More than 100	At least one appointed person At least one first aider for every 50 employed (or part thereof) One additional first aider for every 100 employed
Higher risk e.g. most construction, slaughter houses, chemical manufacture, extensive work with dangerous machinery or sharp instruments	Fewer than five 5–50 More than 50	At least one appointed person At least one first aider One additional first aider for every 50 employed

e.g. from a confined space, employers will need to consider what additional provision may be required.

In any event the arrangements for calling the emergency services should be well known by all particular appointed persons.

Any information relating to the emergency procedures must be made available to any accident and emergency services to ensure that they can prepare their own response plans.

The employer is responsible for putting together an emergency response plan to include such issues as:

➤ Safety equipment and PPE
➤ First aid facilities
➤ Emergency procedures for employees
➤ Procedures for clearing up and safe disposal
➤ Regular safety drills
➤ Any needs of disabled persons.

6.7 Case study

On many occasions supervision is a safety critical element contained in a safe system of work (SSOW). When it fails it may result in dire consequences such as personal and financial losses.

On Friday, 6 February 2004 the HSE issued the following press release:

A construction company was fined a total of £40 000 and ordered to pay £12 983 costs at the Old Bailey after pleading guilty to a breach of health and safety legislation. The case, brought by the Health and Safety

Executive (HSE) against Eugena Ltd, arose following a fatal accident during construction work at St Thomas' Hospital, Lambeth Palace Road, London SE1.

Construction worker Ian Mallon was laying blockwork at a height of about 2.5 metres when he fell from the unguarded edge of a scaffold work platform on 17 June 2001. As a result of the fall Mr Mallon suffered severe head injuries from which he died in hospital several days later. The incomplete scaffold platform used by Mr Mallon and his workmates had not been inspected by a competent person after its alteration.

Eugena Ltd pleaded guilty to a breach of duties under section 3(1) of the Health and Safety at Work Act etc. 1974, in that they failed, so far as was reasonably practicable, to ensure the health and safety of persons not in their employment.

Sentencing the company, the Judge said:

'Sadly it is all too often the case [that] when not adequately supervised corners are cut, a principal contractor must be alive to such risks.'

HSE Principal Inspector, Neil Stephens, said after the case:

'This case has demonstrated that principal contractors simply cannot assume that sub-contractors, if left unsupervised, will act safely. Principal contractors must ensure that they have adequate arrangements in place to supervise the work of their sub-contractors'.

The HSE inspector dealing with the case, Michelle Workman, added:

'If proper supervision had been provided on site before this accident, and the scaffold platform had been inspected prior to use, then this accident could easily have been prevented.'

6.8 Example exam questions for Chapter 6

1. **Identify** EIGHT sources of information that might usefully be consulted when developing a safe system of work. (8)
    (a) **List** the main types of safety signs.
    (b) Provide examples and **describe** their meaning. (6)

2. **Outline** reasons for maintaining good standards of health and safety within an organisation. (8)
3. **Outline** the main health and safety issues to be included in an induction training programme for new employees. (8)
4. (a) **Explain** the meaning of the term 'safe system of work'. (2)
    (b) **Outline** SIX sources of information that may need to be consulted when developing a safe system of work. (6)

7

Principles of fire and explosion

7.1 Introduction

Previous chapters have discussed the overall management issues that need to be addressed if a coherent safety management system is to be established and implemented in an organisation. As this book is primarily focused upon fire safety management and its principles, a key element that must be addressed is ensuring an underpinning knowledge of the principles of fire and explosion; this chapter will concentrate on these areas.

An understanding of the principles are critical to identifying how a fire will start and spread, the former of which are known as primary fire hazards, each of which has the ability to initiate or start a fire, or exacerbate a fire. These elements will be required when completing a fire risk assessment.

Explosion and fire are interlinked in that on many occasions fire occurs after an explosion, it is therefore essential that those involved with fire safety and risk management have a basic knowledge of the causes and properties of explosion, particularly in relation to processes involving gases and dusts.

This chapter discusses the following key elements:

➤ Combustion processes
➤ Ignition of solids, liquids and gases
➤ Fire growth and spread
➤ Explosion and explosive combustion.

7.2 The chemistry of fire

In order to appreciate the differences between fire and explosion it is necessary to look at the definitions:

Combustion – a chemical reaction or series of reactions in which heat and light are evolved.

Fire – chemical reaction brought about by the combining of fuel and oxygen and the application of sufficient heat to cause ignition.

Fuel – can be either a gas, liquid or solid (state). The amount of heat required to release a vapour which allows combustion to take place will depend upon the fuel's state. For example, a block of wood requires a higher level of heat to be applied than petrol.

This is because when heated, combustible materials give off flammable gases or vapours. If the temperature is high enough and a sufficient quantity of oxygen is present, ignition of these gases and vapours will occur, and a fire will result, thus as petrol is already producing vapours it is easy to ignite.

7.2.1 The fire triangle

The 'fire triangle' is a simple representation of the three factors necessary for a fire to start and once started to continue to burn.

All materials have the ability to burn if supplied with sufficient heat to cause the molecules to break down and give off vapour.

Once the vapour or gas is released it is that which ignites, causing more heat to be released, propagating further reactions – the fire process has begun.

As the material that is involved with the combustion or fire decomposes the material that is left has less ability to react, ultimately causing the fire to die down and go out.

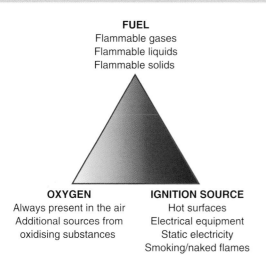

Figure 7.1 Triangle of fire

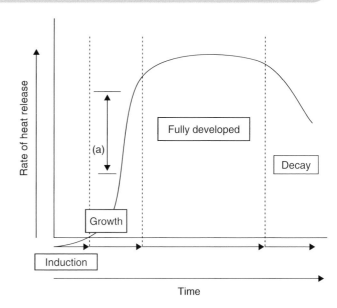

Figure 7.3 Stages of combustion or fire growth (a) = flashover zone

The decomposition of the material in this way is known as pyrolysis and the smoke that can be seen when a fire occurs is in fact unburnt products of pyrolysis included in the vapours given off.

Figure 7.2 illustrates the main elements of the fire process.

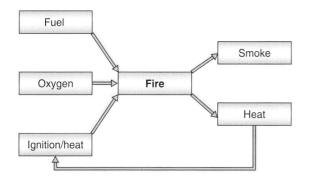

Figure 7.2 Diagrammatic representation of the fire process

7.2.2 Stages of combustion

There are four generally acknowledged stages contained within the process. These start at the **induction** stage where the component parts of the triangle come together and initiate the reactions.

The **growth** stage, where supplied with an uninterrupted level of oxygen or fuel the reactions become rapid and grow in intensity, creating large volumes of smoke (unburnt products of pyrolysis) the time taken in the growth of a fire may be from a few minutes to several hours dependent upon the prevailing conditions. The point at which the fire involves all the combustible materials within the room or area is known as the

'flashover point' (see Fig. 7.3). The time taken to arrive at this flashover point will vary and is likely to involve more than one area or phenomenon, including the size of the room, the surface linings, the availability of oxygen and a variety of complex chemical reactions.

The **fully developed** stage in a fire is whilst the reactions are not as rapid as in the growth stage, the fire continues to burn violently consuming the available oxygen supply and fuel sources. This 'fully developed stage' is characterised by massive flames and very high temperatures (in excess of 300°C). It is in fact at this time that the fire is controlled not by the amount of fuel that it has to burn but by the amount of oxygen it has on which to feed.

Finally the **decay** stage, where having consumed all the available fuel the fire dies down and is eventually extinguished, can be as a direct result of fire service intervention or can occur naturally when there is no further oxygen or fuel to support the combustion process.

During the growth stage there is a period of time prior to reaching the fully developed stage where there is a serious risk of **flashover**. This is due to the layering of hot gases beneath the ceiling and the oxygen concentration in the air being less than normal. It is at this time, as the concentration of gaseous fuel rises sharply in the growth stage, that allowing air to enter causes it to mix within the fuel layer, and with the already existent heat and flame a flashover occurs, a process which is not dissimilar to an explosion.

The fire process is 'exothermic', in other words it releases significant quantities of heat. The amount of heat produced is dependent upon the fuel involved and its location, e.g. whether it is from the building components itself (refer to Chapter 9 on building design and construction) or materials introduced to the building such as plastics, furnishings, etc. As with any chemical reaction, the control of the amounts and levels of the component parts (heat, fuel and oxygen) has a significant bearing on the rate of reaction and the heat output. Thus should a fire involve materials such as polystyrene, audio/video tape or other known high heat releasing materials then the speed at which the fire develops will increase.

The same will be the case if chemicals such as oxidising agents are involved. When heated, oxidising agents give off large amounts of oxygen which can rapidly increase both the growth and the spread of fire. The control of air (oxygen) in ventilation systems and ducts also has an impact upon fire growth and fire spread.

It is appropriate also to consider the effects of a **smouldering** fire which only occur in porous materials such as paper, cardboard, sawdust, fibreboard, etc. (carbonaceous).

Smouldering is the combustion of a solid in air which does not produce a flame. The smouldering process is very slow and can go undiscovered for a very long time; it can, however, produce a large amount of smoke. The smoke must accumulate and reach its lower flammability limit before ignition can occur. Given favourable conditions smouldering will undergo a transition to flaming, such as in the case of a cigarette igniting upholstered furniture. Due to the fact that the mechanism is not fully understood prediction as to when the smouldering to flaming transition occurs is difficult. Following a fire it is also possible that small conglomerations of material can be left smouldering (bull's eyes) which if supplied with sufficient oxygen can cause reignition.

Some chemical reactions also have the ability to extract heat from surrounding materials; these are called **endothermic** reactions, one such reaction can be seen when liquid carbon dioxide (CO_2) is used when tackling a fire using a portable fire extinguisher. The energy required to change the liquid CO_2 into a gas (known as the latent heat of vaporisation) is taken from the surrounding material resulting in the formation of ice on the body of the extinguisher.

7.2.3   Fire initiators

In order for a fire to start there has to be sufficient heat from an initiator or ignition source. Sources of ignition can be found in every workplace and home.

These sources of ignition could be open flames, hot surfaces, electrical sparks (internal or external), electrically generated arcs, friction (machinery), chemical reactions,

Figure 7.4   Heat, fuel and oxygen combine to initiate a combustion reaction

or even the compression of gases. This is not an exhaustive list, the causes and prevention of fires are discussed in the following chapter leaving this chapter to look at the principles of fire and explosion.

Previously the Office of the Deputy Prime Minister (ODPM) produced statistics in relation to the types, numbers, etc. of fire and fire deaths/injuries.

From the statistics it is possible to identify the sources of ignition and the number of occasions that these 'initiators' have been considered to start a fire:

➤ Smokers' materials
➤ Cigarette lighters
➤ Matches
➤ Cooking appliances
➤ Space heating appliances
➤ Central and water heating devices
➤ Blowlamps, welding and cutting equipment
➤ Electrical distribution
➤ Other electrical appliances
➤ Candles
➤ Other/unspecified.

In addition to those sources identified above other common sources of heat in the workplace include:

➤ Electrostatic discharges
➤ Ovens, kilns, furnaces, incinerators or open hearths

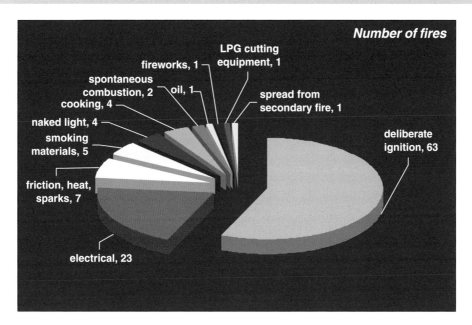

Figure 7.5 Ignition sources 2004–2005

> Boilers, internal combustion engines or oil burning equipment
> Lightning.

7.2.4 Fuel sources

Everything will burn if sufficient heat and available conditions allow. Clearly some items will burn more readily than others, for instance wood shavings or dust will burn more easily than a solid block of wood. These **fuel sources** are generally broken down into three main groups (although these are sometimes subdivided).

Solids

Often referred to as carbonaceous materials (carbon based) these include wood, cardboard, paper, hardboard, soft furnishings such as carpets and curtains and materials such as plastics, foam rubber and even metal. As the structure of solids is based upon tightly formed particles, those, such as metals (unless reactive, i.e. magnesium) are very difficult to break down and require substantial heat sources to be applied for a fire to be initiated.

Some solids are very 'reactive' and may be designated as 'flammable' solids under UN Transport requirements; these will be clearly identifiable by a label/sign.

Liquids

Far more susceptible to supplying fuel for a fire due to their ability to release vapour. Liquids including petrol, paraffin, white spirit, thinners, varnish and paints present

Figure 7.6 Typical flammable solid sign

a significant risk. Chemicals such as twin pack adhesives, acetone and toluene also release vapours and these liquids due to their low **flashpoint** present an even greater risk.

Aerosol containers contain a flammable liquid which is pressurised to a level that it becomes a gas. The containers present a significant risk particularly if they come into direct contact with a heat source.

Flammable liquids are classed as:

> Extremely flammable.

Liquids which have a flashpoint lower than 0°C:

> Highly flammable.

Liquids which have a flashpoint below 21°C but which are not extremely flammable:

> Flammable.

Liquids which have a flashpoint below 55°C but which are not highly flammable.

Figure 7.7 Typical flammable liquid sign

Gases

Flammable gases are common throughout workplaces and include natural gas and liquefied petroleum gas (LPG). The most common types of LPG are stored in pressurised cylinders and include butane and propane (i.e. cooking, heating, plumbers' torches, etc.); in addition dissolved acetylene and oxygen mixtures are utilised in welding operations.

As with the flammable vapours given off by flammable liquids not only do gases present a significant risk of fire they also present an **explosion** risk.

Gases do not always come in containers and can be produced by chemical reactions such as the degradation of waste materials within a refuse dump which produces methane.

Figure 7.8 Typical flammable gas sign

Dusts

Dusts can be produced from many everyday and workplace materials such as wood, coal, grain, sugar, synthetic organic chemicals and certain metals. A cloud of combustible dust in the air can explode violently if ignited. Explosions and explosive materials are discussed later in this chapter.

7.2.5   Oxygen

The final element of the fire triangle that should be discussed is oxygen. Oxygen is contained as an element within the air that we breathe (19.6%) and therefore is available in any area that humans can live.

The supply of oxygen, as previously discussed, can influence fire behaviour and thus recognising how the supply can be enhanced or reduced is vital when considering control factors.

Ventilation and air handling systems can provide an enhanced supply as can the use of oxygen cylinders (medical use and hot work/cutting work).

Chemical reactions can also cause the release of oxygen, particularly chemicals such as ammonium nitrate (fertilisers), sodium chlorate (pesticides), hydrogen peroxide (water treatment, hair care) and chromate (variety of industrial processes).

It should also be noted that oxidisers exhibit highly exothermic reactions when in contact with other substances, particularly flammable substances. Oxidising agents can also increase the combustibility of substances which do not normally burn readily in air, or can lower ignition temperatures to such a point that materials burn more readily and more violently.

Figure 7.9 Typical oxidising agent sign

7.2.6   The chemical process

Conditions

Quite specific conditions are needed for the fire process to start and continue. In order for an evaluation of whether the conditions present the potential for fire, it is essential to gather information as to the likelihood that the conditions will arise.

Flammable materials are provided with information sheets (Materials Safety Data Sheets – MSDS) which are produced by the manufacturer for onward transmission to the suppliers and ultimately the 'end user', as required by both the Chemical Hazards (Information Packaging for Supply) Regulations and HSWA, section 6.

The MSDS will include certain information which will help to analyse the fire risks.

Certain fire data is provided (based upon test conditions):

➤ **Flashpoint** is the lowest temperature at which a substance will produce sufficient vapour to flash across its surface momentarily when a flame is applied
➤ **Firepoint** is the lowest temperature at which the heat from the combustion of the burning vapour is

capable of producing more vapour that once ignited is able to sustain the combustion cycle

➤ **Spontaneous ignition or auto-ignition temperature** is the lowest temperature at which a substance will ignite spontaneously without any other ignition source. A spark is not necessary for ignition when a flammable vapour reaches its autoignition temperature.

Some materials have a spontaneous ignition temperature that is so low that they are capable of igniting under normal conditions at room temperature. One such material is linseed oil which as an organic material reacts with oxygen at room temperature; this is called oxidisation.

The lower the fire-/flashpoints or autoignition temperatures, the more likely will be the chemicals' involvement in supporting a fire at/near ambient or room temperatures and the likelihood that the combustion process will continue and produce a fire condition.

Also included on MSDS is the following information:

➤ **Vapour pressure** is the pressure of a vapour given off by (evaporated from) a liquid or solid, caused by atoms or molecules continuously escaping from its surface. Vapour pressure is measured in units of pressure such as pascals. The higher the vapour pressure of a substance at any given temperature, the more volatile the substance is and therefore the more likely it will be that vapour will be 'given off'. Figure 7.10 shows a small region of a liquid near its surface.

➤ **Vapour density** is the density of a gas, expressed as the mass of a given volume of the gas divided by the mass of an equal volume of a reference gas (such as hydrogen or air) at the same temperature and pressure.

Some of the more energetic particles escape.

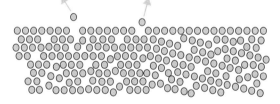

Figure 7.10 Vapour particles escaping from the surface of a substance

The vapour pressure and density affect the substances' rate of reaction (the amount and rate of atoms/molecules escaping) and where/what level the gas will be found (hydrogen sulphide heavier than air, methane lighter than air). This information will enable a variety of differing control measures to be considered.

Table 7.1 Vapour density

	Vapour density	
Hydrogen	1	
Methane	8	} Lighter than air
Ammonia	8.5	
Air	14.4	
Carbon dioxide	22	
Sulphur dioxide	32	} Heavier than air
Chlorine	35.5	

Also included on the MSDS will be information relating to the **flammable limits** or **flammable/ explosive range**; these issues will be addressed in the explosion element later in this chapter.

7.3 Classification of fire

Fires are classified in terms of their types. BSEN 2 identifies five classifications. The classifications assist in recognising not only the type of fire but also the type of extinguishing medium that will need to be utilised to extinguish or suppress the fire.

Class A – fires that involve solid materials, usually of an organic nature, such as wood, cardboard, paper, hardboard, soft furnishings such as carpets and curtains, in which combustion normally takes places with the formation of glowing embers.

Figure 7.11 Class 'A' fires

Class B – fires that involve liquids such as petrol, paraffin, white spirit, thinners, varnish and paints or liquefiable solids such as candles (wax) and fats.

Figure 7.12 Class 'B' fires

121

Class C – fires that involve gases such as LPG (i.e. butane, propane) or those involving natural gas.

Figure 7.13 Class 'C' fires

Class D – fires that involve metals such as sodium, lithium, manganese and aluminium when in the form of swarf or powder.

Figure 7.14 Class 'D' fires

Class F – fires that involve cooking mediums such as vegetable or animal oil and fats in cooking appliances. Such fires are particularly difficult to extinguish as they retain considerable heat allowing the chemical reaction to restart.

Figure 7.15 Class 'F' fires

It is worthy of note that there is no classification for **electrical fires**; this is due to the fact that electricity does not actually involve any fuels which can be extinguished. Fires that involve electrical circuits and appliances, cables, etc. can either be extinguished with a non-conductive extinguishing medium, or the supply can be isolated and the actual material (given its own 'class') can be extinguished.

Figure 7.16 Boxes containing matches being stored adjacent to a light fitting

7.4 Principles of fire spread

Once a fire has started and there is sufficient fuel and oxygen to sustain it, there are three recognised ways in which it can spread within the building: convection, conduction and radiation. There are also the effects of direct burning or heat transfer to take into account.

7.4.1 Convection

Due to its properties hot air rises; this can be seen graphically when smoke from a bonfire rises and disperses within the atmosphere or a fire is started within a grate and rises up through the chimney.

The convection process begins when combustible materials are subject to excessive levels of heat and they give off a vapour which in turn ignites. When these vapours are heated they expand and become less dense than air. As they rise they leave an area of low pressure which is replaced instantly by cooler unheated air. This fresh air is then mixed with the vapour and heated, assisting in the development of greater temperatures. The process is cyclic, continuing to support the fire process.

Convection is the most common cause of fire spread within buildings and structures. During a fire hot gases and vapours (smoke and heated air) will rise vertically through stairwells, lift shafts and service risers to the highest level available. They then form a layer at that height, from which they spread out horizontally until checked.

As the temperature of the smoke and now the toxic gas/vapour layer increases, heat is radiated back downwards and may ignite other combustible materials in the vicinity. Modern research also shows that the unburnt fuel (unburnt products of pyrolysis) in the smoke can reignite on reaching its spontaneous combustion point causing a substantial rise in temperature. This rise in temperature, if provided with a fresh source of oxygen (window glazing fails), has the ability to cause a flashover or explosion.

Clearly there is a substantial difference in a fire starting in a confined area such as a building, in comparison to one that starts in the open air and it is this issue that will need to be addressed when designing buildings that minimise smoke and fire spread enabling people to escape safely.

Convection and the effects of smoke

Although it is important to restrict the spread of fire within a building, it is equally important to consider the speed of spread and effect of the smoke created by the fire as it burns.

As a fire develops it will create large quantities of smoke which will, usually, spread ahead of the fire quickly filling a building. The effect of this is to present a toxic and/or asphyxiant hazard to people within the building. Smoke also reduces visibility and obscures escape routes; this linked to people's natural reluctance to walk into or through smoke can lead to, or increase, panic, which in turn leads to disorientation reducing the chances of safe escape.

Figure 7.17 Smoke spread – convection

7.4.2 Conduction

Conduction is the movement of heat through a material. The ability of conductors to transfer heat varies considerably according to the type of material, e.g. metal is a much better conductor than brick. It should be noted that conduction may occur in solids, liquids or gases; however, in relation to fires within buildings it is most prevalent in solids.

The thermal conductivity (the ability to conduct heat) varies between materials and is a key element in building design and construction, which will be considered in Chapter 9.

A fire in one room can spread to adjacent rooms by heat being conducted through the fabric of the building (walls/ceilings, etc.), especially via metal pipes or frames used in building construction. The heat can then ignite materials in direct contact with the surface, or radiate out from the surface. This can raise the temperature of materials in the adjacent room to their spontaneous combustion temperature, thus spreading the fire.

The relative conductivity of building materials is therefore an important factor in the fire resisting ability of a structure or building. This issue is considered within the Building Regulations Approved Document B.

Figure 7.18 Fire spread through a fire resisting wall by conduction along a steel pipe

7.4.3 Radiation

Radiation is the transfer of heat energy as electromagnetic waves, which heat solids and liquids (but not gases) encountered in its path.

Fire radiation paths do not require any contact between bodies and move independently of any material in the intervening space. If not absorbed by fire resistant material the electromagnetic radiation can radiate through glazing causing fires to spread and involve a number of compartments/rooms. As with the example of a heater or open fire, fires can spread to combustible items left to dry or from building to building when heat from a fire may be radiated to an adjacent building by passing through windows, and igniting combustible contents in the second building.

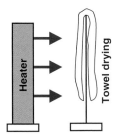

Figure 7.19 Fire spread by radiation from heater to combustible material

Figure 7.20 Fire spread by radiation from one building to another

7.4.4 Direct burning

When combustible materials come in direct contact there is a physical transfer of heat from the ignition source to the material which in turn releases vapours which ignite and propagate the fire. It is true to say that direct burning makes use of one or more of the previously discussed methods of heat transfer; however, it is appropriate to mention this method as a reasonable proportion of fires are started in this way and when

heated to an ignition temperature by coming into contact with a burning material causes fire to spread.

7.4.5 Fire growth

The rate at which a fire grows will depend upon numerous factors and it should be noted that a single factor on its own may not promote fire growth but interreacting factors may develop a fire more rapidly. A rapid growth rate will have an effect upon areas such as the stability of the building and the effectiveness of the emergency plan to ensure that people can leave the building safely.

The fire growth rate is generally recognised as the rate at which it is estimated that a fire will grow; this includes spread of flame over surfaces and behind linings, and within any part of the contents. Fire growth rates may be categorised in accordance with Table 7.2.

Factors that may affect the growth rate include the:

➤ Construction and layout of the building
➤ Ventilation into, throughout and out of the building
➤ Use of the building (including the types of activity being undertaken)
➤ Fire loading within the building.

Construction and layout of the building

How the building or structure is constructed in terms of its materials and the quality in which the materials have been used within the building has an effect upon any potential fire growth rate. Clearly, buildings constructed of wood have the potential to speed the fire growth rate; however, in itself, due to the nature of wood (strengthens when it is burnt) the fire growth rate would be affected more by voids between floors, ceilings and roofs in wooden buildings than the use of wood itself.

The size and layout of a building also has the potential to affect the growth rate. When a building has high ceilings, such as in the case of atria (found in shopping malls), fire growth is likely to be much slower than those buildings with low ceilings, or those that

Table 7.2 Fire growth rates

Category	Fire growth rate	Examples
1	Slow	Open plan office – with limited combustible materials, stored or used
2	Medium	Warehouse – which is likely to have stacked cardboard boxes, wooden pallets
3	Fast	Production unit/warehouse – baled thermoplastic chips for packaging, stacked plastic products, baled clothing awaiting delivery
4	Ultra-fast	Production unit/warehouse – flammable liquids, expanded cellular plastics and foam Manufacturing, processing, repairing, cleaning or otherwise treating any hazardous goods or materials

include basements. On these occasions, reflected radiated heat, due to the lack of height available, will develop the fire more rapidly and raise the growth rate.

The use of sandwich panel walling, containing polystyrene (used as thermal insulation for areas such as cold storage facilities), has historically resulted in rapid fire growth, due to the substantial heat release of the building material, i.e. the polystyrene. The failure of component parts, due to the rapid fire growth rate of the materials, also has the ability to destabilise the panels and cause premature collapse.

Whether included as part of the building construction or as part of the contents of a building, wall and surface lining materials also have a direct effect upon fire growth rates.

Approved Document B of the Building Regulations classifies performance of internal linings. These will be discussed in future chapters; however, the principles involve ensuring that the internal linings should adequately resist the spread of fire over their surfaces and, if ignited, a rate of heat release which should be reasonable in the circumstances.

In this paragraph 'internal linings' means the materials lining any partition, wall, ceiling, or other internal lining, such as plasterboard, plaster wall coverings including paper. Clearly hessian and materials such as polystyrene tiles have the ability to release large volumes of heat and thus would greatly affect fire growth, whereas plaster, plasterboard, etc. release heat very slowly and would not adversely affect the fire growth rate.

Ventilation

As previously discussed the supply of oxygen is critical to the development and spread of a fire. Fire growth rate is inextricably linked to the supply of oxygen and therefore affected by the ventilation of a building. Air conditioning and air circulation systems provide a fire with a ready source of oxygen via forced ventilation, which will aid fire growth. The presence of dust, vapours and fumes within the atmosphere also have the ability to affect the fire growth rate in so far as they provide a rich source of fuel. It should also be noted that appropriate levels of dust in the atmosphere may cause an explosion; this subject is discussed later in this chapter.

When mechanical smoke extraction systems are utilised within a building the effect of fire growth is notably less. The smoke laden air containing particles of fuel and substantial quantities of heat is removed from the atmosphere by the extraction system reducing the speed at which a fire grows, while assisting smoke ventilation and allowing persons clear, smoke-free, escape routes.

Use

The use of a building is directly related to the type of occupancy. Building use can be categorised in the following groups:

➤ Offices and retail premises
➤ Factories and warehouse storage premises
➤ Sleeping accommodation such as hotels, boarding houses, etc.
➤ Residential and nursing homes
➤ Teaching establishments
➤ Small and medium places of assembly – public houses, clubs, restaurants, etc.
➤ Large places of assembly (more than 200 persons) – shopping centres, conference centres, etc.
➤ Theatres and cinemas

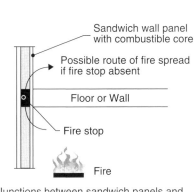

Junctions between sandwich panels and fire separating walls or floors should always be fire-stopped

(a)

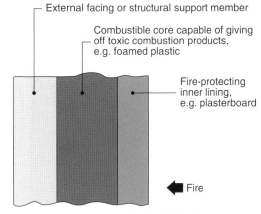

Use wall panels with an effectively fixed fire-protecting inner facing such as plasterboard which does not decompose, disintegrate or shatter in fire

(b)

Figure 7.21   Cross-section of sandwich panel

➤ Healthcare premises
➤ Transport networks.

Clearly, the use to which a building is put will greatly influence the contents within. In addition the quantity of combustible and flammable materials within a building will also reflect its use. The nature of the contents of a building is a key factor in determining the rate of fire growth and spread in the event of fire.

Fire loading within the building

The contents of the building and therefore the growth rate of a fire will vary according to the materials and activities being used or undertaken, for example where a significant amount of dangerous substances or preparations, e.g. substances or preparations that have a fast or ultra-fast fire growth rate or are classified as explosive, oxidising, extremely flammable, highly flammable under the Chemicals (Hazard Information and Packaging for Supply) Regulations 2002, are stored and/or used, the area is considered to be of high fire risk.

When large quantities of readily combustible products are stored or displayed under a large open plan mezzanine or gallery with a solid floor (as in some DIY outlets) there is always a risk of rapid fire growth resulting in flames spreading beyond the edge of the mezzanine or gallery floor, hence posing a threat to life safety, particularly when the occupants of the building are members of the public and are likely to be unfamiliar with the emergency plan (builders' merchants, DIY stores, etc).

Figure 7.22 Example of fire load from storage containers in a warehouse

Not only should the contents and activities be considered in relation to fire growth, but also the arrangements for the storage, of materials with potential for high heat release.

Heat release from materials is measured in megawatts per square metre (MW/m^2). Some materials generate much greater MW/m^2 than others. For example, wooden pallets that are stacked 1.5 metres high are likely to release 5.2 MW/m^2 whereas polystyrene jars packed in cartons stacked at the same height have a heat release of 14 MW/m^2.

Table 7.3 shows some examples of common commodities together with their known fire loading in terms of megawatts per square metre. When considering the fire risk associated with stored materials it is important to think about the heat that may be generated when materials are involved in a fire.

It is also widely recognised that the containers within which some materials are stored add significantly to the fire loading within buildings, most notably bulk warehouses. The Hazardous Installations Directorate report on chemical warehouse hazards states that flammable liquids in plastic, intermediate bulk containers (IBCs) present a very high risk because they inevitably fail in the case of a fire releasing their contents, adding, as a rule of thumb, around 3 MW/m^2 to the total rate of heat release each.

As can be seen from the examples above and the issues discussed in both construction and ventilation, fire growth is not based on any one element but a combination of all. This will need to be taken into account when undertaking fire risk assessments and any subsequent action plans.

Table 7.3 Common commodities together with their known fire loading

Commodity	Heat release MW/m^2
Wood pallets, stack 0.46 m high (5–12% moisture)	1.4
Wood pallets, stack 1.5 m high (5–12% moisture)	5.2
Wood pallets, stack 3.1 m high (5–12% moisture)	10.6
Polyethylene rubbish litter bins in cartons, stacked 4.6 m high	2
Polystyrene jars packed in cartons, compartmented, stacked 1.5 m high	14
Polystyrene tubs nested in cartons, stacked 4.3 m high	5.4
20–25 video cassettes	1

7.5 Explosion

As the terminology surrounding explosion is generally less familiar than that of fire, it is beneficial to define certain terms to clarify future explanations.

7.5.1 Terminology

Explosion – an abrupt oxidation, or decomposition reaction, that produces an increase in temperature, or pressure, or in both temperature and pressure simultaneously.

Explosive atmosphere – flammable substances in the form of gases, vapours, mists or dusts mixed with air under atmospheric conditions, which, after ignition has occurred, combustion spreads to the entire unburned mixture.

Deflagration – a combustion wave propagating from an explosion at subsonic velocity relative to the unburnt gas immediately ahead of the flame (flame front).

Detonation – a combustion wave propagating from an explosion at supersonic velocity relative to the unburnt gas immediately ahead of the flame (flame front).

BLEVE – boiling liquid expanding vapour explosion – an explosion due to the flashing of liquids when a vessel with a high vapour pressure substance fails.

CGE – confined gas explosion – explosion within tanks, process equipment, sewage systems, underground installations, closed rooms, etc.

UVCE – unconfined vapour cloud explosion – a vapour/gas explosion (deflagration or detonation) in an unconfined, unobstructed cloud.

7.5.2 The mechanism of explosion

Explosions can be caused by nuclear reactions, loss of containment in high pressure vessels, high explosives, runaway reactions, or a combination of dust, mist or gas in air or other oxidisers. This chapter concentrates on the latter examples.

Both dust and gas explosions are very similar in nature; when a volume of flammable mixture is ignited it results in a rapid pressure increase and fire moving through the atmosphere or cloud.

A dust explosion occurs when a combustible material is dispersed within the air forming a flammable cloud, this allows the flame to propagate through it.

A gas explosion follows a very similar principle when gas and oxygen are premixed within explosive limits which depend upon the supply of oxygen and the concentration of the fuel. If either of these is too high or too low then explosion will not occur.

7.5.3 Deflagration, detonation and explosive atmospheres

An extremely dangerous situation will occur if a large combustible mixture of fuel and air (gas or dust) is formed in a cloud and ignites. The time period from the release of the fuel mixture in air through to ignition can range from a few seconds to tens of minutes. The amount of fuel involved can also vary. In the case of gas the amount can vary from a few kilograms up to several tons.

The pressure generated by the explosion's combustion wave will depend on how fast the flame propagates through the explosive mixture and how easily the pressure can expand away from the cloud, which is governed by confinement.

When a cloud is ignited the flame can propagate in two different modes through it. These modes are:

➢ Deflagration
➢ Detonation.

The most common mode of flame propagation is deflagration. During deflagration the flame front travels at subsonic speeds through the unburned gas; typical flame speeds do not reach higher than 300 m/s. The pressure developed in front of the flame explosion may reach values of several bars.

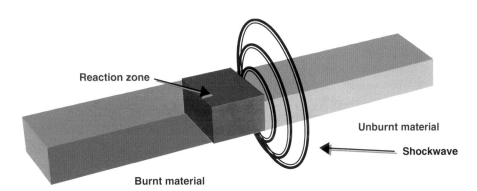

Figure 7.23 Detonation wave through a flammable mixture

In contrast the speed of travel of the flame front during a detonation is supersonic. In a fuel/air cloud a detonation wave will move at speeds of between 1500 and 2000 m/s and the peak pressure in front of the flame can reach 15 to 20 bar. This pressure front when enclosed can cause substantial damage including the collapse of structures.

In an explosion of a fuel/air cloud ignited by a spark, the flame will normally start out with a velocity of the order of 3–4 m/s. If the cloud is truly unconfined and unobstructed (i.e. no equipment or other structures are engulfed by the cloud) the flame is unlikely to accelerate to velocities of more than 20–25 m/s, and the overpressure will be negligible if the cloud is not confined.

In a building or other enclosed space when the mixture is burning the temperature will increase and the fuel/air mixture is likely to expand by a factor of up to 8 or 9, accelerating the flame front to several hundreds of metres per second.

If the ignition source is from a weak source, e.g. a hot surface or a spark, the explosion will initially start as a slow burning deflagration. Due to obstructing objects and confinement, the deflagration can accelerate and become fast burning. When a deflagration becomes sufficiently rapid, a sudden transition from deflagration to detonation may occur.

If this transition occurs, very high pressure loads, up to 50 bar, can be reached locally and severe damage can be expected within the compartment. If a detonation has been established in the compartment it may also propagate into any unconfined cloud outside, potentially creating a UVCE.

A deflagration propagating into a large truly unconfined and unobstructed cloud will slow down and the pressure generation will normally be negligible. A detonation, however, will propagate through the entire cloud at a high velocity and cause severe blast waves. The possibilities of transition to detonation will mainly depend on the:

➤ Type of fuel
➤ Size of cloud
➤ Enclosure conditions, such as obstructing objects and confinement.

Explosive limits and the flammable/explosion range

The explosive limits or flammable/explosion range refers to the range in which the concentration of a flammable

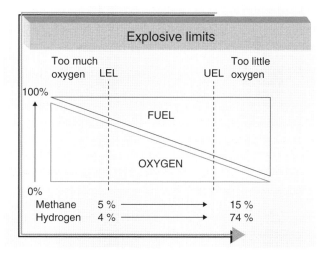

Figure 7.25 Illustration of explosive limits

Figure 7.24 Explosions at the Buncefield oil depot

substance in air may cause an explosion. The range is provided with two limits which are:

➤ **Lower explosion limit (LEL)** – the minimum concentration of vapour in air below which the propagation of flame will not occur in the presence of an ignition source. Also referred to as the lower flammable limit or the lower explosive limit.

➤ **Upper explosion limit (UEL)** – the maximum concentration of vapour in air above which the propagation of flame will not occur in the presence of an ignition source. Also referred to as the upper flammable limit or the upper explosive limit.

Managing an explosive atmosphere requires that the explosive range between the UEL and LEL, where explosion can take place, is avoided and that the concentration of gases, vapours and dusts is kept outside this range. The mechanisms relating to how this is achieved can be found later in the chapter.

These limits are significant in that they provide key information (via MSDS) that will assist the risk assessor in considering the risks and any risk control measures required.

7.5.4 Explosive conditions

In relation to dust any solid material that can burn in air will do so at a rate that increases in direct proportion to the increased surface area, therefore, dust, fibres, etc., are more likely to cause an explosive atmosphere if they are extremely fine.

The types of materials that often cause dust explosions include:

➤ Coal and peat
➤ Metals such as iron, zinc, aluminium
➤ Natural organic materials such as grain, linen, sugar, etc.
➤ Processed materials such as plastics, organic pigments (paint), pesticides, etc.

Each of the above materials has the ability to cause an explosion, though all combustible materials have the potential to react in a similar manner to a flammable gas when mixed with the correct proportion of air (or other

supporter of combustion). The severity will be based upon the heat release of the substance and the size of particles.

Determining the amount and likelihood of creating an explosive atmosphere will depend upon the following:

➤ The presence of a flammable substance
➤ The degree of dispersion of the flammable substance (this will vary dependent upon its state – dusts, gases, vapours and mists)
➤ The concentration of the flammable substance in air within the explosive range

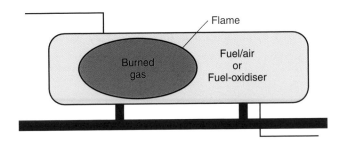

Figure 7.27 Confined explosion within a tank

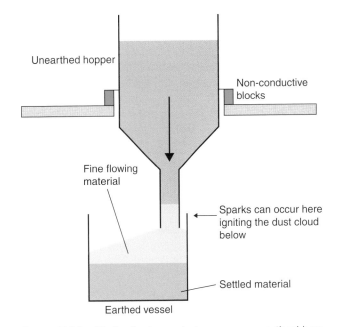

Figure 7.28 Static discharge between an unearthed hopper and an earthed vessel having the potential to ignite dust

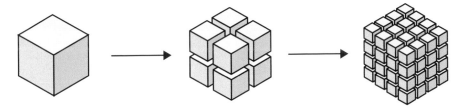

Figure 7.26 The smaller a substance is divided the greater the surface area

➤ The amount of explosive atmosphere sufficient to cause injury or damage by ignition.

The consequences of both gas and dust explosions will depend upon the environment in which the mixture (dust or gas cloud) is contained or, in the case of the gas cloud, what it engulfs. A gas explosion will be classified in relation to the environment in which the explosion takes place, for example **confined gas explosions** which occur within vessels, pipes, tunnels or channels, or unconfined gas explosions commonly referred to as **unconfined vapour cloud explosions**, which occur in process plants and other unconfined areas.

Once the determination of the explosive atmosphere has been established it is then necessary to consider whether there is a presence of an ignition source which will precipitate an explosion; such ignition sources could be:

➤ Hot surfaces
➤ Naked flames and other hot gases
➤ Mechanically generated sparks
➤ Electrical equipment
➤ Static electricity
➤ Lightning
➤ Radio frequency (RF) – electromagnetic waves
➤ Ionising radiation.

Once ignited an explosion may have secondary repercussions, particularly in the case of dust explosions. This

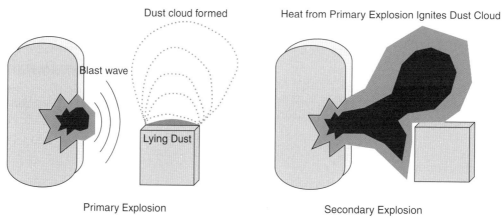

Figure 7.29 Primary and secondary explosions

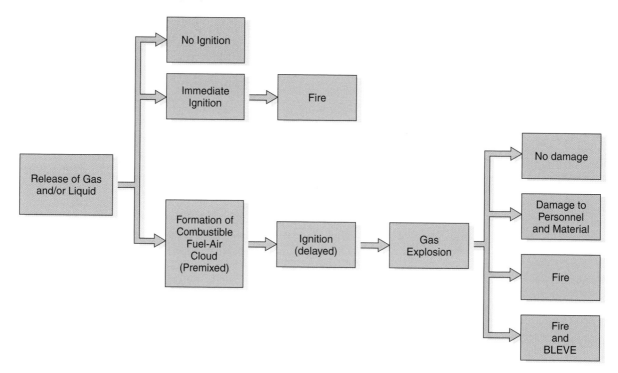

Figure 7.30 Diagram depicting typical consequences of the release of combustible gas into the atmosphere

can be broken down into primary and secondary explosions. The concentrations required for a dust explosion are generally not seen outside process vessels and thus the most significant dust explosions start from within a piece of equipment, e.g. mixers, hoppers and silos.

The first explosion is known as primary causing a rupture of the vessel releasing the flammable gas/air mixture into the atmosphere. With dust generally suspended in the air around the process equipment or dust lying undisturbed within the building a secondary explosion occurs.

A similar chain of events can be seen in relation to gas explosions, particularly when a gas explosion involves pressurised containers. The consequences of gas explosions range from no damage to total destruction and can lead to fires and BLEVEs as indicated in Figure 7.30.

7.5.5 Principles of explosion management

The principles of explosion management can be broken down into two discrete areas: those of control and mitigation, each will be dealt with separately. The Dangerous Substances and Explosive Atmospheres Regulations (DSEAR) require the application of a hierarchy of control measures to manage the risk of accidental explosion.

DSEAR regulation 6(4)

(a) the reduction of the quantity of dangerous substances to a minimum;

(b) the avoidance or minimising of the release of a dangerous substance;

(c) the control of a release of a dangerous substance at source;

(d) the prevention of the formation of an explosive atmosphere, including the application of appropriate ventilation;

(e) ensuring that any release of a dangerous substance that may give rise to risk is suitably collected, safely contained, removed to a safe place, or otherwise rendered safe, as appropriate;

(f) the avoidance of –
  (i) ignition sources including electrostatic discharges; and
  (ii) adverse conditions which could cause dangerous substances to give rise to harmful physical effects; and

(g) the segregation of incompatible dangerous substances.

Control

Critical to the management of explosive atmospheres is the avoidance or reduction of potentially explosive materials within an atmosphere.

The substitution of flammable substances by inert materials or limiting the concentrations of the flammable substances to avoid their explosive range must be considered at the top of any explosion management hierarchy.

Such controls may be the replacement of a fine dusty material by a less dusty granular material or reducing the flammable gas to the absolute minimum. Limiting the concentration to avoid the explosive range with mechanical systems linked to ventilation which may be actuated via gas or flow detectors (including alarms) should be considered. In the case of combustible liquids the objective should be to reduce the concentration of any mist formed, below the lower explosion limit, which in turn will ensure it is sufficiently below its flashpoint to prevent explosion.

An alternative mechanism may be the use of adding inert gases, e.g. nitrogen and carbon dioxide, utilising water vapour or inerting using a powdery substance such as calcium carbonate. With appropriate dispersal these materials can prevent the formation of an explosive atmosphere, which we term inerting.

Design and construction – equipment, protective systems and system components

When considering the types of equipment, protective systems and components that will contain flammable substances, reasonably practicable steps should be made to keep the substances enclosed at all times and the materials of construction should be non-combustible.

Where necessary leak detection systems should be fitted and particular attention should be given to the following areas:

➤ Joints
➤ Piping
➤ Areas that may be subject to impact
➤ Areas that may be subject to hazardous interactions with other substances.

The detection systems should provide advanced warning of any leakage from the equipment, systems or components, so that appropriate steps can be taken to prevent the consequential build-up of any flammable atmosphere.

Dilution by ventilation

While quite effective with gas and vapour in relation to dusts, ventilation is of limited effectiveness and provides sufficient protection only when the dust is extracted from the place of origin and deposits of combustible dust can be prevented.

In order to prevent the formation of an explosive atmosphere from the dispersion of dust in air or by equipment, it is appropriate to design conveying and removal systems to an approved standard. Such equipment will remove the levels of dust within the enclosure which will help to avoid primary explosions; however, secondary explosions could still easily occur if the lying dust is sufficiently agitated leading to entrainment in the conveying and removal systems.

The avoidance of ignition sources that may precipitate an explosion for either gas or dust in the atmosphere provides a relatively good control measure; however, in relation to dusts, accumulations and moisture content must be managed effectively as together they have the potential for self-heating and therefore self-ignition, thus the management of temperature and moisture in relation to dust must be considered.

Mechanical inputs can also produce either glowing sparks or hot spots and while the sparks are not sufficiently energetic to provide ignition repeated contact may run the risk of igniting a dust cloud.

The removal of any foreign objects from process streams and the use of non-sparking or spark-proof equipment (intrinsically safe) must be considered in either gaseous or dusty atmospheres.

Ensuring that any electrical equipment is subject to regular maintenance (planned preventive maintenance) must also be seen as a key area to prevent ignition sources initiating an explosion.

Electrostatic sparks from static electricity must also be minimised and the following should be considered:

➤ The use of conducting materials for equipment, plant, etc. to avoid charge build-up
➤ The earthing of any equipment that may become charged
➤ The earthing of workers
➤ Earth non-conducting materials via an earth rod through the storage vessel.

With regard to electrostatic discharges if there is any doubt earthing should take place.

Mitigation

In relation to dust the best way to contain a primary explosion is to ensure that the process equipment is strong enough to withstand it. Dust explosion pressures are usually within range of 5–12 bars. Designing the plant as though it were a pressure vessel is likely to cause it to be very expensive and beyond what is reasonably practicable. It is therefore quite often that designers will resort to explosion venting.

Explosion venting is one of the most effective ways to relieve pressure; however, it is often difficult to size

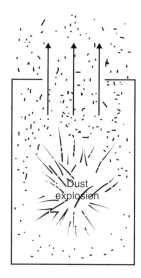

Figure 7.31 A vented dust explosion

the vent correctly to ensure that sufficient pressure relief is available as it must allow sufficient outflow of the burnt fuel and air to relieve the pressure being generated by the heat of the explosion.

There are a wide variety of differing designs of venting dependent upon processes undertaken. These can be simple panels that are ejected, vent covers (attached to process vessels with clips and rubber seals), and hinged doors that can withstand explosions or where necessary redirect the explosion.

The vent area will depend upon the volume of the enclosure, the enclosure's strength, the strength of the vent cover and burning rate of the dust cloud.

There are a number of hazards caused by venting which will need to be taken into consideration, these are:

➤ Emission of blast waves from the vent opening
➤ Ejection of flames from vent opening
➤ Fireballs can be ejected
➤ Emission of solid objects (parts of the vessel, vent covers, etc.)
➤ Reaction forces on the equipment, induced by the venting process
➤ Internal venting may also lead to secondary explosions.

Consideration must be given to the location of any explosion relief panel or venting, this must be considered at the design stage.

If venting an explosion cannot be achieved, **explosion suppression** may be considered. Any suppression unit must be permanently pressurised, fitted with a large diameter discharge orifice and any valve required to operate the discharge mechanism should be of high speed which is quite often achieved via a small detonation charge.

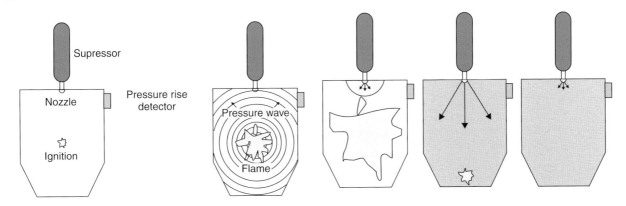

Figure 7.32 Explosion suppression system in operation

The trigger of such a suppression system is likely to be via either a pressure sensor or flame sensor, each will have systems to prevent false alarms and actuation.

In addition to a suppression system an alternative may be control by the addition of a liquid; however, there are environmental implications where water and dust residues have the potential to contaminate the environment.

Housekeeping in relation to the removal of dust accumulations must be considered although as previously discussed this should be undertaken with careful consideration to eliminate the possibility of secondary dust explosions. Installing good dust extraction systems and making arrangements for spilt dust to be removed immediately using vacuum systems or explosion-proof vacuum cleaners will assist in the management of dust control; however, those operated by individuals rely heavily upon human factors and the level of competence, training and instruction.

Ultimately, as is the case with all hierarchies of control the provision of suitable personal protective equipment may be considered.

Figure 7.33 Warning sign for places where explosive atmospheres may exist

Plant layout and classification of zones

Any building or plant where there is the potential for gaseous or dust explosions must be designed upon similar lines to those in which explosives are manufactured,

Table 7.4 Showing the zones for gases and vapours

Zones for gases/vapours	
Zone 0	A place in which an explosive atmosphere consisting of a mixture with air of flammable substances in the form of gas, vapour or mist is present continuously or for long periods or frequently. **Note:** In general these conditions, when they occur, arise inside containers, pipes and vessels, etc.
Zone 1	A place in which an explosive atmosphere consisting of a mixture with air of flammable substances in the form of gas, vapour or mist is likely to occur in normal operation occasionally. **Note:** This zone can include, among others, the immediate vicinity: ➤ Of zone 0 ➤ Of feed openings ➤ Around filling and emptying openings ➤ Around fragile equipment, protective systems, and components made of glass, ceramics and the like ➤ Around inadequately sealed glands, for example on pumps and valves with stuffing-boxes.
Zone 2	A place in which an explosive atmosphere consisting of a mixture with air of flammable substances in the form of gas, vapour or mist is not likely to occur in normal operation but, if it does occur, will persist for a short period only. **Note:** This zone can include, among others, places surrounding zones 0 or 1.

Table 7.5 Showing the zones for dusts

Zones for dusts	
Zone 20	A place in which an explosive atmosphere in the form of a cloud of combustible dust in air is present continuously, or for long periods or frequently. **Note:** In general these conditions, when they occur, arise inside containers, pipes and vessels, etc.
Zone 21	A place in which an explosive atmosphere in the form of a cloud of combustible dust in air is likely to occur in normal operation occasionally. **Note:** This zone can include, among others, places in the immediate vicinity of, e.g., powder filling and emptying points and places where dust layers occur and are likely in normal operation to give rise to an explosive concentration of combustible dust in mixture with air.
Zone 22	A place in which an explosive atmosphere in the form of a cloud of combustible dust in air is not likely to occur in normal operation but, if it does occur, will persist for a short period only. **Note:** This zone can include, among others, places in the vicinity of equipment, protective systems, and components containing dust, from which dust can escape from leaks and form dust deposits (e.g. milling rooms, in which dust escapes from the mills and then settles).

stored, etc. These should be located away from other buildings and actual parts of the plant should be as remote from one another as is possible. Ideally buildings should be of single storey in nature but kept as low as possible and the explosion prone part of any process should be as high as possible, ideally on the roof to minimise the possibility of building collapse.

Where any hazardous part of the plant is located within a building the area should be reinforced and protected from the rest of the area by a blast wall. As discussed previously the area should be vented to avoid damage (structural) from any overpressure.

Escape routes and other emergency response planning must take into account the explosive nature, as should any electrical equipment.

Under the Dangerous Substances and Explosive Atmospheres Regulations 2002 there is a requirement to identify hazardous contents (containers and plant) to ensure that the selection of the correct equipment and systems can take into account the level of and likelihood of there being an explosive atmosphere.

7.6 Case study

Flixborough disaster 1974

One of the most serious accidents in the history of the chemical industry was the explosion at about 16:53 hours on Saturday 1 June 1974 at the Nypro (UK) site at Flixborough, which was severely damaged. Twenty-eight workers were killed and another 36 others were injured as a direct result of the explosion and subsequent fire. It is recognised that the number of casualties would have been more if the incident had occurred on a weekday, as the main office block was not occupied.

Outside the plant, 53 persons were reported injured with 1821 houses and 167 shops suffering damage ranging from major (required rebuilding) to broken glazing from the pressure waves.

The overall cost of the damage at the plant itself together with the damage outside was estimated at over £75 million.

The cause of the Flixborough explosion was a release of about 50 tons of cyclohexane, due to failure of a temporary pipe. The flammable cloud was ignited about 1 minute or so after the release. A very violent explosion occurred. The blast was equivalent to an explosion of about 16 tons of TNT.

Five days prior to the explosion, on 27 March 1974, it was discovered that a vertical crack in reactor No. 5 was leaking cyclohexane. The plant was subsequently shut down for an investigation. The investigation that followed identified a serious problem with the reactor and the decision was taken to remove it and install a bypass assembly to connect reactors No. 4 and No. 6 so that the plant could continue production.

During the late afternoon on 1 June 1974 the temporary 20 inch bypass system ruptured, which is extremely likely to have been caused by a fire on a nearby 8 inch pipe. This resulted in the escape of a large quantity of cyclohexane, which formed a flammable mixture and subsequently found a source of ignition.

At approximately 16:53 hours there was a massive unconfined vapour cloud explosion which caused extensive damage and started numerous fires on the site.

Eighteen fatalities occurred in the control room as a result of the windows shattering and the collapse of the roof. No one escaped from the control room. The fires burned for several days and after ten days those that still raged were hampering the rescue work.

The characteristic of the gas explosion at Flixborough is that the dense fuel (cyclohexane) was able to form a huge flammable gas cloud and that the confinement and obstructions within the plant caused high explosion pressures.

The HSE's subsequent report identified the following failings in technical measures:

➤ A plant modification occurred without a full HAZOP assessment of the potential consequences. Only limited calculations were undertaken on the integrity of the bypass line. No calculations were undertaken for the dog-legged shaped line or for the bellows. No drawing of the proposed modification was produced

➤ With regard to the use of flexible pipes no pressure testing was carried out on the installed pipework modification

➤ In relation to maintenance procedures there appeared to be ad hoc recommissioning arrangements

➤ Those concerned with the design, construction and layout of the plant did not consider the potential for a major disaster happening instantaneously

➤ The plant layout was poor in relation to the positioning of occupied buildings

➤ Control room design did not take into account structural design to withstand major hazards events

➤ The incident happened during start-up when critical decisions were made under operational stress. In particular the shortage of nitrogen for inerting would tend to inhibit the venting of off-gas as a method of pressure control/reduction

➤ Operating procedures were lacking particularly with regard to the number of critical decisions to be made

➤ In relation to inerting the atmosphere reliability/back-up/proof testing appeared inadequate.

7.7 Example NEBOSH questions for Chapter 7

1. (a) **Explain** the meaning of the term 'flash point'. (2)
   (b) **Outline** the precautions to take to reduce the risk of fire when using flammable solvents in the workplace. (6)

2. A major hazard on a refurbishment project is fire.
   (a) **Identify** THREE activities that represent an increased fire risk in such a situation. (3)
   (b) **Outline** the precautions that may be taken to prevent a fire from occurring. (5)

3. **Outline** the measures that should be taken to minimise the risk of fire from electrical equipment. (8)
   (a) **Identify** two sources of oxygen which can be found in the workplace. (2)
   (b) State the ways in which the potential sources of oxygen supply to a fire can be reduced. (4)
   (c) **Define** the terms 'explosive atmosphere'. (2)

4. **Define** the following terms:
   (a) Deflagration (2)
   (b) Detonation (2)
   (c) Upper flammable limit (UFL) (2)
   (d) Lower flammable limit (LFL). (2)

5. **Explain** with the use of a diagram the method of operation of an automatic explosion suppression system. (8)

6. **Describe** the three zones that may be identified within a workplace that has flammable gases and vapours that may produce an explosive atmosphere. (8)

7. **Identify four** methods of heat transfer and explain how each can cause the spread of fire. (8)

8 Causes and prevention of fire

Identifying sources of fuel, ignition and oxygen will undoubtedly assist in preparing a fire safety management strategy based upon a risk assessment approach. However, these areas are only hazards (something that has the potential to cause harm) and it will take human intervention in one form or another to change the hazards into risk, be it accidentally or deliberately. Determining the different causes of fire in the workplace is the next stage in the management process, as having identified the primary sources of harm and how they are caused will enable a robust fire safety management plan to be produced to prevent a fire starting.

This chapter will also include specific reference to causes of fires within construction and maintenance work with the inherent hazards and risks such operations bring, together with the preventive steps to manage them.

This chapter discusses the following key elements:

➤ Common causes of major fires
➤ Fire risks associated with flammable, combustible and explosive substances
➤ Fire risks associated with common workplace processes and activities, including those associated with construction and maintenance operations
➤ Arson
➤ Measures to prevent fires.

8.1 Accidental fires

The term accidental fire refers to all fires other than those which have been deliberately or maliciously started. There are a wide range of causes of fires within the workplace. These will to a certain extent reflect the use to which the workplace is put. It is also useful to consider causes of fires in vehicles as in many organisations a workplace may be a vehicle, such as in the case of a long distance lorry driver working for a haulage company.

Using the current statistics available it can be seen that that the common causes of major accidental fires in the workplace fall under the broad headings of:

➤ Electrical appliances and installations
➤ Cookers, associated cooking equipment and installations
➤ Naked lights and flames
➤ Heaters and heating systems
➤ Chemical and LPG (hazardous materials)
➤ Smokers and smokers' materials
➤ Waste and waste management systems
➤ Other significant causes.

8.1.1 Electrical appliances and installations

Outside deliberate fire setting, fires that are caused by electrical appliances and installations are the most common cause of fires in both industry and the home. There are a variety of different ways that electricity flowing through equipment and installations can cause a fire, these include:

Overloaded wiring – where the electric current flowing in the wires exceeds the rating of the cables. The wiring

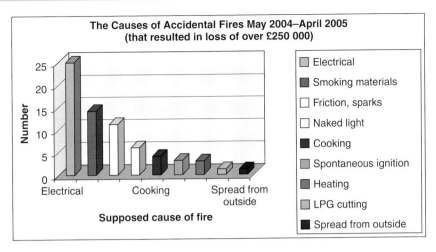

The Causes of Accidental Fires May 2004–April 2005
(that resulted in loss of over £250 000)

Legend:
- Electrical
- Smoking materials
- Friction, sparks
- Naked light
- Cooking
- Spontaneous ignition
- Heating
- LPG cutting
- Spread from outside

Figure 8.1 Causes of accidental fires in industry 2004/2005

heats up and melts the insulation and can set fire to flammable material nearby. Commonly found in flexible cables or cords, or consumer panel/power board leads. Over recent years there have also been a number of fires involving electrical extension leads, where the cable has remained fully wound on the cable drum. The flow of electricity and the tightness of the fully wound cable prevent heat dissipation causing overheating and a fire to start.

Loose wiring connections – the current flowing through the wiring encounters resistance at the connection and generates heat. This can start a fire in the wall at the back of a power point or socket, in a wiring junction box in the ceiling, above light fittings or inside a switch/fuse board or consumer panel.

Electrical 'arcing' (or sparking) – this generally occurs where wiring insulation has been damaged by an external occurrence perhaps due to poor location (e.g. under floor coverings). This deterioration to the insulation allows the copper conductors inside the cable to touch one another briefly, or to just make contact with the metal case of an appliance.

A small current will then flow from the 'live' conductor to the neutral or earth wire, or to the earthed metal case. The current will initially be too small to blow a fuse or to trip a circuit breaker, but because the contact area is also very small (a few strands of wire) the heat produced at this point can reach sufficiently high temperatures to melt or vaporise metals such as copper, brass or sheet metal. The localised heating or energy release will ignite combustible materials in close proximity and start a fire.

The above-mentioned arcing faults are 'uncontrolled' whereas arc welding makes use of this energy release and is an example of 'controlled' arcing, which will be discussed later in this chapter in fire hazards during construction and maintenance work.

Figure 8.2 Electrical equipment involved in a school fire

Poorly maintained equipment, unauthorised use or maintenance of electrical systems, components and equipment, lack of formal and informal checks and inspections can all be directly linked to causes of fires within the workplace.

Fires have been known to have been caused by vermin such as rats eating through electrical cables within basement areas due to a lack of vermin control procedures.

8.1.2 Cookers, associated cooking equipment and installations

Cookers, cooking equipment or other equipment or installations used for the heating of food provide a range of sources of ignition and fuel have the potential to initiate and exacerbate a fire.

Electrical fires associated with cooking fall under the category already covered in the previous paragraph and

Figure 8.3 A typical thermostatically controlled food servery

it is not intended to recover the aspects here; however, failing thermocouples and other heat controlling devices can also cause fires. These generally occur due to a fault and a lack of preventive maintenance and testing, for example using portable electrical appliance testing known as PAT testing. This PAT testing only checks the safety from 'electrocution risk' and it is possible that a damaged heat controlling device will not be identified by such testing.

The failure of the heat controlling device is then likely to allow the overheating of the food or substance being cooked which may combust and catch fire. In instances where oil is being used to cook food, chips, for example, this raising of the oil's temperature releases sufficient vapours to cause the oil to reach its flash and then fire point.

Gas cooking systems incorporate both an ignition source, by way of a pilot light, and a substantial fuel source, by way of the gas supply used in the cooking process. A number of fires are caused each year from faulty systems where the pilot light fails and gas is released to mix with the air until an explosive concentration is reached and any alternative ignition sources such as a light switch are capable of causing an explosion or fire.

Regardless of the type of cooking device used, if left unattended an overheated cooking receptacle and/ or food that is allowed to burn dry will produce sufficient heat for a fire to start.

8.1.3 Naked lights and flames

By their nature lighting units emit heat; the amount of heat will generally depend upon two factors, the wattage or energy consumption and the type of light, such as halogen lamps, incandescent units or fluorescent strip lighting.

As in the case of all fires, the lights themselves will need to be positioned close to or touching a combustible material, or in the case of flammable mixtures in air, provide sufficient heat or an ignition source to ignite the vapours.

Halogen lighting (high heat output) is becoming more popular within industry and it used to light both internal and external signage. If the signage itself is combustible and the halogen unit has been installed incorrectly or poorly maintained the lighting unit may come into contact with the signage starting a fire.

Halogen desk lamps are also in evidence in offices and workshops to provide additional task lighting. These units if poorly positioned and left unattended have the ability to start a fire.

As in the case of all electrical systems lighting systems have the ability if poorly maintained or incorrectly installed to provide an ignition source for any available combustible material.

The inappropriate use of lighting units such as in the case of inspection lamps used in vehicle workshops is also known to be a cause of fires in the mechanical and engineering sectors of industry.

Naked flames are a constant source of ignition within industry. The flame generating devices such as gas torches, welding units and cutting equipment (oxy-acetylene) account for a number of fires throughout industry each year. Fires starting from the use of these pieces of equipment are generally caused by human error when using them in close proximity to other combustible or flammable materials, poorly maintained equipment or in appropriate use. These issues will be discussed later in section 8.2.

8.1.4 Heaters and heating systems

Fires caused by heaters and heating systems can be divided into two key areas; those that are caused by faulty systems including poor design, inappropriate installation or lack of planned preventive maintenance; and those that are caused by inappropriate use or misuse.

Poor design and installation can allow the heat produced from the system to come into direct contact with combustible materials, perhaps such as wall linings through which the system's ducting is routed. As previously discussed with regard to cooking equipment, a lack of maintenance may allow heat controlling devices to malfunction. In addition if moving or rotating parts are included in a heating system any bearings may run dry or seals fail due to a lack of maintenance.

Inappropriate positioning of combustible or flammable items in close proximity to heaters and parts of heating systems may prevent air circulation, causing a heat build-up, or if these materials are in direct contact or within range of any radiated heat a fire may start.

8.1.5 Chemical and LPG (hazardous materials)

The Chemicals (Hazards Information and Packaging for Supply) Regulations classify substances representing certain hazards. If a substance has been classified under CHIP as being dangerous for supply, anyone supplying it must conform to the requirements concerning the provision of hazard information to those who may subsequently come in contact with it. Until the CHIP Regulations were brought in the only legal requirement covering the provision of information on dangerous substances from the supplier was contained in section 6 of the HSWA, where suppliers were required to pass 'relevant information' to customers.

The exact nature of this information was never elaborated upon, so the CHIP Regulations were introduced to fill this gap, leaving the suppliers and customers in no doubt as to the information required to be provided to users.

The CHIP Regulations lay down requirements for the packaging and labelling of these dangerous chemicals. The packaging requirements are straightforward, requiring only that the container for these types of substances shall be suitable for their contents, and shall be sealed when supplied.

The labelling requirements are that a package containing a dangerous substance or preparation bear a label of an approved design, and be of a specific size relative to the size of the package, containing relevant information.

The classifications of dangerous substances under CHIP 2002 are illustrated in Table 8.1.

Fires that involve chemicals and LPG are generally caused by inadequate arrangements for their safe use, transportation, storage and disposal. The lack of appropriate levels of training (and where appropriate levels of supervision) when using chemicals may lead to reactive chemicals being mixed causing a fire or explosive cocktail, or allowing the release of flammable vapours which can find an ignition source to start a fire. Mishandling or poorly connecting LPG gas cylinders can have a similar effect.

LPG gas cylinders can represent a serious risk if they are involved in fire and although they are less often seen as the direct cause of a fire they can produce devastating effects when involved. Poorly maintained equipment and installations are also seen as a significant risk when using LPG and can cause a release that can in turn result in a fire or explosion.

Poor storage and transportation arrangements are also responsible for causing fires involving both chemicals and LPG. Storing chemicals such as cleaning agents which may include acids and oxidisers in the same cupboard may precipitate a reaction if leakages and poor storage arrangements allow the chemicals to come into contact with one another.

The same can be said of disposal arrangements when substandard procedures allow reactive chemicals, flammable mixtures and LPG cylinders to come into contact with sources of ignition or with other reactive chemicals.

Fires can also start from poorly planned dismantling or disposal of equipment containing residues of flammable liquids or when dealing with spillages as part of environmental spill protection procedures.

Table 8.1 Classifications of dangerous substances under CHIP 2000

Explosive	Solid, liquid, pasty or gelatinous substances and preparations which may react exothermically without atmospheric oxygen thereby quickly evolving gases, and which under defined test conditions detonate, quickly deflagrate or upon heating explode when partially confined.
Oxidising	Substances and preparations which give rise to a highly exothermic reaction in contact with other substances, particularly flammable substances.
Extremely flammable	Liquid substances and preparations having an extremely low flash point and a low boiling point and gaseous substances and preparations which are flammable in contact with air at ambient temperature and pressure.
Highly flammable	The following substances and preparations, namely – (a) Substances and preparations which may become hot and finally catch fire in contact with air at ambient temperature without any application of energy, (b) Solid substances and preparations which may readily catch fire after brief contact with a source of ignition and which continue to burn or to be consumed after removal of the source of ignition, (c) Liquid substances and preparations having a very low flash point, or (d) Substances and preparations which, in contact with water or damp air, evolve extremely flammable gases in dangerous quantities.
Flammable	Liquid substances and preparations having a low flash point.

Figure 8.4   Van fire involving cylinders

8.1.6   Smokers and smokers' materials

The majority of current workplaces are non-smoking and with the advent of legislation that bans smoking in public the number of fires in workplaces caused by smokers and their discarded smoking materials is set to be reduced still further if not eradicated.

As is very often the case where prohibition of practices such as smoking is established the consequences are to drive the prohibited practice 'underground'. Boiler rooms, stores and other out of the way places are then utilised for the prohibited practice, which if not checked can lead to increased risks, particularly as there are no formal arrangements for disposal of smoking materials.

8.1.7   Waste and waste management systems

Waste materials and a lack of formalised waste management systems are very often a major cause of fires within the workplace. Arson is a very real threat in today's society and the issues associated with it will be discussed later in this book. Accidental fires are also started from poor management of waste, including the lack of a formalised policy in removal from a building and onward transportation from the site. A lack of staff awareness as to the priority of combustible and flammable material waste management is also responsible for poor management of waste, for example the poor disposal of rags contaminated with linseed oil by persons working within the furniture industry or solvent contaminated materials in the printing industry being discarded inappropriately.

As previously discussed within the section on chemicals and LPG poor disposal arrangements can contribute significantly to the fire risks within a building. Storing waste containers and skips in close proximity to buildings can allow any resultant fire to develop and penetrate buildings though windows, eaves and lightweight wall claddings.

Figure 8.5   Poor storage arrangements for flammable materials

8.1.8   Other significant hazards

Electrostatic discharges from machinery, the decanting of solvents and other processes which are poorly managed due to a lack of procedures, training and use of inappropriate materials can also cause fires.

The heating and magnetic forces produced by the high currents of a lightning strike can cause structural damage to buildings (direct effects), and the associated electric and magnetic fields can induce transients which may damage or disrupt electrical equipment (indirect effects). Either the direct or indirect effects of a lightning strike can cause fires and possibly explosions.

Poorly maintained mechanical equipment can run the risk of causing a fire when a lack of grease allows the bearings to run dry causing friction which turns into heat that may be sufficient to cause a fire. Equally, excessive levels of contaminated grease on machinery are also known to have been responsible for causing a fire.

The selection of incorrect plant and equipment for use in hazardous areas where a flammable or explosive atmosphere can also have the potential to cause a fire or explosion.

8.1.9 Vehicle fires

Accidental fires involving vehicles can occur for a number of reasons, these are generally due to:

➤ Faulty wiring looms or components
➤ Fuel leakage
➤ Discarded smokers' materials
➤ Faulty HT systems (ignition systems)
➤ Hot exhaust contact with combustible materials
➤ Vehicle collisions or accidents.

Modern vehicle interiors are largely composed of polymers, plastics and other synthetic materials – all of which are particularly combustible giving off flammable vapours. The smoke and fumes from the outbreak of fire are highly toxic and can be deadly if inhaled.

Poorly conducted and managed refuelling or recharging arrangements for vehicles can also cause fires. Overcharging of batteries (e.g. forklift trucks) can release hydrogen which has a very wide explosive range and refuelling a petrol driven vehicle in close proximity to ignition sources also has the potential to cause a fire.

Vehicle fires can also be a primary source for a secondary fire in a building, e.g. delivery/haulage vehicle within a distribution depot, or a forklift truck within a warehouse, where a fire starting from a vehicle quickly spreads within a building causing the whole building to be involved.

8.2 Causes of fire relating to construction and maintenance

Many of the causes of fire detailed above can relate to work involving construction and maintenance operations. Equally the preventive measures that may be adopted to minimise the risk of a fire occurring in construction and maintenance operations may be equally valid across many sectors of industry. This section of the book will look at specific causes of fires in construction and maintenance operations.

A large proportion of fires started within the construction sector fall under the following key headings:

➤ Arson
➤ Electrical
➤ Hot work
➤ Flammable and combustible substances.

As arson is dealt with in a section on its own within this section the first area to be addressed will be electrical causes of fire.

8.2.1 Temporary electrical installations

While similar in nature to electrical systems utilised within permanent structures those used in construction and maintenance operations are subject to potentially more arduous or hostile environments.

Some common electrical faults in construction and maintenance operations that pose fire hazards include:

➤ The overloading of electrical sockets and systems in site accommodation
➤ The incorrect use of flat twin and earth cable as extension leads instead of suitable flexible cable
➤ Electrical cables or lighting laid on or near combustible material (frequently in roof and ceiling voids)
➤ Mechanical damage to cables, often as a result of inappropriate routing of cables
➤ The intentional defeating of safety devices, such as fuses or circuit breakers (use of metal objects such as nails)
➤ The accumulation of combustible rubbish against temporary distribution boards and transformers
➤ Unauthorised make-shift cable joints made without correct proprietary connectors.

In addition to the above issues fires have been started by inappropriate location of temporary site and task lighting particularly when using halogen lighting systems in close proximity to flammable or combustible materials.

Figure 8.6 Typical hot works include welding

8.2.2 Hot works

Generally the term 'hot work' applies to the use of open flames, fires and work involving the application of heat by means of tools or equipment.

More specifically hot work includes any works involving the use of naked flames, such as when welding or brazing, or when undertaking plumbing work with gas torches. Hot work also includes the use of grinders which create high temperature sparks and also includes the use of hot air guns for sealing materials such as flooring.

The use of naked flames and tools that apply heat, in themselves will not necessarily be the cause of fire, but the inappropriate use of the equipment or lack of control of combustible materials, flammable vapours, etc. when undertaking hot work operations have been the start of many fires within the construction and maintenance sector.

Sparks, for instance, falling on unprotected wood or behind panelling or onto wood shavings or dust can ignite the material. It has also been known that the heat generated by hot air guns when laying flooring materials can ignite combustible items upon which the flooring is being laid particularly as the heat generated in the process is then sealed in.

It is often the case that sparks of heat generated in such ways may go unnoticed for long periods of time and given the correct conditions, perhaps opening up sheeted areas, etc., will provide additional oxygen supplies to complete the fire triangle. Fires can start in this way as long as 1–2 hours have elapsed after the works have been completed.

Open fires such as bonfires to burn site rubbish are to all intents and purposes banned in construction operations due to fire risk; however there are a variety of different ways that naked flames can start a fire.

Figure 8.7 Oxy-fuel equipment on a construction site

Operations such as the laying of bitumen in flat roofing operations requires a bitumen boiler which is powered by LPG and can pose a significant risk if not managed effectively; this also falls under the title of hot work operations.

The use of bitumen or tar boilers can present a significant risk. Such fires are invariably due to:

➤ Poorly maintained equipment (thermostat faulty or poorly adjusted)
➤ Poor position of the boiler such as on the roof itself with no fire resistant materials between it and the roof
➤ Poorly supervised units allowing the boiler to boil over or boil dry
➤ Attempting to move boilers while the burner is still alight
➤ Overfilling of the boiler allowing the boiler to boil over
➤ Use of LPG cylinders without the appropriate safety features such as flashback arrestors.

The use of oxy-fuel equipment such as acetylene and propane are inevitably responsible for a number of fires in construction and maintenance operations. Many of the key causes have already been discussed in this chapter such as poor maintenance, use in close proximity to combustibles, flammables, etc.

The additional risks that the use of such equipment brings arise from such issues as poor handling and storage arrangements for cylinders of acetylene.

Acetylene is an extremely flammable gas. It is different from other flammable gases because it is also unstable. Under certain conditions, it can decompose explosively into its constituent elements, carbon and hydrogen. This decomposition can be more readily brought about from mishandling, dropping, etc. of the cylinder causing damage to the internal mass. Uncontrolled leaks from the acetylene cylinder regulator or hoses can also have the same effect.

Additionally the contamination of oxygen supplies (cylinder, valve, regulator, hoses, etc.) by oil or grease also has the capability of causing an explosion.

8.2.3 Flammable and combustible substances

The use, storage, transportation and disposal of flammable and combustible substances within construction and maintenance operations bring with them a wide variety of different hazards and associated risks.

While the industry is attempting to reduce the use of flammable substances as a matter of overall risk management many proprietary substances used in the work cannot be substituted.

Flammable substances

Risks arise from a lack of competency of those using flammable chemicals in that they are unaware of either the flammable properties of the substance or any reactions that may occur as part of a mixing process. Chemicals used for degreasing metals are often solvent based as are a large number of paints, varnishes, etc. and if they are used within an area with ignition sources such as hot work operations, fires and possibly explosions can occur.

Poor storage arrangements such as storing incompatible substances, failing to control ignition sources, lack of designated storage facilities, poor ventilation, etc. all have been responsible for causing fires on construction sites and during maintenance operations.

Inadequate arrangements for the safe disposal of flammable waste products such as aerosol containers, paint cans, mastics, together with poor arrangements for removal and disposal of contaminated fuels such as diesel, petrol, etc. are also known to have contributed to fires in construction operations.

Combustible substances

The amount of combustible material encountered at any one time on a construction site will vary dependent upon the nature of the project and the schedule of the works.

Poor management of waste combustible materials is recognised as being a major factor when considering fire risks on a project. Poor housekeeping in relation to the removal of waste, or allowing a build-up of combustible waste such as used cement bags are easily identifiable as sources of fuel, which when taking into account the sources of ignition available, increase the fire risk greatly.

Often the lack of appropriate waste removal arrangements during refurbishment or maintenance operations allows a build-up of highly combustible waste materials to be readily available for any source of ignition to start a fire, for example allowing the debris from dismantling an old flat roof to build up. The roofing materials (wood, roofing felt, bitumen) are highly combustible and failing to remove the debris/waste from either the roof itself or adjacent to the building presents a significant fire risk.

Demolition operations during construction also contribute greatly to the fire and explosion risk with many combustible and flammable substances used, produced or released during the process, such as:

➤ Explosive charges
➤ Fine dust levels
➤ Wood and other carbonaceous materials
➤ Gas supplies
➤ Electrical supplies.

In general the risks associated with fire during construction and maintenance operations are similar in nature to those risks in other areas of industry. However, due to the type of work, the variety of operations being undertaken simultaneously and the materials in use or being produced the risks appear to be greater.

8.3 Arson

Arson can be defined as the deliberate or wilful act of setting fire to a building or item of property, be it an industrial property, dwelling house, car or any similar item. Arson has become the largest single cause of major fires in the UK.

At its worst, arson can lead to loss of life and significant financial consequences as in the case of any fire, but persistent and pervasive minor arson attacks also establish a strongly detrimental (and visually harmful) tone to deprived areas and communities. In some areas, arson now accounts for 70% of all fires, rising to a staggering 82% of fires in certain inner city localities.

The number of arson fires, both in their own right and as a proportion of the total fires in the UK, attended by the fire service has more than doubled over the past decade and now stands in excess of 85 000 per year. The number of casualties, as a direct consequence of arson attacks, over the same period of time, has shown a similar increase in rate.

8.3.1 Causes of arson – why people commit it

Arson is a complex and serious crime which has a wide variety of causes that include:

➤ Arson associated with other criminal acts
➤ Arson associated with a grievance

Figure 8.8 Arson attack on a school

➤ Fraudulent arson
➤ Arson associated with economic or political motivation
➤ Arson associated with mental instability.

Arson associated with other criminal acts

This generally takes the form of an act of criminal damage or similar opportunist vandalism. It may also be associated with an attempt to conceal or destroy evidence of another crime such as a break-in, burglary, or sabotage of the equipment in the premises or the premises itself. Young people are very much associated with this form of fire setting.

Arson associated with a grievance

This may result from the fear of unemployment possibly due to company relocation, lack of advancement or promotion, salary grievances, or having been humiliated in front of co-workers. At a personal level employer/employee conflicts such as jealousy of a work colleague's promotion or success or revenge against a superior or employer are also known to cause arson attacks.

Fraudulent arson

This type of arson may be committed in an attempt to defraud the insurers by an owner of a business employee setting fire to their own premises or property. This may be property in the workplace or at home. Fraudulent fires in industry and commerce do not always involve setting fire to the financial records, they may, for example, involve an attempt to destroy out-of-date (or out-of-fashion) stock, with a claim subsequently being made for more modern items as part of a 'new for old' policy.

Arson associated with economic or political motivation

Arson attacks may be associated with industrial action (strikes), industrial sabotage, campaigns of pressure groups (e.g. animal rights) and more recently as a result of terrorist activities.

Arson associated with mental instability

It is true to say that pyromania is rare; however, the feelings associated with hatred and jealousy, the desire to attract attention, together with frustration and sexual perversion have all led to fire raising and arson. The would-be heroes who light fires in order to be able to 'discover' them and assist in the rescue of their colleagues and involvement in fire fighting operations also come into this category.

8.3.2 Arson – Influencing factors and management

There are a number of factors that influence the likelihood of an arson attack which also serve to provide information on how such attacks can be managed.

When arson occurs

Arsonists, in common with other criminals, do not like to be seen. They often attack at night, under the cover of darkness.

Statistics reveal that:

➤ 49% of all fires occur at night
➤ 68% of arson fires occur during the hours of darkness
➤ 84% of 'major' arson fires (damage valued at over £50 000) occur during the hours of darkness.

Given the above statistics security lighting is undoubtedly a cost-effective way of reducing the incidence of arson attacks. In many premises such lights may be operated by passive infrared (PIR) detectors and not only provide illumination but also produce an element of surprise. PIR is also seen as being more acceptable in built-up areas due to the nuisance caused by permanent illumination.

These lighting systems should also be enhanced by intruder alarms, which ideally should be supported by audible alarms which when sounding often put off a 'would-be' arsonist.

Ensuring that the arsonist is visible is also an important element in other forms of security precautions as part of an overall strategy. For example, palisade welded mesh fencing or in the case of construction, Herras fencing, is preferable to a wall or solid fence, as intruders inside the grounds or site are rendered visible.

Figure 8.9 Typical security lighting

Arson also often occurs during tea or lunch break times, when few staff are present, again reducing the likelihood of the arsonist being seen. This can be combated by introducing tighter entrance control measures and raising staff awareness.

Buildings/sites at risk

When considering the initial design and layout of a building, those with areas that allow a person to linger (an arsonist) unseen present a high risk as do those that have access to the roof. These areas can be designed out at this stage if appropriate thought is applied. These issues should also be taken into account when considering temporary sites such as in construction work, where layouts of site accommodation and building operations, stores, vehicle pounds, etc. must also take into account security against arson.

In many cases the employers or occupiers of a building have had little or no input or control over the design of the premises in which they work, therefore basic measures to prevent access to the roof or other areas of the building must be considered. Simple controls such as the inclusion of a dense, thorny hedgerow at the perimeter of the site may assist in preventing access and the environment of the premises may also be enhanced by such an introduction.

Priority should be given to keeping arsonists out, especially when the premises are not occupied. Attention must be paid to primary control measures like the siting and security of windows, doors, locking mechanisms, fences and gates.

Weak points in building perimeter protection that should be considered may include letter boxes, air vents and louvered windows.

The poor management of access keys can also add to the risk and therefore the security programme must include accountability for keys and proper authorisation for their issue. A register should be maintained of all key/access issues and all keys/cards accounted for at the end of each period of work.

Entry to isolated or less used parts of the premises, such as storerooms and warehouses, should be restricted and monitored. In addition suitable arrangements must be taken to identify legitimate visitors or, where appropriate, restrict the areas to which members of the public have access.

Where buildings or sites are unoccupied at night or at weekends, serious consideration should be given to the installation of CCTV as it has a high deterrent effect. It should be noted that CCTV systems which are not monitored have limited value.

The equipment used must be of a high quality and be installed to suit the prevailing conditions and should incorporate suitable recording equipment. Specialist advice should be sought before installation to ensure the correct system is fitted.

Frequent, but irregular visits by mobile security patrols outside normal working hours can be a deterrent against arson and similar crimes, if there is no 24-hour security presence on the site.

If used, security staff should be vigilant and take note of such crimes in the neighbourhood, this is a particularly good indicator of a security company's competency if they have such information readily available. Keeping a log of such incidents may help in assessing the likelihood of an arson attack occurring.

Perhaps one of the most cost-effective measures against the arsonist is an alert and motivated workforce. All staff should be reminded to challenge those they do not know. It may be that a simple 'Can I help you?' may be sufficient to deter a potential arsonist.

Figure 8.10 CCTV with infrared lights reduce the likelihood of arson attack

Figure 8.11 Risk of arson may increase when companies are targeted by political activists

Table 8.1 An arson risk assessment model (an example – government crime reduction toolkits)

Step 1 Study the vulnerability of the building: (a) Externally (b) Internally	Look at the building and what goes on within it. Note the possible ways in which fires could be started deliberately. Identify the vulnerable points both inside and outside the building and in the external areas within the building perimeter. In addition, consider the area in which the business is located in order to assess the likelihood of an arson attack in the neighbourhood.
Step 2 Identify the fire hazards: (a) All possible sources of ignition (b) Flammable liquids and gases, combustible materials (including waste), furniture or furnishings and combustible elements of the structure (c) Structural features that could lead to the spread of fire	A key element of the arson risk assessment is to identify, and reduce as far as is practical, the sources of ignition and combustible materials that are available to the opportunist arsonist. Although it is recognised that these cannot be eliminated completely, steps can be taken to eliminate or reduce the threat (see step 4). Steps should be taken to identify voids, unprotected ducts, unstopped gaps around services and similar features.
Step 3 Identify the people who could start fires deliberately: intruders, visitors and members of staff. Also consider the people that will be affected, especially anyone with a disability	All staff should receive appropriate training so as to be aware of the danger of arson and the threat that it presents to life and jobs. Everyone should take part in regular fire drills and be aware of the need to assist people with any form of disability.
Step 4 Eliminate, control or avoid the threat	Where possible, action should be taken to remove potential sources of ignition, flammable liquids and combustible materials from the workplace. It may be possible, for example, to replace a flammable solvent with a non-flammable one with similar properties. Checks of the premises should be made last thing at night, especially when contractors have been present. A fire risk assessment should be undertaken and appropriate action taken as necessary.
Step 5 Consider whether the existing security provisions are adequate or need improvement	Ensure that the best use is made of existing security measures before considering new, complex or expensive installations or procedures. For example, many intruders enter buildings through windows or doors that are left insecure so ensure that a check is made at the end of each day, other security measures that should be considered include: ● Perimeter protection ● The strength of the building envelope ● Access control ● The detection of intruders ● Security lighting ● CCTV systems ● Staff relations ● Awareness of activities of pressure groups who could target the premises.
Step 6 Consider whether the existing fire safety provisions are adequate or need improvement	Much can be done, often at little cost, to reduce the threat of arson and limit the horizontal and vertical spread of fire; effective compartmentation is a key element in reducing the damage caused by fire. The installation of a sprinkler system that will not only sound the alarm but will automatically fight the fire is a further advance in protection.
Step 7 Allocate the risk category and record the findings	Allocating the risk category need not involve complex mathematical formulations. A simple low, medium or high categorisation for each part of the premises may be sufficient.
Step 8 Prepare a business continuity plan	The business continuity plan should have a clearly defined purpose. Key members of staff should be identified and their roles defined. Key contractors should be listed with their contact points. Provision should be made for staff welfare as well as practical steps to ensure that the effect on business operations is minimised. A copy of the plan should be kept off the site.
Step 9 Carry out a periodic review of the assessment	The assessment should be reviewed if the nature of the business, the number of staff, the materials used or the character of the neighbourhood changes significantly.

Business operations and activities

The type of operations or activities that an employer or occupier undertakes can also be a target for an arsonist.

Those companies who undertake work that makes use of animals as part of research, or who hold personal, secure data that may be used against a person or organisation, are potentially at risk. It may be that specialist tools and equipment are in use or a particular piece of plant or machinery is stored which could be stolen and sold on for profit, any of which may be attempted to be hidden by the use of fire.

Gathering information about pressure groups that may pose a threat and liaising with local police and fire services may assist in managing the threat of arson. Information is often kept by insurance companies, who may also provide guidance with regard to prevention, some of which may be required for insurance to be offered.

Taking simple steps such as ensuring that stock and valuable items are secured out of the away and outside general view, particularly those on transient sites such as in the construction sector, also assist in managing arson as arson attacks often follow acts of petty theft and vandalism.

Minimising the availability of combustible materials to the arsonist, managing items such as skips and other refuse containers and securing flammable chemicals that could be used as accelerants are all part of the arson risk reduction measures that a company could adopt to prevent this risk of arson and subsequent losses that would be likely to occur following an attack.

8.4 Prevention of fire

The preventive strategies for reducing the risk of fire and explosion within the workplace can be equally applied across the majority of industry sectors as illustrated from the preceding section on arson prevention and control. The principal causes of fire generally only vary in very specific detail and more often than not relate to hazardous operations and sites, e.g. COMAH sites, otherwise the causes remain constant, particularly when it revolves around 'human factor'.

This section will therefore cover general guidelines with specific issues discussed under each of the various headings.

The overall preventive and protective arrangements for each element covered should be included in a comprehensive prevention strategy by way of ensuring that a written policy is formalised and communicated. The written policy should cover the following:

➤ A statement of the specific policy that is being covered, e.g. hot works

➤ An organisational section detailing clearly who is responsible and for what, e.g. who is responsible for securing a contractor to undertake testing of electrical apparatus, or the role of the employees for undertaking visual checks

➤ Specific arrangements for undertaking risk assessments and implementing controls such as identifying the number of fire wardens required and how they are to be trained

➤ Details of any monitoring or measuring of performance should also be included such as the signing off of a permit to work system, whether they were completed adequately and operated effectively

➤ How the elements of the individual policies will be reviewed and audited, to identify effectiveness or waste. For example, gas heating equipment may no longer be in use but may still be included on an inspection aide memoir and the policy still appears to cover gas equipment and persons are still being trained in the safe handling of gas cylinders.

8.4.1 Electrical appliances and installations

Electrical fires rate only second to arson in relation to providing the primary hazard and cause of fire in the workplace. A range of management steps should be taken to minimise the risks associated with electrical appliances and systems and are likely to be included in the management system detailed above.

Electrical appliances and apparatus should only be procured from a reputable (ideally approved) supplier. A risk assessment should have been completed prior to procurement to confirm the equipment is going to be safe and fit for its intended purpose.

Intrinsically safe – apparatus or system

European Harmonised Standard EN50 020

A protection technique based upon the restriction of electrical energy within the apparatus and in the interconnecting wiring, exposed to an explosive atmosphere, to a level below that which can cause ignition by either sparking or heating effects. Because of the method by which intrinsic safety is achieved it is necessary that not only the electrical apparatus exposed to the explosive atmosphere, but also other (associated) electrical apparatus with which it is interconnected, is suitably constructed.

Figure 8.12 Electrical supply testing

Figure 8.13 Portable appliance testing

For example, if electrical equipment is to be used in a hazardous area where flammable substances could be or have been used/stored then it should be intrinsically safe.

Both temporary and permanent electrical supply installations should be installed in accordance with the latest addition of BS 7671 (IEE Wiring Regulations and the Electricity at Work Regulations 1989). Installations, particularly if of a temporary nature, for example in construction works, must be inspected regularly and tested at intervals no greater than those specified by the installation company. It should be noted that if temporary installations are in position for a greater period than 3 months on a construction site then they should be subject to test in line with HSE guidance every 3 months or when they have been altered.

Electrical cabling should be protected against damage by appropriate routing away from potentially hazardous areas such as pedestrian routes or vehicular routes.

Portable appliance testing (PAT)

Portable electrical equipment should be subject to regular inspections and tests, the time frames of which will be dependent upon the nature of use. Ideally portable electrical equipment should carry durable labels which display that it has been inspected and tested and is in a satisfactory condition.

Table 8.2 summarises the HSE's guidance on the type and frequency of inspections and tests for portable electrical appliances in offices and hotels, etc.

Electrical equipment used in more hostile environments such as in construction and mountainous works are tested more frequently.

Electrical installations, whether temporary or permanent, together with electrical equipment and apparatus, should be protected by a range of measures including residual current devices (RCDs) which protect human beings from electrical shock, fuses and miniature moulded case circuit breakers (MCCBs) which provide protection to apparatus from overheating and short circuits. Arrangements should be in place to ensure that all electrical work to installations and/or electrical equipment is undertaken by only trained, qualified and competent electricians who are familiar with the systems involved.

Electrical safety systems such as those detailed above will also assist in reducing the risk of fire caused by damage to installation cabling caused by vermin; however, these systems should be supported by active management to control vermin for the risk of fire from these areas to be avoided or minimised.

All staff members who are to make use of electrical equipment should be familiarised with the hazards associated with electrical equipment, any specific design issues, limitations and signs and symptoms of deterioration or wear.

Documented systems to provide evidence of the testing, inspection, etc., together with a defect reporting system, should be considered to assist in managing electrical safety.

8.4.2 Cookers, associated cooking equipment and installations

Management systems that include those issues covered under electrical safety must be considered when using

Table 8.2 Summary the HSE's guidance on the type and frequency of inspections and tests for portable electrical appliances in offices and hotels, etc.

Equipment/environment	User checks	Formal visual inspection	Combined inspection and testing
Battery operated (less than 20 volts)	No	No	No
Extra low voltage (less than 50 volts AC): e.g. telephone equipment, low voltage desk lights	No	No	No
Information technology: e.g. desktop computers, VDU screens	No	Yes, 2–4 years	No if double insulated – otherwise up to 5 years
Photocopiers, fax machines: NOT handheld. Rarely moved	No	Yes, 2–4 years	No if double insulated – otherwise up to 5 years
Double insulated equipment: NOT handheld. Moved occasionally, e.g. fans, table lamps, slide projectors	No	Yes, 2–4 years	No
Double insulated equipment: handheld, e.g. some floor cleaners	Yes	Yes, 6 months–1 year	No
Earthed equipment (Class 1): e.g. electric kettles, some floor cleaners	Yes	Yes, 6 months–1 year	Yes, 1–2 years
Cables (leads) and plugs connected to the above. Extension leads (mains voltage)	Yes	Yes, 6 months–4 years depending on the type of equipment it is connected to	Yes, 1–5 years depending on the type of equipment it is connected to

Figure 8.14 Gas cooking installation showing fixed fire fighting equipment

electrical cookers and cooking equipment. In addition to the management systems detailed above establishing a planned preventive maintenance scheme (PPM) should take into account heat controlling devices such as thermocouples as it is unlikely that the portable appliance testing will entirely eliminate the risks from such equipment.

Any gas cooking equipment and installation should be installed by a member of the Council of Registered Gas Installers (CORGI). Gas installations to appliances should be by fixed piping and/or protected/armoured flexible tubing. In the case of temporary cooking supplies and installations (e.g. in construction operations) LPG gas cylinders must be located outside buildings and protected from unauthorised interference.

Ideally as technology advances gas systems and appliances should 'fail to safe in the event of an emergency, i.e. fire'. It is likely that gas supplies to appliances and installations will therefore be linked to alarm/detection systems that will automatically isolate the gas supplies upon actuation. It is worthy of note that gas supplies should not be able to be re-energised after an emergency in such a way that gas supplies are allowed to 'free flow' with the presence of an ignition source to control the gas release.

With latter day installations such as in hotels and restaurants with large kitchens, an additional safety measure can also be introduced such as the provision of fixed fire fighting installations. These will be covered in more detail in later chapters.

As in the case of electrical testing, any systems should be subject to regular inspection and maintenance

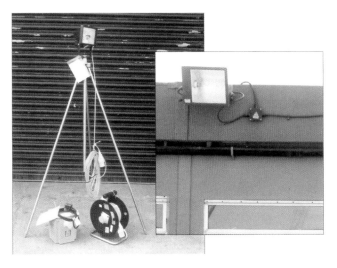

Figure 8.15 Halogen lighting unit fitment

by a qualified competent engineer in line with the manufacturer's or installer's maintenance scheme.

8.4.3 Naked lights and flames

Avoiding the use of naked flame within the workplace is the most effective way of managing fire risks from this source. Replacing hot work and flame generating devices such as oxy-fuel equipment must be seen as the first step in significantly reducing the risks from this area. In relation to practical operations the use of compression plumbers' fittings as opposed to the soldered fittings is a good example of where such replacement negates the risk entirely.

Where the use of naked flames cannot be avoided, formal operating procedures, safe systems of work, additional levels of supervision and adequate levels of regular formal monitoring may serve to reduce the risk.

The burning of rubbish in the workplace or on a construction site should be avoided at all costs both in so far as the fire risks inherent with the operation but also to ensure compliance with the Environmental Protection Act which requires management of omissions into the atmosphere.

A risk assessment as to the requirements for higher heat emitting lighting units such as halogen lamps should be undertaken prior to their procurement. If such units are required, control measures including limiting their use to the confines of areas that do not have combustible materials or flammable vapours present, should be considered.

Ensuring that higher risk lighting is subject to planned preventive maintenance, particularly items such as their retaining hinges and brackets, will ensure that the units remain upright and away from surfaces which could catch fire.

8.4.4 Heaters and heating systems

The preceding paragraphs detail the hazards associated with electrical and gas systems which are also common to those sources of energy used by heaters and heating systems. As has already been discussed, the poor design, inappropriate installation and lack of planned preventive maintenance are all causes of fire. It is therefore appropriate to ensure that systems are designed, installed and subject to planned preventive maintenance.

Systems should be designed in such a way that if a hot product is able to heat areas such as ducting, which runs through the building, the areas should be lined with fire/heat resistant materials to prevent them coming into contact with combustible materials and therefore ensuring competent designers are appointed is a significant risk management step.

The installation of any heaters and heating systems must be undertaken by competent engineers installing to the design specification. It is also likely that these engineers will be retained to undertake the planned preventive maintenance of the system to ensure that it remains safe and fit for its intended purpose.

The maintenance programme must take into account specifically rotating or moving parts which must be kept lubricated to ensure they do not run dry.

It is likely that if designed correctly, combustible materials will not come into contact with any component parts of a heating system; however, given the human factors involved, physical barriers, awareness training and regular inspections are likely to be needed to ensure that combustible or flammable materials do not come into close proximity with the systems or prevent the correct air circulation by obstructing ventilation grilles, etc.

8.4.5 Chemicals and LPG (hazardous materials)

The vast proportion of chemicals used in the workplace are classified under CHIP and therefore fall under the Control of Substances Hazardous to Health Regulations (COSHH). These chemicals together with liquefied petroleum gas (LPG) may also fall under the Dangerous Substances and Explosive Atmospheres Regulations (DSEAR), each of which requires the effective management and control of risk.

When considering the management of chemicals and LPG, formalised procedures must be produced, implemented and monitored in relation to their use, storage, transportation and disposal.

Appropriate training in all areas including the handling of chemicals and LPG gas cylinders is critical to securing the safety of all those in the workplace and others who may be affected.

Figure 8.16 External tank facility

Table 8.3 Minimum separation distances

Quantity stored *litres*	Distance from occupied building, boundary, process unit, flammable liquid storage tank or fixed ignition source *metres*
Up to 1 000	2
1 000–100 000	4
Above 100 000	7.5

The principles of safety relating to both chemicals and LPG are similar in many ways and it is intended to consider these overall principles for each hazardous material as one, with specific areas of difference being highlighted.

Substitution

Ideally the use, storage, etc. of flammable chemicals and substances and LPG should be avoided where reasonably practicable. The same can be said for using low flashpoint liquids. Other chemicals, substances or liquids, which are either non-flammable or have a higher flashpoint, may provide suitable safer alternatives.

It must also be noted that reactive chemicals whose properties in themselves are non-flammable, but may react violently even explosively when mixed together, should also be considered in relation to substituting them for less reactive materials.

Storage

Where substitution cannot be accommodated, the provision of safe storage must be considered as a key element in reducing the fire and explosion risks from the hazardous materials.

Ideally, arrangements should be made for storage of flammable liquids, gases and substances to be stored well away from other processes and general storage areas. This is most effectively achieved by a physical distance. An alternative approach may be to provide a physical barrier such as a wall or partition, combined with a storage system or container.

If the storage facility is to be located within a building and the hazardous materials are to be handled within the area around the store fire resisting separation from other areas of the building or workplace should be considered.

The arrangements for storing flammable substances and LPG will depend upon the storage facilities available, the amount of hazardous materials and the nature of the materials.

External storage arrangements – when storing large quantities (50 litres or more) ideally an external storage area will be provided that will include the following considerations:

➤ Segregation from other processes and storage (flammable substances should be stored separately from other hazardous substances and materials such as oxidisers and LPG, etc.):
  ➤ By physical distance
  ➤ Fire wall
➤ Security against unauthorised access, locking mechanisms, physical walls and fences
➤ Positioned in a well-ventilated area and on impervious ground
➤ Take into account the potential heat on the containers from a fire within or outside the premises' boundary; and a fire within the storage facility on buildings, plant and people inside or outside the premises
➤ Positioned away from sources of ignition (or where electrical supplies are intrinsically safe)
➤ Environmental protection arrangements:
  ➤ Bunded facility (110% of contents)
  ➤ Interceptors to capture potential release

The acknowledged safe separation distances are detailed in Table 8.3.

The arrangements for the storing of larger quantities (more than 300 kg) of LPG gas cylinders are very similar to those required for flammable liquids.

In addition to providing adequate levels of signage on site, formal systems of work will need to be

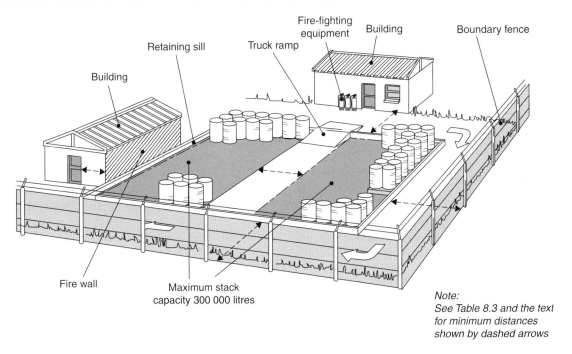

Figure 8.17 Example of a well laid out external storage facility for flammable chemicals – HSG 140

developed and implemented that:

➤ Keep rubbish and anything combustible well away from the cylinders, and keep weeds and grass in the vicinity cut down (avoid using a chlorate-based weedkiller, as it can be a fire hazard in itself)

➤ Prevent any electrical equipment, vehicles, bonfires or other sources of ignition near the cylinders

➤ Prevent smoking when changing cylinders

➤ Prevent people not involved with the storage or installation gaining access to it, particularly children (construction sites particularly)

➤ Keep vehicles well away from the installation (fire and impact hazard)

➤ Protect against accidental damage, ensuring that pipework is properly routed and supported. In the case of underground routing that schematic diagrams of the pipe routes are drawn up, to avoid putting anything in the ground which may damage the pipework

➤ Enable reports of any equipment failure or damage to the supplier without delay.

Internal storage – where flammable substances can be stored within specially designed separate buildings, safe by position, the same control measures as those required for outdoor storage, with the additional requirement that the building should generally be constructed of non-combustible material and the roof of the building be of lightweight materials, may be used. The roof would open readily to release the effects of an explosion, therefore acting as explosion relief.

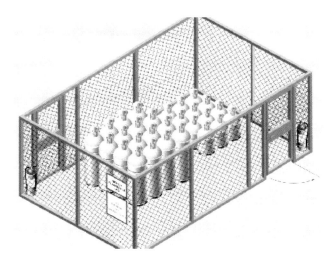

Figure 8.18 External storage arrangements for LPG

Ventilation of any building used for a flammable material store must also be considered; the ventilation apertures or bricks must be so positioned to allow air flow within the building/store and must be positioned at both high and low level. The apertures must not be sited so that they allow ventilation to other buildings or to external areas that could contain ignition sources. The number of apertures (based upon the total floor area in per cent) required will depend on the substance stored, e.g. 1% flammable liquid and 2.5% flammable gases.

Additional measures would be required for stores that cannot be positioned in separate buildings, particularly the fire resistant separation from other parts of the building as shown in Fig. 8.19.

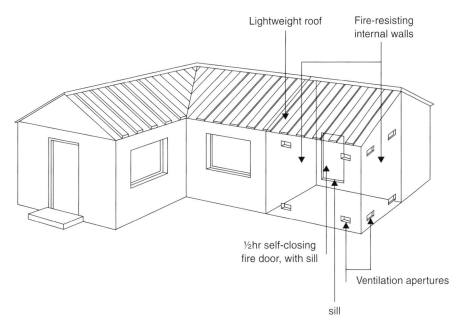

Lightweight roof

Fire-resisting internal walls

½hr self-closing fire door, with sill

Ventilation apertures

sill

Figure 8.19 Example of suitable flammable storeroom in a building

The storage arrangements for smaller amounts of flammable substances and LPG will vary dependent upon the processes for which they are being used. All containers that are to be used to store flammable liquids should be designed and constructed to meet the standard of being suitable for the purpose.

Containers also need to be of a performance tested type, in that they should be robust and have well-fitting lids and spill resistant tops to prevent leakage if knocked over. There are specific standards available for such containers and their packaging to comply with transport legislation.

To ensure that no chemical or physical interaction occurs which might cause leakage, the material, from which the containers are made, needs to be compatible with the properties of the liquid. In addition, containers may need protection against corrosion, e.g. by painting, and plastic containers can often suffer degradation by natural light, which can be reduced by suitable shading of the container.

LPG cylinders can also suffer degradation from corrosive atmospheres particularly in relation to some chemical operations, therefore the use and storage of the cylinders should reflect the workplace conditions and be stored outside.

Storage within the actual workplace where the flammable substance or LPG is to be used must be considered at length. Ideally external storage arrangements should be made; where this is impracticable the following must be considered:

➤ Keep the quantities of flammable materials and LPG stored and used to an absolute minimum

➤ Storage arrangements should be made available for flammable liquids which are separate from other substances including LPG – up to 50 litres may be stored within the work area if enclosed in a fire resisting cabinet or bin

➤ Securing mechanisms should be provided to any storage that allow only authorised access

➤ The storage facility must be provided with spill protection

➤ Flammable liquid containers should always be replaced in the store following use and such containers should be provided with lids to protect against vapour release and spillage

➤ LPG storage (up to 300 kg) should be ideally in open air but must be with good ventilation if stored within a workplace (see Fig. 8.18)

➤ Proprietary safety containers (such as dash pots used in the printing industry) should be fitted with self-closing lids for flammable liquids

➤ Flammable liquids should not be dispensed or decanted within a store or adjacent to a store. Arrangements for such operations should include the provision of ventilation and extraction systems

➤ The storage facility should be located away from ignition sources

➤ Signage denoting the flammable storage facility should be displayed, including 'no smoking'

➤ An inventory of the substances should be kept including as a minimum the name of the substance, type and quantity kept

➤ Arrangements for the safe storage and disposal of items such as rags, cloths, etc. which may have been used to apply the liquids or mop up spillage.

For example, metal containers with tight fitting self-closing lids and regular removal (end of day/shift) from site.

Figure 8.20 LPG cylinder storage facility

Training – as discussed at the start of this section, training is a key element for ensuring effective management of an organisation's fire safety and in particular with regards to flammable substances and LPG. Any training programme must cover the arrangements for the safe handling and storage of these hazardous materials, together with the arrangements for the safe disposal.

An outline of such a programme may be as follows:

➤ The types of flammable substances in use, their properties and hazards
➤ General procedures for safe handling of substances and operation of plant
➤ Use of personal protective equipment and clothing
➤ Housekeeping arrangements
➤ Reporting of incidents and faults, including spills and minor leaks
➤ Specific instructions on individual items of equipment, plant and processes
➤ Emergency procedures, including raising the alarm, and use of appropriate fire fighting equipment, deployment of spill kits, etc.

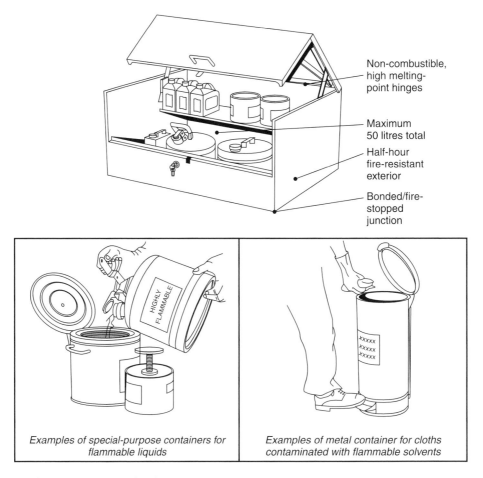

Non-combustible, high melting-point hinges

Maximum 50 litres total

Half-hour fire-resistant exterior

Bonded/fire-stopped junction

Examples of special-purpose containers for flammable liquids

Examples of metal container for cloths contaminated with flammable solvents

Figure 8.21 Internal flammable storage facility

Ventilation of storage areas

Ventilation has been discussed as part of the storage requirements; however, all areas where flammable liquids are handled should be adequately ventilated to ensure that any released vapours are diluted to a safe level.

Whichever ventilation system is used it needs to be capable of providing a minimum of six complete air changes per hour. The system's capacity should ensure that the amount of vapour in the atmosphere is not only diluted to well below its flammable limit, but also reduced to a level below any relevant workplace exposure limit (HSE's guidance note EH 40 gives advice on workplace exposure limits).

Dependent upon the concentrations and volumes in use good, natural, well-positioned ventilation may be adequate, but where this cannot be achieved, mechanical ventilation and/or local exhaust ventilation (LEV) is almost certain to be needed.

Positive pressure mechanical systems can be used to force air into an area; diluting and venting any potential build-up of flammable vapours. Any area provided with a positive pressure mechanical system will also need to be provided with sufficient openings to release the atmosphere created. LEV systems are designed to capture airborne contaminants such as flammable vapours, filter them and ventilate outside a building to a place of safety in the open air.

Any vents provided for the release of pressure or contaminants should be a minimum of 3 m above ground level, at least 3 m from building openings, boundaries and sources of ignition, and away from building eaves and other obstructions where they could become trapped. The ducting taking the flammable vapours away should be arranged so that they cannot condense and collect at low points within them.

Ventilation/fume cupboards – in certain operations (e.g. solvent-based spraying operations or decanting of flammable liquids) booths or cabinets are required to control the build-up of flammable vapours. In such instances the airflow rate into all openings in the enclosure should be sufficient (about 1 m/s) to prevent vapours entering the work area. These ventilation booths, cabinets and other enclosed equipment are generally designed to ensure that the vapours are kept below 25% of their LEL.

As the outcome or failure of any ventilation system or part of the system may result in a build-up in vapour concentrations with the potential for both fire and explosion, the system must be provided with an alarm system to provide warning.

There are also a number of storage systems that are combined with booths and cabinets which use LEV for vapour control. Regardless of the system used each must be subject to rigorous maintenance testing and examination regimes to ensure that flammable vapours are controlled.

Any electrical system used as part of the ventilation or extraction systems must meet rigid electrical standards and as in the case of all such equipment be intrinsically safe. As discussed earlier, classifying hazardous or potentially explosive atmospheres is not only a legal requirement under DSEAR but has been widely used to determine the extent of hazardous zones created by flammable concentrations of vapours.

Over recent years the process of classifying hazardous areas in such a way has been extended from its initial purpose of selecting fixed electrical equipment for use in the area, to helping to identify and eliminate potential ignition sources, including portable electrical equipment, vehicles, hot surfaces, etc. from flammable atmospheres.

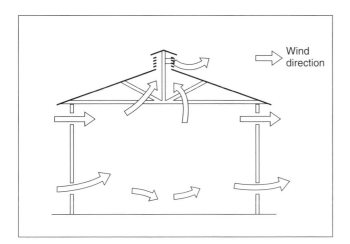

Figure 8.22 The use of wall and roof vents together, give good natural ventilation

Figure 8.23 Example of a fume cupboard

Figure 8.24 Purging tanks during decommissioning

This **zoning** has assisted, for example, in identifying that vehicles such as forklift trucks used for transporting containers of flammable substances, which operate within classified areas/zones inside and outside storage buildings, must be protected to an appropriate standard – particularly during storage operations, when the highest probability of a release from a container occurs while it is being mechanically handled. Therefore an unprotected forklift truck may well provide a source of ignition should a breach of containment occur.

The management of works that involve the removal, replacement or disposal of any part of a plant or system that has previously held flammable liquids, materials or LPG must be planned and undertaken with great care, to avoid fires and explosion.

Many such works will be carried out under the Construction, Design and Management Regulations, utilising specialist contractors undertaking works as part of a permit to work system – PTW (see section on construction and maintenance) to ensure the highest level of management control.

Various techniques can be employed to ensure that flammable substances are managed effectively including purging or inerting the system prior to undertaking operations. Ensuring that residues are removed and that any supply system has been isolated from both power sources and process materials must also be undertaken as part of the systems of work and the PTW.

Disposal arrangements for contaminated flammable materials and aerosol cans, empty flammable liquid containers or gas cylinders must also be considered.

8.4.6 Smokers and smokers' materials

As with each of the previously mentioned sections establishing a policy in relation to formalising the standards, who is responsible and for what, with practical preventive control measures and the training and awareness of staff and all those who may be affected, are the primary steps to take in the management of smokers and their discarded smokers' materials.

Figure 8.25 Metal bin for discarded smokers' materials

Many organisations have effective smoking policies based upon making arrangements for designated smoking areas and prohibiting smoking in all other areas. Less successful are those policies based upon a complete ban; however, this only serves to drive smokers into out of the way, secluded locations such as plant room, storage areas, toilets and emergency escape staircases with no disposal arrangements, where a small fire has a chance to grow undetected.

While the provision of smokers' booths strategically located away from combustible materials and any potential flammable atmospheres, together with the provision of facilities to discard used smokers' materials (ash trays metal bins, etc.), provides practical solutions to the management of this risk, this cannot always be secured in places of work.

Places of public entertainment and public assembly will undoubtedly benefit from being smoker-free zones as the ban on smoking in public areas takes effect; however, it is likely that premises such as nursing homes will still need an effective management strategy to reduce the risk of fire from smoking materials. Such arrangements are likely to include:

➤ Ensuring smoking only takes place in designated areas by establishing a monitoring regime
➤ Provision of metal or glass ash trays or bins for discarded materials
➤ Ensuring when ash trays are emptied (before leaving the area at night, for example) that checks are made to ensure all material is extinguished before placing into external bins, which are separate from combustible waste

➤ All soft furnishings within the smokers' areas are fire retardant (chairs, curtains etc.).

As part of an overall management strategy, including welfare provision for their employees, many successful organisations also encourage staff to give up smoking by providing advice, guidance and support from both internal and external occupational health advisers and by running effective campaigns to raise people's awareness.

8.4.7 Waste and waste management systems

Given the high number of fires that involve waste materials, some 160 000 each year, establishing a waste management system must be seen as a high priority for reducing the risk from fire. Any waste management systems should include effective arrangements for reducing the total amounts of waste stored at any location and ensuring that any residual waste is stored in a way that minimises the risk of fire.

As a large proportion of external waste fires are started deliberately, reviewing the security arrangements as part of a cohesive arson reduction plan must be considered.

An effective strategy for the management of waste is likely to cover:

➤ Conducting a review of the types and volumes of waste produced
➤ Segregation of different types of waste, e.g. combustible, flammable, hazardous
➤ Formal arrangements are put in place for waste collection that prevent a large build-up of waste and establish a monitoring programme to ensure that it works
➤ A monitoring programme that makes sure that rubbish is placed in the containers or designated areas and not left on the ground nearby
➤ Ensuring that any potentially hot materials such as smokers' materials or hot ashes, etc. are cooled and placed within a metal waste container
➤ Wheeled waste bins should be sited 6 metres form buildings and plant (10 metres if plastic) and ideally be fitted with lockable lids and chained to prevent them from being wheeled out of position
➤ Loose combustible waste (e.g. bin bags) should not be stored within 6 metres of a building or plant and be a minimum of 2 metres from a boundary or perimeter fence
➤ If a waste compactor is used ensure that the safe system of work includes turning off when the building is not occupied and the unit is subject to a planned preventive maintenance regime

➤ Raising staff awareness of the need to manage waste responsibly.

As in the case of all arson prevention arrangements, waste management should take into account the provision of adequate security lighting, CCTV systems, fencing and possibly security guards.

8.4.8 Other significant hazards

Electrostatic discharges – the risk of fires and explosions attributed to electrostatic discharges that arise from machinery such as the Neoprene belt on a conveyor unit or from processes such as the decanting of solvents can be managed effectively by selecting the correct equipment, managing the environment, e.g. humidity, by the practice of grounding metallic objects, and replacing insulating materials with static dissipative materials.

Where the presence of a flammable atmosphere is possible, electrostatic discharges must be controlled. The three key methods available are:

➤ The equipotential bonding and earthing of conductors, which may also include operatives
➤ The minimisation of electrostatic charge generation
➤ The maximisation of charge dissipation.

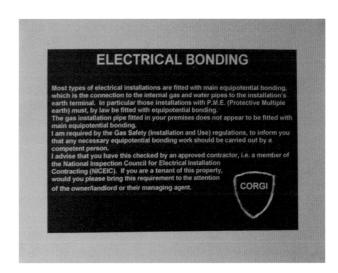

Figure 8.26 Typical electrical bonding notice located on a conveyor belt boiler

In some instances, the above measures, either on their own or in combination, may not be entirely adequate and they should then be supplemented by appropriate safe systems of work and design features to avoid any discharges.

Formal procedures for the safe use, handling, etc. of any article or substance that can effectively cause a discharge of static electricity must be recorded and staff trained to follow them.

The provision of lightning conductors is the main control measure for preventing fires and explosions occurring from lightning strikes. Lightning conductors' function is to attract a lightning discharge, which might otherwise strike a vulnerable part of a structure and to convey the current safely to earth.

There are a variety of types of lightning protection systems available and will be selected upon the level of risk. Structures with inherent explosive risks, e.g. explosives factories, fuel stores, dumps and tanks, invariably need the highest possible class of lightning protection system. Details of the requirements can be found in BS 6651 Code of practice for protection of structures against lightning.

Electro-mechanical equipment – to combat the fire risks from poorly maintained mechanical equipment, e.g. bearings running dry causing friction which can in turn ignite excessive levels of contaminated grease, all such equipment should be included on a planned preventive maintenance regime, which includes a cleaning and where necessary a degreasing programme.

Establishing a procurement policy, which includes a pre-procurement risk assessment, will ensure that correct selection of plant and equipment for use in hazardous areas (where a flammable or explosive atmosphere has the potential to cause a fire or explosion) will minimise the risk relating to such equipment. BS EN 50020:2002, Electrical apparatus for potentially explosive atmospheres – Intrinsic safety 'i', specifies the construction and testing of intrinsically safe apparatus, intended for use in potentially explosive atmospheres and for associated apparatus, which is intended for connection to intrinsically safe circuits which enter such atmospheres. This document may well form the basis from which an organisation establishes its procurement policy for such equipment.

Any intrinsically safe equipment will also need to be used by competent persons and included both within an inspection and testing programme undertaken by competent engineers and in a planned preventive maintenance programme.

8.4.9 Vehicles

Road-going vehicles are subject to testing as part of Ministry of Transport requirements; however, such 'MOT' testing provides only a 'snapshot' test of the vehicle and while failures of electrical components, exhausts, etc. are included the possibilities of electrical wiring defects which may lead to an uncontrolled ignition source may go unnoticed.

Figure 8.27   MOT test in progress

Ensuring that road-going or site-based vehicles are subject to regular inspections and servicing will assist in identifying components and systems that could lead to a fire before it occurs.

Establishing formal procedures for refuelling or recharging of vehicles as part of a safe system of work and ensuring that staff are provided with training to raise their awareness of the risks involved will reduce the likelihood of fire starting from such operations.

As previously discussed within the hazardous substances section the control of flammable vapours must be considered when refuelling operations take place and thus rudimentary controls should be observed such as:

➤ Ensuring that refuelling/recharging is only undertaken in designated areas remote from ignition sources and with good ventilation
➤ No smoking policy is strictly observed and monitored
➤ The use of mobile communications (telephones/radios) is prohibited
➤ Warning signage is displayed
➤ Vapour recovery systems (petrol) are incorporated into the system when bulk tanker delivery takes place
➤ Provision of appropriate fire fighting equipment
➤ Emergency isolation arrangements
➤ Emergency action plans.

When considering the recharging of batteries for vehicles such as forklift trucks and other mechanical handling devices overcharging cut-out safety devices should be installed and maintained to prevent any potential for the release of hydrogen. Such operations will also require many of the aspects covered when refuelling petrol driven vehicles such as good ventilation.

LPG powered vehicles are becoming more prevalent for site-based operations; formal systems of work for refuelling will also need to be provided to ensure that all those involved with the operation have standards from which to work.

The disposal of used tanks, batteries and cylinders must also be taken into account when considering fire and explosion risks associated with vehicles and appropriate arrangements made to ensure that they are disposed of as hazardous waste as part of the organisation's environmental strategy.

8.4.10 Construction and maintenance

A large proportion of construction operations are undertaken under CDM Regulations which require the provision of a construction phase health and safety plan which incorporates the arrangements for the management of fire safety. As part of the requirements fire risk assessments and method statements also need to be prepared for the works. Larger maintenance projects may also fall under CDM with the same requirements.

On longer-term projects an independent fire safety plan may also be created to ensure that as work progresses fire safety management is addressed at each phase.

In order to reduce risk associated with projects of any size it is vital to retain strict control of the building contractors. The CDM Regulations 2007 provide a framework for managing contractors which is appropriate for all building work whether or not the works strictly fall within the scope of the regulations.

Contractor management

An employer/client is liable for the actions of the contractor while working on the client's site. Any unsafe work practice could lead to the client being involved in compensation claims if anyone is injured or to enforcement action if a breach of safety legislation is involved.

There are a number of inherent risks to be considered when employing contractors. In terms of fire safety management these risks arise from the actions of the contractors during the work itself and the subsequent impact upon the fire safety arrangements within the building following the contracted works.

In order to reduce the liabilities of the employer/client it is essential they carefully select competent contractors and then exercise sufficient management control over the project to ensure the contractor employs safe working practices throughout the duration of the contract.

The selection procedure should be flexible enough to ensure the amount of work required in each specific case is kept in proportion to the degree of risk inherent

Fire safety risks associated with contractor within buildings:

The actions of the contractors during the work:
➤ Obstruction of the means of escape in case of fire or other emergency
➤ Wedge open fire doors
➤ Isolation of safety critical services, e.g. fire detection and alarm systems
➤ The introduction of combustible and highly flammable materials
➤ The introduction of sources of ignition, e.g. hot work
➤ Poor housekeeping allowing combustible and highly flammable materials to come into contact with ignition sources
➤ Breach and reduce the effectiveness of security arrangements
➤ Fail to adhere to normal work patterns and procedures, e.g. they often move around the building, work in remote locations and omit to sign in or out.

Impact upon the fire safety arrangements within the building following the contracted works:
➤ Breach fire compartmentation walls, ceiling and floors and service risers
➤ Create voids between floors and service risers
➤ Fail to reinstate safety critical systems, e.g. fire detection and alarm systems
➤ Fail to replace safety critical information signs
➤ Use substandard materials, e.g. non-fire resisting glass in fire resisting elements of construction.

in the contract. The selection process should take place in two stages:

➤ Initial selection stage – to appear on the list of approved contractors
➤ Tender stage – against the specific needs of the contract.

Initial selection stage

A large number of major client organisations have quality assurance (QA), which requires the initial selection or pre-qualification of suppliers and contractors. The successful contractors are placed on an approved list from which contractors are then selected for specific contract requirements.

Where organisations do not have formal QA systems, they would be well advised to adopt the same principles to ensure proper selection of competent contractors.

The selection process would require the contractor to supply relevant information regarding a number of areas of their business management in order for the client to form a judgement of their ability to properly carry out the work.

The common areas which are considered are usually concerned with ensuring the contractor is able to meet the technical requirements of the contract. These may include, where relevant, such items as:

➤ Financial standing
➤ Quality assurance procedures
➤ Technical experience of previous work of a similar nature
➤ Technical skills of contractor's employees
➤ Technical procedures relevant to the contract
➤ Technical method statements
➤ References from previous clients
➤ Industrial relations record
➤ Environmental record.

Safety performance

In addition to information regarding the contractor's technical ability to carry out the project works, the client should seek information to allow them to judge the safety management procedures and practice of the contractor.

Therefore, safety issues should form an integral part of the selection process. The client may seek information on relevant areas, such as:

➤ Contractor's safety policy and/or procedures manual
➤ Employees' safety handbook/safety instructions
➤ Contractor's safety management system and organisation
➤ Safety personnel – competence
➤ Safety representatives/committees
➤ Contractors safety record:
  ➤ *Safety event statistics including; incidents, accidents and fires, etc.*
  ➤ Safety-related work references
  ➤ Enforcement actions taken against them
➤ Membership of relevant professional bodies
➤ Training procedures – sample records
➤ Maintenance procedures – sample records
➤ Risk assessment procedures – generic and specific
➤ Sample assessment records for the type of work
➤ Method statements – generic or specific from previous similar work
➤ Relevant insurance policies.

The client would review **all** of the information presented to form a judgement on the competence of the contractor. In reality it is unlikely that everything provided by the contractor will be perfect; however, a balanced judgement should be possible based on the information provided. If this is not the case the contractor should not be approved and the contractor should be informed of the reasons why.

Tender stage selection

At this stage the client will be selecting a specific contractor from those appearing on the approved list and will require specific details from the contractor as to how they intend to meet all of the contract specification, including the safety requirements of the project works.

To a large degree the response from the contractor will depend on the information provided to them by the client. In order to formulate a comprehensive tender the contractor will require specific information about the technical requirements of the contract, which is often dealt with very well. However, they will also require information about the safety conditions of the contract and any risks which the client has identified, which they will need to consider during the project works.

The types of information the contractor will need will include, as necessary, the following:

➤ A clear definition of the project
➤ Identification of any specific risks presented by the client premises, plant, substances, work processes or operations which the contractor will need to consider
➤ Identities of persons with health and safety responsibilities
➤ Requirements for specific responsibilities within contractor staff, e.g. stipulating for a site safety officer
➤ Risk control measures required by the client
➤ Health and safety performance standards required
➤ Limits on the contractor's actions and areas of invitation
➤ Client and joint management arrangements, reporting and consultation procedures
➤ Permit to work procedures to be adopted
➤ Monitoring arrangements to be implemented by:
  ➤ The client
  ➤ The contractor
  ➤ Jointly.

Additional information:

➤ Site rules
➤ Welfare facilities provided
➤ Administration procedures
➤ Accident reporting procedures
➤ Notification of other employers or contractors on site
➤ Procedures for formal release and receipt of information.

Figure 8.28 Site rules poster

The client can then judge the response from the contractor to make the final selection and award of the contract. The considerations during the final selection process will be based on the information provided by the contractor which should include, as necessary:

➤ Safety plan for the project
➤ Risk assessments for the project works
➤ Method statements for the project works
➤ Response to the issues raised by the client
➤ Evidence of the resources to be allocated to the project, such as:
  ➤ Plant and equipment
  ➤ Access procedures/equipment
  ➤ Materials
  ➤ Safety equipment
  ➤ Numbers and skills of personnel and subcontractors
  ➤ Employment procedures
  ➤ Emergency equipment and procedures
  ➤ Time allocated for completion of the works
➤ Nominated project management personnel, including health and safety responsibility
➤ Procedures for checking the competence of subcontractors
➤ Defined safe systems of work
➤ Permit to work systems
➤ Specific emergency procedures.

8.4.11 Managing the contract

Once the contract has been awarded and the work commences, the client will still need to exercise adequate control over the project works to ensure the contractor does, in fact, maintain the health and safety standards required.

In order to do this the client would need to establish joint management procedures with the contractor. The nature and degree of control required must be kept in proportion to the size and complexity of the project, and will depend on the extent of risk involved in the work and the nature of the work site.

The management procedures should be determined at the planning stage of the project and form part of the conditions of the contract.

The areas that will need consideration will include:

➤ Management of the project:
  ➤ Contractors procedures
  ➤ Client procedures
  ➤ Joint procedures.
➤ Project monitoring procedures for the contractor and client are held jointly and require agreed protocols for:
  ➤ Communication and reporting procedures
  ➤ Procedures for dealing with non-compliance issues
  ➤ Suitable welfare arrangements
  ➤ Emergency procedures for fire, medical and other emergency.

8.4.12 Safety conditions of the contract

In order to make these conditions enforceable through the contract, any joint management arrangements and specific requirements for site controls should be detailed in the contract specification, either as specific conditions included in the contract or by referring to standard safety conditions in a separate document.

Failure to notify the contractor of such details, which may have a cost element, could result in conflict during the contract works and, possibly, claims for additional payment at a later date.

The type of safety conditions or rules which could be imposed include:

➤ A requirement to comply with all legal obligations
➤ Restrictions or controls on the activities of the contractor
➤ Provision for access and egress – client and contractor staff
➤ Specific site rules, e.g. management of fire doors, smoking, etc.
➤ Working times/hours
➤ Site services and facilities to be provided by the client or the contractor, in particular safety critical systems that must be maintained during the contract, e.g. fire detection and alarm systems
➤ Security arrangements

- Restrictions on equipment:
  - Type and use
  - Loans
  - Hire
- Requirements for maintenance of equipment
- Control of hazardous substances
- Housekeeping and site cleanliness/tidiness
- Storage requirements or restrictions
- Welfare arrangements
- Accident reporting and investigation procedures – the client should know as soon as possible of any unsafe event on their site
- Permit to work procedures to be followed
- Sanctions which may be imposed in the event of poor performance.

During the contract works the client should exercise sufficient monitoring of the contractor's activities to ensure compliance with the conditions. Again, the extent of the monitoring will be dictated by the scope of the work and the degree of risk inherent in it. The monitoring arrangements may, therefore, range between simple observation of the contractor's activities at one end of the scale, to formal inspection and monitoring procedures at the other. The client would need to determine the extent of the monitoring activities they wish to employ at the planning stage and inform all internal personnel involved of the procedures.

Many of the potential problems during the project works can be avoided by establishing effective communication channels between all of the parties involved in the works. Once again, the type and extent of communications established would depend on the nature of the project, but the communication channels which may be needed would include accident/incident reporting procedures, to ensure the client is fully aware of all accidents and dangerous occurrences.

8.4.13   Performance review

In order to confirm the contractor has in fact carried out the work to a satisfactory standard and met the safety conditions, the client will need a formal procedure for reviewing the performance of the contractor against the contract specification and conditions.

In the case of the safety conditions of the contract, this will involve both effective monitoring of the work in progress and a formal review on completion of the work.

Monitoring the contractor's performance

As already stated, the degree of monitoring required would be dictated by the nature of the work. However, no matter how simple or low risk the work may be there will always be a need for monitoring in some form. The monitoring may be carried out by the project management team or by the local staff in the area of the works.

In either case the staff carrying out the monitoring will need access to information regarding the extent of the work to be carried out and the methods of work to be followed in order to properly judge the actions of the contractor. This may require the staff given the task of overseeing the work to be provided with copies of the contractor's method statements.

The procedure should include reporting channels for non-compliance issues and formal feedback of performance standards.

All client staff should be encouraged to observe the activities of contractors and report any circumstances they feel to be unsafe to the relevant person within their business area.

Formal review

On completion of the works, or at predetermined periods during contracts of long duration, the client should carry out a formal review of the performance of the contractor against the requirements of the contract and their own method statements.

This should include the health and safety performance of the contractor as well as the technical aspects of the works.

The formal review of performance should be used in determining the competence of the contractor to continue the contract and when letting future contracts.

8.4.14   Electrical and gas supplies, installations and equipment

As with all electrical supplies and installations those relating to construction operations should be installed in accordance with the IEE Wiring Regulations. Arrangements should be established that all contractors on site ensure that their electrical equipment is portable appliance tested in accordance with HSE's guidelines in HSG141. The frequency of test differs from that in an office, hotel, etc. environment in that the arduous environment in which the equipment is used reduces the time frames between tests as detailed in Table 8.4.

Care must also be taken in terms of routing cables and the provision of temporary electrical supplies to ensure that they are kept free from water and plant/vehicle movements.

Temporary lighting, particularly halogen units, should be installed and located away from combustible/flammable materials. Ideally, any electrical plugs should be removed when not in use.

The control of use and installation of gas supplies on a construction site must also adhere to the same requirements as those on permanent sites. Installation must be by a CORGI registered gas fitter and it is

Table 8.4 Electrical safety on construction sites

Equipment/application	Voltage	User check	Formal visual inspection	Combined inspection and test
Battery-operated power tools and torches	Less than 25 volts	No	No	No
25 V portable hand lamps (confined or damp situations)	25 volt secondary winding from transformer	No	No	No
50 V portable hand lamps	Secondary winding centre tapped to earth (25 volt)	No	No	Yearly
110 V portable and handheld tools, extension leads, site lighting, movable wiring systems and associated switchgear	Secondary winding centre tapped to earth (55 volt)	Weekly	Monthly	Before first use on site and then 3 monthly
230 V portable and handheld tools, extension leads and portable floodlighting	230 volt mains supply through 30 mA RCD	Daily/every shift	Weekly	Before first use on site and then monthly
230 V equipment such as lifts, hoists and fixed floodlighting	230 volt supply fuses or MCBs	Weekly	Monthly	Before first use on site and then 3 monthly
RCDs Fixed		Daily/every shift	Weekly	*Before first use on site and then 3 monthly
Equipment in site offices	230 volt office equipment	Monthly	6 monthly	Before first use on site and then yearly

prudent that the main or principal contractor ensures that the fitter is from an approved and registered list.

A large proportion of hot work is undertaken utilising LPG cylinders and thus controls in relation to the cylinders are included within the section on hot work.

8.4.15 Hot work

Ideally, the need for hot work should be eliminated by the use of alternative methods, e.g. bolting component parts rather than welding. However, where this cannot be achieved a dedicated area remote from fuel sources such as combustible waste materials, flammable vapours and combustible parts of the building structure should be considered.

It may be possible to establish a 'safe haven' by erecting a temporary fabrication (fab) shop on site where the hot work may be undertaken. This will also assist in the management of ill-heath issues such as weld fumes as dedicated LEV can be introduced into the fab shop.

These temporary fab shops comprise:

➤ Non-combustible floors or ceilings or alternatively floors and walls with flexible protective coverings that conform to LPS1207

➤ Where protective flexible coverings are installed they are installed by a competent person to ensure that appropriate overlapping is achieved
➤ Electrical supplies may be intrinsically safe dependent upon the hot work processes being undertaken, e.g. the use of LPG
➤ Storage of combustible materials within the fab shop should be prohibited
➤ Extraction systems to remove airborne contaminants should be installed
➤ Detection systems (heat or rate of rise) may be installed
➤ Fire fighting equipment (fire extinguishers) should be provided.

Hot work permits as part of a 'permit to work' system may be adopted for work within the fab shop. However, such systems must be adopted on all other areas of site unless there is minimal risk of damage to surrounding property.

Hot work permits are formal management documents which support the control of the safe systems of work or methods of work. Permits themselves should not be issued as general or blanket authorisations as this can lead to complacency when permits are issued as a matter

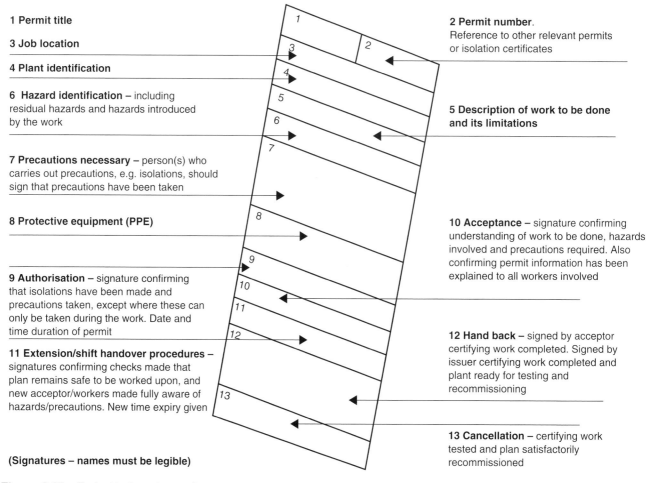

1 Permit title

3 Job location

4 Plant identification

6 Hazard identification – including residual hazards and hazards introduced by the work

7 Precautions necessary – person(s) who carries out precautions, e.g. isolations, should sign that precautions have been taken

8 Protective equipment (PPE)

9 Authorisation – signature confirming that isolations have been made and precautions taken, except where these can only be taken during the work. Date and time duration of permit

11 Extension/shift handover procedures – signatures confirming checks made that plan remains safe to be worked upon, and new acceptor/workers made fully aware of hazards/precautions. New time expiry given

(Signatures – names must be legible)

2 Permit number. Reference to other relevant permits or isolation certificates

5 Description of work to be done and its limitations

10 Acceptance – signature confirming understanding of work to be done, hazards involved and precautions required. Also confirming permit information has been explained to all workers involved

12 Hand back – signed by acceptor certifying work completed. Signed by issuer certifying work completed and plant ready for testing and recommissioning

13 Cancellation – certifying work tested and plan satisfactorily recommissioned

Figure 8.29 Typical hot work permit

of course and not for high risk operations such as hot works.

Hot work permits should normally include as a minimum:

> The nature and location of the hot work to be undertaken
> The proposed start time and duration of the hot work
> The limitation of time for which the permit is valid
> The person in direct control of the hot works
> The authorised signing off (completion and checks) of the works.

Separate permit forms may be required for different tasks such as hot work and entry into confined spaces and therefore different information and records are likely to be required.

The hot work permit used as part of the safe system of work will also provide details on checks that will need to be made prior to the commencement of the work such as the area being cleared of loose combustible material, if work is to take place on one side of a

wall or partition the opposite side must be examined for combustible material, the need to provide fire resistant coverings where work is being undertaken adjacent to immovable wooden structures. In addition the checklist may well include the isolation of existing detection systems and the requirement for the provision of fire fighting equipment and competent persons to operate it.

A key element that will also be included in the hot work permit will be the sign-off following an examination of the area (at least one hour after cessation of hot works) to ensure that there are no fire risks remaining.

A sample of a hot work permit can be found in the Appendix 8.1, together with a sample hot work checklist at Appendix 8.2.

The issues relating to highly flammable liquids and LPG have been covered at length in preceding sections of this chapter; however, quite specific to construction/maintenance operations is the use of items such as tar boilers, which are invariably powered by gas cylinders.

Tar boilers (and similar equipment) should remain at ground level remote from the building where reasonably practicable. However, where the risk assessment

Figure 8.30 Tar boiler used to heat bitumen

identifies that the movement of hot tar presents a greater risk the following should be provided:

➤ An insulated non-combustible base must be provided for the tar boiler to be placed upon
➤ The boiler should be sited in such a way that spilled material can be easily controlled
➤ Hazardous material (flammable substances) must be removed from within close proximity
➤ The equipment must be supervised by a competent supervisor in particular relating to the bitumen level, temperature and lid securing devices
➤ Gas cylinders should be secured in a vertical position, fitted with a regulator and flashback arrestor and be sited at least 3 metres from the burner
➤ Arrangements to ensure the lit boiler is not left unattended
➤ Residual materials should be removed prior to signing off of the hot work permit.

8.4.16 Demolition

All demolition operations fall under the Construction (Design and Management) Regulations (CDM) and therefore require a systematic approach to the operations. Specific details relating to demolition will appear in the construction phase health and safety plan, the fire plan, demolition risk assessments and method statements.

To combat the high risk of fire and explosion when undertaking demolition works, e.g. disruption and ignition of buried gas services or dismantling of tank structures that may have contained flammable residues, a formal system of work should be adopted.

A demolition survey to identify any potential fire and explosion risks together with other risks must be undertaken. This should assist in:

➤ Locating underground services
➤ Identifying whether such services are in use or are disused
➤ Determining how supplies can be disconnected or otherwise made safe (usually undertaken by a representative of the local supply company)
➤ Identifying the flammable and/or explosive concentrations that may be contained within any tanks' structure
➤ Determining methods of residue removal or how the materials will be made inert.

Any demolition work must be conducted by a competent company ensuring that all persons involved with the work have received adequate training, are familiar with the method statement and operate to a permit to work system.

The general requirements for provision of fire fighting equipment must also be considered together with the provision of gas monitoring systems that can detect the presence of flammable vapours.

8.4.17 Arson and site security

The majority of issues relating to arson have been discussed previously within the chapter; however, the instances of arson on construction sites are significant, particularly from unauthorised access, especially by children.

As secure fencing is required under construction legislation, arrangements for monitoring the fencing/hoarding around the site should be included as part of any control measure.

Secure storage of flammable liquids, LPG and any substantial volume of combustible materials must be considered, particularly when the site is closed.

On some sites, particularly those vulnerable to arson, gathering details relating to the area's history is an essential part of site security.

Precautions that need to be taken may include:

➤ Significant physical anti-climb fencing/boarding
➤ Security lighting
➤ CCTV monitoring
➤ Regular out of hours security controls or a permanent presence
➤ Liaison with local authorities including the fire and police service.

Securing items such as expensive plant and tools away from view, together with the management of combustible waste (secure skips), also reduces the risk of arson as any potential thief/arsonist will be either unaware of the items to steal or unable to fuel a fire due to the lack of combustible materials.

Construction and maintenance operations on schools often present a higher level of risk due to the presence of children. There have been instances of fires being set to construction works on schools where signage denoting the presence of flammable mater-ials, as part of the arrangements for providing information to staff and members of the emergency service, has enabled the fire setters to be provided with information of fuel supplies. Signage should therefore be considered in liaison with the police and fire service.

8.5 Case study

Recent arson attack at a school in North West England – estimated loss £1.2 million

The fire was discovered at 00.39 hours by the school janitor. The largest school block, which housed 16 teaching rooms, the library, main office, pastoral offices, the head teacher and deputy's offices together with the staff room, was almost completely destroyed. The block also housed both the history and geography departments which were also completely wiped out. The modern languages, mathematics, English, special educational needs and RE departments lost a large proportion of their resources.

The devastation and trauma was summed up by the head teacher: 'The first reaction is shock and numbness, followed by total disbelief and then realisa-tion that 25 years of resources have gone; that all the carefully collected photographs, booklets and artefacts from all over Europe have gone; that all the paperwork for the administration of public examinations has gone, and all the school text books and personal belongings have gone.'

The timing of the fire was particularly unfortunate; as Year 9 SATS were to be held later in the week and GCSE examinations were due to begin within the next month. The pressure placed upon heads of each subject was enormous as they were required to contact examination boards to discuss what arrangements could be made for loss of coursework and pupils' revision material. There

was considerable torment facing many pupils who had lost work as part of their studies, some of which was ready to be examined by the examination board.

As an interim measure the burnt-out classrooms were replaced by temporary accommodation units and the school had a derelict building at its centre for over a year, becoming a demolition site and then a building site, before restoration works had been completed. These circumstances were obviously not conducive to market-ing the school, and pupil recruitment and the sixth form suffered in particular. This had a massive effect on the school budget resulting in a large deficit, from which the school has struggled to recover.

Clearly insufficient preventive and protective measures were in place, which allowed the arsonist to carry out the attack; the post-fire arrangements included a comprehensive arson and fire risk assessment with an action plan to prevent recurrence.

8.6 Example NEBOSH questions for Chapter 8

1. **Outline** FOUR common causes of fire in the workplace. (8)
   (a) **Explain** the dangers associated with liquefied petroleum gas (LPG). (4)
   (b) **Describe** the precautions needed for the storage, use and transportation of LPG in cylinders on a construction site. (4)
3. **Identify** the preventive measures which can be applied to reduce the possibilities of arson taking place. (8)
4. **Outline** the measures that should be taken to minimise the risk of fire from electrical equipment. (8)
5. A major hazard on a refurbishment project is fire.
   (a) **Identify** THREE activities that represent an increased fire risk in such a situation. (3)
   (b) **Outline** the precautions that may be taken to prevent a fire from occurring. (5)
6. **List** the principles that should be considered for the safe storage of flammable liquids in containers. (8)

Appendix 8.1   Example hot work permit

	Hot Work Permit	**No. ....**
Part A:	Issued to:... Date:................................... Time... I certify that the above person has been contracted to undertake hot works:............................. From...../...../.....at..........hrs until...../...../....at..........hrs For the purpose of... .... Signed...Position..Date...../...../....at..........hrs	

Part B

I certify that the following precautions have been taken: (contractor to complete)

Fire Protection

	Yes	No	N/A
• Where fire detection systems are installed, they have been isolated (in work area only)	Yes	No	N/A
• Where sprinklers are installed they have been isolated (in work area only)	Yes	No	N/A
• Personnel involved with the work and providing the fire watch are familiar with the means of escape and method of raising the alarm/calling the fire brigade	Yes	No	N/A

Precautions within 10 metres (minimum) of the work

	Yes	No	N/A
• Combustible materials have been cleared from the area. Where materials cannot be removed, protection has been provided by non-combustible or purpose-made blankets, drapes or screens	Yes	No	N/A
• Flammable liquids have been removed from the area	Yes	No	N/A
• Floors have been swept clean and are free from combustible material and dust	Yes	No	N/A
• Combustible floors have been covered with overlapping sheets of non-combustible material or wetted and liberally covered with sand. All openings and gaps (combustible floors or otherwise) are adequately covered	Yes	No	N/A
• Protection (non-combustible or purpose-made blankets, drapes or screen) has been provided for walls, partitions and ceilings of combustible construction or surface finish	Yes	No	N/A
• All holes and other openings in walls, partitions and ceilings through which sparks could pass have been fire stopped or otherwise protected	Yes	No	N/A
• Combustible materials have been moved away from the far side of walls or partitions where heat could be conducted, especially where these incorporate metal	Yes	No	N/A
• Enclosed equipment (tanks, containers, dust collectors etc) have been emptied and tested, or is known to be free of flammable concentrations of vapour or dust (see atmospheric test results below)	Yes	No	N/A

Equipment

	Yes	No	N/A
• All equipment to be used for hot work has been checked and found in good order	Yes	No	N/A
• Gas cylinders (when in use) have been properly secured and are fitted with appropriate protection devices (flash back arrestors, etc.)	Yes	No	N/A
• Fire extinguishers have been serviced in accordance with manufacturers' instructions	Yes	No	N/A

167

Results of atmospheric test: NB. Attach certificate of test results of this PTW
form

Test no.........................Date......./......./.......   Result.............% oxygen

The following precautions will be taken during the works:

1. A trained person not directly involved with the work will provide a continuous fire watch during the period of hot work
2. At least two suitable extinguishers are immediately available. Personnel undertaking the work and providing the fire watch are trained in their use
3. Emergency arrangements to stop work, make safe area and evacuate have been established (if alarm not set off by hot work processes)

Part C	**Acceptance** I have read and understood this permit and will ensure that all the precautions detailed in Part B above have been, or will be taken. Signed..Position...Date...../...../.....at.....hrs
Part D	**Completion** I declare that the permitted hot work has been completed..........................on...../...../....at.....hrs Signed..Position... Date...../...../.....at.....hrs
Part E	**Request for an extension** The work has not been completed and an extension is requested Signed......................................Position...Date...../...../.....at.....hrs
Part F	**Extension** I have re-examined the area and confirm that the permit may be extended to..../...../...../at.....hrs Signed..Position...Date...../...../.....
Part G	**Cancellation of Permit** I hereby declare that this permit to work is cancelled and that all specified precautionary measures have been withdrawn or reinstated. Signed......................................Position...Date...../...../.....at.....hrs
Part H	**Fire Watch** The work area and all adjacent areas to which sparks and heat may have spread (such as floors below and above, and areas on other side of walls) have been inspected and found to be free of fire following completion of work (minimum of 1 hour following cessation of works) Signed..Position...Date...../...../.....at.....hrs
Part I	**Hand back of area** I hereby confirm that the works have been completed satisfactorily, the area is safe and the detection/sprinkler system are fully functional Signed..Position...Date...../...../.....at.....hrs

NB a copy of the Method Statement and risk assessment for the works should be attached to this form

Appendix 8.2 Example hot work checklist

HOT WORK CHECKLIST

Fire Protection

1. Where sprinklers are installed they have been isolated (in work area only). ☐
2. Where fire detection systems are installed, they have been isolated (in work area only). ☐
3. A trained person not directly involved with the work will provide a continuous fire watch, in the work area and those adjoining areas to which sparks and heat may spread, during the period of hot work operations and for at least one hour after it ceases. ☐
4. At least two suitable extinguishers or a hose reel are immediately available. Both the personnel undertaking the work and providing the fire watch are trained in their use. ☐
5. Personnel involved with the work and providing the fire watch are familiar with the means of escape and method of raising the alarm/calling the fire service. ☐

PRECAUTIONS WITHIN 10 METRES (MINIMUM) OF THE WORK

6. Combustible materials have been cleared from the area. Where materials cannot be removed, protection has been provided by non-combustible or purpose-made blankets, drapes or screens. ☐
7. All flammable liquids have been removed from the area. ☐
8. Floors have been swept clean and are clear of combustible/flammable material. ☐
9. Combustible floors have been covered with overlapping sheets of non-combustible material or wetted and liberally covered with sand. All openings and gaps (combustible floors or otherwise) are adequately covered. ☐
10. Protection (non-combustible or purpose-made blankets, drapes or screens) has been provided for: ☐
    – Walls, partitions and ceilings of combustible construction or surface finish. ☐
    – All holes and other openings in walls, partitions and ceilings through which sparks could pass. ☐
11. Combustible materials have been moved away from the far side of walls or partitions where heat could be conducted, especially where these incorporate metal. ☐
12. Enclosed equipment (tanks, containers, dust collectors, etc.) has been emptied and tested, or is known to be free of flammable concentrations of vapour or dust. ☐

EQUIPMENT

13. Equipment for hot work has been checked and found in good order ☐
14. Any gas cylinders in use for the work have been properly secured. ☐

(The person carrying out this check should tick the appropriate boxes)

Appendix 8.3 Construction phase fire safety checklist

CONSTRUCTION PHASE FIRE SAFETY CHECKLIST

	Yes	No
Has a site coordinator been appointed by the main contractor?	☐	☐

Has the Fire Safety Coordinator:

	Yes	No
– formulated a fire safety plan?	☐	☐
– ensured that staff are familiar with the plan and understand it?	☐	☐
– monitored compliance with the fire safety plan, especially with regard to hot work permits?	☐	☐
– established a regime of checks and inspections of protection equipment and escape routes?	☐	☐
– established effective liaison with security contractors or staff?	☐	☐
– written records of checks, inspections, maintenance work, fire patrols and fire drills?	☐	☐
– carried out a fire drill and analysed the results?	☐	☐
– checked the arrangements and procedures for calling the fire brigade?	☐	☐

Large Projects

On large projects, has the Fire Safety Coordinator:

	Yes	No
– appointed fire marshals and/or deputies, trained them and delegated responsibilities to them?	☐	☐
– provided site plans for the emergency services detailing the escape routes, fire protection equipment and facilities for the fire brigade?	☐	☐

Fire Safety Plan

Does the fire safety plan detail:

	Yes	No
– the organisation and responsibility for fire safety?	☐	☐
– the site precautions?	☐	☐
– the means for raising the alarm in case of fire?	☐	☐
– the hot works permit scheme?	☐	☐
– the site accommodation, its use, location, construction and maintenance?	☐	☐
– the points of access and sources of water for the fire brigade?	☐	☐
– the control of waste materials?	☐	☐
– the security measures to minimise the risk of arson?	☐	☐
– the staff training programme?	☐	☐

Emergency Procedures

	Yes	No
– Is the means of warning of fire known to all staff?	☐	☐
– Is it checked routinely and can it be heard in all areas above background noise?	☐	☐
– Are fire instruction notices prominently displayed?	☐	☐
– Are the fire brigade access routes clear at all times?	☐	☐
– Have specified personnel been briefed to unlock barriers when the alarm sounds?	☐	☐
– Have signs been installed indicating fire escape routes and the positions of fire protection equipment?	☐	☐

Fire Protection

Have measures been taken to ensure the early installation and operation of:
- escape stairs (including compartment walls)? ☐ ☐
- lightning conductors? ☐ ☐
- automatic fire alarms? ☐ ☐
- automatic sprinkler systems? ☐ ☐
- hose reels? ☐ ☐
Are fire dampers and fire stopping provided at the earliest opportunity? ☐ ☐
Is steelwork protected as soon as possible? ☐ ☐
Are adequate water supplies available for fire fighting purposes? ☐ ☐
Are all hydrants clear of obstruction? ☐ ☐

Portable Fire Extinguishers

Are adequate numbers of suitable extinguishers provided? ☐ ☐
Are sufficient personnel trained in their use? ☐ ☐
Are extinguishers located in conspicuous positions near exits? ☐ ☐
Are carbon dioxide extinguishers in place adjacent to electrical equipment? ☐ ☐
Do all mechanically propelled site plant carry suitable extinguishers? ☐ ☐
Have procedures been implemented for the regular inspection and maintenance
of extinguishers? ☐ ☐

Site Security against Arson

Are adequate areas of the site, including all storage areas, protected by hoarding? ☐ ☐
Is security lighting installed? ☐ ☐
Has closed circuit television (CCTV) been installed? ☐ ☐
If CCTV is in position are the screens monitored and/or recorded? ☐ ☐
Is the site checked for hazards at the end of each work period, particularly
where hot work has been in progress? ☐ ☐

Temporary Buildings

Has the contractor made application for a fire certificate (if applicable)? ☐ ☐
Is the firebreak between the temporary building and the structure undergoing work
more than 10 metres? ☐ ☐
If the firebreak is les than 6 metres:
- is the temporary building constructed with materials which will not significantly
  contribute to the growth of fire? ☐ ☐
- is the building fitted with an automatic fire detection system? ☐ ☐

Recommendations

9 Fire protection in buildings

The design, construction, layout and furnishing of buildings has a key role to play in any fire safety management strategy. The building and how it is furnished has a huge impact upon not only whether a fire is likely to start, given the furnishings within, but also if the building itself can withstand a fire long enough for persons to safely escape in the event of a fire.

As the building has such a key role to play in relation to fire safety and risk management it is imperative that a knowledge of the Building Regulations Approved Documents, British Standards and official guidance is included within any course of study. Each of these areas is covered within this chapter and will have a significant impact on the preventive and protective control measures identified within a 'suitable and sufficient' fire risk assessment (see Chapter 14).

9.1 Definitions

Throughout this section reference is made to some technical words which have the meanings provided in the following definitions:

Building – any permanent or temporary building but not any other kind of structure.

Fire resistance – the ability of a component of a building to maintain its integrity and stability when exposed to fire for a given period of time.

Means of escape – a structural means whereby a safe route is provided for persons to escape in case of fire, from any point in a building to a place of safety, clear of the building, without outside assistance.

Flashover – In relation to the behaviour of fire in buildings and other enclosed spaces flashover is the point at which the whole room or enclosure where the fire started becomes totally involved in fire. It is caused by the radiated feedback of heat. Heat from the growing fire is absorbed into the upper walls and contents of the room, heating up the combustible gases and furnishings to their autoignition temperature. This build-up of heat in the room triggers flashover.

> This chapter covers the following areas relating to the fire protection of buildings:
>
> ➤ Legislative requirements
> ➤ Construction; preventive and protective measures
> ➤ Means of escape
> ➤ Fire detection and alarm systems
> ➤ Means of fighting and suppressing fire
> ➤ Access and facilities for the fire and rescue service.

Figure 9.1 Temporary building

Backdraught – an explosive reaction that occurs when a fire in an enclosed space, which has died down due to insufficient oxygen, is provided with large quantities of oxygen.

Fire compartment – a building or part of a building, comprising one or more rooms, spaces or storeys, constructed to prevent the spread of fire to or from another part of the same building.

Dead-end – an area from which escape is only possible in one direction.

Inner room – a room that can only be accessed through another room.

Storey exit – an exit that provides egress from a compartment floor into a protected stairway or final exit.

Protected route – a corridor or stairway that is separated from the rest of the building by a minimum degree of fire resistance (normally 30 mins) and is a 'fire sterile' area, i.e. contains no ignition sources or combustible material. A protected route will contain or lead to a final exit.

Final exit – a door that gives access to a place of ultimate safety outside the building.

9.2 Fire protection (preventive and protective measures)

Fire protection of buildings, the preventive and protective measures that will protect persons in the event of a fire, fall into two broad categories referred as passive and active protection.

9.2.1 Passive fire protection

Passive fire protection is based on the principle of containment; the compartments of the building are constructed so that if a fire should occur, it will be restricted to one area. For example, fire doors should prevent the spread of smoke and flames from lobbies, stairwells and lift shafts.

Another example of passive fire protection is the design of escape routes, which should not incorporate combustible wall, ceiling or floor linings. Fire dampers should be installed in ducts where they pass through compartment walls, and holes in such walls around cables and other services should be fire stopped.

Doors and shutters in compartment walls should be able to withstand the effects of fire for the same period of time as the walls themselves.

9.2.2 Active fire protection

Active fire protection systems may detect or extinguish a fire, with a water sprinkler or inert gas flooding installation performing both functions. An automatic fire detection installation will detect heat or combustion products of a fire in its early stages and raise the alarm. Such systems should be monitored remotely when the building is not occupied to allow the fire brigade to be summoned without delay, thus reducing the damage. A sprinkler installation will release water from the heads nearest the flames with flow switches raising the alarm in a similar way to a conventional detection system.

Active systems also include those that assist in compartmenting the fire such as fire door release mechanisms, fire shutters and mechanical damping systems. In addition other systems may be actively used for smoke extraction.

Neither passive nor active fire protection measures can be installed and then forgotten; they require regular inspection and maintenance. Service contracts should be established with accredited contractors for installed equipment but the fire safety manager should also ensure that regular inspections are made of escape routes, fire doors and housekeeping standards and that suitable records of such inspections are kept.

9.3 Legislative requirements

The design and construction of buildings in the United Kingdom must achieve reasonable standards of fire safety and resistance from a fire starting inside the building or a fire spreading from outside the building. In addition, all buildings must be designed and constructed in a way that protects the occupants from the fire and ensures their safe evacuation. In England and Wales the legal requirements are set out in the Building Regulations 2000 (as amended 2006) and the Building Regulations (Scotland) 2003 for Scotland.

Figure 9.2 Front cover of Approved Document B

Table 9.1 Functional requirements of the Building Regulations 2000

Title	Regulation (functional requirement)	Summary
B1 Means of warning and escape	*The building shall be designed and constructed so that there are appropriate provisions for the early warning of fire, and appropriate means of escape in case of fire from the building to a place of safety outside the building capable of being safely and effectively used at all material times.*	This section outlines the principles of horizontal and vertical escape within buildings and provides detailed guidance of how the requirement of B1 may be met.
B2 Internal fire spread (linings)	1) *To inhibit the spread of fire within the building the internal linings shall* (a) *Adequately resist the spread of flame over their surfaces; and* (b) *Have, if ignited, a rate of heat release which is reasonable in the circumstances.* 2) *In this paragraph 'internal linings' means the materials lining any partition, wall, ceiling or other internal structure.*	This section provides definitions and classifications for the linings of ceilings and walls. It also identifies those areas within buildings that should be finished to the standard classifications.
B3 Internal fire spread (structure)	1) *The building shall be designed and constructed so that, in the event of fire, its stability will be maintained for a reasonable period.* 2) *A wall common to two or more buildings shall be designed and constructed so that it adequately resists the spread of fire between those buildings. For the purposes of the sub-paragraph a house in a terrace and a semi-detached house are each to be treated as a separate building.* 3) *To inhibit the spread of fire within the building, it shall be sub-divided with fire-resisting construction to an extent appropriate to the size and intended use of the building.* 4) *The building shall be designed and constructed so that the unseen spread of fire and smoke within concealed spaces in its structure and fabric is inhibited.*	This section defines the load bearing elements of the structure and describes the nature and importance of fire resisting compartmentation. It includes guidance on how openings within compartment walls and floors should be designed and managed.
B4 External fire spread	1) *The external walls of the building shall adequately resist the spread of fire over the walls and from one building to another, having regard to the height, use and position of the building.* 2) *The roof of the building shall adequately resist the spread of fire over the roof and from one building to another, having regard to the use and position of the building.*	This section provides detailed guidance relating to the fire resistance of external walls, space separation between buildings to minimise the risk of fire spread through radiation and discusses the requirements of roof coverings.
B5 Access and facilities for the fire service	1) *The building shall be designed and constructed so as to provide reasonable facilities to assist fire fighters in the protection of life.* 2) *Reasonable provision shall be made within the site of the building to enable fire appliances to gain access to the building.*	This section exclusively deals with access and facilities for the fire service when attending fires at the building. It details the requirements for the provision of fire mains, vehicle access, access to buildings for fire fighting personnel and the venting of heat and smoke from basements. This guidance is provided to allow the fire service to deal with serious incidents in a way that minimises the risk to the building and themselves.

Both of these Building Regulations lay down certain 'functional requirements' that all buildings must satisfy. In the regulations, these functional requirements are nothing more than brief descriptions of the function of various aspects of the building, for example the Building Regulations require that all buildings have 'adequate drainage'. The specific design and construction methods used to achieve the functional requirements of the Building Regulations are, in the UK, left to those who design and construct buildings. So in the example given, the design and construction of the drainage for a building may be of any type, *but* it must be adequate.

In the same way, the regulations relating to the fire safety of buildings are not in the main prescriptive but outline certain functional requirements, for example the Building Regulations require that all buildings have 'appropriate means of escape in case of fire'.

The Building Regulations are enforced by local building control authorities, which work under the auspices of the government department responsible for planning. It is the local building control authorities who normally decide whether or not a particular approach to satisfying the functional requirements of the Building Regulations is adequate.

In order to provide some helpful guidance to designers and constructors, the DCLG issues detailed guidance on how the functional requirements of the Building Regulations may be met. This detailed guidance takes the form of 'Approved Documents'; these documents have the same status as all approved Codes of Practice.

In the case of fire safety DCLG have issued 'Fire Safety – Approved Document B' to the Building Regulations 2000 (Revised 2006).

The Building Regulations are divided into various 'parts'. Part 'B' of the regulations deals with the main fire safety requirements for all buildings.

There are two volumes:

➤ Approved Document B (Fire Safety) – Volume 1: Dwellinghouses (2006 Edition)
➤ Approved Document B (Fire Safety) – Volume 2: Buildings other than dwellinghouses (2006 Edition).

Part B of the regulations is further subdivided into the following five sections:

➤ B1 Means of warning and escape
➤ B2 Internal fire spread (linings)
➤ B3 Internal fire spread (structure)
➤ B4 External fire spread
➤ B5 Access and facilities for the fire service.

Table 9.1 provides a summary of the functional requirements of the regulations relating to fire safety.

In addition to 'Approved Document B' the Building Regulations also provide direction on 'access to and use of buildings' in 'Approved Document M' for all persons, but particularly those with disabilities.

The Workplace (Health Safety and Welfare) Regulations, Disability Discrimination Act and Discrimination (Employment) Regulations all require arrangements to be established to ensure all persons can safely gain access and use facilities and do not differentiate these requirements from those required in the event of an emergency.

Thus due consideration must be made when accessing the requirements for those with disabilities.

9.4 Building construction and design – preventive and protective measures

Designing and maintaining a 'safe' building is the first step to ensuring reasonable levels of fire safety for the building and its occupants in the event of a fire. Without a safe building, all the efforts of management to provide systems that prevent fires occurring or ensure emergency evacuation will be of little or no effect. The fundamental features that provide a basis for both building and life safety in all buildings are:

➤ Elements of structure
➤ Compartmentation
➤ Internal linings.

9.4.1 Elements of structure

An element of structure is defined as being part of a building, which *supports* the building, i.e. a *load-bearing* part, for example:

➤ Any part of a structural frame (beams and columns)
➤ Any load-bearing wall (other than part which is only self-load-bearing)
➤ A floor or any element that supports a floor.

A roof structure is not considered as an element of structure unless the roof provides support to an element of structure or which performs the function of a floor.

The elements of structure should continue to function in a fire. They should continue to support and maintain the fire protection to floors, escape routes and access routes, until all occupants have escaped, or have been rescued. In order to achieve this load bearing elements of structure are required to have a demonstrable standard of fire resistance.

The degree of fire resistance for any particular building depends upon its use, size and location.

Elements of structure are required to have specific fire resistance in order to:

➤ Minimise the risk to the occupants, some of whom may not evacuate the building immediately
➤ Reduce the risk to fire fighters who may be engaged in fire fighting or rescue operations
➤ Prevent excessive fire damage and collapse of the building
➤ Prevent excessive transfer of heat to other buildings and structures.

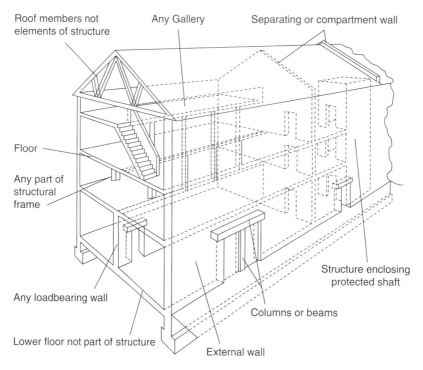

Figure 9.3 Elements of structure

Junctions

In a building that is designed to offer some resistance to fire it is crucial that the joints between the various elements of structure do not present any weak spots in the fire protection.

It is therefore critical that the junctions of all of the elements of structure that form a compartment are formed in such a way that fire is prevented from passing through the join for a period at least equal to the period of fire resistance of any of the elements it joins. In some situations, for example roofs in terraced buildings and buildings that abut each other, the compartment walls extend through the roofs and walls forming the compartment.

9.4.2 Compartmentation

Compartmentation is the subdivision of the building into compartments. Each compartment separated from others

Figure 9.4 Party wall between terraced properties

Figure 9.5 Party wall between two adjacent buildings

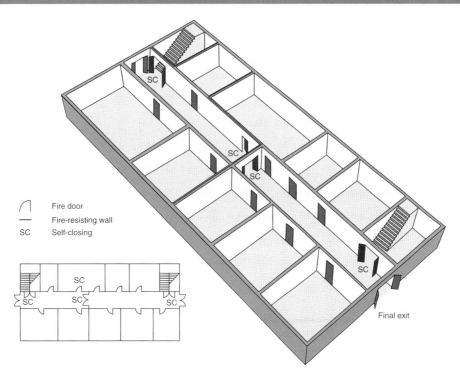

Fire door
Fire-resisting wall
SC   Self-closing

SC

SC
SC

SC

Final exit

SC   SC   SC   SC   SC

Figure 9.6 Example of the horizontal subdivision within a floor which limits fire spread and protects the means of escape

by walls and/or floors, thereby restricting the growth and spread of fires in buildings.

Effective compartmentation limits the extent of damage caused by the heat and smoke from a fire, which, in turn, will have direct and significant implications for the business continuity and resilience of occupants of the building.

Compartmentation is also used as a means of preventing fire spread between adjacent buildings. Compartmentation can be achieved horizontally within a floor area or vertically between floors. Compartmentation is also used to create areas of relative safety for occupants escaping from fire.

Horizontal compartmentation

In a single storey building or on any one level of a multi-storey building, compartmentation can be applied:

(a) To meet travel distances requirements (see later)
(b) To enclose specific fire hazards
(c) To assist progressive horizontal evacuation
(d) To assist a phased evacuation
(e) To separate areas of different:
  ➤ Occupancy
  ➤ Risk category
  ➤ Standards of fire resistance or
  ➤ Means of escape

(f) Where individual compartments are too large and exceed the limit for the standard of fire resistance proposed
(g) Where it is desirable or necessary for the occupants to stay in a building involved in a fire for as long as possible, for operational or safety reasons, e.g. an air traffic control centre; intensive therapy unit, the control centre of an oil rig.

Vertical

In multi-storey buildings, each storey of any non-domestic building should be a separate compartment. Each compartment should be capable of sustaining the total destruction of the compartment involved without permitting the fire to spread to other floors. This vertical compartmentation also protects occupants of the building who might have to pass the storey involved in fire while escaping. Finally vertical compartmentation also provides a degree of protection to fire fighters working on storeys immediately above or below the fire.

Fire resistance

Compartment walls and other elements of structure are normally required to have a degree of resistance to fire. Obviously any element of structure that was unable to resist the passage of a fire would rapidly collapse. There have been some notable instances where the lack of fire resistance in the structural elements of buildings has led to rapid fire spread and rapid collapse of a building resulting in significant loss of life.

It is therefore vital that the structural elements within all buildings are designed and constructed in a way that:

➤ Limits and contains fire spread
➤ Ensures structural stability for appropriate periods of time
➤ Ensures adequate means of escape in case of fire.

There is a British Standard test for fire resistance contained in BS 476 Fire resistance of elements of structure. To pass the test, elements of structure must maintain their integrity and stability when exposed to fire for a given period of time.

Part 20 of BS 476 categorises the elements of construction into three main groups:

(a) Load-bearing elements that have a fire resistance
(b) Non-load-bearing elements that have a fire resistance
(c) Elements that make a contribution to the fire resistance of a structure.

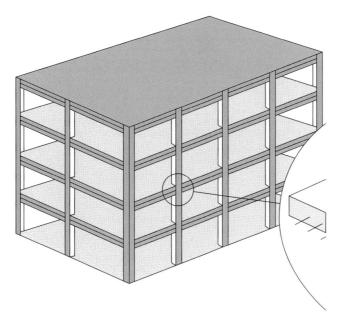

Figure 9.7 Vertical compartmentation in a multi-storey building

The Building Regulations Part B – Section 2 outlines the minimum periods of fire resistance for the structural elements of buildings. The period of fire resistance is given in time. The minimum period that an element of structure can have is 30 minutes; this can be raised to 4 hours in certain circumstances.

9.4.3 Classes of building construction

The Building Regulations categorise buildings into one of the following into three classes of construction:

Class 'A' – complete non-combustible construction, i.e. elements of structure, floors, walls. Supporting structure of brick or concrete
Class 'B' – traditional construction, i.e. non-combustible walls with combustible floors
Class 'C' – combustible construction, i.e. timber floors and walls.

Figure 9.8 Structural beams having undergone fire tests

Figure 9.9 Class A – complete non-combustible construction, i.e. concrete or brick floors and walls

Figure 9.10 Class B – traditional construction, i.e. non-combustible walls with combustible floors

Figure 9.11 Class C – combustible construction, i.e. timber floors and walls

Purpose groups

The degree of fire resistance and other measures that Approved Document B may 'require' for the main elements of construction depends to a great extent upon the purpose to which the building is put. Buildings are categorised into seven 'purpose groups':

1. Residential domestic
2. Residential institutional
3. Office
4. Shop and commercial
5. Assembly and recreational
6. Industrial
7. Storage (including warehouses and car parks).

Appendix 9.1 provides a more detailed breakdown of the types of premises in each purpose group.

In addition to the use of the building, the degree of fire resistance for any element of structure, including compartment walls and floors, is the size of the building; the particular dimensions that building control officers, designers and builders need to consider are:

➤ Height of the building
➤ Total floor area
➤ Volume of each compartment.

9.4.4 Materials of construction

Fire resistance is often achieved in buildings due to the inherent qualities of the building materials used. Architects and builders select a variety of materials for both esthetic and practical reasons. Some of the common materials used in construction are:

➤ Brick and concrete
➤ Steel
➤ Plasterboard
➤ Glass
➤ Steel sandwich panels.

Brick and concrete

When using brick or concrete blocks for construction adequate fire resistance is achieved by ensuring the joints at walls and ceilings are sound and providing sufficient vertical stability by the provision of piers and/or corners. When mass concrete is used it is reinforced with steel which provides the necessary stability.

Steel

Steel is used because it is light, strong and to a degree flexible. The major disadvantage of using steel for the

Figure 9.12 Typical building materials

Figure 9.13 Example of a building with an all glass exterior

elements for structure is that it has a low melting point and will lose 60% of its strength at temperatures in the region of 600°C. The temperatures in fires in buildings often reach 1000°C and therefore it is important that the steel components of a building are protected against the heat from any fire to prevent early collapse of the structure. Methods for protecting structural steel include:

➤ Encasing in concrete
➤ Enclosing in dry lining material, e.g. plasterboard
➤ Coating with cement-based materials
➤ Coating with intumescent materials.

Plasterboard
Plasterboard achieves its fire resistance because it is made from non-combustible material, commonly gypsum. A wall made from a 12 mm thickness of plasterboard which is adequately sealed at the joints will achieve 30 minutes' fire resistance. The disadvantage of plasterboard is that it has little strength or load-bearing capacity. Its durability relies on the strength of its supporting structure (normally wooden or metal stud work) and its protection from mechanical damage.

Glass
The use of glass in buildings is becoming more widespread with the development of glass production technology which has resulted in glazing that has a variety of specific applications, for example:

➤ In internal doors as vision panels
➤ As internal and external doors
➤ As partitions and compartment walls

➤ In roofs, floors and ceiling
➤ In escape and access corridors.

The stability of glass elements of structure relies totally on the systems that support the glass, for example the beading, seals and fixings used.

When assessing the fire resistance of glass it is important to find evidence of its compliance to the required fire resistance. Fire resisting glazing should be marked with a permanent stamp which indicates at least the product name and manufacturer. The mark should be entirely visible and legible.

It should be noted that there are many different proprietary types of fire resisting glass available, many of them with similar sounding names. The main glass types are as follows:

➤ Non-insulating glasses:
  ➤ Integral wired glass
  ➤ Laminated wired glass
  ➤ Monolithic 'borosilicate' glass
  ➤ Monolithic 'soda-lime' glass
  ➤ Laminated clear 'soda-lime' glass
  ➤ Ceramic glass
  ➤ Laminated safety ceramic glass
➤ Insulating glasses:
  ➤ Intumescent multi-laminated soda-lime glass
  ➤ Intumescent 'gel-filled' glass
➤ Partially insulating glasses:
  ➤ Intumescent laminated glass
➤ Radiation control glasses:
  ➤ Coated monolithic 'soda-lime' glass.

The glazing system requirements for each of these glasses are very different and any change in the glass type without a change in the glazing system has the potential to reduce the performance as low as 10% of the required level in many cases, i.e. 3 minutes instead of 30 minutes. It is critical that the method of installation and the material and design of the construction being glazed fully complies with the glass manufacturer's recommendations.

Steel sandwich panels

Lightweight sandwich panels are being increasingly used in buildings; they are often constructed with combustible plastic core material which is included to provide thermal insulation. Lightweight sandwich panels combine the strength of the external material with the insulation properties of the inner core. Therefore they have become popular as a building material that enables simple and rapid erection. Unfortunately this type of panel has been implicated in the rapid fire spread and early collapse of a number of large buildings. It is crucial that any cavities or concealed spaces that may be created when using sandwich panels are adequately protected against concealed, internal fire spread.

9.4.5 Concealed spaces

Concealed spaces in buildings provide easy routes for fire to escape both horizontally and vertically. Fire spreading in the concealed spaces in a building presents significant risks due to the fact that it can develop spread without being detected. Concealed spaces may also allow a fire to move through fire compartmentation. Access to concealed spaces is always, by their very nature, limited, therefore, even if a fire is discovered before it has developed sufficiently to affect other parts of the building, the fire service is often faced with difficulties in bringing it under control.

Concealed spaces are found in numerous locations in a building including:

➤ Roof spaces
➤ False ceilings
➤ Service risers
➤ Behind decorative panelling
➤ Cavity walls
➤ Floors
➤ Raised floors for computer suites.

It is for those reasons that cavities in buildings should always be provided with barriers that resist the concealed spread of fire (see Fig. 9.14). It is particularly important that cavity barriers are provided at those locations where the cavity passes through a compartment wall or floor.

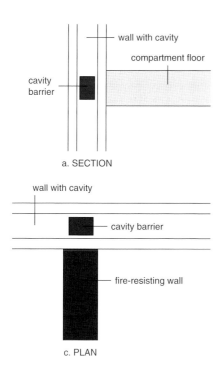
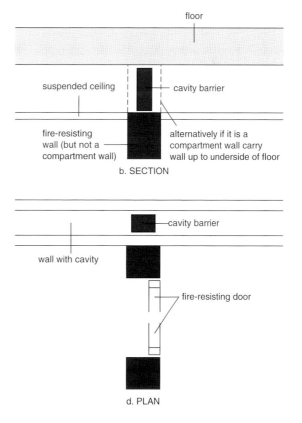

Figure 9.14 The provision of cavity barriers

Typical locations where cavity barriers should be in position are as follows:

➤ At the junction between an external cavity wall and an internal compartment wall or floor
➤ In any ceiling void where a long corridor (over 12 m) has been subdivided to limit fire or smoke spread
➤ Between a compartment wall and the underside of the floor above, i.e. above any false ceiling.

Every cavity barrier must be constructed to provide at least 30 minutes' fire resistance and should be tightly fitted and mechanically fixed in position in such a way so as not to be affected by the:

➤ Movement of the building due to subsidence
➤ Collapse, as a result of a fire, of any services penetrating the barrier
➤ Failure in a fire of any construction into which they abut.

It is often the case that during alterations to a building or its services, fires are started in concealed cavities and develop and spread without easily being detected. Alterations that require a fire resisting enclosure are often made without consideration of their effectiveness, being negated due to the presence of a concealed cavity.

9.4.6 Openings in compartmentation

Compartmentation is vital for the safety of the building and its occupants. However, a building, which has compartment walls and floors with no openings, has very limited use. There are always occasions when a compartmentation wall or floor must be 'breached' in order for occupants to move around and the building to be fitted with services. It is when these necessary breaches are made in compartmentation that it is absolutely vital that the compartmentation maintains its integrity against fire spread.

The Building Regulations Approved Document limits the openings permitted in a compartment floor or walls to those for:

➤ Doors and shutters with the appropriate fire resistance
➤ The passage of pipes, ventilation ducts and other services
➤ Refuse chutes of non-combustible construction
➤ Atria designed in accordance with specific rules
➤ Fully enclosed protected shafts
➤ Fully enclosed protected stairways.

The two types of openings in compartment walls and floors that are the most difficult to manage once a building is occupied are doors with the appropriate fire resistance

Figure 9.15 A typical fire resisting door

and the passage of pipes, ventilation ducts and other services. Time and time again breaches in compartmentation that occur as a result of inadequate fire risk management have allowed relatively small fires to spread and develop in tragic proportions.

Breaches of compartmentation are found to routinely occur by fire doors being faulty or being wedged open and contractors' works involving the routing of new services through compartment walls or floors being carried out without the necessary reinstatement of the integrity of the compartment.

Fire doors and shutters

Fire doors – fire doors are provided not only to allow passage through a fire compartment wall but also to protect persons escaping from the heat and more importantly smoke generated from a developing fire.

In addition to the fire resistance of the door and its assembly, it is also vital, in order to protect escape routes and prevent smoke damage to the building, that the door limits the spread of smoke. This is achieved by ensuring that the door effectively self-closes and is fitted with two types of seal; an intumescent strip that expands when it becomes hot and forms a fire resisting seal around the door and a cold smoke seal which is normally in the form of bushes or a felt material, which prevents the movement of cold smoke.

The fire resistance of doors is certified through a testing procedure laid down in BS 476. Fire doors can only achieve a certified rating in conjunction with its assembly, i.e. door frame hinges, door handles, glazing, etc. It can be understood therefore that the fire resistance on any particular door relies as much on the entirety of the completed assembly as on the door itself.

A Code of Practice for fire doors with non-metallic leaves (BS 8214:1990) states that all fire doors should be

Table 9.2 Examples of the marking regime suggested in BS 8214:1990

Core colour	Background colour	Fire resistance in minutes	Colour code meaning
Red	White Yellow Blue	20 30 60	Intumescent seals are required as part of the door assembly at the time of installation
Green	White Yellow Blue	20 30 60	No additional intumescent seals need to be fitted when installed
Blue	White	20	No additional intumescent seals need to be fitted when installed.
	White	30	Intumescent seals are required as part of the door assembly at the time of installation.

permanently marked with their fire resistance rating at the time of manufacture. This marking confirms that that particular door (and associated assembly) has passed the necessary BS 476 test. The marking should indicate, not only that the door provides a degree of fire resistance but also whether or not heat activated (intumescent) seals need to be incorporated when the door is installed to enable it to achieve the require period of resistance.

It can be seen that doors indicated with a blue core on a white background can serve to provide either 20 minutes' fire resistance if installed without additional intumescent strips or 30 minutes if additional strips are added.

Once fire doors are installed within a building it is necessary to manage their continued effectiveness. The inspection of fire doors is a key component of all fire safety management systems. Those who find themselves responsible for managing fire safety in buildings will need to be aware of the common requirements for fire doors which are summarised below.

Fire resistance	Doors should display evidence of conformity to BS 476, normally by the inclusion of a colour coded plug in compliance to BS 8214. In general all fire doors and their assemblies must be capable of resisting fire for 30 minutes. The notable exception to this is for fire doors that have been provided to break up a long (over 12 metres) corridor, and doors protecting dead ends, in which cases only 20 minutes' fire resistance is required. Any glass fitted in the door should be similarly fire resistant and should be securely fitted with fire resistant beading.
Fire and smoke stopping	The vast majority of fire resisting doors are required to have both fire and smoke stopping capability. Fire stopping is achieved by fitting intumescent strips either to the door or the frame. When the door assembly gets hot the strip expands and forms a seal which prevents the passage of fire. Smoke stopping is achieved by fitting Neoprene strips to the door. These will prevent the passage of 'cold' smoke through the door assembly prior to the action of the intumescent strip. Notable exclusions from the requirement to have cold smoke seals are doors to external escape routes and an unprotected lift shaft.
Well fitting	All fire doors are tested in their assemblies and as such when in situ should be well fitting. Fire doors that develop gaps between them and the door frame will not perform their functions.
Self-closing	All fire doors should be fitted with self-closing devices that are maintained so that the door closes positively against the door stops and where appropriate latches effectively. The notable exceptions to this are those doors that are normally kept locked shut, e.g. doors to cupboards or service ducts.Where it is considered a hindrance to have a fire door closed at all times it may be held open by an automatic release mechanism that is actuated by an automatic detection and alarm system.
Door furniture	Hinges that a fire door is hung on should display a CE mark and should normally be made of a material that has a melting point above 800°C. Locks handles, etc. should not be so fitted as to compromise the integrity or stability of the door assembly.
Signage	All fire doors in places of work must be suitably indicated, on both sides, with signs that comply to the Health and Safety (Safety Signs and Signals) Regulations 1996, or BS 5499 Part 1, indicating either: ● That they are to be kept closed when not in use ● That they are to be kept locked when not in use ● That they are held open with an automatic release mechanism – keep clear.

Fire shutters – there are occasions when it is not practicable to protect an opening in a compartment with normal doors. In order to protect a large opening in compartment walls and floors there are a number of systems that can be applied including:

➤ Fire resisting roller shutters (FRRSs)
➤ Sliding doors in large scale industrial applications
➤ Fire curtains in, for example, the proscenium arch of a theatre.

The principles of operation remain the same for all these systems. However, the most common method of protecting large openings is with FRRSs. FRRSs are commonly fitted to protect openings, for example:

➤ In kitchen walls
➤ In party walls between two premises
➤ Between floors in shops fitted with escalators in corridors
➤ Protecting atriums or escalator wells
➤ Protecting vehicle entry openings
➤ Entrances to retail stores from shopping malls
➤ Escalator hood shutters.

Figure 9.16 Fire shutter

FRRSs must be capable of being opened and closed manually by the fire service in the direction of the approach when positioned across fire fighting routes. Power operated FRRSs must be provided with a Declaration of Conformity and following installation should be CE labelled in accordance with the EC Machinery Directive. Power operated FRRSs also fall within the scope of the Supply of Machinery (Safety) Regulations 1992. Where an FRRS is held in the open position there should be controlled descent when the automatic self-closing device operates.

There must be a planned system of maintenance with items maintained in an efficient state, kept in efficient working order and in good repair. It is critical to provide safe operation when it is intended to connect to a remote smoke control or fire alarm system, which may result in activation and operation while there are persons in the vicinity of the opening.

Typically, FRRSs are tested/assessed for any period between 30 and 240 minutes depending on what classification the manufacturer wishes to achieve for his product.

Traditionally FRRSs have been manufactured from steel; recent innovation has seen the introduction of new products made from reinforced glass fibre matting. Aluminium is often used in security rolling shutters but melts at circa 660°C. In a standard fire test it is likely that the temperature will exceed this melting point in less than 10 minutes and hence it is not an appropriate material for FRRSs.

All FRRSs should be fitted with an automatic self-closing device. In cases where self-closing would be a hindrance to normal use, they can be held open by a local heat detection release mechanism such as a fusible link particularly where a FRRS has been permitted across a means of escape route. All doors installed in compartment walls which need to be held open in normal use, should be fitted with an automatic release mechanism which is connected directly to a local or remote smoke detection system and not rely on local fusible link for operation. The smoke detection system, if local, should have the detectors mounted on both sides of the compartment wall and should activate both visual and audible warnings and cause the door or shutter to close automatically without any delay.

Passage of pipes, ventilation ducts, chimneys and other services

In the same way as fire spreading through concealed cavities presents a serious risk, buildings are also vulnerable to fire spreading through compartment walls and floors as a result of openings in them that provide no resistance to fire. When buildings are designed, built and managed it is vital that when pipes, ventilation ducts or other services pass through a compartment wall or floor they are provided with adequate 'fire stopping'.

In many cases services and the holes in compartments are 'stopped' with cement mortar or other non-combustible material. However, proprietary fire stopping and sealing systems are now widely available and are often used to seal the area around a breach in the compartmentation. These proprietary systems, such as intumescent pillows, offer a relatively inexpensive and flexible method to control fire spread. Other common fire stopping materials include:

➤ Gypsum-based plaster
➤ Cement- or gypsum-based vermiculite
➤ Mineral or glass fibre
➤ Ceramic-based products
➤ Intumescent materials including pillows, mastics, etc.

Those responsible for conducting fire risk assessments will need to pay particular attention to alterations made to buildings which often result in significant breaches of compartment walls as a result of the retrofitting of services such as communication cabling.

In addition to the stopping around pipes, ducts, etc., where there is the potential for fire to spread within ducting, fire dampers should be fitted to maintain the integrity and form part of the fire resisting compartment as shown in Figure 9.18.

Internal linings

Once the elements of structure are built and (as importantly) maintained, consideration must be given to the lining materials used on the walls and ceilings within the building.

The nature and properties of materials used to line the interior of a building can have a direct influence on the safety of occupants in a fire situation, regardless of how well the building is constructed.

The Building Regulations Part B – Section 2 outlines the minimum requirements for the surface spread of flame of lining materials, i.e. the speed at which fire spreads along the surfaces of walls and ceilings.

Tests have been performed to ascertain if wall and ceiling linings play a part in the time taken to fail

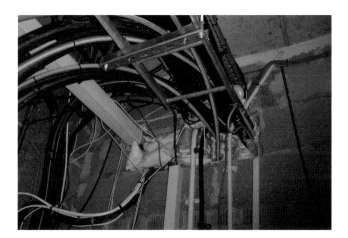

Figure 9.17 Fire stopping with proprietary intumescent pillows

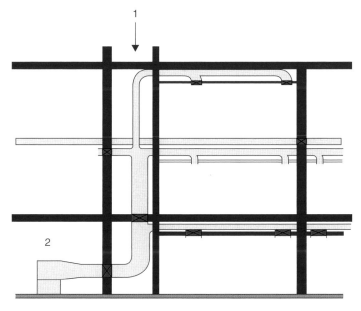

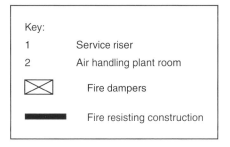

Key:
1   Service riser
2   Air handling plant room
⊠   Fire dampers
▬▬   Fire resisting construction

Figure 9.18 Fire dampers in fire resisting compartmentation

The Stardust fire

The importance of lining material was tragically highlighted in a fire that swept through a building used as a discothèque in Ireland in 1981. The fire killed a total of 48 young people that night. Nearly all were from the local area, many from the same family. A memorial park serves as a reminder of what happened.

The fire started in a partitioned balcony area where two seats were alight when it was discovered. The seating was raised to enable people in that area of the club to see the floor shows. The seating was filled with highly combustible and toxic polyurethane foam and it was covered with combustible vinyl covering.

The walls of that area of the nightclub were decorated with carpet tiles which are normally used for floor coverings.

Attempts by security staff to put out the fire failed. The flames very quickly spread into the main part of the club where ceiling tiles and wall coverings caught light, generating huge quantities of hot thick toxic black smoke.

People trying to escape the rapid spread of fire were crushed in the stampede for the exits, some of which were found to be locked. Some collapsed where they were standing from inhalation of the toxic smoke.

Escape routes were further restricted because many of the windows were barred for security reasons. One group of young people were trapped in the men's toilets because the windows had bars. They were very fortunate to be rescued by fire fighters.

The subsequent inquiry resulted in full-scale tests being conducted in order to ascertain how the fire could have spread so quickly that it could claim so many lives. During the course of the tests it became apparent that the combination of the combustible furniture and the flammable ceiling and wall covering combined to cause conditions where the fire spread faster than had ever before been known to be the case.

The fire highlighted a number of valuable lessons not least the importance of limiting combustible materials on ceilings and walls.

---

structurally and reach flashover temperature. Table 9.3 is a summary of the test results.

These tests along with experience from fatal fires demonstrate that it is absolutely essential to restrict the speed of fire along internal surfaces. The degree to which flame spreads across the surface of a lining material is classified by a physical test which is detailed in British Standard BS 476 Part 1.

The three BS 476 classifications used by the Building Regulations Approved Document for lining materials are:

➤ Class '0' – Any totally inorganic material such as concrete, clay, fired clay, metal, plaster and masonry

➤ Class 1 – Combustible materials which have been treated to proprietary flame retardants, flame retardant decorative laminates

➤ Class 3 – Wood, plywood, hardboard, fibreglass.

The nature of the internal linings of a building along with the size and integrity of any fire compartments are fundamental factors that affect not only the safety of the building but more importantly the ability of the occupants of a building to escape in case of fire.

Areas where people circulate such as lobbies and corridors and all routes used to escape from fire must be to the highest standard, i.e. have the slowest surface spread of fire rating. Large rooms may be to the

Table 9.3 Table showing comparative tests of fire spread

	Construction of walls	Construction of floors	Time until flashover	Time to structural failure
Test 1	Cellulose fibre 12.5 mm wallboard lining on timber battens	Timber	4 minutes	6 minutes
Test 2	Plasterboard on timber battens	Plasterboard on timber battens	23 minutes	33 minutes

(Source: BRE Digest 230, 1984; BS EN ISO 1182 2004)

inter-mediate standard but small rooms may be to the lower standard of Class 3 classification due to the shorter travel distances.

Table 9.4 summaries the Building Regulations' requirements for the use of the various classes in a number of locations.

Table 9.4 Classification of surface linings

Location	National class*
Small room – area not more than: 4 m² in residential buildings 30 m² in non-residential buildings	3
Domestic garages – area not greater than 40 m²	
Other rooms	1
Circulation spaces in dwellings	
Other circulation spaces	0

Notes:
  (i) Any material/product classified as non-combustible when tested in accordance with the requirements of BS 476 Part 4:1970.
  (ii) Any material which, when tested to BS 476 Part 11:1982, does not flame or cause any rise in temperature on either the specimen or furnace thermocouples.
  (iii) Any material which is totally inorganic, such as concrete, fired clay, ceramics, metals, plaster and masonry containing not more than 1% by weight or volume of organic material. Any material conforming to the above requirements can be classified as Class 0 in accordance with Approved Document B of the England and Wales Building Regulations 2006.
* BS 476 Part 7:1987 details the method for classification of the surface spread of flame of products (and its predecessor BS 476 Part 7:1971). The test is a classification system based on the rate and extent of flame spread and classifies products 1, 2, 3 or 4 with Class 1 being the highest classification.

9.4.7 Preventing external fire spread

It is important that the potential spread of fire from a building to a neighbouring structure is restricted. The first attempts to limit the spread of fire from one building to another followed the Great Fire of London in 1666 when a small fire in the city spread to disastrous proportions due to the complete lack of fire resistance or fire separation of buildings. Since that time it is a requirement that buildings are constructed from non-combustible material, have limited openings (such as windows) close to adjacent buildings and have external surfaces that are themselves resistant to radiated heat from adjacent properties.

There are two ways that fire can spread between buildings, i.e. by the flames from a fire in one building directly impinging on an adjacent building or by radiated heat from the building on fire igniting the adjacent building. Experience has shown that in both these cases fire spread is aided by flying embers of burning material being blown from the fire onto surrounding buildings.

For all buildings that are built within 1 m of the boundary to the property, flame spread is the main mechanism. Beyond this distance, the mechanism for fire spread is assumed to be radiation. Fire spread from building to building by radiation is dependent upon the:

➤ Size and intensity of the fire
➤ Distance between and building on fire and any neighbouring structures
➤ Extent of the building surface capable of transmitting heat.

The risk of fire spread by radiation is affected by a number of factors relating to space separation between buildings, the construction of the external walls and the combustibility of roof coverings.

Figure 9.19 External fire spread

Space separation
When planning a building or assessing the fire risk of an existing building the distance between it and the next nearest building should be calculated using the distance to the boundary of the property. If this is not appropriate, for example when two buildings are on the same site, a 'notional' boundary must be used which is normally half the distance between the two buildings. Although there may not actually be another building on that boundary at the time of planning or assessment it allows consideration of how much fire resistance and/or unprotected areas may be appropriate in the walls of the building being considered.

Construction of the external walls
The nature of the building materials and the numbers of windows or other openings obviously affect both fire spreading to and from a building. It is therefore necessary to ensure sufficient fire resistance of external walls and to limit the amount of the surface area which is not protected by fire resistance.

187

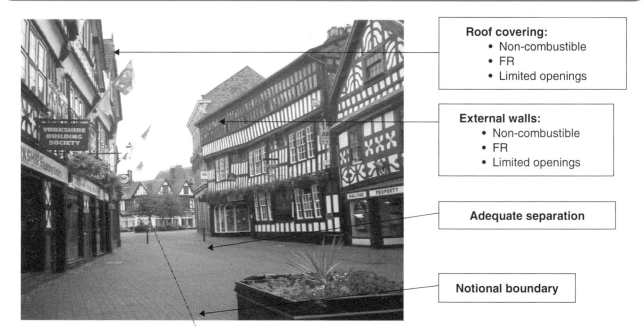

Roof covering:
- Non-combustible
- FR
- Limited openings

External walls:
- Non-combustible
- FR
- Limited openings

Adequate separation

Notional boundary

Figure 9.20   Preventing external fire spread

Fire resistance – the degree of fire resistance of the external building will vary with the size and purpose group of the building but as a minimum will normally provide fire resistance for at least 1 hour.

Unprotected areas

To prevent fire spread by radiation from one building to another, there would be no openings in the exposed face of the building. This would of course be impractical; however, the openings in external walls that present a risk to adjacent properties must be limited. Limiting the size of any openings has the effect of limiting the size of any source of radiated heat that may come from that opening. For example, a large opening will allow a large amount of flame to develop outside the opening which will in turn radiate large qualities of heat. The smaller the opening, the less the flame, the smaller the amount of radiated heat.

The combustibility of roof coverings

Any surface that is pitched at an angle of less than 70° from the horizontal should be considered as a roof. The most significant risk of fire spread via the roofs of buildings is not only radiated heat but also flying embers of burning material settling on the roof. It is important therefore that the roof covering offers an appropriate degree of fire resistance. As with the construction of external walls the amount of fire resistance and the amount of unprotected openings (e.g. roof lights) will

depend upon the space separation between the roof and any boundary likely to present a risk.

9.5   Means of escape

Means of escape is defined by the Approved Document as a 'structural means whereby a safe route is provided for persons to escape in case of fire, from any point in a building to a place of safety, clear of the building, without outside assistance'.

This definition provides a basis for solutions to fire safety planning in buildings and other structures. However, there are a few prescriptive rules governing the provision of the means of escape in particular situations. It is useful to bear in mind that the means of escape provision varies from building to building depending upon such factors as the nature and size of the building or structure, the use of the building and the nature and disposition of the occupants. For example, the means of escape in a traditionally built multi-storey hotel will need to be of of a far higher standard than the means of escape from a single storey office.

When considering what standard of means of escape may be appropriate in a given situation it is necessary to consider the factors relating to the:

➤ Evacuation time
➤ Evacuation procedures
➤ Occupancy

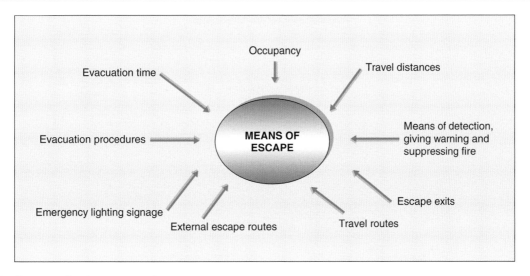

Figure 9.21 Factors affecting means of escape

➤ Travel distances
➤ Escape exits
➤ Travel routes
➤ External escape routes
➤ Emergency lighting signage
➤ Means of detection, giving warning and suppressing fire.

It is important to understand how quickly it is necessary to evacuate people from a building, not least because it allows a realistic calculation of the numbers and width of exits required to achieve evacuation within the time available.

In published document PD7974 the British Standard Institute suggests that evacuation time can be seen as the sum of a number of key elements including the time taken from ignition to detection and from detection to the sounding of an alarm. When planning buildings it is important for architects to understand how quickly they need to evacuate people in case of fire. PD7974 describes this as the required safe escape time (RSET). The RSET includes time taken to:

➤ Detect the fire
➤ Sound the alarm
➤ Occupants to recognise the alarm
➤ Occupants to respond to the alarm
➤ Travel time.

It can be seen from Figure 9.22 that the safe evacuation time is divided into free movement time and travel time.

The overall available time for occupants of any building to escape is described as the available safe escape time (ASET). The difference between RSET and ASET is the margin of safety between the time when evacuation is complete and conditions within the building become untenable.

The available safe escape time is therefore the time from ignition to complete evacuation of the building. The time it actually takes a person to move from any occupied part of the building to a place of ultimate safety will vary considerably due to the physical condition of the occupants as well and the distance they have to travel before reaching safety.

It is sometimes useful to consider the ASET in four simple phases:

Phase 1 – alert time from fire initiation to detection/recognition
Phase 2 – pre-movement time taken by behaviour that diverts an individual from the escape route(s)
Phase 3 – travel time to physically get to an exit
Phase 4 – flow time, i.e. how long it takes for the occupants to move through the various stages of the escape route. Doorways are invariably the least efficient element with the longest flow time and restriction on the route.

Research has shown that Phase 1 can be as long as two-thirds of the total evacuation time required. The most important variable to consider when assessing means of escape provision is the construction of the building. If the building is less likely to burn or allow rapid fire spread the occupants will have more time to evacuate and therefore a better chance of escape. (Chapter 10 explores reasons why people may delay reacting to a fire alarm.)

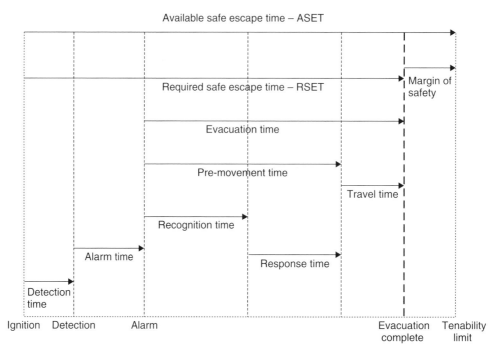

Figure 9.22 Available safe escape time

Generally accepted evacuation times are:

Class 'A' construction – 3 minutes
Class 'B' construction – 2.5 minutes
Class 'C' construction – 2 minutes.

These are not rigid, and can be increased or reduced according to the circumstances of each building, but they give a reasonable benchmark for most situations. Certainly anyone managing a building or assessing the fire risks associated with a building will normally need to ensure that evacuation to a place of safety for all occupants is achieved in under 3 minutes, except in the most exceptional case when the safety of the occupants during evacuation is ensured by additional engineered control measures such as pressurised staircases or the provision of refuges.

9.5.1 Evacuation strategies and procedures

As a part of fire risk assessment for non-domestic buildings, the effectiveness of the procedure for evacuation will need to be assessed. There are a number of strategies that can be adopted to ensure the safe evacuation of buildings. The strategies depend upon the magnitude and nature of risk from fire. In general there are two distinct strategies:

➤ Total evacuation
➤ Progressive evacuation.

Total evacuation strategy

It is most common for smaller, simple buildings to adopt a strategy that results in the total evacuation of the building. In these cases all of the occupants of the building will immediately move to a place of ultimate safety outside the building at risk, by a process of either simultaneous evacuation or phased evacuation.

Total simultaneous evacuation – the most common way to achieve simultaneous evacuation is that upon the activation of a single call point or detector an instantaneous warning from all fire alarm sounders is given as the signal for an immediate evacuation. This is referred to as a single stage evacuation.

However, even when it is considered to evacuate the whole building simultaneously there are occasions where an immediate wholesale, uncontrolled evacuation would present its own, significant risks. For example, in a theatre or other place of public entertainment it is important that an evacuation is achieved without creating panic among the audience who may be seated in dark unfamiliar surroundings. In this case and in some other circumstances it will appropriate to adopt a two-staged evacuation.

In a two-staged evacuation, there is an investigation period before the fire alarm sounders are activated. This approach is adopted for facilities such as halls of residence (student accommodation) and places of public entertainment. A staged evacuation ensures that confirmation of a fire can be verified prior to running the risk of total uncontrolled evacuation. Typically the sequence of events for two-staged evacuation is as follows.

1. An exclusive alert is given to staff (this may be coded or a discreet visual or audible signal or message)
2. The situation is then investigated and the presence of a fire verified (this is a limited period, and, if verification cannot be confirmed in the prescribed time, full evacuation will commence)
3. The evacuation signal is then broadcast and simultaneous evacuation commenced if:
   ➤ A fire is confirmed or
   ➤ The agreed investigation and verification period lapses without the alarm being cancelled or
   ➤ Confirmation of a fire is received from more than one source.

Total phased evacuation – phased evacuation is a common approach that is adopted in high-rise premises, where the floors are separated by fire resistant construction. It may also be adopted in a similar way to the progressive evacuation for hospitals and residential care homes (see below). In these cases simultaneous evacuation would be impractical due to the numbers of people involved, the travel distances and the limited numbers of stairs and exits.

In a phased evacuation it is common that the first people to be evacuated are all those on the storeys immediately affected by the fire, usually the fire floor and the floor above; the remaining floors are then evacuated, two floors at a time, at phased intervals.

In order to safely adopt a phased approach it is necessary to provide and maintain additional fire protection measures and supporting management arrangements. It is often the case that those with physical impairments, either temporary or permanent, are moved to a place of relative safety as part of the personal emergency escape plan (PEEP) (see Chapter 10).

Progressive evacuation strategy
As an alternative to total evacuation, it is sometimes necessary to consider a progressive evacuation strategy. Progressive evacuation is typically adopted in hospitals where fire-resisting compartments are provided to allow the evacuation of non-ambulant patients on a phased basis whereby they are moved into adjoining compartments prior to assisted removal from the building.

There are two types of progressive evacuation procedures that can be adopted:
Progressive horizontal evacuation – progressive horizontal evacuation is the process of evacuating people into an adjoining fire-resisting compartment on the same level, from which they can later evacuate to a place of ultimate safety. For example, when evacuating hospital patients from surgical theatres, they will be initially moved to an adjacent fire compartment within the building prior to an eventual removal, if necessary.

Zoned evacuation – a 'zoned' evacuation is achieved by moving the occupants away from the fire affected zone to an adjacent zone. An example of this would be a shopping centre where the occupants would be moved to the adjacent smoke control zone while the fire affected zone was brought under control.

9.5.2 Occupancy

One of the most significant factors that influence the design standards of the means of escape in any type of building regardless of the method of evacuation adopted is the nature, distribution and mental and physical state of the occupiers. Approved Document B provides guidance on appropriate aspects of means of escape based on purpose groups.

These purpose groups take into account the types and density of the occupants that should normally be planned for in various buildings.

> In schools the occupants are assumed to be awake, in the main young and able bodied, subject to discipline and very familiar with the building they may need to escape from. Whereas in a hotel, the occupants are often asleep, not disciplined, potentially under the influence of alcohol and not familiar with the building that they may need to escape from. For this reason the physical condition, state of consciousness and their expected reactions must be considered at the design stage of any building. Moreover, the means of escape should be reviewed at the time when any alterations are made to a building.

Appendix 9.1 summarises the purpose groups that are used in ADB. It can be seen that the categorisation into purpose groups takes account of the physical condition, state of conciseness and expected reactions of the occupancy of a building. However, occupancies vary throughout the life of a building and it is important that these factors are considered when developing and managing evacuation procedures once the building is occupied.

When considering the occupancy of a building and how it may impact upon the requirements for the

Figure 9.23 The occupancy of sports stadium

provision of the means of escape thought must be given to the following factors:

➤ Number of occupants
➤ Distribution
➤ Density.

Number of occupants

The number of occupants for an existing building with a reasonably fixed population may be ascertained by questioning the responsible person who owns or occupies the building. For buildings such as theatres or cinemas, the number of seats provided should be counted. However, there are many buildings where it is necessary to assume or estimate the total number of occupants.

It is important to consider the totals that may be reasonably expected, for example in school halls although the normal use may be for sports where small numbers are present the space may also be used occasionally for assemblies or fundraising social events where there may be large numbers of pupils and/or parents attending.

Distribution

The distribution of occupants within a building is also critical to the safety of occupants when assessing escape routes. In addition to the total numbers of people who may occupy a building, where those people are located within the building has a significant affect on the requirement for their means of escape. For example, the means of escape from a retail premises situated at ground floor level will be of a lesser standard than if the premises were situated in a basement, or on the second floor of a department store.

Consideration must also be given to those people who work in remote parts of buildings such as storerooms, boiler rooms or lift plant rooms. When assessing the means of escape for the building a competent assessor will ensure the escape arrangements for all persons in the building are considered despite their location throughout the building.

Distribution is clearly linked with the purpose group and exact type of premises, e.g. factory, shop, etc., and the nature and type of the occupancy of the building.

Density factor

When planning the means of escape during design (unoccupied premises), or for those with variable populations such as places of public entertainment, it may be difficult to calculate the maximum numbers who will be present. In addition, there will be occasions when it will be necessary to determine the maximum numbers of people who should be allowed to occupy a particular space in any particular purpose group.

In order to achieve a realistic calculation as to how many may or should be occupying a certain space reference must be made to 'density factors'. The density factor may be defined as 'the available floor space per person'. The Approved Document B provides numeric values to assist in calculating how many persons may be present (see Table 9.5). In order to calculate the density factor for a particular building or area, it is necessary to know:

➤ The use
➤ The floor space.

To assess the density factor, the *usable* floor space must be calculated. The usable floor space *excludes permanent features*, i.e. stairs, toilets, lifts, escalators, corridors and other circulation spaces.

Consider the example in Figure 9.24.

Because the parts of the building are put to different uses, it is necessary to use different density factors.

In order to calculate the number of people, it is first necessary to divide the floor area by the floor space allowed per person, i.e.:

If density factors for Figure 9.24 are:

$$\text{No. of people} = \frac{\text{area of room or storey}}{\text{floor space per person}}$$

(a) Shop floor (main sales) – $2\,m^2$ per person
(b) Office – $6\,m^2$ per person
(c) Dining room and canteens – $1\,m^2$ per person

Table 9.5 Density factors given in Approved Document B

Type of accommodation	Density factor m²
Standing spectator areas, bars without seating and similar refreshment areas	0.3
Amusement arcade, assembly hall (including a general purpose place of assembly), bingo hall, club, crush hall, dance floor or hall, venue for pop concert and similar events	0.5
Concourse or shopping mall	0.7
Committee room, common room, conference room, dining room, licensed betting office (public area), lounge or bar (with seating), meeting room, reading room, restaurant, staff room or waiting room	1.0
Exhibition hall, studio (film, radio, television, recording)	1.5
Shop sales area, skating rink	2.0
Art gallery, dormitory, factory production area, museum or workshop	5.0
Office	6.0
Shop sales area, kitchen, library	7.0
Bedroom or study bedroom	8.0
Bed-sitting room, billiards or snooker room or hall	10.0
Storage and warehousing	30.0

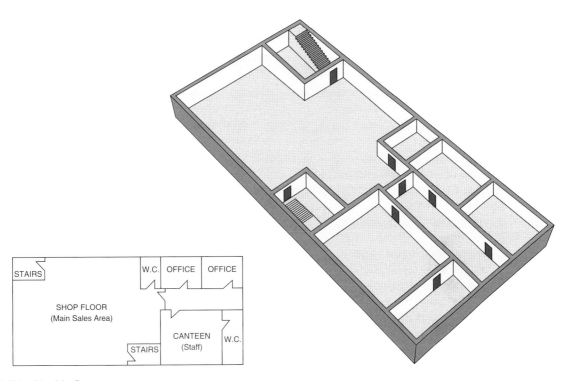

Figure 9.24 Usable floor space

and the relevant floor areas are:

(a) shop floor – 20m × 30m = 600m²
(b) offices – 8m × 4m = 32 × 2 = 64m²
(c) canteen – 8m × 10m = 80m².

This is the *total* floor area – not the usable floor area. It is therefore necessary now to deduct the floor area of the permanent features in the area where they are situated, i.e.:

(a) WC in shop – 10m²
(b) Stairs in shop – 15m² × 2 = 30m²
(c) WC in canteen – 10m².

Therefore, the total theoretical number of people for which escape routes must be designed is:

(a) Shop $= \dfrac{560}{2} = 280$

(b) Offices $= \dfrac{64}{6} = 11$

(c) Canteen $= \dfrac{70}{1} = 70$
    Total = 361 people

Figure 9.25 A room with a final exit to an unenclosed area outside

The number, width and disposition of exits can then be provided and should be designed to allow 361 people to escape within the required time.

As an alternative to calculating a theoretical number is it perfectly acceptable to use real data from real situations. In this case the data that is used to determine the requirements of the means of escape should be based on the highest average occupancy density.

9.5.3 Travel distances

Having first considered the physical state and reactions of the occupants and how these can be related to build-ing type, it is important to look at the process of escape and the maximum distances people can be expected to travel to escape from fire.

Obviously the distances people travel to get to safety has a direct bearing on the time taken to evacuate in the case of fire. It is important to remember that travel distances will vary depending upon the type of building, the use of the building and the occupancy of the building.

Two crucial notions to help understand and assess travel distances are that of a 'room of origin' and a 'final exit'. The room of origin is the term applied to the room where a fire may break out. For all buildings, all rooms may be considered as rooms of origin and although the likelihood of a fire occurring in some rooms is greater than

in others for the purposes of assessing the adequacy of the means of escape, the travel distances from all rooms must be considered.

In some cases it will be possible to escape directly from a room where a fire may start to a final exit to a place of ultimate safety which will normally be an open area in the open air. In these cases the only distance that is relative is the distance within the room.

Stages of travel – in most building designs it is neither practical nor desirable to have final exits from all the rooms where a fire may start. For most situations it is necessary to consider the process of escape in four distinct 'stages':

Stage 1: escape from the room or area of fire origin
Stage 2: escape from the compartment of origin via the circulation route to a final exit / entry to a protected stair-way / to an adjoining compartment offering refuge
Stage 3: escape from the floor of origin to the ground level
Stage 4: final escape at ground level.

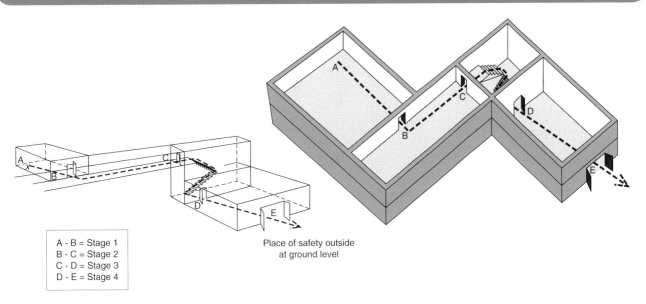

A - B = Stage 1
B - C = Stage 2
C - D = Stage 3
D - E = Stage 4

Place of safety outside
at ground level

Figure 9.26 Component parts of an escape route

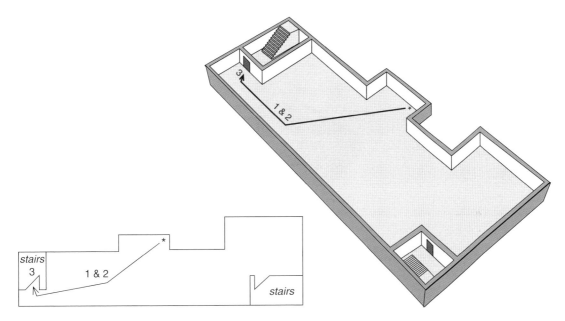

Figure 9.27 The stages of escape in open planning

Figure 9.26 illustrates the four stages of escape in a common traditional building layout. Whereas Figures 9.27 and 9.28 show how the various stages of escape in open planned and cellular planned internal layouts are applied.

In buildings with simple internal layouts, the stages are relatively easy to identify. However, in more complex buildings where there may be phased evacuation, for example, Stage 2 may only be to a place of safety on the same level, referred to as a 'refuge'.

In high-rise buildings it may not be practicable or desirable to commence total evacuation so the design must allow for a phased evacuation. Structural elements and internal fire systems must have sufficient resilience to restrict fire spread, so that people most at risk can

evacuate first, while others can be evacuated later, once the fire fighting teams arrive.

This type of scenario clearly demonstrates the interaction of escape, containment and extinguishment, with communication in the central role. Phased evacuations may go through Stages 1 and 2, with Stage 3 held in reserve, or Stages 1 to 4 (the total evacuation) of some occupants while others are put on 'alert' in readiness for evacuation and moved to safety when practicable.

Stage 1 Travel – from the room of origin

When considering escape from a room, the speed of fire spread needs to be considered and compared with

195

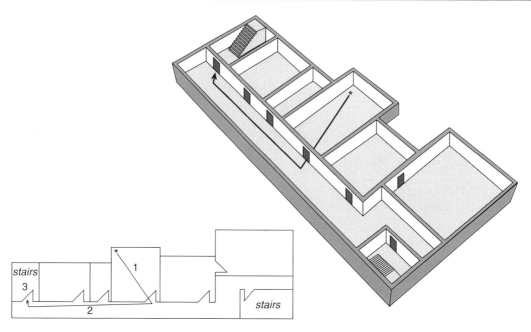

Figure 9.28   The stages of escape in cellular planning

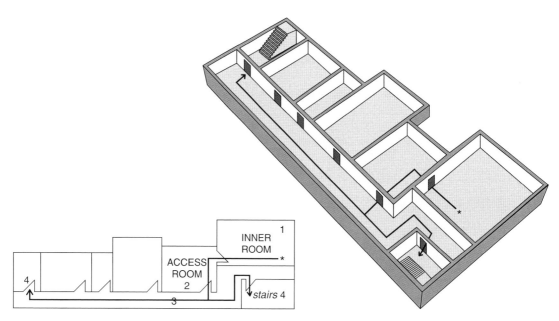

Figure 9.29   Inner room and access room

the speed the occupants can leave. Because it can be difficult to predict the rate of fire growth spread, it is important to ensure that the occupants of the room become aware of the fire as soon as possible.

In a large room more than one exit may be required, so that the occupants are never too far from an exit. The maximum distance they should have to travel is referred to as the 'Stage 1 Travel Distance'. The actual distance

in metres will vary depending upon the nature of the building and its occupants.

There are circumstances where immediate access to an exit route is not possible and egress from an 'inner room' has to be made via an 'access room' – see Figure 9.29.

In this case both rooms (1 and 2) are treated as part of the Stage 1 escape, and it is necessary to ensure that

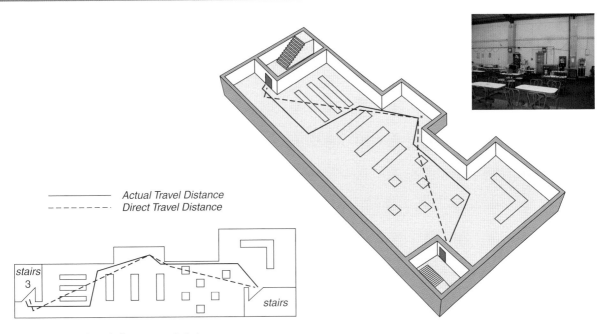

Actual Travel Distance
Direct Travel Distance

Figure 9.30 Actual and direct travel distances

the occupants of the inner room are aware of conditions in the access room. Commonly this is achieved by providing glazing in the wall and door separating the two rooms, or providing automatic fire detection in the access room. It may also be necessary to restrict the use of the access room to reduce the risk of outbreak from fire, which would prevent the escape from the inner room. For example, egress from an inner room should not be through a high risk area such as a kitchen or workshop.

Stage 2 Travel – from the compartment of origin to a place of relative safety

The next stage of the escape process is from the compartment where the fire ignited. This is usually via a circulation route leading to:

➤ A final exit
➤ A protected stairway or
➤ An adjacent compartment that provides a refuge.

The occupants need to reach a place of safety, either by leaving the building completely or by being isolated from the fire by effective fire containment measures and fire resisting construction.

The building should be designed so that any compartmentation affords the occupants sufficient time to escape from the compartment of origin before being overcome by fire and smoke. However, it is difficult to calculate precise times because people, buildings and fires vary, and most codes and legislation specify distances based on experience of past fires and past experiences.

Figure 9.31 Extended travel distance due to fixed chairs, tables, etc.

A designer therefore must work from first principles and make an assessment of the life risk and feasible travel distances.

Table 9.6 gives some typical values for travel distances that can be applied at a preliminary design stage, or where there is no controlling legislation or guidance.

The travel distances must be calculated to take account of the internal layout for furniture and fittings.

Special risks associated with very tall buildings (i.e. over 10 storeys) or deep basements (i.e. more than one level) require special attention.

Table 9.6 Suggested total actual travel distances

Purpose group	Escape – single direction	Escape – more than one direction
Residential institutions (care homes, etc.)	9 m	18 m
Hospital wards (high dependency patients)	15 m	30 m
Hotels, flats other residential	18 m	35 m
Offices	18 m	45 m
Schools (with seating – halls, etc.)	15 m	32 m
Shops	18 m	45 m
Factories and warehouses	25 m	45 m
Assembly (primarily for disabled persons)	9 m	18 m

Table 9.7 Minimum number of exits

No. of occupants	No. of exits
1–60	1
61–600	2
601+	3

Table 9.8 Minimum widths

No. of occupants	Width of exits
1–60	750 mm
51–110	850 mm
111–220	1050 mm
221+	1050 mm plus 5 mm/person
NB wheelchair users require a minimum of 900 mm width	

In addition to the requirement to limit travel distances, it is also necessary to consider the flow of people through the exits from buildings. The Approved Document B provides guidance on the number and widths of exits that are required to provided an adequate means of escape.

In practice this means that when planning or assessing the means of escape for a particular situation, there must be adequate numbers and widths of exit as well as adequate directions and distances of travel.

Minimum number of escape exits
In terms of the numbers of exits that are required, this is based on a simple correlation between the numbers of people likely to need to escape and the number of doors.

The figures within the tables should be used in conjunction with the travel distance figures and should also be used with the same caution.

Minimum widths of escape routes and exits
In terms of the width of the escape exits this again is a simple correlation between the numbers of persons who are likely to make use of the exits and the width of the doors and corridors.

Although it is important to consider the number and widths of exits when considering the travel distances (particularly when considering the adequacy of alternative routes that result in increased travel distances), it is also important to consider the appropriate evacuation time.

It is also essential to take into account those that may need to make use of wider exits, corridors and open plan offices with designated routes, such as those using wheelchairs, walking aids, etc.

The use of refuges
Stage 2 of an escape route normally ends when the occupants reach an adjoining fire resisting compartment for refuge, or a protected stairway which leads to ground level. Descent to ground level is Stage 3 travel, but this may not take place immediately. It may be more appropriate for people to take refuge in another part of the building separated from the fire by a series of compartment walls.

The Codes of Practice for escape for disabled persons and guidance on homes for the elderly, and the guidelines for new and existing hospitals (NHS Firecodes),

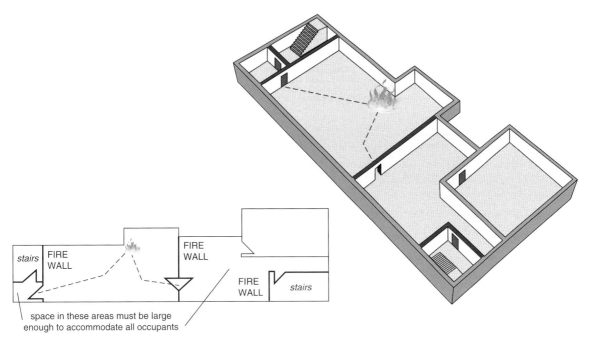

Figure 9.32 Concept of 'refuge'

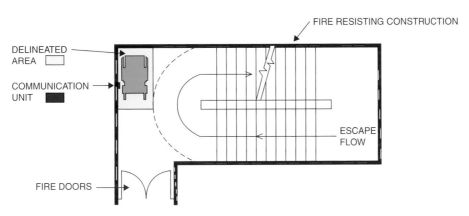

Figure 9.33 A refuge in a stairway

all recognise the importance of the concept of a refuge (see below).

At its simplest, it means that the disabled, infirm or those of limited mobility will only have to move a short distance within the building to another fire resisting compartment or protected zone to await further evacuation.

In the case of disabled persons in shops and offices, the concept is limited to the provision of fire-protected 'refuge areas' on each floor level adjoining lifts and stairs which serve the building.

In other premises, buildings are subdivided into subcompartments, which afford fire containment and reception areas for those evacuating as illustrated above.

In addition to the provision of a fire resistant lobby area (or similar) with a safe route from the storey exit, a

satisfactory refuge may also be an area in the open air such as:

➤ Flat roof
➤ Balcony
➤ Podium.

Each of which must be protected or remote from the fire; they must also both have a satisfactory means of escape.

Stage 3 Travel – out of the floor of origin

If the occupants of a building are on a level other than the ground floor, the next stage is vertical escape travel to ground level. Even if the evacuation plan involves the use of refuge and fire resisting compartments on the floor of origin, vertical evacuation may still be necessary if the fire is not brought under control.

The advantage of Stage 3 travel is that those escaping should by this stage have reached a position where they are protected from fire for a period of 30 minutes or more. They should not encounter further fire hazards en route to ultimate safety.

For most of the occupants vertical evacuation will be by staircases. Normal accommodation access lifts should never be used in a fire situation as the occupants may become trapped or be taken to a floor that is at risk from the fire itself.

For disabled people it may only be possible to evacuate using a specially designed lift on an independent electrical supply, with the potential for fire warden control and communications between the lift car and the evacuation control point.

Stage 4 Travel – final escape at ground level

Stage 4 travel is from the foot of the staircase to the outside. The stairs should not all converge into one common area at ground level, otherwise a single incident can block all escape routes.

The final exit and external design of a building also have to be considered in escape planning. It must be possible to leave the building and to get to a safe distance. The volume of people that may escape from a building governs their need for a readily identifiable assembly point or transfer area. Where large numbers of people may be involved, it will be necessary to plan these areas, so that access for the emergency services is not compromised.

9.5.4   Escape exits

Many assessments of rate of flow of persons through exits have been made following tests and simulation, particularly those using Paris firemen in 1938 and 1945 and on the London underground system.

The generally accepted rate is 40 persons per minute per unit (0.75 m) exit width. This is an average figure and the actual figures varied between 20 and 170.

Exit widths – the width of one unit corresponded to the average shoulder width and was determined to be 525 mm. Two units of 525 (i.e. 1050 mm) was the requirement for two people to travel through, shoulder-to-shoulder.

However, there will be overlapping between groups of people and so further units of width required need only be an additional 450 mm.

Hence, three units of exit width = 525 + 525 + 450 = 1500 mm and four units would be 525 + 525 + 450 + 450 = 1950 mm.

Calculating exit capacity

Where two or more exits are provided from either a storey exit or on ground floor, it should be assumed that one of the exits may be compromised by fire thus preventing the occupants from using it. It is therefore essential that the remaining exit or exits are of sufficient width to allow all persons to escape in the available time.

Therefore when considering the adequacy of the means of escape for a building it is necessary always to discount the largest available exit in any given situation.

For example, if a social club room is expected to accommodate 150 people it will need to have two separate exits of a minimum width of 1050 mm each, or three exits of 850 mm each. It can be seen in this example that the capacity of the exits after one of the largest exits has been discounted is still sufficient to comply with the guidance in the ADB regarding exit widths.

Alternative escape routes – the 45° rule

When considering the adequacy of the exit capacity it is also necessary to understand the term 'separate exits'. Exits can only be regarded as separate if, from any point in the room or space in question, they are at least 45° apart.

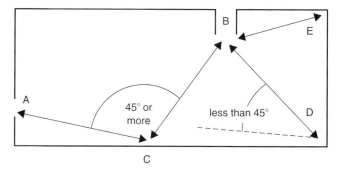

Figure 9.34   The 45° rule applied in rectangular space

Alternative escape routes need to be sited to minimise the possibility of all of them being unavailable at the same time. Alternative escape routes should be either:

(a)   45° or more apart (see Options 1 and 2 below) or
(b)   If less than 45° apart, separated from each other by fire resisting construction.

Option 1 – alternative routes are available from C because angle ACB is 45° or more, and therefore CA or CB (whichever is the less) should be no more than the maximum distance for travel given for alternative routes.

Alternative routes are not available from D because angle ADB is less than 45°. There is also no alternative route from E.

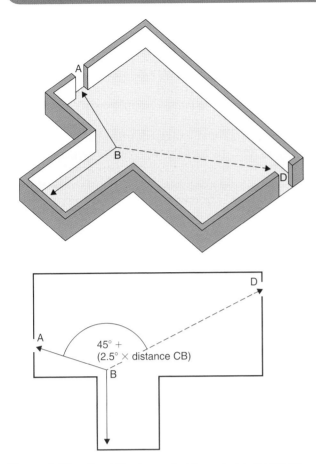

Figure 9.35 The 45° rule applied to an irregular space

Option 2 – travel distance in dead-end condition – C–B.

Angle ABD should be at least 45° plus 2.5° for each metre travelled from point C.

CBA or CBD (whichever is less) should be no more than the maximum distance of travel given for alternative routes, and CB should be no more than the maximum distance for travel where there are no alternative routes.

Doors on escape routes

The time taken to pass through doors while making an emergency evacuation can be critical. There have been many cases where the death toll in fires has been significantly increased as a result of problems with the doors on escape routes. All doors on escape routes should therefore be readily openable at all times when the building is occupied.

To ensure that doors on escape routes do not cause unnecessary delays they must satisfy the following criteria:

Door fastenings – escape doors should either not be fitted with any locking device, or where security is required they should be fitted with simple fastenings that can be easily operated from the direction of escape.

The fastenings must be operated without the use of a key.

It is permissible to fit a door with a simple turn latch on the side from where escape may be needed and for the latch to be operated by a key on the other side, e.g. a hotel bedroom.

The method of operating doors on an escape route must be indicated with 'how to open' instructions that comply to the Health and Safety (Safety Signs and Signals) Regulations 1996.

Direction of opening – if it is reasonably practical, all doors on exit routes should be hung so that they open in the direction of escape.

If it is likely that more than 60 people will use the door, or it provides egress from a very high risk area with the potential for rapid fire spread or is at the foot (or near the foot) of a staircase, the door *must* open in the direction of escape.

The same requirement must also be applied if the door is on an exit route in a building used for public assembly, public entertainment or conference centre/hall.

Scope of opening – all doors on escape routes should open to a minimum angle of 90° and should swing clear of any changes in floor level.

Any door that opens onto an escape route should be recessed so as not to obstruct the escape route.

Vision panels – vision panels are required where doors are provided to subdivide long corridors (also aids compliance with the WHSW Regulations).

Revolving and automatic doors and turnstiles – revolving doors, automatic doors and turnstiles *must not* be used as escape doors unless:

➤ They are of the full width of the escape route
➤ They fail to safe and open outward from any position of opening
➤ They fail to safe in the event of power failure.

There are sufficient non-revolving, automatic or turnstile doors immediately adjacent.

Securing mechanisms for emergency escape doors

In some circumstances securing mechanisms other than panic devices may be fitted to emergency escape doors. The securing mechanism must be capable of being easily and immediately opened from the inside without the use of a key to facilitate prompt escape.

There are a number of products available that meet this requirement; some of the more common are described below.

Panic devices – include a mechanism consisting of a bolt head or heads which engage with a keeper in the surrounding door frame or floor for securing the door when closed. The device can be released when the bar (positioned horizontally across the inside face of the door) is moved anywhere along its length in the direction of travel in an arc downwards.

Escape mortice deadlocks – this type of mechanism is widely used. On the inside there is a simple turn snib which unlocks the bolt and on the outside the mechanism is key operated. When the door remains unlocked a door closer will be required to hold the door in the closed position.

Cylinder mortice deadlocks – this type of mechanism (Yale) is satisfactory providing the cylinder has a thumbturn, which can be installed on the inside of the door. The fire authority does not recommend the type which is key operated on both sides for any class of emergency escape door, unless there is a separate method of releasing the lock in an emergency.

Non-deadlocking mortice nightlatches – this type of mechanism is also suitable; a normal door handle, a push handle (Fig. 9.37) or a thumbturn provides an easy means of escape. Care should be taken to avoid

Figure 9.36 An example of a panic device

Figure 9.37 Example of a push handle

specifying any type that permits the handle on the inside to be locked. Some models allow the latchbolt to be retained in the unlocked position. If this facility is used a door closer may be necessary to hold the door closed.

Break glass tube/panel – this is simple to operate in an emergency. The bolt securing the door in the frame is held in position by the glass tube or panel. The bolt is spring loaded so that immediately the glass is broken it withdraws from the frame and leaves the door open. This type of device is usually supplied with a hammer on a chain that is fixed alongside it to ensure there is no need to search for an implement with which to break the glass.

The mortice version of this type of lock should not be used on doors which are fire resisting because the glass panel would adversely affect the fire performance of the door. Both types can be opened without the need to break the glass. The glass tube can be removed once a padlock is opened but access is only possible from the side of the door on which the bolt is mounted. The glass panel type can be fitted with a special strike plate that allows the door to be opened from either side by the use of the key.

The glass panel is not affected. Both models can provide emergency escape from both sides of the door. There are reservations about these devices and careful consideration should be given to their use. In situations where smoke has filled the area it would be difficult to locate the bolts, particularly if they are fitted at the top of the door. It is, therefore, recommended that if they have to be used they are fitted no higher than 1 metre from the floor level. This system is especially not recommended where the public have access.

Break glass locks – breaking the glass or smashing the fragile plastic dome allows access to a handle or turn. Both locks can provide escape from either side of the door. Some can be used by key holders as a pass door from either side of the door. Also by use of a special key the latchbolt can be retained in the withdrawn position allowing the door to be pushed open until such time as the latchbolt is released and the door closed. In

Figure 9.38 An example of a break glass tube

this situation it will be necessary to install a door closing device to retain the door in the closed position. As above this system should not be used where public have access.

Electromagnetic locks – these units provide a great amount of flexibility in the control of escape doors. They must be wired into an automatic fire warning/smoke detection system and they must have a local override such as a break glass call point adjacent to which operating instructions would be displayed. Free entry and exit can also be time controlled or operated from a remote station. This type of hardware is relatively easy to fit, reliable and, compared with panic bolts, fairly unobtrusive. It is simply an electromagnet which, while it is energised, holds the door closed. When the power is cut, which could be effected by the alarm system, the magnet releases and the door is free to open. If the door needs to be opened in the normal course of events, a key switch on the frame could turn off the current. The system must fail in the safe/open position.

> ### Keys in boxes
>
> The use of these products is not acceptable where emergency escape doors are concerned. Where they are installed it is not unusual to see the keys missing from the boxes. In the event of the need to escape from a smoke filled area it may not be easy to find the box, smash the glass, remove the key from among the broken glass, find the lock, insert the key and open the door. In view of the alternatives available there is no need to rely on this method of providing emergency exit. (There are some certificated premises with this arrangement, which is not retrospective.)

9.5.5 Protection of escape routes

The aim, when planning or assessing escape routes, is to comply with the requirements of the law and ensure occupants have adequate means of escape. Although the law is not prescriptive and requires only compliance to functional requirements, the guidance provided on how to achieve adequate means of escape is helpfully detailed and allows the easier planning or assessment of various means of escape. The guidance contains a number of specific principles, which should be complied with in most cases.

As is the case with all the guidance relating to fire safety it is not necessary to follow it exactly, provided that the functional requirement of achieving adequate means of escape is achieved in other ways.

The following key points with regard to escape routes should be considered:

➤ Access to storey exits
➤ Separation of circulation routes from stairways
➤ Storey/floors divided into different uses
➤ Storey divided into different occupancies
➤ The height of escape routes
➤ Separation of high risk areas
➤ Protection of dead-ends
➤ The subdivision of long corridors.

Access to storey exits – a storey, or floor, with more than one escape stair should be planned so that occupants do not have to pass through one stairway to reach another, although it is acceptable to pass through one stairway's protected lobby to reach another stair.

Separation of circulation routes from stairways – any stairway and any associated exit passageway should not form part of the primary circulation route between different parts of the building at the same level. This is because self-closing fire doors can be rendered ineffective as a result of constant use, or because some occupants might regard them as an impediment; and wedge them open or remove the self-closing devices.

The exception to this is where the doors are fitted with an automatic release mechanism operated in conjunction with the automatic fire detection and alarm system.

Storeys divided into different uses – where a storey contains an area ancillary to the main use of the building, for the consumption of food and/or drink, such as a canteen or mess room, then:

➤ Not less than two escape routes should be provided from each such area
➤ One of the escape routes should lead directly to a storey exit without entering the remainder of the storey, a kitchen or an area of special fire hazard.

Storeys divided into different occupancies – where a storey is divided into separate occupancies (i.e. where there are separate ownerships or tenancies of different organisations):

➤ The means of escape from each occupancy should not pass through any other
➤ If the means of escape includes a common corridor or circulation space, then either it should be a protected corridor, or a suitable automatic fire detection and alarm system should be installed throughout the storey.

Figure 9.39 Example of abuse of fire compartmentation by the use of door wedges

Height of escape routes – all escape routes should have clear headroom of not less than 2 m except in doorways. This is important not only to negate ceiling fittings causing obstruction to people escaping but also to provide a reservoir that will keep smoke above head height.

The separation of high risk areas – all rooms and areas containing high fire risks, for example boiler rooms, kitchens, tea rooms or rooms in which high risk processes are carried out, should be separated from the remainder of the building by fire resisting construction of at least 30 minutes and in the case of boiler rooms this should be increased to a minimum of 1 hour. Access to any such rooms should not be provided from doors that open directly onto a staircase used for means of escape.

The protection of dead-ends – in a dead-end situation occupants seeking to evacuate the building can only do so by passing doors to other rooms. If fire has broken out in one of the rooms through which the occupants in a dead-end need to pass there is a high risk that their route will be blocked by smoke (Fig. 9.40).

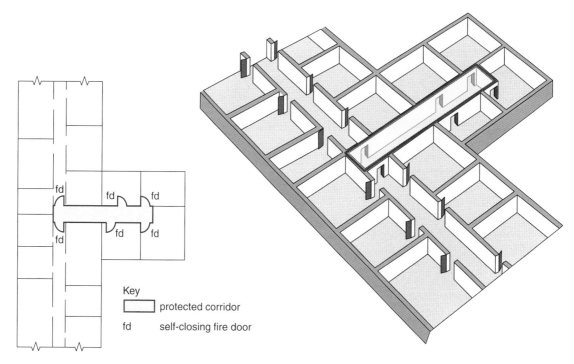

Key

☐ protected corridor

fd self-closing fire door

Figure 9.40 Protection of a dead-end

Therefore all dead-end corridors that are longer than 2 metres must be constructed of fire resisting construction including doors and door assemblies that provide at least 30 minutes' fire separation.

Subdivision of corridors – if a corridor provides access to alternative escape routes, there is a risk that smoke will spread along it and make both routes impassable before all occupants have escaped.

To avoid this, every corridor more than 12 m long which connects two or more storey exits should be subdivided by self-closing fire doors, so that the fire door(s) and any associated screen(s) are positioned approximately mid-way between the two storey exits. Any doors to the accommodation that would allow smoke to bypass the separating door should be self-closing.

Corridors connecting alternative exits are illustrated in Figure 9.41.

If alternative escape routes are immediately available from a dead-end corridor, there is a risk that smoke from a fire could make both routes impassable before the occupants in the dead-end have escaped. To avoid this:

➤ Either the escape stairway/s and corridors should be protected by a positive pressurisation system conforming to BS 5588-4 for smoke control or

➤ Every dead-end corridor exceeding 4.5 m in length should be separated by self-closing fire doors (together with any necessary associated screens) from any part of the corridor which:
 ➤ Provides two directions of escape or
 ➤ Continues past one storey exit to another.

When assessing the adequacy of a means of escape scheme the competent assessor will need to balance a number of factors as discussed elsewhere in this chapter. However, there are some situations that will be encountered where assessors will have to finely balance assessments of what is reasonable in any particular situation.

In order to assist these assessments, it will be helpful to consider characteristics of travel routes that are normally considered acceptable or unacceptable.

Acceptable means of escape – the following means of escape are generally deemed to be acceptable for all buildings:

➤ A door that leads directly to the outside, or *final exit*
➤ A door that leads to a protected staircase, or *storey exit*

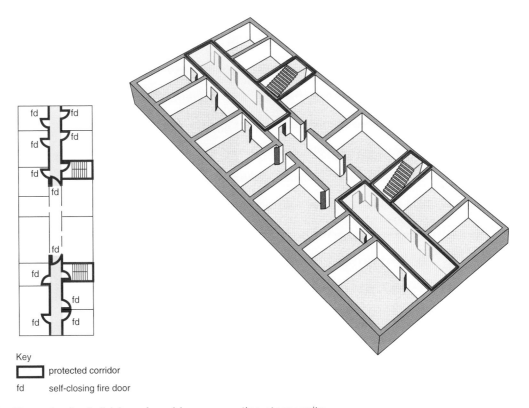

Key
▭ protected corridor
fd self-closing fire door

Figure 9.41 Example of subdivision of corridors connecting storey exits

➤ A door or doors that lead to another compartment enclosed by fire resisting construction – see *progressive horizontal evacuation* above

➤ An open staircase (accommodation stairs) where the length of the stair is included in the total travel distance

➤ Any ramps conforming to BS 8300, at a gradient of no more than 1:12

➤ Any moving walkways where the distance is included in the total travel distance and the walkway is designed not to stop suddenly

➤ Any wicket doors and gates (except from high risk areas), provided that:

> ➤ They are not to be used by members of the public

> ➤ Not more than 10 persons are expected to use them in an emergency

> ➤ They provide an opening at least 500 mm wide, with the top of the opening not less than 1.5 m above the floor level and the bottom of the opening not more than 250 mm above the floor level.

Generally unacceptable means of escape – the following are not normally acceptable as means of escape, but they can be used in certain situations where their reliability can be demonstrated to the appropriate authorities:

➤ Lifts, except for an appropriate evacuation lift for the evacuation of disabled people in an emergency

➤ Fixed ladders, except those in plant rooms which are rarely used and accommodate fewer than 10 able bodied people

➤ Portable ladders and throw-out ladders, e.g. rope ladders from upper rooms

➤ Appliances requiring manual operation/manipulation, e.g. fold-down ladders

➤ Power operated or manually operated sliding doors, except where they are designed to 'fail safe' in the open position on loss of power

➤ Security grilles and shutters (roller, folding or sliding), loading doors, goods doors, sliding doors and up-and-over doors, unless they are capable of being easily and quickly opened

➤ Wicket doors and gates at exits from high risk areas

➤ Escalators

➤ Wall and floor hatches

➤ Window exits.

Stairways

A fundamental aspect of any means of escape in multi-storey buildings is the availability of sufficient numbers of adequately wide, protected escape stairways.

Figure 9.42 Escalators are generally unacceptable as a means of escape

Figure 9.43 Typical modern escape stairway

Number of escape stairs needed is determined by the following:

➤ The number of people in the building
➤ Whether independent stairways may be required for mixed occupancies
➤ Whether a single stair is acceptable – in certain low rise (and low risk) buildings
➤ Width needed for escape – dependent upon whether simultaneous or phased evacuation strategies are employed.

In some larger buildings, additional fire fighting stairs may be provided for access for the fire and rescue service.

Protection of escape stairs – escape stairs should ideally be compartmented (enclosed), providing a minimum of 30 minutes' fire resistance, to ensure that they can fulfil their role as areas of relative safety during a fire evacuation.

However, it may be possible that an unprotected stairway (e.g. an accommodation stair) may be used as an internal escape route to a storey or final exit, provided that the travel distance and the number of people that may use it is very limited. This unprotected stairway should not link more than two floors and be additional to another stair being used for escape purposes; it should also not be able to compromise a dead-end route.

The provision of **fire rated lobbies or corridors** is used to provide additional protection to an escape stairway where it is the only stair in a building (or part of a building) of more than one storey (other than the ground floor) or the staircase extends above 18 m. The provision of lobbies and corridors are also required as part of phased evacuation strategy.

The protected lobbies or protected corridors should be provided at each level, except the top storey, including all basement levels or where the stair is a fire fighting stair.

A smoke control system may be considered as an alternative to the provision of any protected lobbies and corridors.

As in the case of escape routes generally a protected stairway needs to be free of potential sources of fire. Therefore, facilities that may be incorporated in a protected stairway are limited to toilets or washrooms, so long as the room is not used as a cloakroom.

A reception desk or enquiry office area at ground or access level may be included in an escape staircase enclosure if it is not in the only stair serving the building or part of the building and is limited to a maximum of 10 m².

Any cupboards or service riser shafts should be enclosed with fire resisting construction.

Special fire hazards – in the case of a building containing a special fire hazard, a protected lobby of not less than 0.4 m² should be provided between the hazard area and the escape stairway. In addition the lobby should have permanent ventilation, or a mechanical smoke control system to protect the area from the ingress of smoke.

Single staircase – ideally all buildings should have alternative escape routes and therefore a single staircase is not an option; however, a single staircase may be suitable for workplaces of low or normal fire risk (see Fig. 9.44). In such cases people on each floor should be able

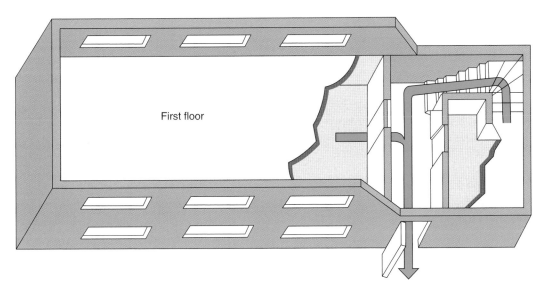

First floor

Figure 9.44   Protecting a single staircase

to reach it within the appropriate travel time. Any single staircase must:

➤ Be constructed to FR30 standard
➤ Serve no more than three floors above, or one floor below ground level
➤ Access the stair (other than top floor) by means of a protected lobby/corridor
➤ Be of sufficient width to accommodate the number of persons
➤ Lead directly to open air.

9.5.6 External escape routes

External escape stairs

Where more than one escape route is available from a storey or part of a building one or more of the routes may be by way of an external escape stair, provided that there is at least one *internal* escape stair from every part of each storey and that the external stair(s) meet the following recommendations.

If the building (or any part of the building) is served by a single access stair, the means of escape may be improved by the provision of an external escape stair provided that the following measures are incorporated:

➤ All doors to the external stair should be fire resisting and self-closing, except that a fire resisting door is not needed at the head of any stair leading downwards where there is only one exit from the building onto the top landing
➤ Any part of the external walls within 1.8 m of (and 9 m vertically below) the flights and landings of an external escape stair should be of fire resisting construction, except that the 1.8 m dimension may be reduced to 1.1 m above the top level of the stair if it is not a stair up from a basement to ground level.

Any part of the building (including any doors) within 3 m of the escape route from the stair to a place of safety should be provided with fire resisting construction

➤ Glazing in areas of fire resisting construction should also be fire resisting and should be fixed shut
➤ Where a stair is greater than 6 m in vertical length, it should be protected from weather conditions.

If more than one escape route is available from a storey, or part of a building, one of those routes may be by way of a flat roof, provided that all of the following conditions are met:

➤ The route does not serve a building where the occupants require assistance in escaping, or part of a building which is for public use

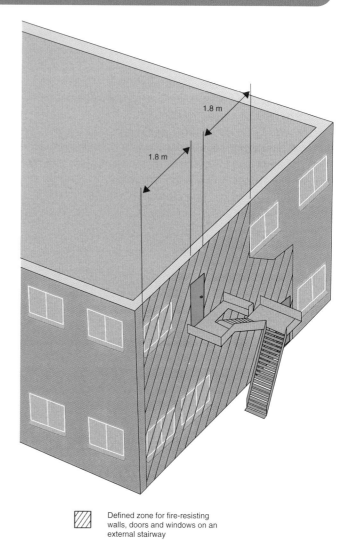

Defined zone for fire-resisting walls, doors and windows on an external stairway

Figure 9.45 Protecting an external escape stair

➤ The flat roof is part of the same building from which escape is being made
➤ There are no ventilation openings of any kind within 3 m of the escape route
➤ Any wall, including a door or a window in the wall, within 3 m of the escape route has at least 30 minutes' fire resistance for integrity from the inside – and there is no unprotected area below a height of 1.1 m measured from the level of the escape route
➤ Any roof hatch or roof light forming part of the roof within 3 m of the escape route has at least 30 minutes' fire resistance for integrity from the underside
➤ The route is adequately defined and guarded by walls and/or protective barriers
➤ The route across the roof leads to a storey exit or an external escape route.

Figure 9.46 Metal fire escape above non-FR glazing – note open window

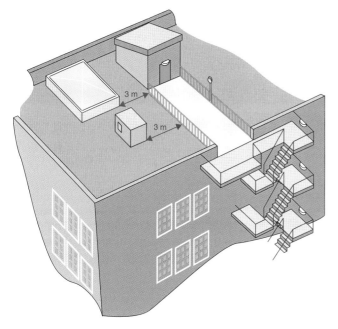

Figure 9.47 Protecting an escape across a roof

9.5.7 Escape facilities for disabled

People

Even with extended distances (where additional means of support are included), most disabled people can be expected to be able to reach a place of relative safety without assistance. In some cases, e.g. wheelchair users, people may be unable to negotiate stairs unaided. The following additional measures can be taken to aid the evacuation of disabled people.

Means of warning – alarm is commonly given by an electronic sounder, a mechanical bell, or by people shouting. Most people, including those who have impaired hearing, will hear such an alarm, or will be responsive to the subsequent actions of other people in the building.

There are circumstances where additional measures might be needed, for example in very noisy areas, or where a person might be isolated from others or asleep. In such cases visual alarms or personal vibrating pagers linked to the alarm system should be considered.

Escape time – many disabled people cannot move quickly. Disabled people may need to be escorted or even carried to a safe place, and they may need to rest for a while as they make their escape.

In order to minimise the amount of time taken to escape, the following matters should be considered:

➤ Clear signage indicating escape routes should be provided, maintained and kept free from obstruction
➤ Escape routes should be kept clear of obstructions.

Vertical escape – unless a lift or platform lift in a building is an evacuation lift, which is extremely unlikely, it must not be used for escape in case of fire because power failure could occur while it is in use, trapping any occupants. A lift that is not an escape lift should have signs at all landings advising people not to use it in the case of an emergency.

A stair lift, also being susceptible to power cuts, commonly obstructs the stairway when in use (and not folded away), and should therefore not be used for means of escape.

Particularly in an older building, a route provided only for means of escape might well include a stairway or other unsuitable feature.

The route of travel does not end at the external door but at a safe place away from the building. Sometimes it may be unrealistic for these external routes to be fully accessible, for example at the rear of a building where the floor level might be substantially lower or higher than adjacent ground level.

It follows that there will be circumstances where a disabled person would be unable to escape unaided and might need to find a place of relative safety to await assistance, or might simply need to rest for a while in a place of relative safety before continuing. In such circumstances, refuges should be provided.

Refuges – a protected stairway is an internal stairway intended for use as an escape route (it may also be in everyday use). The protection takes the form of an enclosure of fire resistant walling and doors.

A protected stairway, and a final exit from a building leading to an escape route to a place of safety where that route includes a stairway, should be provided with refuges (unless the building has no more than two storeys plus one basement storey, each under 280 square metres floor area).

A refuge:

➤ Should be provided at each point where an escape route leads into a protected stairway, or an external stairway
➤ Could be in the stairway, or in a protected lobby or corridor leading to a stairway. Where two protected stairways serve a storey, the whole storey could be effectively divided into two refuges using fire resistant walling and doors, giving two alternative refuges
➤ Could be in the open air on a flat roof or balcony, protected by fire resistant walling and flooring as necessary
➤ Where serving an external stairway, could be external, protected by fire resistant walling as necessary
➤ Should not obstruct the flow of other people escaping
➤ Should be at least 1400 mm × 900 mm, and perhaps larger depending on the number of wheelchair users it might need to serve.

Escape via an evacuation lift

Lifts should not be used for general evacuation, but they may be used for the evacuation of disabled people. If an evacuation lift is provided, it should be designed and installed in accordance with appropriate standards (BS 8300 and BS EN 8 1-70).

Evacuation lifts are generally placed in protected stairways. Where they are not in a protected stairway, there should still be a refuge available.

Figure 9.48 A typical sign indicating a refuge point

Evacuation by stairs

Evacuation of disabled people can take place on any evacuation stair if:

➤ The stair is not less than 900 mm in width
➤ The stair should be wide enough to manoeuvre a wheelchair around any bends.

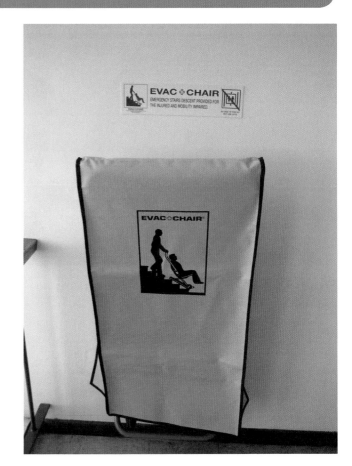

Figure 9.49 Evacuation chair provided to assist non-ambulant persons to escape in case of fire

The management plan of a building should specify the procedure necessary for carrying disabled people up or down stairs where this is required. Identified staff should have this responsibility and be given suitable training for the task.

9.5.8 Smoke control and ventilation

Protected means of escape including stairways provide a degree of smoke control within a building. In some cases it is necessary to provide additional smoke control to protect vulnerable stairways, for example a single stairway serving a number of upper floors. In this case it may be enough to provide lobbies at each landing to reduce the possibility of the stairway becoming smoke logged. In general there is no need to provide smoke ventilation in these circumstances; however, there are some occasions where ventilation is required to protect the means of escape.

In the case of large buildings with many occupants, e.g. shopping malls, smoke control and ventilation play a

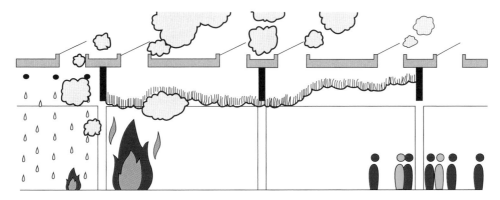

Figure 9.50 Sprinklers that moderate fire growth and smoke can be vented from ceiling reservoirs

major role in ensuring that the extended travel distances are kept free from smoke to allow safe evacuation.

Smoke controls are in general passive and comprise reservoirs provided at high level which allow the smoke from a fire to accumulate without affecting people within the mall. Once in the reservoirs the smoke can then be vented clear of the building.

Smoke ventilation can be achieved either naturally or mechanically.

Natural ventilation – is achieved by providing openings from the building to fresh air; the building or space is ventilated by the natural movement of the air. In order for this natural ventilation to be effective sufficient vents must be provided to allow air to flow in and out. The openings for natural ventilation can be permanent, as in the case of multi-storey car parks, manual, as is the case with openable windows in an escape stairway, or automatic, for example at the head of an escape staircase.

Mechanical ventilation – is achieved by either forcing air in, forcing air out or by a combination of both methods. When using mechanical means to ventilate smoke, it is obviously important to ensure the equipment is capable at operating at the temperatures that may be expected. A common use of mechanical ventilation is pressurisation. With this system the area that needs to be protected against smoke seepage is pressurised, often at the time of the actuation of a fire alarm. This pressurised area then establishes leakage of air through door seals and other gaps which prevent the backflow of smoke into the pressurised area. Some fire fighting stairways and other escape stairways have this additional protection.

9.5.9 Emergency lighting

There are broadly two types of emergency lighting; that which is provided for use when the normal supply

fails and emergency escape lighting, which provides illumination for the safety of people leaving a building to escape a potentially dangerous situation. BS 5266 Part 1: 2005 defines the acceptable standard for emergency lighting both if the normal supply fails and for people evacuating a premises.

Emergency lighting – in addition to emergency escape lighting it may be appropriate to provide other forms of lighting which will provide illumination if supplies fail. This lighting will enable manufacturing process to be shut down safely, assist people to find fire fighting equipment and fire alarm call points, or provide illumination of disabled toilets, etc.

Emergency escape lighting – emergency escape lighting is required along the escape routes from all places of work, including plant rooms and storage facilities where it is foreseeable that persons may be required to work and where there is insufficient ambient light to safely use the means of escape.

Figure 9.51 Emergency lighting unit

Requirements for luminance and response
The important aspect of an emergency system is that it provides sufficient lighting to enable the means of

Table 9.9   Provision of emergency escape lighting

Areas needing emergency escape lighting
Each final exit door including both internal and external routes
All escape routes including those that may assist persons making their way from the building to the assembly point
Intersections of corridors (should also assist in illumination of emergency exit signage)
Any change in direction or floor level, including temporary partitions, ramps, steps or ladders
All stairways to ensure each flight and landing is provided with adequate illumination
Windowless rooms and WCs that exceed 8 m²
Disabled WCs
Areas needing emergency lighting
Fire fighting equipment and fire alarm call points – fire points
Hazardous plant and equipment – needing isolation in the event of an emergency
Lift cars
Halls and other areas exceeding 60 m²
Fire alarm panels

escape, etc. to be used at all times. The amount of light that falls on a surface is referred to as 'luminance'. Luminance is measured in lux (lx). The standard that is normally considered acceptable for escape routes is between 0.2 and 1 lx.

For example, for routes that are normally clear as in the case of corridors and stairways, the minimum luminance on the centre line of the route at floor level should be 0.2 but 1 lux is considered preferable.

In all cases emergency lighting should activate within 5 seconds of the failure of power to the normal lighting circuit. The emergency lighting will then continue to operate for a specified period of time, typically 3 hours, but sometimes less.

Types of luminaries

The required level of luminance for **emergency escape lighting** can be provided in a number of ways using a variety of light fittings. Individual emergency lighting units can be self-contained, integrated into a standard light fitting or provided as an integral part of fire escape signage. The choice of fitting will be dependent upon the:

➤ Size and nature of the space to be illuminated
➤ Use of the building
➤ Type of occupancy.

For some larger buildings the power for emergency lighting may be provided by a generator or central battery system. In all other cases emergency lighting is provided by self-contained units. Emergency lighting luminaries are categorised according to the following aspects of their design:

➤ Whether they are illuminated under normal conditions (referred to as being maintained) or only operated on the failure of the normal power supply (non-maintained)
➤ Whether they form part of a normal light fitting (referred to as combined) or are free standing (satellite)
➤ Whether the emergency illumination is provided by the same bulb or filament that is illuminated under normal conditions (compound).

BS 5266 Part 1: 2005 categorises emergency lighting units by their type, mode of operation, facilities and duration of emergency lighting it will provide. Table 9.10 summarises the various aspects of the categorisation of emergency escape lighting units and their codes.

Using a combination of the codes and some prose suggested by BS 5266, individual emergency lighting units are now described by the various aspects of their design in relation to each of the categories. The full description of a particular unit can then appear as a code. This allows easy specification of units during the design stage of a system and also allows the characteristics of individual units to be easily identified once they have been installed. For example, a luminaire conforming to BS 5266:2005 that is self-contained, which only operates when the normal power supply fails, has a local test facility and is designed to operate for 3 hours will be described as in Figure 9.52.

The specification, design and installation of emergency lighting systems require competence in fire safety, physics and electrical engineering. Calculations relating to the level of lux falling on the centre line of the means of escape that take into account the natural luminescence of the surrounding building and decorating materials are complex.

It is of course possible to apply 'rules of thumb' to the provision of emergency lighting; however, once a system has been installed it must be certified by a competent engineer that it fully satisfies the requirements of BS 5266:2005.

Table 9.10 Categorisation of emergency escape lighting

	Type
X	Units that are self-contained, i.e. they have an integral rechargeable battery
Y	Units that are provided with a central supply
	Mode of operation
0	Non-maintained – not illuminated continuously
1	Maintained – illuminated continuously
2	Combined non-maintained – a luminaire containing two lamps, at least one of which is energised from the emergency supply not illuminated continuously
3	Combined maintained – a luminaire containing two lamps, which are maintained from the emergency supply
4	Compound non-maintained – self-contained unit that can provide power to a satellite unit not illuminated continuously
5	Compound maintained – self-contained unit that can provide power to a satellite unit illuminated continuously
6	Satellite – individual unit that is powered from a compound unit
	Facilities
A	The unit includes a local test device
B	The unit includes a device for remote testing
	Duration of operation in minutes
60	Has a duration of 1 hour
120	Has a duration of 2 hours
180	Has a duration of 3 hours

X	0	A this unit has a local test facility	180

Figure 9.52 An indicative label classification for emergency escape lighting

Once a system has been installed it is necessary to ensure that its continued operation is effectively monitored. Those who find themselves responsible for the means of escape within buildings will need to ensure that the emergency escape lighting system:

➤ Is certified as complying with BS 5266
➤ Is subject to routine testing and inspection:
  ➤ Daily to ensure maintained lamps are illuminated
  ➤ Monthly to ensure each unit activates on a failure of the power supply

Figure 9.53 Maintained emergency exit box

Figure 9.54 Maintained blade emergency light – double sided

Figure 9.55 Non-maintained satellite emergency light

➤ Six monthly to ensure that the unit maintains emergency operation for at least 1 hour
➤ Three yearly (and after the first three yearly test, annually) to ensure each unit operates for its full design period.

The monitoring of emergency lighting systems needs to form part of the overall management of fire safety.

9.5.10 Signs and signage

Fire safety signs and signing systems form an integral part of the overall fire safety strategy of a building and are fundamental to the communication of good fire safety management information. Clearly visible and unambiguous signage is essential for speedy escape, particularly in buildings where many of the occupants might be unfamiliar with the building layout.

All fire safety signs must comply with the Health and Safety (Safety Signs and Signals) Regulations 1996 or comply with British Standard BS 5499 Part 1. Signage in hospitals will also need to comply with the Health Technical Memorandum way finding (supersedes HTM 65).

Where a fire risk assessment identifies the need for a sign, the sign should be displayed prominently, conspicuously and appropriately having regard to the environment and occupancy profile of the building.

Fire safety signs should not be sited such that they are overridden with other types of public information or property management signs, and should be consistent in style and design throughout the building.

Signage will also need to be displayed that gives information and instructions relating to:

- Escape routes
- Fire doors
- The location of extinguishers
- The location of manual fire alarm call points
- The location of evacuation assembly points
- Fire action signage.

Escape routes

Every doorway or other exit providing access to a means of escape, other than exits in ordinary use (e.g. main entrances), should be distinctively and conspicuously marked by an exit sign.

All fire exit signs should be illuminated under normal conditions (signs that are not internally lit or backlit should be lit by primary or secondary lighting). Internally lit or backlit signs should remain illuminated in the event of power failure.

Photoluminescent signs will assist in providing clear visibility in all lighting levels, particularly those that are lower light such as in basements and plant rooms. The signs are manufactured with a material which absorbs and stores energy from daylight or artificial light and thus glow in low light/dark conditions.

Signs should be provided in stairways to identify the current floor and the final exit.

Figures 9.58 to 9.59 some of the more common safety and fire safety signs for use in buildings.

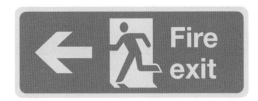

Figure 9.56 Photo luminescent fire exit sign

Sign, Colour & Pictogram	Description, Uses and Conventions
Safe Condition!	Safe condition signs are used to indicate escape routes, emergency exits and operation of devices related to Escape Routes emergency door operations, etc.
FIRE EXIT	Used without an arrow above the '**Final Fire Exit**' door leading directly to safety. Usually at the door leading outside to fresh air.
FIRE EXIT	Used to direct a person to a fire exit indicating that the direction of tracel is - **straight on from here or straight on and up from here**
FIRE EXIT	Used to direct a person to a fire exit indicating that the direction of travel is - **straight on and down from here**
FIRE EXIT	Used to direct a person to a fire exit, indicating that the direction of travel is - **left from here**
FIRE EXIT	Used to direct a person to a fire exit indicating that the direction of travel is - **right from here**

Sign, Colour & Pictogram	Description, Uses and Conventions
FIRE EXIT	Used to direct a person to a fire exit indicating that the direction of travel is - **up and left from here**
FIRE EXIT	Used to direct a person to a fire exit indicating that the direction of travel is - **up and right from here**
FIRE EXIT	Used to direct a person to a fire exit indicating that the direction of travel is - **down and right from here**
FIRE EXIT	Used to direct a person to a fire exit indicating that the direction of travel is - **down and left from here**
FIRE EXIT FIRE EXIT	**General:** Example of directional signs indicating a route that is suitable for Disabled Persons with mobility problems - leading directly to a fire exit or a disabled refuge. Idicating that the direction of travel is - **right or down from here etc.**

Figure 9.57 Typical emergency escape signs

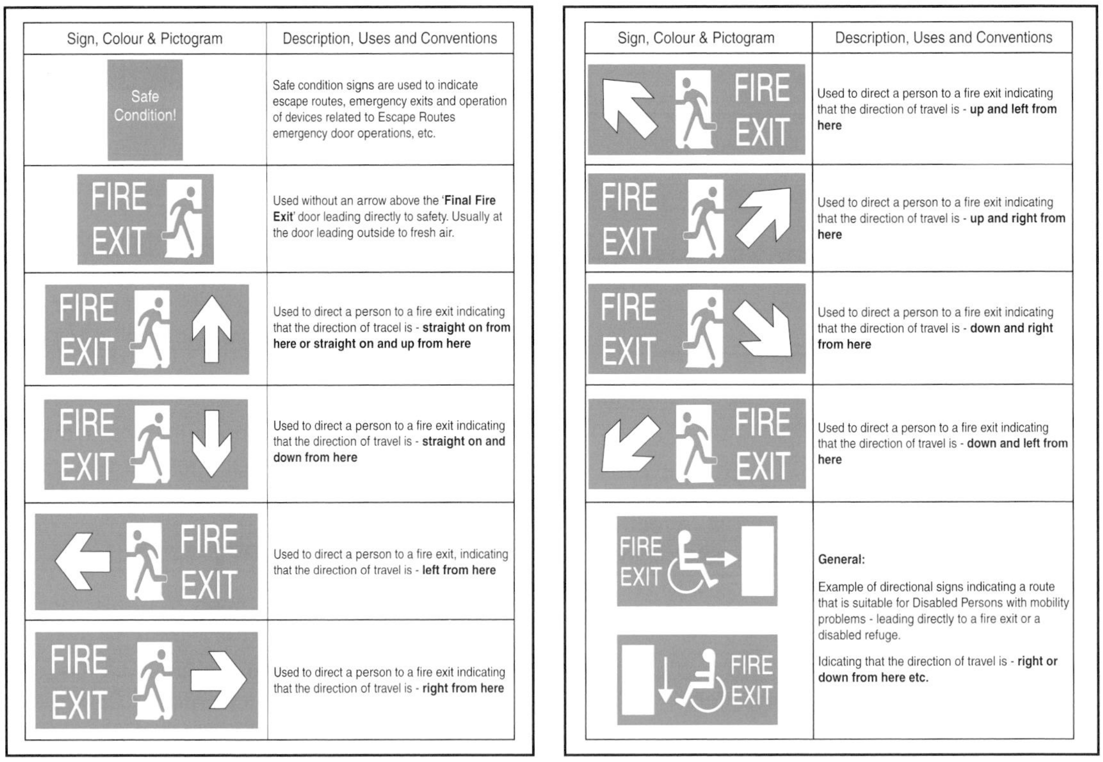

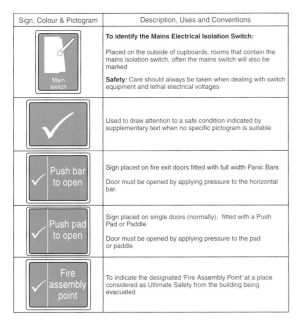

Sign, Colour & Pictogram	Description, Uses and Conventions
Main switch	**To identify the Mains Electrical Isolation Switch:** Placed on the outside of cupboards, rooms that contain the mains isolation switch, often the mains switch will also be marked **Safety:** Care should always be taken when dealing with switch equipment and lethal electrical voltages
✓	Used to draw attention to a safe condition indicated by supplementary text when no specific pictogram is suitable
Push bar to open	Sign placed on fire exit doors fitted with full width Panic Bars Door must be opened by applying pressure to the horizontal bar.
Push pad to open	Sign placed on single doors (normally), fitted with a Push Pad or Paddle Door must be opened by applying pressure to the pad or paddle
Fire assembly point	To indicate the designated 'Fire Assembly Point' at a place considered as Ultimate Safety from the building being evacuated

Figure 9.58 Green 'safe condition' signs

Sign, Colour & Pictogram	Description, Uses and Conventions
Fire door keep shut	Identifies a Fire Door that is a 'Self Closing Fire Door' that **MUST** be kept shut and unobstructed at all times Positioned at eye level on both faces of each leaf of self-closing fire doors Note: Door may be signed 'Fire Door Kept Closed'
Fire door keep locked	Identifies a Fire Door that **MUST** be kept shut and unobstructed at all times Used on fire doors **without a 'Self Closing' devices**, i.e. Cleaner's Cupboards, Stores and Plant Rooms or Service Ducts
Automatic fire door keep clear	Identifies a 'Self Closing Fire Door' that is held open on electro magnetic devices which **MUST** be kept clear and unobstructed at all times. Obstruction may impede the closing of the door or shutter, which is released on activation of the fire alarm. Placed at eye level on doors or shutters, which close automatically in the event of a fire
Fire exit keep clear	Keep area clear of obstructions that may impede escape on an escape door or route Found on escape routes where onstructions may occur, e.g. on the external face of a fire exit door so that vehicles do not obstruct
Fire action	Fire Action Notices – provide information and instruction on what to do on discovering a fire, or on hearing the fire alarm. Positioned at fire alarm call points or to accompany fire extinguishers, etc.

Figure 9.59 Blue 'mandatory' door signs

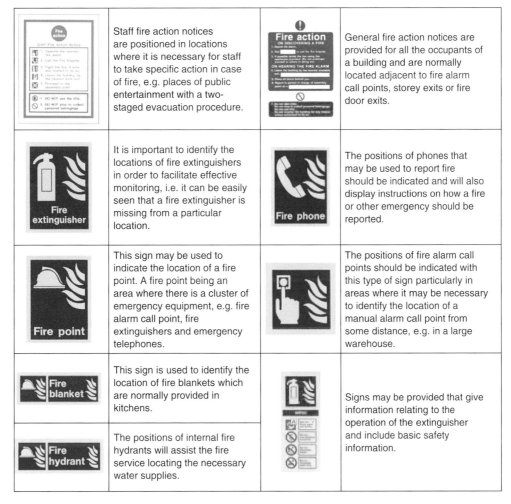

Fire action Notice	Staff fire action notices are positioned in locations where it is necessary for staff to take specific action in case of fire, e.g. places of public entertainment with a two-staged evacuation procedure.	Fire action	General fire action notices are provided for all the occupants of a building and are normally located adjacent to fire alarm call points, storey exits or fire door exits.
Fire extinguisher	It is important to identify the locations of fire extinguishers in order to facilitate effective monitoring, i.e. it can be easily seen that a fire extinguisher is missing from a particular location.	Fire phone	The positions of phones that may be used to report fire should be indicated and will also display instructions on how a fire or other emergency should be reported.
Fire point	This sign may be used to indicate the location of a fire point. A fire point being an area where there is a cluster of emergency equipment, e.g. fire alarm call point, fire extinguishers and emergency telephones.		The positions of fire alarm call points should be indicated with this type of sign particularly in areas where it may be necessary to identify the location of a manual alarm call point from some distance, e.g. in a large warehouse.
Fire blanket	This sign is used to identify the location of fire blankets which are normally provided in kitchens.		Signs may be provided that give information relating to the operation of the extinguisher and include basic safety information.
Fire hydrant	The positions of internal fire hydrants will assist the fire service locating the necessary water supplies.		

Figure 9.60 Red 'fire equipment' signs and composite 'fire action' signs

Table 9.11 Typical actions required to maintain the means of escape

Item	Action	Example of time period
Escape routes	Check that all escape routes are clear and free from obstruction and combustion storage.	Daily/weekly
Fire doors	Check fire doors are in a good state of repair and are not being compromised by being wedged open. Check that the operation of fire doors that are held open with automatic mechanisms operate correctly.	Daily/weekly Monthly
Fire exit doors	Check that all doors that are required to open to provide emergency egress operate correctly.	Monthly
Emergency signage	Check the signage is in place and has not been obstructed by alterations in the layout of the building or its contents.	3 monthly
Escape lighting	Check the operation of the emergency escape lighting.	3 monthly
Training	Train staff and visitors to make them aware of their responsibilities not to negate the means of escape, for example not wedging fire doors open or blocking fire exits.	When necessary for visitors. On induction and annually for staff
Competent testing and inspection	Establish and manage a system whereby all engineered aspects that support the means of escape, e.g. emergency lighting is periodically inspected and tested by demonstrably competent persons.	Annually
Control of building works	Establish and manage a system whereby any work conducted by in-house staff or contractors does not: ➤ Negate the means of escape during the course of the building works ➤ Result in breaches of the fire compartmentation	At all relevant times

9.5.11 Management actions required to maintain means of escape

Persons responsible for fire safety within buildings will need to ensure that the means of escape within their area of responsibility is maintained so as to be available at the time when it is required.

It is normal for the elements that comprise the means of escape of a building to form part of a more comprehensive fire safety management system. A comprehensive system will include the competent testing of fire safety systems including systems that detect and give warning of fire.

Table 9.11 outlines those actions that managers and responsible persons will need to ensure occur together with examples of typical time frames.

9.5.12 Fixed fire fighting systems

Fixed fire fighting systems (FFS) provide active protection for a building, its contents and occupants. There

have been some notable incidents where the fixed fire fighting system in a building has been extremely successful in controlling and extinguishing a fire that the fire service has not been able to deal with.

At the design stage architects may consider the provision for FFS as a compensatory feature to provide additional protection to a large space or vulnerable part of a building. FFS are occasionally 'retro' fitted to existing buildings to mitigate a historic weakness or reduce the perceived fire risk of a change of use or circumstance.

The benefits of providing active FFS include:

➤ Increased size of compartments permitted under the Building Regulations
➤ Reduced insurance premiums
➤ High levels of protection for valuable assets, i.e. building fabric or contents
➤ Increased life safety.

The question of whether a fire suppression system such as sprinklers should be fitted will depend on several factors all of which should be identified during the risk assessment process. Such factors will include:

➤ Likelihood of the event
➤ Consequences or outcome of the event
➤ Location of the building/s
➤ How accessible the buildings are
➤ Vulnerability to intruders through the perimeter of the site
➤ Whether there is public access to the site
➤ How good the security system is
➤ Vulnerability of the construction materials to catch fire
➤ The space and separation arrangements for waste disposal and storage
➤ Previous history of vandalism and arson
➤ The time it takes the fire and rescue service to reach the buildings and fight the fire
➤ The availability of a good water supply for the system (sprinklers or other water-based systems).

In order to operate effectively any FFS must be able to detect a fire, activate and then suppress the fire. Systems are available that utilise CO_2 or other gas and dry powder; these are discussed later. The most common FFS are:

➤ Automatic water sprinklers
➤ Drencher systems
➤ Flooding and inerting systems
➤ Water mist systems.

These systems are discussed below.

Automatic water sprinklers

Those who are responsible for fire safety in the workplace should be aware that the relevant British Standard is BS 5306 Part 2: 1990: Fire extinguishing installations and equipment on premises.

Automatic water sprinkler systems will provide efficient means of fire control throughout most parts of a building ensuring:

➤ Detection of a fire at an early stage
➤ Control of fire growth, fire spread, heat and smoke generation, by delivering water to the seat of the fire
➤ The provision of a local alarm system operation, and confirmation of the alarm at the central control room, and
➤ If appropriately arranged, the transmission of an alarm to the fire service.

Figure 9.61 shows the basic elements of any sprinkler system, which are:

Water supplies – normally there should be two independent and reliable water supplies for a non-domestic sprinkler system. It is usual for a town's water mains to be supported with a purpose built water reservoir/tank storage facility on site.

Pipes – pipes are used to move the water from the various water supplies to the control valves of the system and then onto the range pipes in which are fitted the individual sprinkler heads.

Control valves – the control valves are fitted between the water supply and the range pipes and control the

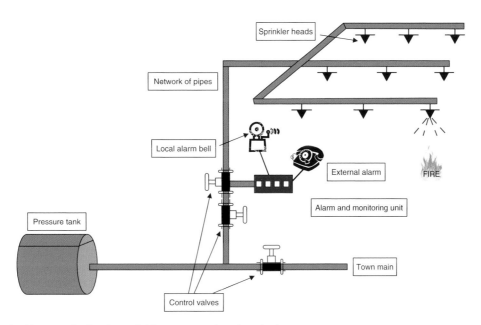

Figure 9.61 Typical layout of a basic sprinkler system showing the key components

flow of water through the system. Most control valves are designed in such a way that a small amount of pressure in the range pipes will hold back a greater amount of pressure in the supply pipe. This is simply achieved with the use of a differential valve which has the range pipe side much bigger than the water supply side.

Alarm and monitoring unit – this unit is fitted immediately above the control valve and monitors the pressure in both sides of the control valve. An alarm is activated if the pressure in either the supply pipes or the range pipes drops or a flow of water is detected in the system, which indicates the operation of a sprinkler head.

Local/external alarm – once the alarm monitoring unit has detected a situation that triggers an alarm it may activate a local and external alarm. The local alarm which indicates that there is a flow in the system is commonly achieved by the use of a Pelton wheel where a small quantity of water that is flowing through the range pipes is diverted and turns a mechanical bell situated near the sprinkler control valve location. The alarm can be raised externally through an automatic link to a commercial, alarm receiving centre (ARC).

Sprinkler heads – sprinkler heads have two main components, the head itself and a fusible link. When a head has operated the water from the range pipe hits a deflector plate which causes the water to be sprayed in any number of different patterns and/or directions. The shape of the head and the size of the hole for the water is dependent upon the location of the sprinkler head and fire risk being protected. The most common type of fusible link is a glass bulb which is filled with a liquid. Each glass bulb has a small air bubble to allow for the normal range of temperature rises in the atmosphere, i.e. when there is a temperature increase the liquid in the glass bulb will expand and compress the air bubble. However, when there is a significant rise in the temperature within the bulb the air bubble will compress as the liquid continues to expand and will fracture the glass bulb itself thereby allowing water to flow and thus operating the entire system. Glass bulbs are colour coded in relation to their operating temperature, see Table 9.12.

Table 9.12 Colours and operating temperatures of sprinkler heads

Colour	Temperature at which it will activate °C
Orange	57
Red	68
Yellow	79

Figure 9.62 A variety of sprinkler heads in use

To ensure effective operation, sprinkler heads must always be installed in the condition in which they are supplied by the manufacturer; they must not be painted or decorated in any way. The choice of sprinkler head will depend upon:

➤ Sprinkler head pattern
➤ Sprinkler head spacing
➤ Location of thermally sensitive element relative to the ceiling
➤ Clearance below the sprinkler
➤ Potential for a shielded fire to develop.

Types of automatic water sprinkler systems

Wet, dry and alternate – automatic water sprinkler systems can be designed to be wet, dry or alternate. The range pipes of wet systems are always charged with water and therefore will provide the most rapid response to a fire. However, wet systems can only be provided in locations that are not subjected to freezing temperatures. Sprinkler systems in shopping malls and offices are typically wet systems. In contrast a dry system is filled with pressurised air. Dry systems are therefore appropriate where the temperature is likely to fall below freezing, for example in multi-storey car parks. The disadvantage of a dry system is that there can be a delay in reacting to a fire situation while the air is driven out of the range pipes by the advancing water. To minimise this disadvantage an alternate system may be used on those occasions where, in the winter, the system may be exposed to freezing temperatures but in the summer in a temperate climate the range pipes can be filled and thereby achieve a more rapid response time.

Pre-action sprinkler system – this is a combination of a standard sprinkler system and an independent approved system of heat or smoke detectors installed in the same area as the sprinklers. Like dry pipe systems the pipes are filled with air but water is only let into the pipes when the detector operates. Pre-action systems are used to prevent accidental discharge of water from the sprinkler pipework following mechanical damage.

Domestic sprinklers – the vast majority of fire deaths and injuries happen in the home. Over the last 15 or so years the thrust for improving fire safety in the home has been centred around the provision of smoke detection and although domestic smoke detectors do save lives and are absolutely vital to reduce the risk of death and injury in the home, it has recently become obvious that early warning alone is not sufficient to reduce deaths in the home. The focus now for increasing fire safety in the home is moving towards the provision of residential sprinklers.

Although a relatively new development in the UK residential sprinklers are more common in America. In one town in Arizona, all new homes have been required to be fitted with sprinklers for the last 15 years. As a result, residential sprinklers now protect over 40% of the dwellings in this town. A recent report showed that in the past 15 years there has been:

➤ No fire deaths
➤ 80% reduction of fire injuries
➤ 80% reduction in property damage
➤ 95% reduction in water usage for fire control.

In the UK domestic sprinkler systems should be designed and fitted to comply with BS 9251:2004: Sprinkler systems for residential and domestic occupancies. These must be designed in such a way that they are able to cope with the expected size of fire in the particular building.

Drenchers

Drencher systems are installed to protect against radiated heat or to flood a high risk process with water for guaranteed and rapid suppression. In essence a drencher system is a range of pipework that resembles the range pipes of a sprinkler system. Drenchers may be either open or sealed. Open drenchers are operated simultaneously by the opening of the main valve, while the sealed type is individually actuated in the same way as a sprinkler head. Sealed drenchers differ little from sprinkler heads except in the shape of the deflector plate.

While a sprinkler system protects a building from internal fire, drenchers are located on roofs and over windows and external openings to protect the building from damage by exposure to a fire in adjacent premises.

The controlling valves for drencher systems are located in accessible positions on or near ground level but away from the adjacent fire risk. The valves are secured against unauthorised interference, protected from frost and clearly indicated.

Drenchers normally operate on the alternate system, and are more economical in the use of water than open drenchers, since only those heads that operate are required, and the pressure is maintained more efficiently. There are three main types of drencher system:

➤ Roof drenchers
➤ Wall or curtain drenchers
➤ Window drenchers.

Roof drenchers – roof drenchers have a deflector rather similar to that of a sprinkler head. From the roof ridge they throw a curtain of water upwards which then runs down the roof (Fig. 9.63). All parts of the roof and any skylights, windows or other openings must be protected.

Wall or curtain drenchers – wall or curtain drenchers throw water to one side only of the outlet in the form of a flat curtain over those openings or portions of a building most likely to admit fire. A special use for

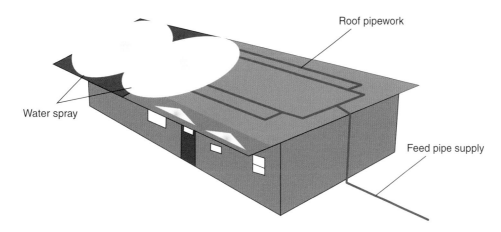

Figure 9.63   The operation of a protective roof drencher system

this type of drencher is on the stage side of a theatre proscenium arch to protect the safety curtain.

Window drenchers – as their name implies, window drenchers are used to protect window openings. They are placed horizontally level with the top of the window providing a curtain of water to protect the glass. From the tail of the deflector, a jet is thrown inwards onto the glass near the top of the window, while two streams are directed at an angle of 45° to the lower corners.

Flooding and inerting systems

Flooding and inerting systems operate by literally flooding an area with a particular extinguishing media. These systems use a variety of media including dry powder and various gases. Extinguishing powder systems have a number of advantages to offer in the form of cost, ease of maintenance, efficiency and reliability. Gas flooding systems commonly use carbon dioxide, nitrogen, argon or a combination of these gases.

Typical flooding systems will consist of a central storage tank or banks of cylinders where gas is stored under pressure, and a piping system conveys the gas to the point(s) of discharge. Special nozzles are used to facilitate the even distribution of the gas. Flooding systems can either flood a total compartment or building or be used to flood the immediate area of a fire risk. Release of the fire-fighting medium is either achieved manually or automatically. Automatic systems use various methods of detection including:

➤ Temperature sensitive wire – where the insulation of a wire melts completing a circuit
➤ Heat detectors – which operate at a predetermined temperature

➤ Gas-filled stainless steel tubing – where a fire is detected through increased pressure in the tube by the expansion of the gas
➤ Pressurised plastic tube – which melts releasing its pressure and triggering the system.

Small fixed flooding systems are increasingly used in kitchen extraction hoods and ducts where they are effective in dealing with fire involving fatty deposits.

Larger systems protect such risks as industrial quench tanks as well as a range of risks where flammable liquids are sprayed, stored or mixed. Flooding and inerting systems are fitted in such locations as:

➤ Computer suites
➤ Archive storage areas
➤ Historic buildings
➤ Electrical switch rooms
➤ Kitchens
➤ Gas turbines.

Figure 9.65 Detail of a fixed flooding system fitted to a kitchen range

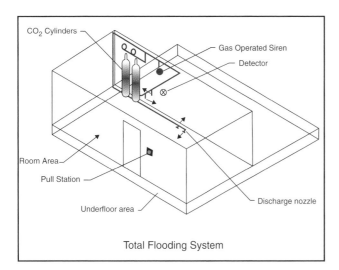

Total Flooding System

Figure 9.64 A typical layout of a flooding system

Figure 9.66 Typical warning sign to a computer room fitted with a flood system

The design specification of such systems is often based on specialist experience and will usually be governed by factors such as the type of fire, whether the protected area is occupied or not, the volume of space to be inerted, etc.

Note: halon gases were widely used in flooding systems because they are extremely efficient at extinguishing fires by chemically inhibiting flame propagation. Two specific halons (halon 1211 bromochlorodifluoromethane (BCF)) and 1301 bromotrifluoromethane (BTM)) are very effective for many types of fire. However, because of their adverse effects on the depletion of the ozone layer, their production was discontinued in 1994 as a result of the 1987 Montreal Protocol. Halon systems can only be specified if they use recycled gases. Alternatives are being developed, for example HFC-227ea fire suppression agent is considered to be an

An important feature of any flooding system must be the method by which the system is prevented from being operated while the space is occupied. In all cases the extinguishing media will harm and may kill the occupants of any confined space in which it operates. Systems are often designed to be operated when the space they protect is unoccupied. Other systems provide pre-warning to occupants or rely on manual operation.

acceptable replacement for halon 1301. HFC-227ea has a zero ozone depleting potential, a low global warming potential, and a short atmospheric lifetime.

Water mist systems

Fixed water mist systems have begun to be introduced in recent years. They can quickly extinguish large fires in enclosed spaces because the heat that is generated by the fire rapidly vaporises the water mist spray droplets into steam. However, small fires that have not yet heated an enclosure or are in large open areas cannot be easily extinguished by water mist sprays unless the fire is literally within the range of the spray.

There are currently about 1000 systems in use in the UK that use water mist to flood a compartment when a fire breaks out. There is some debate regarding the effectiveness of these systems for life safety. In common with all flood systems these can be fitted to protect the whole compartment or just specific fire risks within the compartment. These systems are currently used in both domestic and industrial applications often as a compensatory feature to enable normal design parameters to be increased. There is currently no British Standard for their design or application.

Figure 9.67 High pressure water mist system in operation

9.5.13 Portable fire fighting equipment

Fire extinguishers are usually designed to tackle one or more classes of fire. All extinguishers that conform to current regulations are coloured red. In order to differentiate the specific type of extinguisher they display a colour coded panel which should be at least 5% of the body of the extinguisher. The size of the extinguisher provided for any particular location depends upon the fire risk and the persons who may need to use it. There are seven basic types extinguishing equipment:

➤ Water
➤ Aqueous film forming foam (AFFF)
➤ Foam
➤ Dry powder
➤ Carbon dioxide
➤ Wet chemical
➤ Fire blanket.

Water – water extinguishers are most commonly found in offices and other places of work where the combustible material is carbonaceous, i.e. wood, paper, plastic, etc. Water extinguishes the fire by cooling. Modern extinguishers are of the stored pressure type where the pressure is supplied by a small CO_2 cylinder within the extinguisher, in much the same way as a soda siphon; alternatively the body of the extinguisher can be pressurised. The water is applied to the fire through a discharge hose which allows the water to be played on the fire without having to bodily move the extinguisher backwards and forwards. The pressure within the extinguisher is indicated on the neck of the cylinder.

AFFF – AFFF extinguishers are almost identical to the water type extinguisher apart from the fact that they have a foam additive in the water (aqueous film-forming foam) that increases the efficiency of the extinguisher

and therefore creates a quicker knockdown with fewer media. AFFF extinguishes the fire by cooling.

Foam – foam extinguishers again are similar in construction to water type extinguishers, the foam being expelled from the body of the cylinder by stored pressure. The foam is created by the premixed foam being aerated through a special attachment at the end of the discharge hose. Because the foam is less dense than water it floats on the burning surfaces of oils and other flammable liquids. It extinguishes the fire by smothering, i.e. by preventing air getting to the fuel.

Dry powder – dry powder extinguishers are constructed as shown in Figure 9.69. The dry powder is expelled from the extinguisher by the stored pressure in a CO_2 cylinder. When the CO_2 is operated, the dry powder is transformed into a liquid state and discharges through the hose to the nozzle. Dry powder extinguishes fire by smothering, i.e. by forming a barrier between the burning fuel and the air. It is the only extinguishing media that can successfully deal with flowing fuel fires.

Carbon dioxide (CO_2) – CO_2 extinguishers are constructed as shown in Figure 9.70. They contain only CO_2, which has been pressurised to such an extent that it is a liquid. When the operating lever is depressed, the liquid CO_2 moves out through the discharge tube. At the end of the tube is a horn which is provided to slow the rush of pressurised CO_2 and direct it at the fire. These extinguishers work by smothering, i.e. they replace the immediate atmosphere of air around a fire with an inert gas that is one that does not support combustion. CO_2 is the only medium that can be used to safety fight fires

Figure 9.68 Pressure gauge on the neck of a water extinguisher

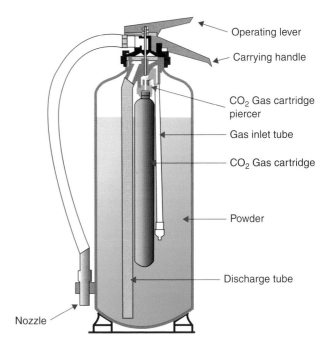

Figure 9.69 Dry powder extinguisher

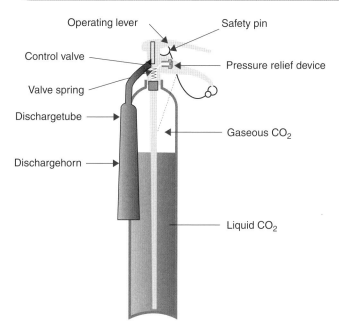

Figure 9.70 Carbon dioxide extinguisher

Figure 9.71 Wet chemical fire extinguisher

that involve electrical equipment; however, they have a number of disadvantages:

➤ When they operate they are very noisy and as a result can distract the operator
➤ They only operate for a limited period; the smaller ones may last only 12 seconds
➤ They do not cool the fire and therefore fire can easily reignite unless the CO_2 is periodically reapplied
➤ They do not work in an open area where draughts may cause the CO_2 to disperse.

Wet chemical – wet chemical fire extinguishers have been developed specifically for use on fires involving cooking oils and fats. They are the most effective type of extinguisher for this type of fire risk. The extinguishers work by discharging the wet chemical agent, through a spray type nozzle, onto the surface of the burning oil or fat. The wet chemical agent reacts with the burning oil or fat to form a 'suds-like' blanket across the fuel surface. This 'suds-like' blanket extinguishes the fire by excluding the air, and by preventing the release of flammable vapours.

Table 9.13 provides a summary of information relating to the various types of extinguishers and media.

Siting of portable extinguishing equipment

Extinguishers should be provided in numbers and sizes recommended by BS 5306 – Fire extinguishing installations and equipment: on premises: Part 3: Code of Practice for selection installation and maintenance of portable fire extinguishers. Those provided should also satisfy legal and insurance requirements.

In general, extinguishers and fire blankets should be positioned at locations where they are conspicuous and readily accessible for immediate use.

Extinguishers for general protection should be:

➤ Near stairways
➤ In corridors or landings
➤ Close to exits.

Extinguishers should not be located:

➤ Where a potential fire might prevent access to them
➤ In concealed positions behind doors, in cupboards or deep recesses
➤ In positions where they might cause obstruction to exit routes or be damaged by trolleys or vehicles
➤ Over or close to heating appliances
➤ Where they may be subjected to extremes of heat or cold
➤ Where they are exposed to wet or damp atmospheres.

Extinguishers provided to deal with special fire risks should be sited near to the fire risk concerned, but not so near as to be inaccessible or place the operator in undue danger in case of fire.

Where large undivided floor areas necessitate positioning appliances away from exits or outer walls, they should be installed on escape routes. They should be positioned so that it is not necessary to travel more than 30 m from a fire to an extinguisher.

Extinguishers should be mounted on brackets so that the carrying handle is 1 m above the floor for large extinguishers and about 1.5 m for smaller ones. Potentially dangerous siting should be avoided. For example, foam

Table 9.13   Summary of the types and uses of portable fire fighting extinguishers

Extinguisher/ medium	Best for	Danger	How it works
Water Panel colour	Wood, cloth, paper, plastics, coal, etc. Fires involving solids	Do not use on burning fat or oil or on electrical appliances	Mainly by cooling burning material.
Dry powder Panel colour	Liquids such as grease, fats, oil, paint, petrol, etc. **but not on chip or fat pan fires**.	Safe on live electrical equipment, does not penetrate the spaces in equipment easily and the fire may reignite. Does not cool the fire very well and care should be taken that the fire does not reignite	Knocks down flames and smothers the fire.
Aqueous film forming foam (AFFF) Panel colour	Wood, cloth, paper, plastics, coal, etc. Fires involving solids Liquids such as grease, fats, oil, paint, petrol, etc. **but not on chip or fat pan fires**.	Do not use on flammable oil fires	Forms a fire extinguishing film on the surface of a burning liquid. Has a cooling action with a wider extinguishing application than water on solid combustible materials.
Wet chemical Panel colour	Cooking oils and fats	These extinguishers are generally not recommended for home use	The wet chemical agent reacts with the burning oil or fat to form a 'suds-like' blanket across the fuel surface. This 'suds-like' blanket extinguishes the fire by excluding the air, and by preventing the release of flammable vapours.
Carbon dioxide CO_2 Panel colour	Liquids such as grease, fats, oil paint, petrol, etc. **but not on chip or fat pan fires**.	Do not use on chip or fat. This type of extinguisher does not cool the fire very well and you need to watch that the fire does not start up again Fumes from CO_2 extinguishers can be harmful if used in confined spaces: ventilate the area as soon as the fire has been controlled	Vaporising liquid gas which smothers the flames by displacing oxygen in the air.
Fire blanket	Fires involving both solids and liquids. Particularly good for small fires in clothing and for chip and fat pan fires **provided the blanket completely covers the fire**.	If the blanket does not completely cover the fire, it will not be able to extinguish the fire	Smothers the fire.

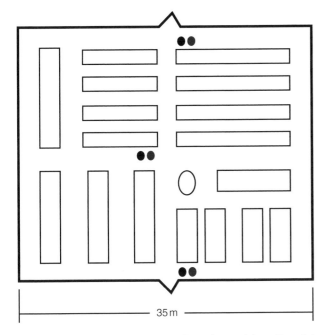

Figure 9.72 Plan drawing of siting of portable extinguishing equipment

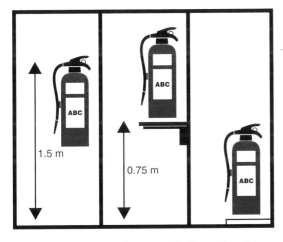

Figure 9.73 Mountings for portable fire extinguishers

extinguishers should not be positioned next to molten salt baths and water extinguishers should not be sited near high voltage equipment. Care should also be taken to ensure that a heavy extinguisher does not itself cause injuries by being dislodged and falling onto limbs or bodies. Extinguishers installed under conditions where they are subject to dislodgement should be installed in specifically designed brackets.

When necessary, the positions of extinguishers should be indicated by signs, and there should be a record (conveniently on a plan) of the type, number and location of the extinguishers within the building or workplace.

In some situations, e.g. in schools or police custody suites, extinguishers may need to be provided in recesses or secured against unauthorised use.

Managing portable extinguishing equipment

As is the case with all fire protection systems portable fire fighting equipment must be the subject of active monitoring. Persons responsible for fire safety at work will need to ensure that extinguishers are provided, sited, inspected and tested in line with the manufacturers' recommendations; this must include an annual test and inspection by a competent person. In addition, extinguishers should be periodically inspected to ensure they are in place, are full and in good condition and have indeed been tested by a competent person within the relevant period.

9.5.14 Fire detection and alarm systems

It is a requirement of the RRFSO and MHSWR that employers have adequate emergency arrangements. For employers or occupiers of a building, an essential element of ensuring people can safely evacuate is to provide appropriate and reasonable arrangements for detecting fire and raising the alarm.

In some cases, for example in small detached offices, it may well be reasonable to rely solely on the occupiers to detect a fire and raise the alarm to others verbally. In this way a reasonable level of safety for the occupants could be said to have been achieved. For slightly larger premises it may be considered necessary to enhance this very basic system with an electrically operated fire alarm that is operated by manual call points.

However, for larger premises or where the owners/occupiers or responsible persons wish to provide a higher level of building, some form of automatic system will be utilised. According to Home Office statistics, 67% of all fires in business premises occur at times when the buildings are closed, i.e. after 6pm and at weekends.

Installing an automatic system that both detects and raises the alarm will significantly increase both life and building safety and can bring with it the following benefits:

➤ Early detection of fires in unoccupied parts of the building, e.g. store and boiler rooms
➤ Early warning to the occupants of a fire, to enable effective escape
➤ The operation of other protective devices, such as:
  ➤ Automatic door closers or stairway pressurisation systems
  ➤ Closing down ventilating and air conditioning plant
  ➤ Bringing fire control systems into operation

➤ Opening ventilators or starting fans for smoke control
➤ Opening doors or ventilators
➤ Operating door release mechanisms, etc.
➤ Early notification to the fire service, enabling an early start to fire fighting operations and reduced fire spread.

9.5.15 Types of automatic fire detection and alarm systems

Fire alarm systems may be installed in buildings to satisfy one, or both, of two principal objectives, namely protection of life and protection of property. Table 9.14 outlines the types available, their primary purpose and in which areas of a building they are installed. The standard for fire detection and alarm systems in buildings is contained within BS 5839 Part 1.

BS 5839 uses a system whereby fire detection and alarm systems are categorised according to their operating method and purpose.

Obviously a system that is intended to provide protection for life will have a knock-on effect to the safety of the property and vice versa. It must be stressed that BS 5839 does not specify what types of systems must be used for any particular building. The type of system selected and installed will need to be identified as an outcome of the fire risk assessment for the premises as required by the RRFSO.

9.5.16 Methods of detection

Manual – systems that allow for the detection of fire by people will be provided with suitable manual call points. The call points must:

➤ Be located so that egress cannot be made from a floor or building without passing one
➤ Be conspicuous
➤ Comply to BS EN 54-11 type A, i.e. be operated by a 'single action'
➤ Be protected from accidental operation.

Table 9.14 Fire detection and alarm systems

Category	Type	Purpose	Where detection system is installed
M	Manual systems	Provide a means of manually communicating the presence of a fire	Small low risk premises
L	The prefix L indicates fire detection systems designed for the protection of life		
L1	Automatic	Earliest possible warning of fire, so as to achieve the longest available time for escape	Throughout all areas of the building
L2	Automatic	Early warning of fire in specified areas of high fire hazard level and/or high fire risk	Installed only in defined parts of the building
L3	Automatic	To give a warning of fire at an early enough stage to enable all occupants, other than possibly those in the room of fire origin, to escape safely	Installed only in defined parts of the building
L4	Automatic	Providing warning of smoke within escape routes	Those parts of the escape routes comprising circulation areas and circulation spaces
L5	Automatic	Location of detectors is designed to satisfy a specific fire safety objective (other than that of a Category L1, L2, L3 or L4 system)	Specified protected areas
P	The prefix P indicates those systems that are provided for the protection of property		
P1	Automatic	To minimise the time between ignition and the arrival of fire-fighters	Throughout all areas of the building
P2	Automatic	Early warning in areas in which the risk to property or business continuity from fire is high	In defined parts of the building

Figure 9.74 Proprietary device for protecting a manual fire alarm call point from accidental operation

Figure 9.75 An ionisation smoke detector

Figure 9.76 A typical heat detector

Automatic – fire may be detected automatically in a variety of ways and apparatus designed for the purpose needs to be sensitive to at least one of the particular phenomena that are associated with fire, i.e. the presence of heat, smoke or flame.

Different detection methods may be incorporated into the same system so that the most suitable detection method can be chosen for any location. In addition to their method of operation automatic detectors can be classified as being either point detector (i.e. equipment that detects a fire at a particular location or area detectors that are capable of detecting a fire over a wide area).

Some point detectors are designed to have the dual function of both detection of fire and sounding an alarm. Traditionally these types of detectors have had a mainly domestic application but are increasingly being used for workplace fire alarm systems.

Smoke detectors – the most common form of detection is the point smoke detector. There are two types of detection unit which are:

➤ **Ionisation units.** These detect all types of smoke that contain small particles, but they respond most rapidly to smoke caused by fires such as a chip pan fire (very sensitive). The detector in these units consists of a small amount of radioactive material that detects any invisible smoke particles that are floating in the air. When particles in the smoke are detected, the alarm is sent to the panel.

➤ **Optical units.** These detect all types of smoke, but are ideal for smoke with larger particles, such as burning furniture. In this unit regular pulses of ultraviolet light are sent to a detector; if the light is prevented from reaching the detector, the alarm will be transmitted to the panel.

Heat detectors – these detectors operate by sensing heat in the environment. They operate in two distinct ways. Some are designed to operate at a given temperature whereas others detect the rapid 'rate of rise' of temperature that is associated with a developing fire. They are often used in a kitchen area, where smoke detectors would continually actuate because of the normal processes in a kitchen area.

Beam smoke detectors

Beam smoke detectors are designed to detect the obscuration caused by smoke over a wide area, typically atria and warehouses. They usually consist of a combined transmitter and receiver unit with a remote reflective element, or a separate transmitter/receiver unit. They are normally positioned high on walls at either side of a monitored area. In this way the system can detect smoke from fires that might occur over a large area. Typically, an infrared beam is transmitted along this length. In the event of smoke passing through the beam, the receiver measures the resultant attenuation.

Aspirating smoke detectors

Aspirating smoke detection systems draw the air from a monitored area via pipework to a remote detection unit where the air is sampled for the presence of smoke. The sampling pipework usually contains predetermined

Figure 9.77   Beam detector

holes through which the air is drawn. Aspirating smoke detection systems are commonly used for applications such as laboratories or computer rooms, they can also be used to monitor individual computer cabinets for the first signs of overheating cables or components.

Linear heat detecting cable (LHDC)

Linear heat detectors comprise heat detecting cables which respond to temperature along their length. There are two types of cable:

➤ Integrating cable, where heat distribution along the length is summed and averaged such that the resultant signal given does not necessarily equate to the highest temperature at any point on the length
➤ Non-integrating cable, where heat sensed at any point along the length will be detected and signalled as appropriate.

LHDC is used in a wide variety of applications but is particularly suited where there is a harsh environmental condition, a physical or hazardous maintenance access constraint to the protected area, and/or a requirement to cost-effective install detection in close proximity to the fire risk. The main benefits of LHDC systems are that they are effective in detecting a rise in temperature at any point along their length, and can be used in environments that may be potentially explosive and therefore require the use of intrinsically safe equipment (see Chapter 8).

Flame detectors

Flame detectors operate by recognising the specific bandwidths of light emitted by a fire. They are used in locations where immediate detection of fire is needed such as petrochemical installations.

A form of filtering is normally combined with flame detectors to ensure that static light emitting sources are not mistaken as a fire. Due to their principle of operation, flame detectors need a direct line-of-sight with the fire to allow them to detect it. Therefore, it is often necessary to train a number of detectors at specific areas in order to provide for full coverage. Even then, objects, furniture, etc., introduced after the detectors have been placed, can reduce their detection function. In some cases it is possible to utilise building materials to 'reflect' light from dead areas onto the detector. However, once a flame is seen by the detector, operation is virtually instantaneous. Therefore, flame detectors are particularly useful as part of an explosion suppression system.

Figure 9.78   Flame detector

9.5.17   Fire panels and zoning

Fire alarm panels comprise control and indicating equipment and perform three principal functions:

➤ Automatically monitor and control fire detection and fire alarm devices and their power supply
➤ Indicate fire signals, system faults and locations
➤ Provide a means of manual control to facilitate testing, disabling of devices, triggering and silencing of audible warning systems and resetting of the system following a fire signal.

The complexity and type of the control and indicating required will vary according to the size and type of premises.

When small and less complex systems are installed they are housed within a single control panel or box. In larger multi-building and complex premises it is likely that the control equipment will be provided with a main fire panel with repeater panels located within each building (many such panels are fitted at alternative points of entry for the fire service).

All such systems, including the installation, wiring, power supplies, panelling, etc., must comply with the requirements contained in each of the parts to BS 5839.

A variety of mechanisms are used to display the location of detectors, alarm points, etc. These indicators may be via the use of text display, a light emitting indicator or any such alternative approved device. Whichever device is used each zone should be indicated in such a way as it provides a simple 'at a glance' overview of all zones, and where fitted addressable information for each detection call point unit. At each panel a clear plan indicating the zones that are covered by the particular panel should be displayed (see Fig. 9.80).

In a large proportion of buildings a simple evacuation strategy will be adopted and on the operation of a manual call point or detection by an automatic fire detector, fire alarm sounders will provide warning and indicate the need for a full evacuation of the building.

Larger more complex buildings may well operate a phased evacuation strategy and the evacuation signal, as has already been discussed, may be restricted to a single floor or a limited area within the building. On these occasions other areas may be provided with an 'alert' warning signal.

Zones – to support the above arrangements the building will need to be divided into a number of alarm zones in such a way that the fire alarm sounders reflect an operation of a manual call point within that area or a detector.

In order to achieve effective zoning the following should be considered:

➤ Boundaries of every alarm zone, with the exception of external walls, should be fire resisting (compartment)
➤ Any overlap between signals and alarm zones should not result in confusion (which zone has been alarmed)
➤ A common alarm signal must be provided into each alarm zone for evacuation and a different signal in all the zones for an alert signal.

The manual alarm zones may incorporate more than one detection zone but the detection zones cannot incorporate more than one alarm zone.

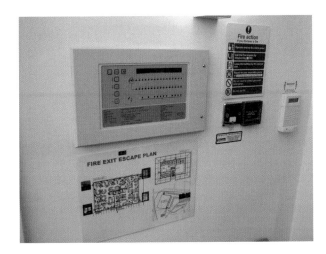

Figure 9.79 Typical fire alarm panel

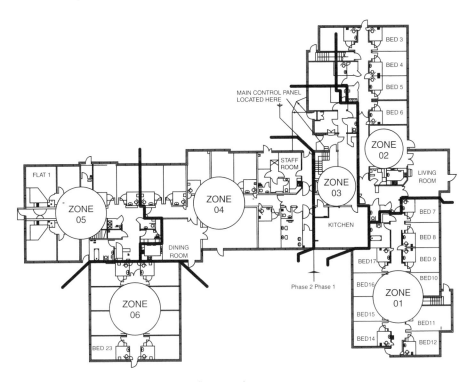

Figure 9.80 Typical zone plan located adjacent to a fire panel

9.5.18 Methods of raising the alarm

In general alarm sounders giving audible signals must be appropriate to the circumstances and must not be able to be confused with any other alarm.

Research has suggested that persons respond very much quicker and more positively to verbal instructions to evacuate a building. However, alarms are most commonly provided by bells, sirens or beeps which may vary in tone and pitch.

In those locations where an audible warning is insufficient, e.g. where there may be persons with impaired hearing or in a noisy working environment, the fire warning system should be augmented with a visual signal. See Table 9.16 for examples of situations, together with appropriate warning arrangements.

The nature and type of audible or other type of alarm signals will depend upon the complexity and the

Figure 9.81   Audible fire alarm

Table 9.16   Examples of various arrangements for raising the alarm in case of fire

Use of building	Alarm arrangements
Cinemas	Coded message to staff followed by a general announcement to public to evacuate
Prisons	Discrete alarm to staff on which they instigate controlled evacuation to secure external location
Hotels	General alarm sufficient to wake sleeping occupants and instigate an immediate evacuation
Large office blocks	Two-stage general warning to evacuate floors affected by fire and warn other floors that an evacuation may be necessary
Factory production line	Audible and visual warnings signal an immediate shut down and evacuation

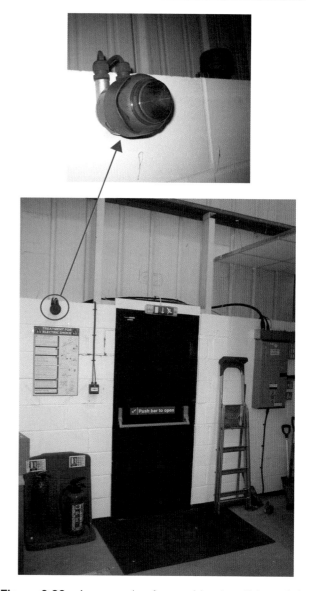

Figure 9.82   An example of a combined audible and visual fire alarm installed in a potentially noisy workplace

occupancy of a building. For example, in care facilities such as hospitals and residential care homes where persons may need assistance to evacuate, the alarm system may not necessarily be required to rouse people from sleep but to ensure that those who are required to assist are alerted. Clearly the audibility requirements will vary and traditionally 65 dB(A) is accepted as the minimal requirement, although 60 dB(A) may well be acceptable in enclosed spaces such as offices (not open plan) and in stairways.

Where members of the public frequent entertainment or shopping facilities where the potential background noise from music, etc. is greater than 80 dB(A) the music should be capable of being muted when the fire alarm signal is actuated. Where persons need to be alerted during periods of sleep (hotels, hostels, etc.) 75 dB(A) at the bed head is recommended, to ensure that they are adequately warned.

It is important that once the alarm signal has begun to sound that it continues until silenced, either due to a false alarm (alarm verifier), or by the fire service.

9.5.19 Reducing unwanted fire signals

In order to limit false alarms the owners, occupiers, or other responsible person with control over a building with a fire-detection and fire-alarm system should ensure that:

➤ The system remains in working order and is properly maintained
➤ Faults are dealt with quickly and efficiently
➤ False alarms are investigated and action taken to solve any problem
➤ Activities which may affect the system (for example, processes which may produce heat or smoke, redecorating or a change in manufacturing processes) are controlled
➤ Maintenance or other work is carried out on the system only by a competent person
➤ Attitudinal, specific advice is sought from:
  ➤ The company that installed the fire detection and fire alarm system
  ➤ The company that maintains the system
  ➤ Your local fire and rescue service.

9.5.20 Fire safety systems maintenance and testing

In addition to being responsible for daily checks of the premises, it is the management's responsibility to ensure that all fire safety equipment is adequately and routinely maintained and tested. The fire safety of the occupants is dependent upon a large number of interrelated features. Failure to maintain any one of the fire safety provisions in effective working order could negate the whole fire safety strategy.

The maintenance of furniture, furnishings, décor and equipment is as important for the safety of occupants as is the maintenance of fire safety equipment. Contents and equipment affect the likelihood of fire occurring, its development and subsequent events. Diligent attention to detail can minimise the risk of fire.

All fire safety installations need to be tested individually, but interdependent fire safety installations need to be tested collectively to demonstrate satisfactory interfacing/interlinking, etc.

Arrangements should be made for all services (including fire detection systems, door control mechanisms, pressurisation systems, evacuation and fire fighting lifts, emergency lighting, standby power systems) to be regularly inspected and maintained.

Alterations and modifications to an existing installation should not be carried out without consultation with the enforcing authority and, where possible, the original system designer or installer (or other qualified persons). This is particularly important where systems are combined and depend upon a sequence of control events.

9.5.21 Fire service access and facilities

The Building Regulations 2000 require that buildings are designed and constructed so as to provide reasonable facilities to assist fire fighters in the protection of life and to enable reasonable access for a fire appliance.

In order to achieve reasonable access for fire appliances and reasonable assistance for fire fighters it is necessary for a building to be provided with:

➤ Sufficient vehicle access
➤ Sufficient internal access
➤ Sufficient fire mains
➤ Adequate means of venting heat and smoke from basements
➤ Adequate means for isolating services.

Sufficient vehicle access

As a general rule, the larger the building the more access is required for fire service vehicles. For small buildings it is normally sufficient for the service to have access just to one face of the building. However, larger buildings require clear access to more than one side of the building. There are two types of vehicles that may require sufficient

Figure 9.83 Fire service access road

Figure 9.84 A high reach hydraulic platform (HP)

access to a building, i.e. high reach (aerial) appliances (e.g. turntable ladders) and normal pumping appliances.

For buildings fitted with protected shafts and internal fire mains (see below) access for pumping appliances must be available near the inlets of the mains.

Fire appliances are not standardised in the UK; the widths, lengths and weights of fire appliances vary from fire authority to fire authority therefore the local building authority will normally inform planners of the exact dimensions that may be required. In terms of load-bearing capacity it is normally considered that roadways capable of supporting 12.5 tonnes per axle are sufficient.

Typical fire service vehicle route specification of a local authority will include dimensions for:

➤ The width of the road between kerbs
➤ The minimum width of gateways
➤ The minimum turning circle between kerbs and walls
➤ Minimum height clearance.

Table 9.17 provides details of the requirements for access according to specific building size and gives some typical dimensions.

Table 9.17 Summary of clearances for pumps and high reach appliances

Minimum clearance/ capacities	Pump	High reach (TL HP)
Road width between kerbs (m)	3.7	3.7
Width between gateways (m)	3.1	3.1
Turning circle between kerbs (m)	16.8	26.0
Turning circle between walls (m)	19.2	29.0
Clearance height (m)	4.0	3.7
Carrying capacity (tones)	12.5	17.0

When assessing the suitability of access for fire service vehicles, the fire exits and assembly points for the building should also be taken into account, since people using these routes can impede fire service access. Assembly points should be located sufficiently far from the premises to minimise interference with the fire service or danger from falling debris.

The indiscriminate parking of cars and other vehicles using the site can seriously obstruct fire service access roads and gates leading to the building. Control and enforcement of parking restrictions can prove difficult, but notices giving clear instructions regarding parking should be in place.

Sufficient internal access

In low rise buildings without deep basements fire fighters are able to gain reasonable access within the building by a combination of their own ladders and other access equipment. In other buildings it is necessary to provide additional internal facilities to enable fire fighters to safely and quickly reach the scene of a fire in order to suppress it.

These additional facilities include fire fighting lifts, stairs and lobbies. It is obviously necessary to protect these facilities from a fire in the building and they are therefore contained within a 'protected shaft'. The plan in Figure 9.85 illustrates a typical layout of a protected fire fighting shaft.

One or more fire fighting shafts are required for buildings that have:

➤ A floor level above 18 metres or a basement below 10 metres

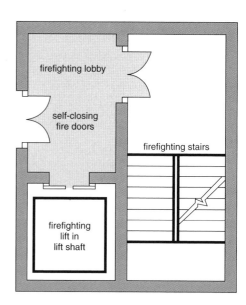

Figure 9.85 Typical layout of a fire fighting shaft containing a fire fighting lift

➤ A floor above 7.5 metres if it is used as a:
  ➤ Shop, or commercial building
  ➤ Public assembly or recreational building
  ➤ Any storage or other non-residential building that is not sprinklered
➤ Two or more basements of over 900 m².

It should be noted that the outlets from the fire mains should be located within the protected fire fighting shaft to enable fire fighters to connect their equipment to the fire main in the relative safety of the protected shaft.

A fire fighting lift is required if the building has a floor more than 18 m above, or more than 10 m below, the fire service vehicle access level.

Sufficient fire mains

Fire mains should be provided in all buildings that require a fire fighting shaft.

Dry rising mains – as the name suggests all dry water mains are fitted vertically within a building. They are operated exclusively by the fire service who when necessary will 'charge' the dry mains with water via inlet valves positioned at ground floor level. Every landing in the building has an outlet valve, sometimes located in a cupboard. Dry rising mains rely solely on the ability of fire service pumps and hoses to pump water up the building. This limits the use of dry rising mains to a maximum height of 50 metres.

Foam mains – in some instances large underground boiler rooms with flammable oil storage are fitted with dry mains that facilitate the application of fire fighting foam from a fire service pumping appliance. The foam main allows fire fighters to apply premix foam, which is aerated and applied at the seat of the fire, from the outside of the building.

Wet rising mains – where a rising main is required to be used for fire service at a height above 50 metres, it is necessary to provide wet risers. Wet risers are similar to dry risers apart from that they are constantly charged with a water supply from a source that is able to provide water at a sufficient pressure to reach the required height.

Falling mains – falling mains are identical to rising mains apart from the fact that they serve floors below the normal level of access. Falling mains are used for extensive basement areas.

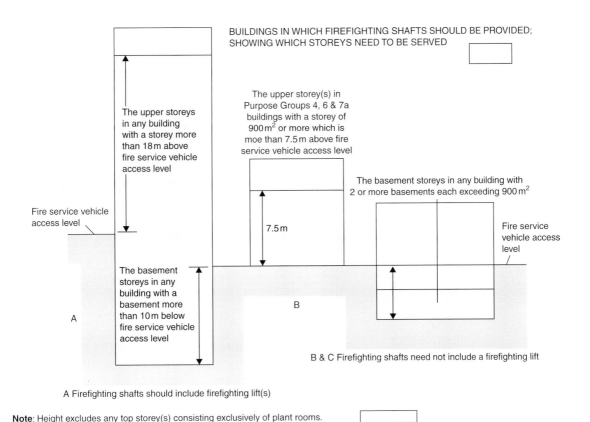

BUILDINGS IN WHICH FIREFIGHTING SHAFTS SHOULD BE PROVIDED; SHOWING WHICH STOREYS NEED TO BE SERVED

The upper storeys in any building with a storey more than 18 m above fire service vehicle access level

The upper storey(s) in Purpose Groups 4, 6 & 7a buildings with a storey of 900 m² or more which is moe than 7.5 m above fire service vehicle access level

The basement storeys in any building with 2 or more basements each exceeding 900 m²

Fire service vehicle access level

7.5 m

Fire service vehicle access level

The basement storeys in any building with a basement more than 10 m below fire service vehicle access level

A

B

B & C Firefighting shafts need not include a firefighting lift

A Firefighting shafts should include firefighting lift(s)

Note: Height excludes any top storey(s) consisting exclusively of plant rooms.

Figure 9.86 Buildings in which fire fighting shafts should be provided

233

Figure 9.87 Dry riser inlet

Figure 9.88 Dry riser outlet

Figure 9.89 Fire hydrant indicator plate

Hose reels – in some larger buildings it is considered advantageous to provide wet rising mains that are fitted with fire fighting hose reels. These are provided to be used by both the occupants of the building and fire service personnel. In reality standard fire service operations generally ignore fixed fire fighting hose reels in buildings and connect their own equipment to rising mains. However, the provision of hose reels does ensure that water for fire fighting is available in greater quantities than portable fire extinguishers.

Water ring mains and hydrants – in certain situations, for example, in remote rural areas or high risk industrial premises the normal water supplied can be augmented with a water ring main which has standard fire hydrant outlets.

Such outlets must remain unobstructed to ensure that the fire and rescue service can make use of them.

Figure 9.90 shows the layout of a building with a dry riser, foam main and hose reels.

Adequate means of venting heat and smoke from basements

The build-up of heat and smoke from a fire in a basement can seriously impede fire fighting due to:

➤ A layer of smoke and heat within the basement which fire fighters need to penetrate before attacking the fire and
➤ The chimney effect of openings in basements which funnels the products of combustion through the normal means of access and egress from a basement.

In order to facilitate fire fighting in basements, all basements that are over 200 m² and have a floor level 3 metres

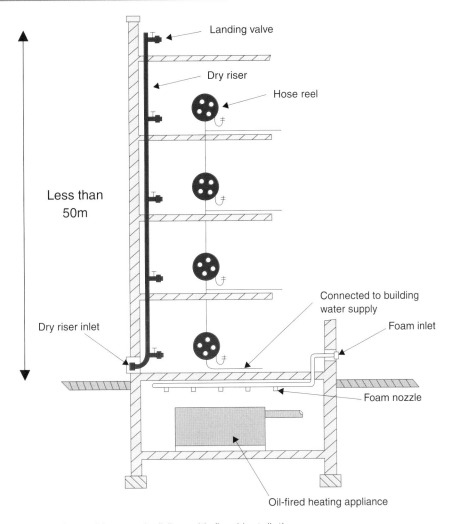

Landing valve

Dry riser

Hose reel

Less than
50m

Dry riser inlet

Connected to building
water supply

Foam inlet

Foam nozzle

Oil-fired heating appliance

Figure 9.90 Cross-section of a multi-storey building with fixed installations

or more below ground level must be provided with smoke outlets. Smoke outlets can be either natural or mechanical. Figure 9.91 shows two ways of achieving natural ventilation from a basement.

If mechanical ventilation is provided for a basement it must be robust enough to continue to function under fire conditions therefore a mechanical system will only be considered as sufficient if:

➤ The basement is fitted with a sprinkler system
➤ The system is capable of achieving 10 changes of air per hour
➤ The system can handle gases at temperatures of 300°C for at least 1 hour.

Adequate means for isolating services
Fire fighters' switches for luminous tube signs, etc. – isolations or cut-off switches to protect fire fighters must be provided for luminous tube signs or other similar equipment designed to work at voltages in excess of 1000 volts AC or 1500 volts DC if measured between any two conductors or 600 volts AC or 900 volts DC if measured between a conductor and earth.

The cut-off switch must be positioned and coloured or marked to ensure that it is readily recognisable by and accessible to fire fighters. The local fire and rescue authority and the Institution of Electrical Engineers provide information on such colouring and positioning.

Gas and other service isolations – many organisations produce documentation relating to the positioning and mechanisms of isolation for a variety of services and processes to assist fire fighters on arrival. Typically these documents consist of floor plans, schematic diagrams and manufacturers' guidance documents (when available). On larger sites arrangements are also made for engineering staff to be on hand to assist the fire fighters to identify and make safe services and systems to minimise the risks to the themselves and those they may be rescuing.

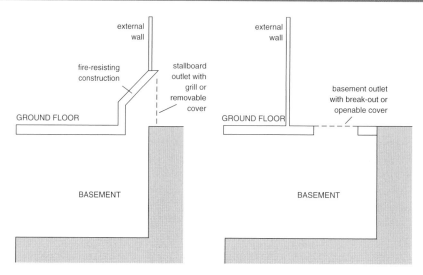

Figure 9.91   Ventilation of a basement by stall board or pavement lights.

Figure 9.92   Fire fighters switch

9.6   Case study

The Summerland leisure complex opened in 1971 and was the largest on the Isle of Man. The Summerland 'family fun centre' housed sports facilities and entertainment venues with the associated retail and food preparation areas.

It was a very large, chiefly single storey building which was considered to be 'state of the art' in terms of innovative use of building materials. The building also had an advanced controlled internal climate. Novel construction techniques had been used, including the use of new plastic materials throughout the building. In order to make the best use of borrowed light combined with some degree of heat insulation the roof had been built of a new Perspex-type material. At that time this novel approach to construction would not have been permitted for buildings on the mainland.

On 2 August 1973, a fire started in a public area where smoking was permitted. The fire quickly developed and spread through the building due to a number of factors including the amount of combustible material used in its construction and breaches in the fire compartmentation by ventilation shafts, which were not fitted with the required fire stopping, passing through compartment walls. It quickly spread through vents which were not properly fire proofed. The fire investigation concluded that the fire possibly started from cigarettes.

When the fire reached the plastic roof of the complex it melted quickly releasing highly toxic smoke into the building and the surrounding area. The roof soon suffered a catastrophic failure and collapsed thereby increasing the oxygen available for the fire and increasing its intensity.

The situation was made worse by the failure of power supplies to safety critical control systems. There was no effective back-up power supply, and inadequate ventilation to allow the occupants to escape.

There were 51 fatalities and over 80 casualties as a result of the fire. Some casualties occurred as a result of being trampled by persons trying to escape the rapidly developing fire while using inadequate means of escape.

An official inquiry after the disaster introduced new fire regulations for the construction industry. Summerland was rebuilt in 1978, and was demolished in 2004.

9.7 Example questions for Chapter 9

1. (a) **Outline** the function 'fire resisting compartmentation' as it relates to fire safety within buildings (8)
   (b) **Identify** THREE ways in which fire may spread from one compartment to another. (6)
   (c) **Discuss** THREE methods that may be used to ensure that the integrity of fire resisting compartments is maintained. (6)

2. (a) **List** the types of fire extinguisher identifying their individual colour code and what class of fire they are used to extinguish. (6)
   (b) **Identify** the most important location for fire extinguishers on a fire escape route. (2)

3. Automatic fire detectors are an important part of a building's fire protection system:
   (a) **Identify** TWO common types of detector and explain how they work. (6)
   (b) **Give** an appropriate siting position for both types you have identified. (2)

4. **Explain** the term 'density factor' and discuss the role this factor has when planning the means of escape from the building. (8)

5. **Identify** the components of a water sprinkler fixed fire fighting system and give an example of the type of premises in which it can be appropriately installed. (8)

6. (a) **Outline** the key principles for the provision of a fire alarm system. (6)
   (b) What is a category M type fire alarm system? (2)

7. As part of your workplace risk assessment, internal self-closing fire doors should be inspected to ensure their adequacy. **Outline** what should be examined. (8)

Appendix 9.1 Classification of purpose groups

Title	Group	Purpose for which the building or compartment of a building is intended to be used
Residential* (institutional)	2(a)	Hospital, home, school or other similar establishment used as living accommodation for, or for the treatment, care or maintenance of, persons suffering from disabilities due to illness or old age or other physical or mental incapacity, or under the age of five years, or place of lawful detention, where such persons sleep on the premises.
(Other)	2(b)	Hotel, boarding house, residential college, hall of residence, hostel, and any other residential purpose not described above.
Office	3	Offices or premises used for the purpose of administration, clerical work (including writing, bookkeeping, sorting papers, filing, typing, duplicating, machine calculating, drawing and the editorial preparation of matter for publication, police and fire service work), handling money (including banking and building society work), and communications (including postal, telegraph and radio communications) or radio, television, film, audio or video recording, or performance (not open to the public) and their control.
Shop and commercial	4	Shops or premises used for a retail trade or business (including the sale to members of the public of food or drink for immediate consumption and retail by auction, self-selection and over-the-counter wholesale trading, the business of lending books or periodicals for gain and the business of a barber or hairdresser) and premises to which the public is invited to deliver or collect goods in connection with their hire, repair or other treatment, or (except in the case of repair of motor vehicles) where they themselves may carry out such repairs or other treatments.
Assembly and recreation	5	Place of assembly, entertainment or recreation: including bingo halls, broadcasting, recording and film studios open to the public, casinos, dance halls; entertainment, conference, exhibition and leisure centres; funfairs and amusement arcades; museums and art galleries; non-residential clubs, theatres, cinemas and concert halls; educational establishments, dancing schools, gymnasia, swimming pool buildings, riding schools, skating rinks, sports pavilions, sports stadia; law courts; churches and other buildings of worship, crematoria; libraries open to the public, non-residential day centres, clinics, health centres and surgeries; passenger stations and terminals for air, rail, road or sea travel; public toilets; zoos and menageries.
Industrial	6	Factories and other premises used for manufacturing, altering, repairing, cleaning, washing, breaking-up, adapting or processing any article; generating power or slaughtering livestock.
Storage and other non-residential +	7(a)	Place for the storage or deposit of goods or materials (other than described under 7(b)) and any building not within any of the purpose groups 1 to 6.
	7(b)	Car parks designed to admit and accommodate only cars, motorcycles and passenger or light goods vehicles weighing no more than 2500 kg gross.

Appendix 9.2   Limitations on travel distance

Maximum travel distance (1) where travel is possible in:

Purpose group	Use of the premises or part of the premises	One direction only (m)	More than one direction (m)
2(a)	Institutional (2)	9	18
2(b)	Other residential (a) in bedrooms (3) (b) in bedroom corridors (c) elsewhere	9 9 18	18 35 35
3	Office	18	45
4	Shop and commercial (4)	18(5)	45
5	Assembly and recreation (a) buildings primarily for disabled people except schools (b) schools (c) areas with seating in rows (d) elsewhere	9 18 15 18	18 45 32 45
6	Industrial (6)	25	45
7	Storage and other non-residential (6)	25	45
2–7	Place of special fire hazard (7)	9(8)	18(8)
2–7	Plant room or rooftop plant: (a) distance within the room (b) escape route not in open air (overall travel distance) (c) escape route in open air (overall travel distance)	9 18 60	35 45 100

10 Safety of people in the event of a fire

The safety of people in the event of a fire in buildings is dependent on having emergency procedures that make full use of the fire safety design features of the building and take account of the behaviour of the occupants when faced with an emergency situation.

Earlier chapters have covered the design of buildings and this chapter examines how people perceive and react to the danger of fire. Only by understanding how people may respond to a fire can effective emergency procedures be developed and implemented that overcome human behavioural problems.

The aim of devising effective emergency procedures is to ensure that the occupants of a building are never exposed to fire effluent or heat or that, if they are, any such exposure does not significantly impede or prevent their escape and does not result in people experiencing or developing serious ill-health effects.

This chapter discusses the following key elements:

➤ Perception and behaviour of people in the event of a fire
➤ The measures needed to overcome behavioural problems and ensure safe evacuation of people in the event of a fire
➤ Emergency evacuation procedures
➤ Assisting disabled people to escape.

10.1 Perception and behaviour of people in the event of a fire

Fire safety in building design is aimed at providing a safe environment for occupants while inside the building. Provision is also made for a safe means of escape for all occupants since a fire emergency usually involves evacuation to a place of safety. Obviously the effectiveness of the means of escape that is provided in any building is reliant upon how they are used at the time of an emergency by individual occupants.

The way an individual occupant of a building will behave to a fire danger is complex. The psychological response of each person is based on their perception of the situation they find themselves in. In order to understand how people perceive the danger of fire it is necessary to consider the principles of sensory perception.

When a person has become aware of an emergency they may react, for example, by spending time thinking about what they should do or by starting to move. This decision will be based upon how seriously they see the risk and how much time they think they may have to evacuate. It can be seen that individual perception is therefore critical to overall escape time.

10.1.1 Principles of sensory perception

The way in which people perceive risk is dictated by individual attitudes, skills, training, experience, personality, memory and their ability to process sensory information; it is the process by which we detect and interpret, i.e. recognise information from our environment.

Detection – is the process of receiving information from the outside world. The detection process involves all the sense organs:

➤ Eyes for gathering visual information
➤ Ears for sensing vibrations in the air, including sound
➤ The nose and tongue are sensitive to certain chemical stimuli
➤ Skin responds to pressure, temperature changes and various stimuli related to pain
➤ The skeletal structure receptors in our joints, tendons and muscles are sensitive to body movement and position.

Interpretation – takes place in the brain. The sense organs send messages to the brain by converting stimuli from the outside world into nervous impulses.

The brain organises the nervous impulses and interprets these as recognisable information about our surroundings such as people, places and events. Recognition and subsequent behavioural responses to what has been perceived is affected by internal factors such as the experience and emotional state of the individual. External factors such as time or money constraints will also have a significant impact on how individuals may interpret information received from their senses.

In order to be able to 'recognise' a situation or object from data received from the senses, the brain will attempt to match a set of data to a previous pattern. For example, a child will learn the look, behaviour, smell and feel of a dog and subsequently interpret anything that looks, behaves, smells or feels like a dog as a dog.

Perception vs reality

Perception varies with individuals, who can interpret sensory data in a number of ways.

The illusion shown in Figure 10.2 is not an inaccurate perception: it is a demonstration of how one perception can be inconsistent with another perception. Recognition of the saxophone player is just as valid as recognition of the young woman. Both are as real and accurate as one another. This example also demonstrates that perception is an active process; humans constantly interpret sensory data to produce recognisable objects and events.

The problems associated with individual perception of reality can be further understood by considering the following key principles in recognition:

➤ There is a tendency to perceive things as complete, filling in the gaps in order to get an overall impression
➤ There is a tendency to perceive objects as constant in size, shape, colour, and other qualities
➤ Sometimes an object that is constant is perceived as variable, for example one moment there appears to be a single object, the next there appears to be more than one.

It is quite unusual for a person to have experience of a real fire or emergency that warrants an immediate evacuation of a building. The normal experience is that of a 'false alarm' therefore it is perfectly understandable that people, in general, do not perceive a serious personal threat when they hear a fire alarm in a building. There have been instances in fatal fires when individuals have failed to perceive the risk of a small fire because the rapid growth and movement of a fire is outside their previous experience and therefore they fail to recognise the magnitude of the risk. If a person underestimates the level of the risk in a fire emergency, they are likely to delay evacuation and thereby increase the risk.

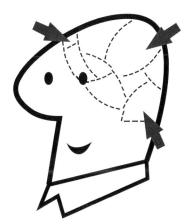

Figure 10.1   Interpreting information

Figure 10.2   Perception can differ

10.1.2 The time required for escape

As discussed in Chapter 9, evacuation can be seen as having four distinct phases:

Phase 1 – alert time from fire initiation to detection/recognition

Phase 2 – pre-movement time taken by behaviour that diverts an individual from the escape route/s

Phase 3 – travel time to physically get to an exit

Phase 4 – flow time, i.e. how long it takes for the occupants to move through the various stages of the escape route. Doorways are invariably the least efficient element with the longest flow time and restriction on the route.

The evacuation time is totally dependent upon the occupants responding to the fire evacuation warning. Once the occupants of a building have been warned of the danger of fire, there is always a delay when a number of initial behaviours can occur before they begin to travel to the exits.

Whereas Approved Document B to the Building Regulations provides guidance on the calculation of flow rates for exit routes, the British Standards Published Document (PD7974 Part 6 2004) – Human factors: life safety strategies – Occupant evacuation, behaviour and condition, provides detailed information relating to phases 2 and 3 of evacuation, i.e. '**pre-movement** and **travel time**'.

When devising a fire evacuation procedure it is useful to consider these two phases of human behaviour. In addition, it is important to understand that there are interactions between pre-movement and travel behaviours for all individual occupants that also impact upon the overall evacuation time.

Evacuation time has two phases:

1) Pre-movement time – the time between recognition and response. This is the time between when the occupants become aware of the emergency and when they begin to move towards the exits.

   In many situations the longest part of the total evacuation time is taken up with this first phase. It is difficult to quantify as it is dependent on the complexities of Human Behaviour.

2) Travel time – the time required for the occupants to travel to a place of safety.

BSI PD 7974-6:2004 p. 2

Pre-movement behaviours

Pre-movement behaviours comprise the recognition by an individual of a fire emergency and their initial response to the situation they have perceived. Recognition that there is a high level of risk acts as a powerful incentive to respond by deciding to evacuate.

During the pre-movement phase occupants will be carrying out their normal activities at the same time as receiving and processing information about the developing emergency situation. In this phase the normal activity of occupants may be active or inactive.

Once the perception process has been completed individuals will make decisions about how they may respond to the information received. Depending upon the perception of risk, occupants of buildings will often embark on activities which do not involve movement

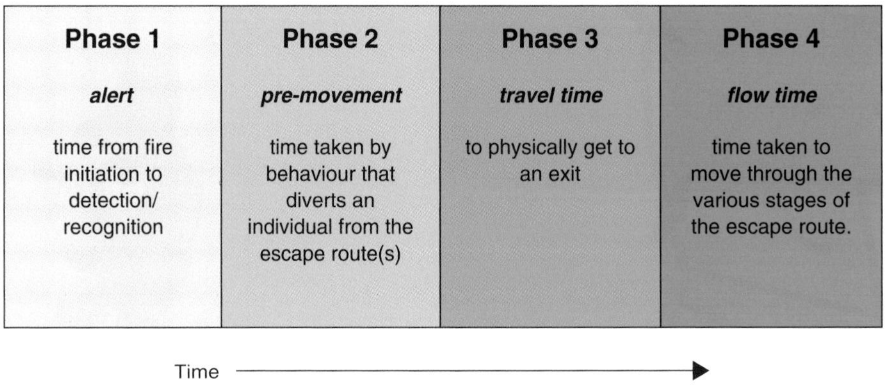

Phase 1	**Phase 2**	**Phase 3**	**Phase 4**
alert	*pre-movement*	*travel time*	*flow time*
time from fire initiation to detection/recognition	time taken by behaviour that diverts an individual from the escape route(s)	to physically get to an exit	time taken to move through the various stages of the escape route.

Time ⟶

Figure 10.3 Four phases of evacuation

to the escape routes. Examples of pre-movement behaviours are:

➤ Completing the activity being undertaken
➤ Trying to verify reality or importance of the warning
➤ Investigation, e.g. to determine source
➤ Safety activities, e.g. stopping machinery
➤ Security activities, locking tills
➤ Alerting others

Figure 10.4 Pre-movement behaviours affected by the job in hand

➤ Gathering together others, e.g. children
➤ Fire fighting
➤ Collecting personal belongings.

Well understood and rehearsed emergency evacuation procedures help individuals maximise their recognition of the fire danger and reduce any of the above, possibly unnecessary and irrelevant, pre-movement behaviours.

Travel behaviours

If people perceive the building as unsafe then the normal response is to leave a building as soon as possible. Whether the decision to evacuate is delayed or taken as soon as possible, once it has been made, individual occupants will begin to travel through the escape routes. Their behaviour in this phase of evacuation is then influenced by such factors as:

➤ Their role
➤ The number of people in the building
➤ Their distribution within the building at different times
➤ Their familiarity with the building
➤ Their familiarity with the route
➤ The characteristics of the occupants and the building.

Table 10.1 provides some examples of how these factors or 'travel time determinants' can influence the travel times of people with various roles.

Table 10.1 The characteristics of the occupants and the building interact and together determine the time required to escape

Travel time determinant	Example
Role of the individual	Those with responsibility for others during a fire, such as parents, elder siblings, nurses, teachers, etc., will delay their evacuation to ensure those they are responsible for are ready and able to escape. Others will have follower relationships and affiliations with such leaders.
Number and distribution of occupants	If there are a large number of people in the building, the travel time will be dependent on the maximum flow capacity of the escape routes. This is particularly relevant at times where the distribution of people in the building is concentrated in certain areas, such as a canteen where there may be normal circulation problems. Crowd flow can cause danger and prohibit safe escape but the flow can be modified by emergency evacuation messages.
Familiarity with the escape route	People nearest to a familiar entrance route typically leave by that entrance. Familiarity means that even when this route is not close, there is a tendency to return to it and use it as the escape route. Regular use of a fire exit route has a strong influence on people's inclination to leave by that fire exit in an emergency. 'Emergency Exit Only' signs, far from encouraging use of an exit in an emergency, may have a detrimental rather than positive effect on travel times. It reduces people's familiarity with the route and reduces their inclination to move towards it in an emergency.
The characteristics of the occupants and the building	People who are wide awake, fit and mentally alert have the potential to escape quickly whereas others who may be less fit or alert will take longer to escape. In some cases individuals may be unconscious or non-ambulant in which case they will need extra consideration when planning emergency evacuation procedures. In the case of secure accommodation, e.g. detention centres and prisons, the occupants of a building may wish to leave faster than the emergency procedure allows.

10.1.3 Characteristics of people influencing safe evacuation

How the occupants of a building may react during both the pre-movement and travel time phases of evacuation are influenced by the characteristics of the people involved. When designing the means of escape from a building it is vital to take into account the number of occupants in a building, together with their density and distribution throughout the building.

However, in addition to these 'occupancy factors' there are a number of other important aspects that influence people's behavior at the time of a fire emergency in a building, which must be considered when developing and managing an emergency evacuation procedure, including:

➤ Sensory condition
➤ Physical condition
➤ State of conciseness
➤ Initial reactions
➤ Stakeholding
➤ Fire and/or heat in the building
➤ Building design features.

Sensory condition
When considering the development and management of emergency evacuation procedures it is often assumed that the full range of senses is available to the full range of occupants of a building. Most fire warning systems rely on an audible warning to initiate evacuation. Once evacuation has been initiated, most exit routes are indicated by visual signs (e.g. a pictograph of a fire exit) and instructions (e.g. push bar to open).

It follows that a sensory disabled person may have difficulties not only in perceiving the risk associated with a fire in a building but may not hear the audible evacuation warning or be able to recognise the way out. It is therefore essential to identify the special needs of any disabled employees when planning your fire safety arrangements and evacuation procedures, for example those with sensory impairments:

➤ People with hearing impairment may not have a clear perception of some types of conventional (audible) warning signals
➤ People with visual impairment may require assistance with moving away from affected areas and locating the designated place of safety.

It is also worthy of note that recent surveys suggest that fewer than 10% of those people suffering with visual impairment can read Braille.

Physical condition
Evacuation studies have been carried out which suggest a general guide to the number of people who might require special assistance in an evacuation. Studies have concluded that, on average, some 3% of people in such buildings cannot or should not evacuate using multiple flights of stairs. Similarly it should be assumed that 3% of the members of a large public crowd may require

Figure 10.5 Buildings should be designed to facilitate human understanding of their role and function

Figure 10.6 Sign for people with hearing impairment

Figure 10.7 Sign for people with visual impairment

assistance to use the means of escape provided for the able bodied.

Figure 10.8 includes individuals who may be affected by disabilities that are not readily apparent to the onlooker, for example some psychological disorders or phobias can affect an individual's ability to effectively use the means of escape which are provided for general use. Although most people within this group may be perfectly capable of using, for example, an external staircase, they should ideally evacuate following, not among, other evacuees.

The empirical evidence suggests that less than 1% of the generally active population, outside institutions, use movement aids such as wheelchairs, walking frames, crutches, etc., and some of this group are able to use stairs unassisted, but at a reduced speed.

With the introduction (as a result of the DDA) of the duties placed on persons who control access to buildings to provide access for those with disabilities, the numbers of people using movement aids in workplaces may be expected to increase.

State of consciousness

An important aspect that affects individual survival from a fire situation is the time delay between discovery of a fire and people beginning to evacuate. In the best cases persons who need to evacuate a building will be awake and alert at the time they are required to react to a fire or an alarm.

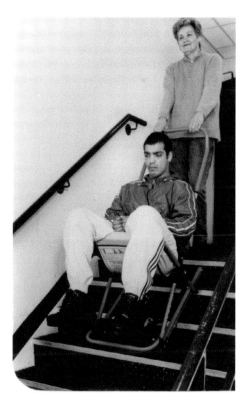

Figure 10.8 Evacuation chair in use

In many fatal fires it is a fact the people are not fully conscious for a number of reasons, including:

➤ Being asleep
➤ Being under the influence of alcohol or recreational drugs
➤ Suffering from a condition that results in confusion
➤ Being under the influence of medication.

Where it is known or can be reasonably foreseen that persons requiring to escape a building in case of fire may have a reduced state of consciousness, consideration must be given to increasing the standard of means of escape, including the provision of adequate detection and warning systems that ensure early warning is given to occupants.

Staff evacuation procedures must also be taken into account, especially when dealing with occupancies such as residential care premises, where total evacuation may be inappropriate.

Initial reaction

Research has shown that the average time to react to the alarm (i.e. start to move) is normally approximately 30 seconds. However, there are occasions where the time taken to initially react to a fire or alarm has been greatly extended, for example:

➤ In hotels where people may be asleep
➤ In offices where people may be engaged with finishing a particular piece of work
➤ In sports halls where people may be changing or engaged in a sporting activity

Enquiries into disasters such as the Summerland fire and the Bradford City Football Club fire indicate that

Figure 10.9 Alcohol affects people's reaction to emergencies

the time between the fire starting and people beginning to evacuate was considerably longer than 30 seconds. The enquiries suggest that early and informative warning of the public is an absolutely critical aspect of any building evacuation management system. When dealing with the evacuation of large numbers of the public it is also vital that there is efficient communications between staff.

Stakeholding

The time that individuals take to respond to a fire situation, the pre-movement behaviour and the time taken to travel can be seriously affected by any stake that the person may feel they have in the outcome of a fire. For example:

Financial stake – if a person is liable to losses of money as a result of a fire they are likely to be motivated to attempt to either fight the fire or salvage valuable assets. Enquiries into fires indicate that having even a relatively small financial stake in the outcome of a fire has led people to lose their lives. There is an example where the evacuation of a restaurant was delayed because the customers had paid for their meal and were determined to eat it rather than respond immediately to a fire alarm.

Moral stake – if it is apparent that there may be loss of life at a fire, individuals will be motivated to attempt to save people who are endangered. It is not only human life that people are willing to delay their own evacuation for, attempts to save animals, particularly family pets, often result in people risking their lives rather than getting to a place of safety.

Legal stake – if a person feels that they may be liable to be subjected to legal action as a result of a fire,

they can be motivated to make attempts to minimise the impact of the fire. There are numerous examples where there have been significant delays in raising the alarm and/or calling the fire service because an individual has accidentally or negligently started the fire for which he knows he may well be sued or prosecuted.

Fire effluent and/or heat

People may be exposed to one or all of the following three main categories of fire dynamics: flames, smoke and heat. In a major fire disaster any one or combination of these exposures affects people's ability to react effectively to the situation.

Flames – exposure can be minimal, i.e. seeing the fire. Greater exposure will cause burns and can be life threatening.

Smoke – there is a strong relationship between visibility and disinclination to move through smoke. The general indication is that visibility has to be reduced before people begin to be strongly deterred.

The majority of people become less inclined to move through smoke when it is described by them as 'thick' or 'black'. This has been estimated to be when visibility through smoke has been reduced to a minimum of approximately 0–5 m.

Case studies of fires have further shown that the floor a person is located on and the thickness of smoke conditions have a major influence on the likelihood of a person using a room as a refuge and waiting for fire brigade rescue.

Heat – there are fire emergencies when people will become aware of an increase in temperature. Exposure to heat is most likely for those who are in the vicinity of the origin of the fire. Heat is a physiological stressor and can be life threatening.

It is important to include the likely effects of exposures to fire effluent and heat in both minor and major fire emergencies when devising evacuation procedures.

Building design features

The building design will determine the travel distances to reach a place of safety and therefore evacuation time. The key characteristics of a building that impact upon people's ability to evacuate in time of an emergency are:

➤ Use of the building
➤ Dimensions of the building, including the number of floors
➤ Layout of the building, for example individuals' pre-movement times are less variable in an open plan setting such as offices, theatres, etc., than in other settings
➤ Building services – including method of detection, provision of warnings and fire safety management systems

Figure 10.10 Having a financial stake in the outcome can affect people's reaction to a fire

➤ Emergency management strategy
➤ Means of escape – safe routes provided for people to travel from any point in a building to a place of safety.

10.1.4 Crowd movement

The problems associated with the behaviour of individuals in the event of fire are increased when large numbers of people are gathered together. Research has highlighted several crucial factors which influence the way a crowd may behave. The behaviour of individuals in a crowd often differs from when those same people are by themselves or in smaller groups. For example:

➤ Individuals in a crowd can be greatly influenced by the actions of others in a crowd, e.g. if one or two people in a crowd take a short cut, then others will tend to follow
➤ Individuals in a crowd are more likely to voice collective frustration at delays caused by excessive queuing
➤ Individuals in a crowd are more likely to be susceptible to panic and once panic starts in a crowd it can quickly spread
➤ In emergencies individuals will become more aggressive in order to escape
➤ Individuals' emotions are often heightened in a crowd, sometimes as a consequence of public entertainment, sometimes by the experience of just being in a large group of people.

In the early stages of an emergency a crowd can be easily influenced by a person who is perceived to be 'an expert'. It is for this reason that it is vital to establish good clear communication with a crowd and provide other good assistance to it in order to give the crowd confidence in the arrangements. This will minimise individuals in the crowd going their own way and leading others into danger.

Flow rates

The time it takes a crowd to flow towards and through an exit reduces as the size of the crowd increases. It is therefore important to understand the flow rate of a crowd of people evacuating a building, particularly when considering the evacuation of large numbers of people in case of fire.

One individual who is able bodied and unimpeded can be expected to travel along an escape route at a speed of around 1.2 metres per second. Therefore to travel, for example, 45 metres will take an individual (in theory) 37.5 seconds (i.e. 45 ÷ 1.2). The speed of unimpeded flow down stairways is in the region of 0.8 metres per second. However, when travelling in a crowd, an individual's progress is impeded.

Many assessments of rate of flow of persons through exits have been made. As a result of these tests, the guidance given to designers in the UK (contained in the Building Regulations ADB) is that 40 persons per minute per 0.75 metre of exit width (see Chapter 9). In other words, a 750 mm exit will allow 60 persons to move through it in 1½ minutes.

Average flow rates of people through given exit widths can be calculated to demonstrate that as the width available for people increases so does their rate of flow through that width. Flow rate is expressed as persons per second per metre (persons/s/m).

Table 10.2 summarises the flow rates of people through various exit widths, assuming that they have 1½ minutes to pass through the exit. The table demonstrates that flow rates increase with the increase of the available escape width.

Figure 10.11 An individual's behaviour changes when in a crowd

Table 10.2 Examples of maximum flow rates of persons using escape widths

Maximum number of people	Minimum width mm	Maximum flow rate persons/s/m
60	750	0.66
110	850	1.22
220	1050	2.44

There are of course a number of factors which impact upon the flow rates of a crowd including:

➢ The clarity of the indication of the direction of travel
➢ Any apparent sign of fire
➢ Degree of perceived urgency of the evacuation
➢ The lighting levels
➢ The underfoot conditions
➢ The characteristics of the crowd, i.e.:
  ➢ Age
  ➢ Mobility
  ➢ Alertness, etc.

Traditionally flow rates have been calculated from historical data and these are now being reviewed in the light of electronic data gathered from computer simulation. The more information individuals have about a fire emergency the greater their ability to respond appropriately. However, influences also include the provision of appropriate design arrangements for areas such as crèches within public buildings, e.g. sports centres.

Crowd dynamics will be directly affected by parents and guardians wishing to ensure that their charges have left the building and it is likely that they will not rely upon the attendants or nursing staff.

In situations where there are large numbers of people or people are within large spaces, e.g. shopping malls, their behaviour is shaped by a number of factors including the following.

Spatial awareness – the degree to which people are able to orientate themselves in unfamiliar surroundings will affect their choices of action in the event of a fire. It is generally accepted that people will leave a building by the route they entered irrespective of the availability of closer alternative exits.

Smoke movement – there have been instances where the movement of smoke within shopping malls has given the false impression that the fire is in one particular direction. It is understandable that people will want to move away from the perceived seat of the fire which has resulted in them moving towards the fire rather than away from it.

Ergonomics – despite the adequacy of the means of escape (size, travel distance) unless it is laid out in a way that people can easily understand and signed in a consistent and logical manner it will slow the evacuation of a building. In order to encourage people to evacuate within a reasonable time the means of escape must reflect good ergonomic principles in such a way that evacuating a building becomes intuitive.

Figure 10.12 Manchester airport fire involving an aircraft on the runway – A quickly developing fire in an enclosed space which resulted in an urgent need to escape and reduced evacuation time

In the early morning of 22 August 1985, a full complement of 130 passengers boarded a British Airways Boeing 737-200 aircraft at Manchester airport, which was scheduled to fly to the holiday island of Corfu.

As the aircraft approached take-off speed, the flight crew heard a loud thump. Believing that a tyre had burst they immediately aborted the take-off run, and informed the control tower of the situation. However, what the flight crew had heard was not a tyre bursting but the port engine partially disintegrating. Some of the engine casing ruptured a fuel tank, and as the aircraft began its emergency deceleration, aviation fuel gushed over the red hot exhaust and ignited.

Unaware of the fire, the flight crew continued to use the reverse thrust to slow the aircraft; this action served only to fan the flames. When the aircraft eventually turned off the runway and stopped, the fire was well established and very intense. Aviation fuel spilled out of the wing tank and formed a flaming lake on the concrete. Because of the orientation of the aircraft when it was bought to rest, the prevailing wind blew the flames towards the body of the aircraft, and within 30 seconds the fire had broken through to the passenger cabin.

The scene inside the aircraft was by that time chaotic, and as black toxic smoke filled the cabin, the crew lost precious seconds struggling with a jammed door. People were overcome by smoke as they struggled in the dark and confusion towards emergency exits, bodies began to block the central aisle. The number of passengers, and the fact that two of the exits were engulfed in flames, further hampered the evacuation. In a further 60 seconds the rear fuselage collapsed.

Although the emergency services arrived at the scene quickly they were unable to save 55 of the passengers, almost all of them were in the rear cabin. Dozens of the survivors suffered different injuries.

A subsequent research project found that although the exit widths were adequate for evacuation under normal conditions, when the need to escape became urgent the understandable panic significantly reduced the flow rates through the doors.

10.2 The measures needed to overcome behavioural problems and ensure safe evacuation of people in the event of a fire

It can be seen from the foregoing sections that there are many factors that affect people's behaviour in a fire emergency. Different and complex combinations of these factors influence the time and direction of movement for a particular incident. In contrast, there are only a few simple measures that need to be taken to overcome the majority of these behavioural problems. Each of the following plays a key role in overcoming behavioural problems with securing the safe evacuation in the event of a fire:

➤ The emergency plan
➤ Detection
➤ Warning signals
➤ Layout of escape routes
➤ Emergency instructions
➤ Rehearsal
➤ Competent staff.

10.2.1 The emergency plan

The key to ensuring that the behavioural problems of people in an emergency situation are minimised is to develop a comprehensive emergency plan. If there is a plan that is well thought out and clearly communicated people will tend to trust it and be willing to play their part. The plan should include, not only the actions that individuals are expected to take in an emergency, but also some arrangements for business continuity.

Figure 10.13 High visibility clothing infers authority in a crowd situation

The emergency plan for the evacuation of a building will be based on the findings of the fire risk assessment (see Chapter 14). In many small buildings the emergency plan may merely be a set of simple instructions to staff as to the actions they are required to take in the event of a fire.

However, for most buildings, particularly those that are shared with others, or have hazardous processes, it is necessary to develop a more detailed and comprehensive plan.

In guidance published jointly by the government and the HSE (*Fire Safety – An Employers Guide*) the following contents of an emergency plan are suggested.

The plan should provide clear instructions on:

➤ The action employees should take if they discover a fire
➤ How people will be warned if there is a fire
➤ How the evacuation of the workplace should be carried out
➤ Where people should assemble after they have left the workplace and procedures for checking whether the workplace has been evacuated
➤ Identification of key escape routes, how people can gain access to them and escape from them to places of safety
➤ The fire-fighting equipment provided
➤ The duties and identity of employees who have specific responsibilities in the event of a fire
➤ Arrangements for the safe evacuation of people identified as being especially at risk, such as contractors, those with disabilities, members of the public and visitors
➤ Where appropriate, any machines/processes/power supplies which need stopping or isolating in the event of fire
➤ Specific arrangements, if necessary, for high fire risk areas of the workplace
➤ How the fire service and any other necessary emergency services will be called and who will be responsible for doing this
➤ Procedures for liaising with the fire service on arrival and notifying them of any special risks, e.g. the location of highly flammable materials
➤ What training employees need and the arrangements for ensuring that this training is given.

In addition to the written description it is often very useful to produce a simple line drawing of the plan layout of the building. If the premises are larger or complex, then it is particularly useful to include a line drawing. This can also help to check fire precautions as part of the ongoing review.

The drawing should include:

➤ Essential structural features such as the layout of the workplace, escape routes structure and self-closing fire doors provided to protect the means of escape
➤ Means for fighting fire (details of the number, type and location of the fire fighting equipment)
➤ The location of manually operated fire alarm call points and control equipment for the fire alarm
➤ The location of any emergency lighting equipment and any exit route signs
➤ The location of any automatic fire fighting system and sprinkler control valve
➤ The location of the main electrical supply switch, the main water shut-off valve and, where appropriate, the main gas or oil shut-off valves.

A clear plan of the building that is available to employees, visitors, contractors and the fire service has a number of important benefits including:

➤ Assisting employers to better understand the emergency plan
➤ Facilitating easy management of the means of escape

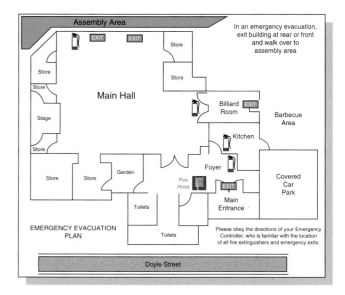

Figure 10.14 Plan of premises must be included in the overall emergency plan

➤ Assisting the safety briefing of employees, visitors, contractors, etc.
➤ Assisting fire service operations at the time of an emergency.

10.2.2 Detection

In order to maximise the amount of time that people in buildings have to escape a fire it is important to ensure that there are adequate arrangements for detecting a fire as soon as one breaks out.

All fires give off heat, light and smoke, each of which can be detected by a variety of means. In some small premises that are occupied around the clock, it may be sufficient to rely on human detection, i.e. people smelling the smoke or seeing the flames. In most cases, however, some form of automatic system is the only means of achieving adequate detection of a fire.

Providing automatic fire detection in zones often results in a quicker identification of the location of the fire and can provide an overview of the extent of fire and smoke spread throughout the building which will aid evacuation and response.

In the case of a multiple storey building the fire detection system is often zoned in floors which allows for appropriate sequential evacuation of the occupants if necessary. Chapter 9 discusses in detail automatic fire detection systems.

10.2.3 Warning signals

Confirmation that there is a requirement for an emergency evacuation is totally reliant on the perception of the person reporting the incident and his or her ability to comprehend the situation and possible impact.

There should be a simple system in place to enable confirmation and activation of the emergency action procedure. This in turn will depend on the communication of reliable information about the risk and/or the level of exposure to hazards.

If the warning is perceived as a precautionary measure the decision to move may be delayed or in some circumstances even ignored. This will be influenced by the experience of any previous unnecessary evacuations. As previously discussed the UK standard for fire warning systems is given in BS 5839-1:2002, which outlines the requirements for these signals.

Audible alarm signals – it is essential that audible alarm systems are sufficient in nature and volume that all those persons for whom it is intended are able to recognise it for what it is. In general a sound level of 60 to 65 dB(A) is considered appropriate. Additional requirements will be required when workplaces have higher background noise levels or where people will need to be roused from sleep.

A critical factor to also consider in the use of audible alarms is the possibility that a workplace may have a number of alarms that could be of the audible type, such as pressure alarms, boiler alarms and burglar/security alarms. Selecting the correct alarm to ensure people know what is expected of them is essential if the response in the event of a fire emergency is to be successful.

Research has been conducted into the effectiveness of providing audible speech warnings incorporating evacuation instructions. Informative fire warning systems are found to be more effective than a conventional fire tone alarm in encouraging a prompt evacuation and also provide specific details of directions to take, etc.

Over recent years, particularly within large places of public assembly such as shopping centres, sports stadia, etc. emergency voice communications (EVC) systems have been utilised; BS 5839 Part 9 provides detailed information of the system requirements. EVCs are intended for specific types of communication; they are not, for example, designed for general use for non-emergency purposes.

EVC systems are intended for:

➤ Use by the management of the building or complex for initial evacuation in the first stages of evacuation, before the fire service arrives. The EVC system

Figure 10.15 Smoke detector

Figure 10.16 Audible warning devices

Figure 10.17 Braille press for help sign

Figure 10.18 Visual alarm combined with an audible sounder

is used for communication between a fire incident controller at a control centre with, for example, fire wardens or fire marshals on various floors or areas of a building or with stewards at a sports venue

➤ Use by the fire service during an evacuation after their arrival at the building or venue; the fire service would normally take over control of evacuation, with an officer at the fire control centre communicating with other officers via the EVC system

➤ Use by the fire service after evacuation to assist fire fighting operations

➤ Use by disabled people in refuge during a fire, to identify their presence and communicate with a person, e.g. a control room operator, at the fire control centre.

Research into the effectiveness of voice warning systems has concluded that, although informative messages are extremely effective, it is even more important to provide an adequate means of escape and support training.

Visual alarm systems – visual alarm systems (beacons) are used to supplement audible warning signals in situations where an audible system is likely to be ineffective, for example where the ambient noise levels in the building would make it difficult to hear an audible signal. Beacons must be provided where ambient noise levels exceed 90 dB(A), or in areas where hearing protection is worn or people with hearing impairments may be present.

There are a number of other occasions where it is appropriate to provide visual warning, for example in radio or television studios or hospitals and theatres where the evacuation of the building's occupants is reliant upon staff and where it can only be done when the occupants remain calm.

Where beacons are required they should:

➤ Be distributed so that they can be seen from all normally accessible locations and under all normal lighting conditions

➤ Flash at a rate between 30 and 130 flashes per minute

➤ Be easily distinguished from other visual signals used in the building

➤ Be bright enough to attract attention, but not so bright as to cause difficulty with vision due to glare

➤ Be mounted at a height of 2.1 m, but no closer than 0.150 m to the ceiling.

As in the case of audible alarms, confusion may be an issue if emergency escape visual alarm systems become confused with other types of visual warning devices.

Fire alarm warnings for people with impaired hearing – impairment of hearing does not necessarily mean that a person is completely insensitive to sound. Many people have a sufficiently clear perception of some types of conventional audible warning signals to require no special arrangements for warning of a fire. However, there may be some occasions where people with impaired hearing rely on others to provide the necessary warning.

If the occupants of a building have impaired hearing it may be appropriate to provide visual alarm systems in the areas occupied by those people. This should include any toilets and welfare facilities. If people with impaired hearing sleep in the building it may be appropriate to provide tactile devices linked to visual alarm devices. These devices may, for example, be placed under pillows and wired into the fire alarm circuits or be triggered by radio signals.

Fire alarm warning signals for people with impaired hearing may also be provided as portable equipment, for example a radio pager or other system using radio communication.

Portable alarm devices – it is possible to provide portable alarm devices to supplement the primary means of giving warning in case of a fire. It is common for these systems to be operated by local radio signals. If this is the case the portable devices should meet the following criteria:

➤ The alarm should operate within 5 seconds of the radio signal being sent and operate for at least 1 minute

➤ Where the alarm device is used for other purposes, the fire signal should have priority over any other signal and be sufficiently differentiated from signals for other non-emergency purposes

➤ Any failure of the system should be identified at the time of the failure

➤ Portable alarm devices must be fitted with a low battery warning system

➤ If the alarm device is capable of being switched off, it should be designed in such a way so as to avoid accidental operation

➤ Any faults at the control equipment for the portable devices should result in an alarm.

10.2.4 Layout of escape routes

The elements of the means of escape, which are provided within a building, are discussed in Chapter 9. However, in addition to the size, number, width and length of escape routes it is vital that the routes are laid out and indicated in such a way that people who need to use them in an emergency are able to readily identify the direction of travel. In order to ensure that people do not hesitate to use the means of escape with which the building is provided, it is necessary to ensure that there is:

➤ Adequate signage, which complies with the Health and Safety (Safety, Signs and Signals) Regulations 1996

➤ Adequate lighting of the escape routes and any signage

➤ Clear information relating to the operation of any security devices throughout the route, for example 'push bar to open' or 'twist knob to open'

➤ Escape routes should be intuitive.

10.2.5 Emergency instructions

It is vital to ensure that the emergency instructions for occupants in a building are clear, consistent, simple and understandable. Emergency instructions should be given on various occasions, for example when staff join an organisation or are relocated within it. For contractors and others the emergency instructions should be provided at the time they enter the relevant building. For members of the public and for others it is normal that the emergency instructions are only given by the provision of notices located at various points throughout the building.

Emergency instructions provided on notices should contain information relating to such issues as the action to be taken when a fire is detected, the action to be taken when the alarm is heard and any special additional instructions that may be required.

A typical example of a fire action notice, which complies to the Health and Safety (Safety Signs and Signals) Regulations 1996, is shown in Figure 10.19. It

Figure 10.19 Typical fire action notice

will be noted that the written instructions are supported by pictograms where possible in order to overcome problems associated with literacy in English.

In some situations it will be necessary to consider the provision of emergency instructions in a number of different languages.

10.2.6 Rehearsal

Irrespective of the quality of the means of escape within a building and the associated systems, it is always necessary to rehearse the emergency procedure. The advantages of conducting fire drills include:

➤ The testing of the systems to ensure they operate as expected

➤ Increasing the familiarisation of the procedure with the occupants of a building

➤ To allow those who hold key roles, e.g. fire wardens and fire incident controllers, to practise their roles

➤ To demonstrate to staff and enforcement bodies that reasonable arrangements have been made to ensure effective evacuation in case of fire.

The frequency with which the rehearsal of a fire evacuation procedure is conducted will be dependent upon the size and nature of the building and its purpose group (see Chapter 9). In general fire evacuation drills should be conducted on a 6 monthly basis and if a building is occupied over a 24 hour period, the fire evacuation drills should be arranged in such a way that staff on each shift are provided with the opportunity to rehearse their actions.

A critical element in the rehearsal for an emergency is the debrief, which if omitted negates the purpose of the exercise as key learning opportunities are lost.

10.3 Emergency evacuation procedures

Decisions relating to the method of emergency evacuation are made at the time a building is designed and based on such factors as the proposed purpose group and the size of the building. Once a building is occupied, it will be necessary to devise detailed emergency evacuation procedures that take into account the design features and fixed installation of a building. Since the introduction of the RRSFO, the person responsible for planning and implementing emergency evacuation procedures is the person identified as being 'responsible' for fire safety management.

When devising emergency procedures the responsible person will need to consider:

➤ The characteristics of the occupants, their disposition within the building, their physical and mental state

➤ The characteristics of the building in terms of its size, use and construction
➤ The physical provisions for means of escape in the building
➤ The circumstances under which it will be necessary to evacuate
➤ What the arrangements are for fighting the fire
➤ What type of evacuation will be appropriate, e.g. phased or full
➤ How the evacuation is to be initiated
➤ What arrangements are needed to call the fire service
➤ What special roles are required to support the procedure, e.g. fire warden and fire incident controllers
➤ What information, instruction and training is necessary to support the procedure.

The fire risk assessment will provide much of this information and it will also give an indication of the time available for the occupants of a building to reach a place of safety. In most cases, the full evacuation of a building should be achieved in the region of 2.5–3 minutes. This time will be extended where a phased/sequential procedure is adopted.

10.3.1 The procedure

The procedure should cover:

➤ What to do on discovering a fire or smell smoke
➤ What to do on hearing the alarm
➤ Roles and responsibilities of staff, e.g. conducting in assisting disabled occupants
➤ Arrangements for calling the fire service
➤ How to save time, e.g. leaving personal belongings behind
➤ Where to evacuate to
➤ Any special precautions that may need to be taken.

Clear concise arrangements should be displayed; they may also be supplemented by additional information such as escape route plans in hotels and licensed premises.

10.3.2 Competent staff

In order to ensure the safe evacuation of people in the event of fire it is not possible merely to rely on building design, adequate means of escape, fire alarms, emergency lighting, etc. In all but the smallest workplaces it will be necessary to have staff that have been trained to assist with emergency evacuation. Specifically those people responsible for fire safety within buildings and outside venues will need to consider the provision of fire wardens/marshals, crowd safety stewards, fire alarm verifiers and fire incident controllers.

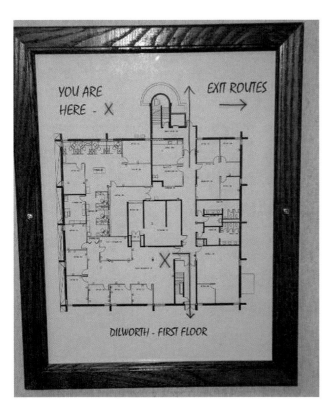

Figure 10.20 Visual representation of an emergency evacuation procedure

Fire wardens/marshals

Role – the terms fire warden and fire marshal are used to describe the same role. It is important to realise that fire wardens and fire marshals have two distinct roles which contribute to the safety of people in fire. First, they have a proactive role that requires them to support the ongoing management of fire safety by carrying out such duties as:

➤ Carrying out an ongoing fire risk assessment while at work
➤ Identifying fire hazards and removing them or reporting them to management, for example:
  ➤ Taking action to reduce the risk of arson
  ➤ Ensuring flammable liquids are stored and used correctly
  ➤ Ensuring sources of ignition are limited or controlled (for example, checking permits to work for any hot work)
  ➤ Monitoring smoking in the workplace
  ➤ Monitoring the build-up of combustible storage and waste
➤ Monitoring fire protection measures, for example:
  ➤ Ensuring fire doors are in good condition and kept locked or closed shut as necessary
  ➤ Ensuring fire fighting equipment is in position, tested and in good condition
  ➤ Ensuring means of escape including corridors and final exit doors are not obstructed particularly with combustible material
  ➤ Ensuring all doors required to provide emergency egress are clearly marked and operate as they should
➤ Knowing what action to take in the event of a fire
➤ Being trained to tackle fire should the need arise
➤ Being competent to assist in the full and safe evacuation of people in the event of a fire.

The specific proactive duties of a fire warden/marshal will need to be devised as a result of the fire risk assessment for the workplace. It is important to record the identification of the individuals fulfilling the role along with their training and their specific duties (see Appendix 10.2).

In terms of the reactive role of a fire warden/marshal, they need to competently respond to emergencies. Different organisations will develop slightly different procedures for emergency evacuation and therefore will require their wardens/marshals to take actions that are tailored to the particular building/organisation.

However, the emergency role of a fire warden/marshal must include the following:

➤ Knowing how to raise the alarm
➤ Knowing how to call the fire service
➤ Knowing where the means of escape for the part of the workplace that they are responsible for is
➤ Being prepared and trained to use fire fighting equipment if it is safe to do so
➤ Assisting the evacuation of people by:
  ➤ Donning a high visibility jacket or waistcoat in order to be easily recognised
  ➤ Assisting disabled staff members in accordance with individual PEEPS
  ➤ Conducting a quick but thorough check or sweep of all rooms including walk-in cupboards, plant rooms and toilets
  ➤ Ensuring heat generating equipment is turned off
  ➤ Closing doors and windows if possible
  ➤ Reporting to the fire incident controller the situation within their area of responsibility)
  ➤ Take a roll call
  ➤ Assist the return to the workplace when the fire service confirm it is safe to do so.

Figure 10.21 A fire warden supervising an emergency evacuation

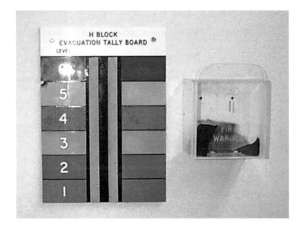

Figure 10.22 Example of an evacuation tally board

Deployment – for most normal low risk workplaces the fire risk assessment is likely to identify a small group of staff who act as fire wardens/marshals. Typically in multi-storey buildings organisations will ensure that there are sufficient fire wardens allocated to each floor to enable their duties to be effectively discharged. It is vital that those responsible for managing fire wardens understand the need to ensure effective arrangements are in place to cover for staff absences. Fire wardens are likely to be away from the workplace for a variety of reasons including holidays, sickness, visits to the doctor, meal breaks. Management will therefore need to ensure that their duties are covered for these periods of absence. It is normally sufficient to nominate and train additional staff and to locally manage their availability in the workplace.

Training – as with all safety training it is important to record the fact that a certain degree of competence has been achieved through the process. Once again the specific content of the training will be informed by the contents of the fire risk assessment for the workplace. However, as a minimum, fire warden/marshal training should include such topics as:

➤ The reasons why managing fire safety is important
➤ Understanding how fire starts, develops and spreads
➤ Measures required to reduce the risk of fire
➤ Measures required to protect people in the event of fire
➤ The specific proactive and reactive role of the fire warden/marshal.

Fire incident controllers (FIC)

Role – fire incident controller is a term used to describe those responsible for coordinating fire wardens/marshals in their proactive and reactive roles. FICs will normally nominate and organise fire wardens and marshals to ensure that fire risk is effectively managed on a day-to-day basis within the workplace. In addition to their proactive role they are also responsible for the control of an emergency evacuation. Their emergency duties will normally include such activities as:

➤ Donning a distinctive high visibility surcoat to ensure they are visible
➤ Establishing control at the fire assembly point
➤ Ensuring the fire service has been called
➤ Collating the information from fire wardens/marshals giving instruction to the people who have evacuated
➤ Liaising with the fire and other emergency services who may attend
➤ Controlling the dispersal or re-entry of the occupants of the building when the fire service say it is safe to do so.

In order to ensure that the FIC's role can be carried out effectively, it is necessary to provide equipment and information at the assembly point at the time of evacuation. This is achieved by the provision of a 'grab bag'. A grab bag is a distinctive holdall that is kept at a position that is normally continually staffed, for example in the reception area of a building. At the time of an evacuation the bag is taken to the assembly point either by a nominated person, e.g. the receptionist, or any other member of staff/fire warden/marshal. Typical of the contents of a grab bag are:

➤ High visibility surcoat displaying the words 'Fire Incident Controller'
➤ Method of communication, e.g. megaphone, portable radio
➤ Pre-printed forms to record the reports from fire wardens/marshals
➤ A copy of the relevant fire emergency plan
➤ Information relating to the building for fire service use including:
  ➤ A plan of the layout
  ➤ The location of service shut-off facilities
  ➤ The location and nature of any specific hazards
  ➤ The details of the fire alarm system
  ➤ Details of any fixed fire fighting installations, e.g. sprinklers.

Deployment – as with fire wardens/marshals, it is important to ensure that a competent FIC is always available

Figure 10.23 A typical grab bag used to carry information and equipment to the assembly area

when required. In order to achieve this adequate management arrangements must be made to ensure that the role is fulfilled in the event of any short- or long-term staff absences. In some cases there may be more than one potential FIC at the fire assembly point. The provision of one high visibility surcoat for the FIC role will ensure that absolute clarity is achieved as to who is fulfilling the role on any particular occasion.

Training – FICs will need to demonstrate competence in their role which will normally be achieved as a result of training and experience. As a minimum it must be expected that an FIC will receive the same training as provided to the fire wardens/marshals and in addition must be given the opportunity to practise their role during fire evacuation drills.

Fire alarm verifier

Role – at the time that a fire alarm warning signal is given it is important to instigate the emergency evacuation of the occupants of the building, as quickly as possible. A balance must be drawn between the risks associated with delaying an emergency evacuation against the risks associated with an unnecessary evacuation, which may include the shutting down of plant and work practices. For example, although it is vital to evacuate a sports stadium in the event of fire, the evacuation itself is not without risk. It is therefore important that there is a degree of confidence that the evacuation itself is necessary. As part of the evacuation procedure of a premise, fitted with an automatic fire alarm, the role of an alarm verifier should be considered. The fire risk assessment will indicate at what stage an emergency evacuation may be initiated and, depending upon the risk, this will either occur prior to a verification of the alarm, at the same time as

the alarm is being verified or as a result of the verification of the alarm. The specific duties of the alarm verifier include:

➤ Attending the fire alarm indicator panel at the time the alarm sounds
➤ Identify the reported location of a fire
➤ Conducting a physical check of the location of the fire to ascertain if there is a fire or a false alarm
➤ Communicating the situation to the fire safety manager or fire incident controller.

Deployment – as with other key roles that assist in the emergency evacuation of people in a fire it is vital that, if the emergency evacuation procedures involve a fire alarm verifier, adequate management arrangements are made to ensure that there is at least one person available to fulfil this role at all times.

Training – in order to be effective in this role it is important that any persons nominated to act as a fire alarm verifier are provided with training and instruction on how to interrogate the fire alarm panel and interpret the information from the panel and any signs of fire at the location indicated.

Stewards/security staff

Role – on occasions when managing the fire safety of large numbers of the public it will be necessary to utilise safety stewards. The specific situation where stewards are used will determine their specific roles but like fire wardens/marshals their roles will be split into both proactive and reactive to ensure crowd safety. Examples of the proactive roles of safety stewards include:

➤ Providing a visible presence and thereby reassuring the public
➤ Providing safety and other information to the public
➤ Giving direction to the crowd and individuals
➤ Identifying and dealing with hazards such as inappropriate behaviour, blocking or obstructing of escape routes.

In emergencies stewards will have specific roles, for example:

➤ To prepare for the emergency evacuation of the venue by opening doors, etc.
➤ To ensure members of the public remain calm and evacuate safely
➤ Assist those who may require assistance
➤ Conduct any fire fighting if appropriate.

It is important that all stewards receive a written statement of their duties, a checklist (if this is appropriate)

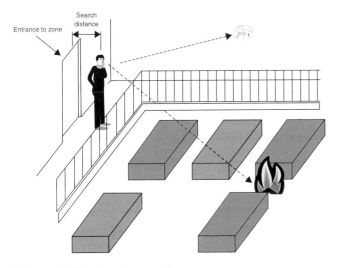

Search distance

Entrance to zone

Figure 10.24 An alarm verifier

Figure 10.25 Crowd control stewards

Figure 10.26 Fire marshal supervising emergency evacuation

and a plan showing the key features of the venue including hazardous areas and the means of escape.

Deployment – the numbers of stewards provided must be based on a risk assessment rather than a precise mathematical formula. The factors that should be considered when assessing the numbers of stewards required at any venue should include:

➤ Previous experience of the type of crowd expected at the particular venue
➤ The state of underfoot conditions including any uneven ground
➤ The presence of obstacles within or around the site that may affect crowd flow rates
➤ The size of the expected crowd
➤ The method of separating crowds from hazards
➤ The deposition of the crowd particularly whether or not they are expected to be sitting or standing.

Irrespective of the numbers of stewards provided, it is most important that an effective chain of command is established. Typically stewards will be controlled by the use of a chief steward (with similar duties to an FIC) supported by a number of senior supervisors who are in turn responsible for the direct supervision of a group of 6–10 stewards. In order to support this chain of command it is vital that good communications exist between the stewards. This is normally achieved by the provision of radio communications.

Training – due to the potential risks associated with managing safety within crowds, it is important that stewards can demonstrate appropriate levels of competence. The level of training will depend on the type of functions to be performed; again it is important to keep a record of any training and instruction provided, including the date and duration of the instruction, the details of

the person giving and receiving instruction and the nature of the instructional training. As a result of the training and instruction stewards will need to be able to demonstrate the following;

➤ An understanding of their responsibilities towards the safety of themselves and others
➤ Conduct pre-event safety checks
➤ Be familiar with the layout of the site particularly regarding means of escape, first aid, welfare and facilities for people with special needs, etc.
➤ Their specific duties, e.g. which areas they are responsible for and what their specific duties may be while the event is in progress
➤ Controlling and directing the crowd under normal circumstances
➤ Controlling and directing the crowd under emergency situations
➤ Recognising crowd conditions, e.g. overcrowding/ frustration
➤ Ensuring gangways and exits are clear at all times
➤ Dealing with disturbances or incidents
➤ Ensuring that combustible refuse does not accumulate
➤ Being familiar with the arrangements for evacuation including coded messages
➤ Communicating with the chief steward in the event of an emergency.

10.4 Assisting disabled people to escape

There is no specific legislation that requires special arrangements for disabled people in the event of fire. However, the Disability Discrimination Act 1995 (DDA)

Table 10.3 Requirements of the Building Regulations that ensure disabled people can have reasonable access and use of buildings

Part	Requirement	Limit of application
M1	**Access and use of existing buildings** Reasonable provision must be made for disabled people to access and use the building and its facilities	Does not apply to: ● Domestic dwellings or ● Parts of buildings used only for maintenance
M2	**Access to extensions to buildings (other than dwellings)** Suitable independent access must be provided where reasonably practical	Does not apply where suitable access to the extension is provided through the building that is extended

requires that employers make what is termed as 'reasonable adjustments' to their premises to ensure that no employee is at any disadvantage. In addition, from 1 October 2004 the DDA placed duties on all those who provide services to the public, e.g. any business premises that customers may visit or any public service buildings such as schools and libraries, to take reasonable steps to change or alter their premises where access for disabled people is either 'unreasonably, difficult or impossible'.

The specific requirements for access and use of buildings by disabled people is detailed in Part M of Schedule 1 to the Building Regulations 2000.

Table 10.3 summarises the requirements of the Building Regulations that ensure disabled people can have reasonable access and use of buildings.

> The Disability Discrimination Act 1995 requires employers to make 'reasonable adjustments' to their premises to ensure that no employee is at a disadvantage. This includes ensuring that disabled people can leave the premises safely in the event of fire.

It is important that the emergency plans for buildings take account of disabled people to ensure that there are reasonable arrangements not only for access and use but also for egress in the case of an emergency. Arrangements will need to be considered for persons whose escape from fire may be made more difficult on the basis of the following:

➤ Aged
➤ Less able bodied
➤ Of impaired mobility

Figure 10.27 Typical impaired vision symbol

➤ Wheelchair users
➤ Impaired vision
➤ Impaired hearing
➤ Learning difficulties or mental illness.

When undertaking an assessment of the needs of disabled persons during an emergency evacuation it will be helpful to consider the following questions:

➤ Is there a need to locate disabled workers to parts of the building where they can leave the building more easily?
➤ Is any special equipment needed, such as an evacuation chair?
➤ Are storage areas provided with any necessary evacuation equipment? Are they easy to access?
➤ In the event of emergencies, are specific members of staff designated to alert and assist disabled persons (evacuation assistants)?
➤ Are visual or vibrating alerting devices provided to supplement audible alarms? Are visual alarms installed in all areas, including toilet facilities?
➤ Do routes and procedures take account of the potentially slower movement of people with disabilities?
➤ Are all disabled people familiar with escape routes and provided with instructions and training in safety procedures? Distribute emergency procedures in Braille, large print, text file and cassette format

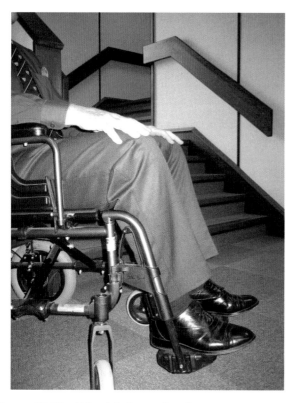

Figure 10.28   Wheelchair user in refuge

Figure 10.29   Escape routes in a typical sports stadium

➤ Are areas of and locations that are safe from immediate danger established?

➤ Are arrangements for people with disabilities included in the written evacuation procedures? Are arrangements periodically reviewed?

➤ Are employees encouraged to make a list of medications, allergies, special equipment, names, addresses and telephone numbers of doctors, pharmacies, family members and friends, and any other important information?

In order to ensure that the emergency evacuation plans for disabled people are effective at the time they are required it will be necessary to provide training for any staff who act as evacuation assistants. Training should include such topics as: evacuation techniques to use, particularly how to carry or assist individuals who use mobility aids; the use of any special evacuation equipment and training in basic sign language to effectively communicate with individuals who are deaf; and the instructions for individuals who use assistance animals.

When developing emergency plans for disabled people it will be useful to obtain advice from the local fire service periodically about such issues as whether people with disabilities should remain in their workplaces, assemble in an area of refuge to await the arrival of rescue workers or immediately evacuate.

Devising emergency arrangements for disabled people is best achieved by consulting with those who know best about the specific issues relating to the physical or mental impairment. Information can be sought from specialist disability organisations; however, when possible and particularly for members of staff, it is considered reasonable to devise a 'personal emergency evacuation plan' (PEEP).

PEEPs are written by the employer in conjunction with the individuals concerned and explain, on a case by case basis, the means of escape arrangements for each individual in case of fire. A PEEP is a very useful tool for identifying exactly what arrangements may be necessary in each individual's case. A common way of gaining information to develop a PEEP for staff is to use a questionnaire. An example is shown in Appendix 10.1.

10.4.1   Evacuating members of the public

The degree to which arrangements need to be made to ensure the safe evacuation of members of the public from a building depends upon the level of risk. In some cases it will be considered to be reasonable to simply display some fire notices within some premises where there is limited public access. At the other end of the scale, for large premises or when large numbers of people are present, additional arrangements, such as public address systems and fire marshals, may be needed.

The numbers, location, physical and emotional state of members of the public must be considered when devising a system for their safe evacuation. Methods that can be considered to safely evacuate crowds of people whether in a building or external venue include:

➤ Limiting the numbers of people allowed in a particular building or venue

Figure 10.30 Visual and audible communication for managing crowd safety

- ➤ Limiting the numbers of people permitted in a particular part of the building or section of seating
- ➤ Providing adequate means of escape that is:
  - ➤ Obvious and well signed
  - ➤ Has good underfoot conditions
  - ➤ Opens in the direction of escape
  - ➤ Opens when under pressure from a crowd, e.g. a push bar to open device
- ➤ Providing adequate communications equipment, e.g. a public address system or personal megaphones for fire marshals
- ➤ Providing an audible/visual safety briefing which includes basic safety information along with instruction of what action will be required in case of emergency
- ➤ Providing additional emergency lighting
- ➤ Providing assistance in the form of competent fire wardens or fire marshals
- ➤ Provision for people with special needs
- ➤ First aid facilities.

10.4.2 The purpose of drills, evacuation and roll-calls

There is no evidence from research to suggest that a fire exit sign will necessarily encourage people to head towards it in the event of a fire, unless the route is already familiar. Occupants may choose to ignore specific fire exit routes and choose familiar routes. It is quite likely that people would often be more inclined to move to a familiar exit which is further away than an unfamiliar exit nearby. Therefore it is vital that where possible

people who may need to use an exit in the event of an emergency are familiar with it.

The purpose of practising an evacuation procedure is to ensure it functions adequately, to ensure all those with a specific role in the procedure are aware of and competent in their role and are able to demonstrate to all parties that arrangements have been put in place to achieve a reasonable level of safety in the event of fire.

The object of a fire evacuation procedure is to practise good evacuation behaviour, so that people do not experience or develop serious health effects associated with being exposed to the effects of fire.

In order to test evacuation procedures it is important to achieve as much realism as possible. The procedure will not be fully tested if people think that the actual emergency escape routes cannot be used.

During a fire evacuation drill in a police station, staff delayed their escape because once at the bottom of a staircase, which was only used for emergency evacuation, they were confronted with a break glass to open fitting on the final exit door. They were reluctant to break the glass, fearing unnecessary damage and so they retraced their steps and left the building through the main entrance.

As a result:

- ➤ The procedures were not tested fully
- ➤ The final exit door and its security fitting remained untested
- ➤ Everyone else who had left their work had, in effect, wasted their time.

10.4.3 Confirming the building is clear

In the event of a fire in a building the fire service will need to know, among other things, whether or not there are still people left in the building. It is important to provide the fire service with clear information regarding the situation relating to people in the building. If the fire service officer on the scene has any doubt as to whether there are persons in a building their first priority becomes the rescue of those people. This will involve committing fire fighters into the building to conduct search and rescue operations and is very likely to delay an attack on the main seat of the fire.

In order to be confident when reporting to the fire service, the fire incident controller will need to know that everyone has either left the building or, if people are still in the building, where they are likely to be found. The two methods used by a fire incident controller are the 'sweep' and 'roll-call'.

Sweep – the sweep technique is simply the application of a systematic and progressive checking of all the areas within a building or within the area of responsibility of a fire warden/marshal. When conducting a sweep of an area it is important to ensure that all areas that may be occupied are swept. This includes any plant rooms where contractors may be working, any walk-in cupboards or storerooms and any toilets or rest facilities. During a sweep of an area the fire warden/marshal should also ensure that where possible and without causing undue delay any doors or windows are closed, any heat generating equipment is turned off and any signs of fire are noted and, if appropriate, dealt with.

In order to ensure that an effective sweep is made of a building at the time of an emergency evacuation it is necessary for the fire safety manager or fire incident controller to allocate specific areas of the building to specific fire wardens/marshals. In the case of multi-storey offices this is often achieved by nominating two to three fire wardens per floor, whereas, in the case of an open plan factory, fire wardens may be allocated to production areas.

When planning the division of a large building to be swept by fire wardens it is important to bear in mind that the area to be swept by one individual must be of a size and nature that will allow the sweep to be conducted and the warden to evacuate within 2½ and 3 minutes of the alarm sounding.

Roll-call – in addition to a sweep of the premises it is often the case that some form of roll-call will be taken at the fire assembly point. The level and nature of the roll-call will be determined by the fire risk assessment and be dependent upon factors such as:

➤ The size and nature of the workforce
➤ The number and nature of any visitors that may be present in the building
➤ The resources required to maintain an accurate roll-call of persons in the building.

Organisations often adopt a minimalist approach to roll-calls, particularly where they have effective arrangements for conducting a sweep of the building. An example of this is where the reception desk of a building will merely record details of visitors, e.g. contractors and clients, to the building and take this roll to the assembly point to ensure that those people not normally in the building have evacuated.

When combined with an effective sweep technique this type of roll-call is seen to provide the necessary level of care for those who are less familiar with the building.

10.5 Case study

On Saturday 11 May 1985 a fire broke out in Valley Parade, home of Bradford City Football Club.

It turned out to be a day that sent shockwaves around the world as fire engulfed the main stand and eventually claimed the lives of 56 supporters; 265 people were injured.

It should have been one of the happiest in the club's modern history. City had won the Third Division championship. Over 11 000 fans were in the ground to watch the players do a lap of honour before the final match of the season against Lincoln City.

At 3:40 smoke was seen raising between the gaps of the wooden benching where rubbish had accumulated in the main stand. There was no attempt to evacuate at this stage, fire fighting equipment was requested by the stewards. The occupants of the stand stood watching the fire; police and stewards were on the scene but there was no appreciation of how fast the fire would spread. There was no perception of the magnitude of the risk.

However, within 4 minutes the flames had grown and were clearly visible from the pitch. It was then that the police on duty in the ground began to evacuate people from one block of the main stand. At this time the game was abandoned and the players left the field.

As the occupants of the block of seating that was on fire began to evacuate, others in the stand took no immediate action and there appeared sufficient time for an orderly evacuation.

However, due to the nature of the construction of the wooden stand, which had a wooden and bitumen covered roof, the fire quickly spread by a combination of convection and radiation until it threatened all the occupants of the main stand and others in the ground and on the field who where within range of the massive amounts of radiated heat.

Within a few minutes the fire totally engulfed the roof of the main stand and the orderly evacuation had turned into an urgent crush to get away from the radiated heat. The clothing of police officers providing assistance to the crowd was seen to spontaneously combust and members of the public were seen to be on fire.

The death toll from the fire finally reached 56 and a further 265 people were seriously injured. Many of the dead were the most vulnerable, either the old or the very young, trapped as they tried to escape from the fire. Some of them were lifelong City supporters, who had waited for many years to see their team return to the top two divisions for the first time since before the war.

The fallout from the fire tragedy hung over the city for months after the disaster. There were some sad and poignant memorial and funeral services over the following 3 weeks; players visited hospitals where the injured were recovering while a disaster fund appeal raised over £4.5 million.

The cause of the fire was said to be the accidental dropping of a match or cigarette stubbed out in a polystyrene cup. Following the disaster an inquest was held by Justice Popplewell which resulted in the introduction of new safety procedures for sports grounds and stadia in the UK including the banning of smoking in stands and the provision for additional means of escape exits.

Following the fire at Bradford City Football Club, the general perception of the risks from fire in sports stadia has increased and as a result all those who design, manage and attend large sporting events are likely to behave differently to the way they would have done prior to May 1985.

10.6 Example NEBOSH questions for Chapter 10

1. What factors may be considered when devising a personal emergency evacuation plan (PEEP) for a disabled person? (8)
2. (a) **Outline** the purpose of conducting a sweep of a building during a fire evacuation. (2)
   (b) **Describe** the actions to be taken to ensure an effective sweep of a building is achieved. (6)
3. **Identify** the factors to be considered when assessing the numbers of fire wardens/marshals that may be required to ensure the full and safe evacuation of a workplace. (8)
4. **Outline** the benefits of undertaking regular fire drills in the workplace. (8)
5. **Discuss** the reasons why people may not respond immediately to an audible fire alarm signal. (8)
6. A cafeteria in a department store where customers queue for service and utilise tables in the cafeteria to eat is busy with customers when the fire alarm system is activated.
   (a) **Identify** the main ways in which customers might react to the fire alarm. (4)
   (b) **Outline** the issues which may need to be considered to deal with the behaviours which may be displayed. (4)

Appendix 10.1   Example personal emergency evacuation plan questionnaire for disabled staff

PERSONAL EMERGENCY EVACUATION PLAN QUESTIONNAIRE FOR DISABLED STAFF

NAME

JOB TITLE

DEPARTMENT

BRIEF DESCRIPTION OF DUTIES

LOCATION

1. Where are you based for most of the time while at work?

2. Will your job take you to more than one location in the building in which you are based?
    YES ☐   NO ☐

3. Will your job take you to different buildings?
    YES ☐   NO ☐

AWARENESS OF EMERGENCY EGRESS PROCEDURES

4. Are you aware of the emergency egress procedures that operate in the building/s in which you work?
    YES ☐   NO ☐

5. Do you require written emergency egress procedures?
    YES ☐   NO ☐

    5a Do you require written emergency procedures to be supported by British Sign Law interpretation?
        YES ☐   NO ☐

    5b Do you require the emergency egress procedures to be in Braille?
        YES ☐   NO ☐

    5c Do you require the emergency egress procedure to be on tape?
        YES ☐   NO ☐

    5d Do you require the emergency egress procedures to be in large print?
        YES ☐   NO ☐

6. Are the signs which mark the emergency exits and the routes to the exits clear enough?
    YES ☐   NO ☐

EMERGENCY ALARM

7. Can you hear the fire alarm/s provided in your place/s of work?
    YES ☐   NO ☐   DON'T KNOW ☐

8. Could you raise the alarm if you discovered a fire?

    YES ☐   NO ☐   DON'T KNOW ☐

ASSISTANCE

9. Do you need assistance to get out of your place of work in an emergency?

    YES ☐   NO ☐   DON'T KNOW ☐

If NO please go to Question 13

10. Is anyone designated to assist you to get in an emergency?

    YES ☐   NO ☐   DON'T KNOW ☐

If NO please go to Question 12. If YES give name/s and location/s

_____

_____

11. Is the arrangement with your assistant/s formal (this is the arrangement written into their job description)?

    YES ☐   NO ☐   DON'T KNOW ☐

    11a Are you always in easy contact with those designated to help you?

    YES ☐   NO ☐   DON'T KNOW ☐

12. In an emergency, could you contact the person/s in charge of evacuating the building/s in which you work and tell them where you were located?

    YES ☐   NO ☐   DON'T KNOW ☐

GETTING OUT

13. Can you move quickly in the event of an emergency?

    YES ☐   NO ☐   DON'T KNOW ☐

14. Do you find stairs difficult to use?

    YES ☐   NO ☐   DON'T KNOW ☐

15. Are you a wheelchair user?

    YES ☐   NO ☐   DON'T KNOW ☐

Thank you for completing this questionnaire. The information you have given us will help us to meet any needs for information or assistance you may have.

Remember, we do not see you as the problem – you are not a safety risk. The problem belongs to us and the building in which you work.

Please return the completed form to:

_____

_____

_____

_____

_____

_____

_____

_____

Appendix 10.2 Example of responsibilities of fire wardens

A. In the event of an emergency or fire

1. Ensure that the fire alarm is raised.
2. If continuous alarm sounds, or when advised of an emergency, ensure that all personnel, including visitors, evacuate the building by the nearest means of escape.
3. Wear a high visibility jacket, or waistcoat, so that you can be easily recognised.
4. Ensure that staff are designated to assist disabled staff members, and that they are removed to a place of safety.
5. Without danger to themselves, search all areas, including toilets and kitchens, to ensure complete evacuation.
6. Without danger to themselves, close all windows and doors, while undertaking the sweep of the building, employing the 'search and cascade technique'.
7. Direct all evacuated personnel to the designated fire assembly point.
8. Use portable fire fighting equipment **ONLY** to aid safe egress.
9. Report to the fire incident controller, identified by the wearing of a high visibility jacket or waistcoat.
10. Prevent any re-entry to the building until given authorisation.

B. In the event of an emergency or fire not in your area

1. On the sounding of an intermittent alarm, alert all staff to be prepared to evacuate, but to remain calm.
2. Stand by and be prepared to receive further instruction.

C. General duties

1. Familiarise themselves with their work location, escape routes and their designated areas of responsibility.
2. Be aware of disabled staff members off duty and ensure that staff are designated to assist them during an evacuation.
3. Familiarise themselves with the location of their designated assembly point.
4. In the event of being away from their place of work, inform their line manager, or deputy warden, so that they assume responsibility in the event of an emergency.
5. Report all defects to line manager, or facilities supervisor, for action.
6. Undertake initial and refresher training of new and existing staff in the fire procedures.
7. Maintain appropriate records.
8. Assist in fire drills.
9. Attend any appropriate training to ensure fire warden duties are carried out effectively.

D. Routine duties – relevant to the designated area of responsibility

Daily
1. Ensure escape routes are clear and available for use.
2. Check final exit fire doors are not improperly locked or blocked.
3. Ensure all rubbish is being removed regularly, and that all materials are stored safely.
4. Ensure portable fire fighting equipment and fire signs are in place.

Weekly – as per checklist
Monthly
1. Check fire doors are in good condition and function properly.
2. Check all electrical equipment is PAT tested in date.
3. Check all fire fighting equipment is in test date.
4. Complete fire log book and report deficiencies.

WEEKLY FIRE WARDEN CHECKLIST

Location:				
Inspected by:				
Appropriate fire exit doors unlocked and opening (open them)	Yes		No	
'Keep locked shut' fire doors all secure	Yes		No	
Fire escape corridors free from obstruction	Yes		No	
Emergency exit signs clearly visible	Yes		No	
Visual check – fire extinguishers in good order and in correct location	Yes		No	
Visual check – fire alarm call points in good order	Yes		No	
Fire escape and 'in case of fire' notices correctly displayed	Yes		No	
Plug sockets not overloaded, and for equipment not in use, the plug removed from plug socket	Yes		No	
Combustible materials stored safely; away from any possible source of ignition, not outside a doorway and not under exit stairs	Yes		No	
Boxes and other stored items not too close to lights	Yes		No	
Evidence of smoking materials in no smoking areas	Yes		No	
Rubbish removed regularly and not allowed to build up	Yes		No	

Areas of concern or which need urgent attention:

Action taken by fire warden:

Signed:		Date:	

(to be retained by fire warden unless action required by another)

If applicable – forwarded for further action to:

Signed:		Date:	

If applicable – forwarded for further action to:

Signed:		Date:	

11 Monitoring, auditing and reviewing fire safety systems

While there is no specific format or guidelines for establishing a 'fire safety management system' detailed within fire safety legislation or official fire safety guidance, there are a variety of management systems that may be adopted as has previously been discussed (Chapter 1). Regardless of the management system used, the measurement of performance forms a key component part of any such system if safety is to be effectively managed.

HSE's *Successful Health and Safety Management* (HSG65) guidance document separates the measurement of performance into distinctly different areas – those of reactive monitoring and those of proactive monitoring; this is the same in BS 8800 'Occupational health and safety management systems – guide'. Each of these differing areas will be discussed within the chapter, particularly in relation to fire safety management. It should be noted that although focusing on fire, in this instance this process may be equally applied to the measurement of safety performance as a whole.

Simplistically put 'that which cannot be measured cannot be managed' as without a knowledge of the strengths and weaknesses of the management system it is not possible to identify any opportunities to improve, or threats that may come about from failing to manage.

The purpose of monitoring and measuring safety performance is to provide an organisation with information on the current status of its policies, procedures, etc., and its progress to its end goal. Besides the legal requirement for reporting certain outcomes (RIDDOR) to the enforcing authorities the RRFSO and MHSW Regulations also require an employer to make arrangements for monitoring and reviewing its preventive and protective measures.

When measuring safety performance, successful organisations use a combination of both **proactive** and **reactive** monitoring techniques.

This chapter discusses the following key elements:

➤ The benefits of monitoring and measuring
➤ Active safety monitoring procedures
➤ Conducting workplace inspections
➤ Auditing fire safety management systems
➤ Reviewing performance
➤ Reactive monitoring.

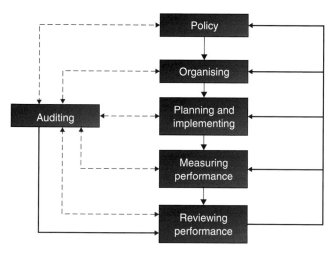

Figure 11.1 Successful management structure and process from HSG65

Figure 11.2 The outcome of a failure to adequately manage fire safety

Proactive – proactive systems monitor the design, development, installation and implementation of management arrangements for workplace precautions and risk control (management systems).

Reactive – reactive systems monitor events that have already occurred such as personal injury accidents, fires, ill health and other evidence of deficient safety performance (these may be false alarms, near misses, etc.) (see Chapter 12).

Organisations that relying solely upon incident, injury/ill-health statistics for their monitoring and measurement of their management system, particularly in relation to fire, are only able to measure management system failures. Crucially due to human nature, there may be underreporting of incidents and accidents for numerous reasons, which have been discussed in Chapter 4.

With regard to reactive analysis it is often the case, particularly in relation to fire, that a small number of incidents or accidents may lead to complacency and a belief that all is well, when this may not necessarily be the case. Statistics themselves are likely only to demonstrate a knowledge of outcomes, i.e. twisted ankle during evacuation and not the cause such as poor flooring conditions presenting a slip and trip risk. It is therefore essential for a combined reactive and proactive approach to be taken when considering safety monitoring or performance measurement systems.

The analysis of both reactive and proactive data will enable a company to analyse its safety performance producing a quantifiable and measurable range of data which will indicate the effectiveness of the management system (see Appendix 11.1 for a sample of performance indicator data).

11.1 Benefits of monitoring and measuring

A key aspect of safety management is to strive for continuous improvement of performance. Therefore it is vital that organisations are able to know how well their systems are performing. Data gathered by monitoring systems will enable an organisation to understand:

➤ Its current position in relation to its safety management
➤ Its progress against its standards
➤ Its priorities for action
➤ The effectiveness of its resource allocation and management
➤ The areas where it may be exposed to excessive risk
➤ Where and how its management systems should be reviewed.

Establishing an effective monitoring and measuring system helps to demonstrate management's commitment to safety objectives in general and will also assist in the development of a positive safety culture, particularly rewarding any positive work undertaken in controlling levels of risk.

In addition, information gathered through the performance measurement enables managers to understand how well they are managing in comparison to their previous experience and the performance of others in similar industry sectors. With safety performance data that can be compared across an industry, the process of benchmarking safety performance can give useful insights to managers. By using comparative data, benchmarking can be conducted both internally and externally. Table 11.1 gives some examples.

When compiling the performance statistics an organisation will need to take into account the mechanisms by which it will gather the information. These should be reflected in the safety policy particularly in relation to reporting and investigation of safety events, incidents, accidents, etc. and its proactive monitoring, inspections, tours and audits.

11.2 Active safety monitoring procedures

11.2.1 Proactive monitoring

The purpose of proactive monitoring is to provide feedback on an organisation's safety performance before a personal injury accident, fire, illness, etc. occurs.

269

Table 11.1 Examples of internal and external benchmarking

Internal	External
Relative position against aims and objectives	Relative position against competitors/similar organisations
Control of hazards and risks arising from an organisation's operations	Relative position against national standards
The efficiency and effectiveness of the safety management system across all parts of the organisation	Learn from other organisations' industry sectors
Safety culture pervading within the organisation	Safety culture comparisons with competitors/similar organisations

Figure 11.3 Fire safety hazards require active monitoring

This monitoring also measures achievement against specific plans and objectives, the effectiveness of the safety management system and compliance with legal and performance standards. In contrast to reactive monitoring, proactive monitoring measures success, reinforcing positive achievement rather than penalising failure and therefore has a direct effect upon safety culture within an organisation, increasing motivation and the desire for continual improvement.

In an organisation that has an effective safety culture, managers, staff, contractors and all persons who may work with or come into contact with the organisation will be included in informal proactive monitoring of workplace precautions and risk control (management) systems during day-to-day working operations.

This informal approach, by undertaking hazard spotting and taking initial preventive and protective steps on the 'shop floor', will have a direct effect upon the safety management within an organisation, which will reduce the overall risks.

In support of an effective safety culture and the informal hazard spotting approach, formal proactive monitoring will be undertaken by supervisors and members of the management team.

There are various forms and levels of proactive monitoring that an organisation may adopt. The exact methods adopted will be dependent upon the size and nature of an organisation, its risk profile and available resources. In most cases a portfolio of approaches is adopted which will generally include:

➤ The systematic inspection of premises, plant, processes and equipment by supervisors, maintenance staff and safety representatives

➤ Direct observation by first line supervisors of work and behaviour to assess compliance with procedures, site rules, etc.
➤ Environmental monitoring and health surveillance to check on the effectiveness of control measures preventing ill health (used to detect early signs of ill health)
➤ Periodic examination of documents and records relating to fire safety systems, permits to work, training records, etc. that would prove to be safety critical and have an effect upon the overall safety management
➤ The completion of regular reports and reviews on safety performance, which may be considered by senior management and the board of directors
➤ The undertaking of regular audit programmes.

The level of monitoring must be considered proportionally to a company's hazard profile and should concentrate on areas where it will produce the most benefit, which in turn will lead to the most enhanced level of risk control.

For active monitoring to be successful there are a number of key factors that must be addressed:

➤ The standards to be applied
➤ Frequency of monitoring
➤ Who will undertake the monitoring
➤ The anticipated results or actions required.

Each of these factors must be included in the formalised policy and procedures contained within the safety management system. It is likely that such issues as the standards and frequency of monitoring will reflect legislation, British/European Standards and industry guidance, e.g. emergency lighting equipment should be checked to ensure all units are in a good state of repair and in good working order on a monthly basis (BS 5266).

The responsibility for undertaking the monitoring will be determined by the type of monitoring that will need to be undertaken and the competencies of those involved. Some anticipated results are obvious, such as in the case of emergency lighting units which should function upon test; others such as how well a permit to work has been complied with, will need greater definition and are likely to reflect the policy and procedures. It is in these instances that an aide memoir or checklist may be usefully applied.

11.2.2 Proactive monitoring methods and techniques

An organisation will need to decide how it will allocate responsibilities for proactive monitoring at the different levels within its management chain and the level of detail that is considered appropriate. It is likely that the decisions will reflect the company or organisation's structure and also its existing management systems, e.g. quality management system, environmental management system.

Each manager should be given the responsibility for monitoring the achievement of company objectives and compliance with standards for which they and their team are responsible, ideally this should be reflected in their own performance indicators. The methods involved will vary from those that will be undertaken within the management chain and those that may be required as part of an independent analysis.

A variety of methods may be used in assessing safety performance including:

➤ **Safety inspections** – these involve the straightforward observation of a workplace and/or activities or equipment within it. They may also include visual checks on documents and records; this is usually carried out by a manager (or on some occasions employee representative), often aided by the use of a checklist with the aim of identifying hazards and assessing the use and effectiveness of control measures
➤ **Safety tours** – similar to safety inspections but which involve unscheduled checks on issues such as means of escape, housekeeping and PPE
➤ **Safety surveys** – which focus upon a particular activity such as hot work or manual handling; they

Figure 11.4 Formal active monitoring system

may also include attitudinal surveys which examine the employees' and management's attitudes towards safety
➤ **Safety sampling** – which involves the sampling of specific areas such as the effectiveness of fire doors, positioning of fire extinguishers; these may be conducted by a range of staff members where specific areas of safety are targeted
➤ **Audits** – involving comprehensive and independently executed examinations of all aspects of an organisation's safety performance against stated objectives
➤ **Health or medical surveillance** – which involves using medical techniques that analyse human performance such as hearing, lung function (particularly with regard to industrial fire teams).

Audits are discussed later in the chapter and due to their similarity, safety inspections and safety tours will be discussed as one.

Safety surveys and safety sampling involve closer scrutiny of areas that may be missed when undertaking an overall inspection or tour and are likely to take a greater proportion of time to prepare for and conduct, due to their specific focus.

Health and medical surveillance, particularly in relation to fire safety management, is likely only to be required by those involved in fire fighting action; however, in general occupational safety and health it is necessary to confirm the effectiveness of areas such as dust and solvent control, where ill-health effects on the human body can have a devastating effect. It is, however, prudent to mention that health surveillance relating to stress must be considered, particularly for those who may be engaged in fire fighting and rescue actions (critical incident stress disorder and post-traumatic stress disorder). Regardless

271

of the type of monitoring that is being undertaken, there are a number of techniques that can be used to gather information, these include:

➤ Direct observation of:
  ➤ Workplace conditions, e.g. obstructed emergency escape route
  ➤ People's behaviour, e.g. smoking in an unauthorised area
➤ Communicating with people to gather information, both fact and opinion
➤ Examining documents, e.g. fire policy or fire plan, contractors' method statements
➤ Examining records, e.g. fire alarm test, emergency lighting testing, training records.

The frequency of formal proactive monitoring will depend upon a number of factors, including legislative and British/European Standards and the level of risk. They may also take into account previous trends and events and reflect the findings of previous monitoring, e.g. if a workplace inspection identifies that fire extinguishers are being moved away from their allocated position then more frequent inspections may be required.

The effectiveness of the proactive monitoring programme will be judged by the reduction of substandard findings and issues identified by whichever technique is used over a given period of time. When completing proactive monitoring programmes to assist trend analysis, many organisations utilise a scoring mechanism, particularly in relation to auditing. The scoring and analysis mechanism is similar to that of the way reactive data is measured and analysed so that graphical evidence may be produced which will then be used to communicate the effectiveness of the safety management systems throughout the organisation.

11.3 Conducting workplace inspections

A system for inspecting workplace precautions is essential in any proactive monitoring programme. On many occasions it can form part of an organisation's arrangements for the planned preventive maintenance of plant (e.g. electrical maintenance and testing) and equipment (e.g. flashback arresters for oxy/fuel systems) which are also covered by legal requirements.

In addition, inspections should include other workplace precautions such as the ability for a fire door to close against its rebates effectively and other aspects of the premises such as the location/obstruction of fire fighting equipment and means of escape.

A suitable inspection programme will address all risks; however, it should take into account those that present a low risk (inspections may be undertaken every month to two months) covering a wide range of precautions, or higher risks that may need more frequent and detailed inspections such as the storage and use of flammable materials (inspections may be undertaken each day or twice daily).

The inspection programme must reflect any specific legal requirements and the risk priorities which are usually identified as part of the risk assessment process.

Particularly, in relation to fire, a fire log book's contents and records will reflect the schedules and performance standards required by not only the law but a variety of British and European Standards. These schedules are likely to be supplemented with inspection forms, checklists or aides-memoires, which will enable a consistency of approach and assist the individuals conducting the inspection to cover all key requirements without the need to remember each element.

Inspections should be undertaken by competent persons who are able to identify relevant hazards, workplace precautions and risk control systems, the standards that need to be met and any shortcomings in the preventive and protective measures. This area will be discussed later within this section.

11.3.1 Recording an inspection

It may well be that fire safety inspections are included as part of a general health and safety workplace inspection programme. It is likely that a safety inspection form will be produced to ensure that key elements are covered by the inspection and to ensure consistency of approach. There are a number of key headings which are likely to be found on a fire safety inspection record, whichever recording mechanism is used. These following items are should be included as a minimum:

➤ Ignition sources
➤ Fuel and oxygen sources
➤ Specific high fire risk activities
➤ General housekeeping
➤ Means of escape
➤ Warning and detection systems
➤ Emergency lighting
➤ Fire fighting equipment
➤ Procedures and notices.

Under each of the headings there will be a number of subquestions to confirm the adequacy of workplace precautions and management controls which can be either answered in a closed question technique (yes or no), or be provided with a score in relation to the level of compliance/non-compliance.

In either of the above cases, where non-compliance is identified, there should be a mechanism for recording

Figure 11.5 Records should be kept in a local fire safety log book

Figure 11.6 Misuse of electrical adaptors

the issue together with recommendations to achieve compliance.

In order to ensure that resources are targeted at the non-compliance issues that present the most significant risk a prioritisation or ranking system may also be applied.

This will require high risk areas to be weighted in such a way that prioritised action is given, for example the misuse of electrical adaptors which present a danger of fire, rather than the provision of signage above fire extinguishers to denote their type (which is indicated on the extinguisher body). Ideally, the inspection record form will also identify who will be responsible for implementing any recommended actions, together with a target date for implementation.

> A systematic approach to an inspection programme is likely to include:
>
> ➤ The preparation of a well-designed inspection form that will help to plan and initiate remedial action – it may also assist in ranking any substandard conditions/deficiencies in order of importance
> ➤ Summary lists of any remedial action required with names and arbitrary timescales to track implementation progress
> ➤ A periodic analysis of completed inspection forms to enable identification of common trends which may reveal system weaknesses
> ➤ The gathering of information to consider changes required to the frequency, type or nature of the inspection programme.

11.3.2 Who will conduct the inspection?

The responsibility for either conducting the inspection itself or nominating persons to undertake the inspection is likely to be part of the fire safety manager's role (a variety of titles may be given to the person who has overall control of fire safety management) in accordance with British Standard 5588 Part 12 and current best practice.

Who is responsible and for what should be clearly identified within the fire safety arrangements section within the policy.

For any inspection to be of value, a trained, competent person or team must conduct it. In larger organisations the role of the fire warden/marshal (whose main function is to assist persons to escape safely in the event of a fire) is extended to undertake part of the inspection process, which may be referred to as a safety tour, as the process involves a physical check of the areas for which they are fulfilling their fire warden/ marshal function.

These safety tours are slightly less formal than a full inspection programme; however, they assist in fulfilling

Table 11.2 Example of an inspection aide memoire

Frequency		Item to be checked
Daily	Morning:	Are escape routes clear? Are the fire exits available for use?
	Evening:	Is the electrical equipment switched off? Has rubbish been disposed of safely? Are all windows shut?
Weekly		Do self-closers on the doors operate correctly? Are fire signs visible? Is there sufficient space between stored materials and sprinkler heads, fire detectors and lights?
Monthly		Are all extinguishers in their correct places and do they appear to be in working order? Are the hoses on the hose reels neatly coiled and the valves easy to turn on?

the main objective which is to monitor the condition of workplace precautions such as the effectiveness of the housekeeping programme and no smoking regime or the management and control of flammable substances. A simple checklist may be created from the details suggested in Table 11.2 (see Appendix 10.2).

Details of such inspections are likely to be retained within the fire log book or fire safety manual so that evidence is readily available for future fire safety inspections, reviews and audits.

On larger sites or where there are a number of buildings that require formal fire safety inspections, it may be the responsibility of the facilities management team or those nominated to take charge of the facilities, to coordinate the fire safety inspection programme. Where fire safety systems are in place it is likely that the facilities management team, or their nominees, will be actively involved in the inspection programme as they are responsible for managing a large proportion of the systems that are in place, for example the fire alarm and detection system, fixed installations (sprinklers, etc.) and fire doors.

The periodic inspections of fire safety systems are likely to be recorded within the fire log book and will be considered as part of the proactive monitoring programme, particularly as the items form a critical component of the workplace precautions or control measures that reduce the overall risk in relation to fire.

11.3.3 Maintenance of general fire precautions

Formal systems need to be adopted to ensure that general fire precautions are maintained in a good state of operation and repair. Table 11.3 gives an overview of a typical inspection regime indicating the equipment to

be inspected, the period between inspections and who should conduct them.

11.3.4 Maintenance of technically complex systems

Further details of proactive monitoring including testing, cleaning and maintenance of items such as gas flood systems, sprinkler systems and smoke and heat exhaust ventilation systems will be found within the manufacturers' guidance documentation (operations and maintenance manuals or construction health and safety file – CDM).

Clearly the proactive monitoring and testing that is required for the above systems will require a high level of technical competence and it is often the case that external contractors are engaged to undertake such work. The responsibilities of the organisation in respect of this external monitoring will be to ensure that documentary records and evidence are maintained to show that such equipment remains safe and fit for its intended purpose.

11.3.5 Using the findings of an inspection

The results of the inspection will provide evidence of the effectiveness of the workplace precautions and controls that have been introduced to reduce the risks from fire and explosion.

As has previously been discussed, the mechanisms for analysing the findings may take a number of forms that may also include statistical analysis, if numerical scales have been utilised, which is often the case when undertaking general health and safety workplace inspections.

Table 11.3   Overview of a typical inspection regime

Equipment	Period	Action
Fire detection and fire warning systems including self-contained smoke alarms and manually operated devices	Weekly	Check all systems for state of repair and operation. Repair or replace defective units. Test operation of systems, self-contained alarms and manually operated devices.
	Annually	Full check and test of system by competent service engineer. Clean self-contained smoke alarms and change batteries.
Emergency lighting equipment including self-contained units and torches	Weekly	Operate torches and replace batteries as required. Repair or replace any defective unit.
	Monthly	Check all systems, units and torches for state of repair and apparent working order.
	Annually	Full check and test of systems and units by competent service engineer. Replace batteries in torches.
Fire fighting equipment including hose reels	Weekly	Check all extinguishers including hose reels for correct installation and apparent working order.
	Annually	Full check and test by competent service engineer.

Figure 11.7   Contractor testing electrical equipment

Regardless of the mechanisms of analysis, actioning the findings of an inspection must be seen as a key management requirement in fire safety management and any report produced must not only detail what has to be undertaken but also provide a guide to the speed of response. On occasions, short-term fixes may need to be taken where the inspection has identified a weakness and significant resources over a long term will need to be considered.

> The information from inspections as with all information gathered from active monitoring systems must be used to inform senior management about the effectiveness of their systems and where improvements can be made. It is therefore vital that the inspection process formalises how the information will be collated and presented to management.

Ideally the close-out actions, when completed, should also be recorded thus providing evidence that may be used to demonstrate legal compliance to enforcement authorities and be considered when conducting a fire safety review or audit.

11.4   Auditing fire safety management systems

A fire safety audit may be described as a periodic (typically annual), systematic and thorough assessment of the implementation, suitability and effectiveness of the fire safety management system. It is a significantly more

Figure 11.8 The senior managers of an organisation must review the outcomes from the active monitoring systems

'in-depth' process than conducting a fire safety inspection or a check on the contents and completeness of a fire safety manual or log book.

It is very likely that during the audit process fire safety auditors will use some or all of the methods for monitoring performance described previously and that audit findings may well include performance indicators such as the results of inspections, and performance measurement data that includes both reactive and proactive monitoring.

The key differences between monitoring performance and auditing are:

➤ It always involves active or proactive monitoring
➤ It is carried out by either independent external auditors or employees without line management responsibility for the site or activities being audited
➤ It is carried out in order to provide an independent input into the safety management review process leading to continual improvement
➤ A scoring or measuring system is always used.

The overall purpose of an audit is to measure performance against a standard, in this case the safety management system against an organisation's policy or a standard such as HSG65 or OHSAS 18001. The results of the audit provide an effective feedback loop that enables an organisation to measure its effectiveness of the safety management and areas of weakness that may require attention.

In general, safety auditing comprises an assessment of parts, or all, of the safety management system. Typically, the safety auditor or audit team may:

➤ Carry out a comprehensive audit of the whole safety management system or

> **Audit**
>
> The structured process of collecting independent information on the efficiency, effectiveness and reliability of the total health and safety management system and drawing up plans for corrective action.
>
> HSG65

➤ Look at a horizontal slice, e.g. an audit of the 'organising' element in the system or
➤ Look at a vertical slice where the arrangements to control a specific hazard, as in the case of fire, are audited in terms of the policy, organisation, planning and implementing, measurement and review processes in relation to fire safety management.

The aim of conducting a fire safety audit is to establish that appropriate management arrangements are in place such as the provision of a formal policy, arrangements for staff training, adequate risk control systems (such as regular fire safety inspections and testing of fixed fire fighting systems) exist and that they are being implemented and that other workplace precautions (preventive and protective measures) such as flammable stores, permits to work and fire doors are in place.

An effective auditing programme will be able to provide a comprehensive picture of exactly how effectively and efficiently the fire safety management system is controlling fire risks. The programme should also identify when and how each of the component parts of the system will be audited. Due to the technical nature of some of the workplace precautions (e.g. detection and alarm systems, sprinklers, etc.) these may be required to be audited on a more frequent basis than other elements of the system and the auditing process should reflect this.

11.4.1 Key stages of an audit

For an audit to be effective it must be based on a sound foundation and will generally include the following key stages:

➤ Agreeing a protocol:
  ➤ Ideally a protocol should be developed and utilised when undertaking audits. This protocol is a documented set of procedures and instructions, which is used to plan and organise the audit. It provides a step-by-step guide for the audit team and forms the framework in which audit is undertaken, providing the necessary guidance to the team on how the audit should be carried out.

- ➤ Scoping the audit:
  - ➤ This is the process of deciding what to audit. The level and detail of the audit will depend upon the size and complexity of the organisation. A multi-site office-based organisation, or an organisation with high risk processes who rely on technical systems to control fire, are likely to have differing audit profiles although there will be similarities in many areas
  - ➤ Auditing is essentially a sampling technique which seeks to obtain evidence across a range of activities of the status of the management system, therefore consideration will be required to judge the correct level and type of sampling that will provide reliable data.
- ➤ Preparation:
  - ➤ Discussion with the auditing team, managers, workforce and representatives regarding the aim objectives and scope of the audit prior to starting is critical to the success of the process
  - ➤ Preparation and agreement of an outline audit plan including approximate timings for interviews, etc.
  - ➤ Preparation of audit 'question sets' to ensure that each area is addressed.
- ➤ Gathering evidence:
  - ➤ As in the case of all active monitoring information, sources for the audit are likely to come from talking to people, undertaking site visits to observe physical conditions activities, etc. and the checking of documents and records
  - ➤ Records of evidence found are recorded against audit 'question sets', which on many occasions are numerically scored

- ➤ It is very likely that if technical systems are to be included in part of the audit process that appropriate qualified engineers may be required to provide additional information on fire safety systems.
- ➤ Report findings:
  - ➤ The audit report should reflect the findings of the assembled evidence upon which the auditors have based the evaluation of the fire safety management system
  - ➤ The findings may be supported with quantitative or qualitative analysis
  - ➤ A close-out presentation may be made to the management team and their representatives by the auditor or audit team.

11.4.2   The report

The report should outline the system's strengths and weaknesses and verify compliance (or otherwise) with standards – including legal compliance. The report should:

- ➤ Assess the adequacy, or otherwise, of the safety management system
- ➤ Accentuate all good practice, highlighting good performance
- ➤ Benchmark against accepted standards or best practice – detail areas of both compliance and non-compliance
- ➤ Highlight areas for improvements
- ➤ Contain sufficient detail and evidence to substantiate the judgements made by the auditors
- ➤ Differentiate between opinion and objective findings
- ➤ Utilise quantitative scoring systems (if these are part of the audit system)
- ➤ Be easy to read.

11.4.3   Auditors and composition of audit teams

Audits may be undertaken by one or more persons. Those chosen as auditors must be competent, whether as individuals or as members of a team. They should be independent of the part of the organisation or the premises that are to be audited. The nature and extent of the audit will determine whether it is undertaken by employees from another part of the organization or by external auditors.

Regardless of whether the audit is to be completed by an internal team or an external consultancy or audit team, the involvement of managers, representatives and employees in preparing for the audit will not only assist in devising and implementing the audit programme, it may also assist in completing the audit if undertaken internally.

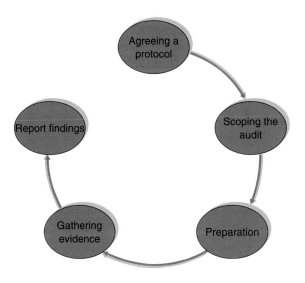

Figure 11.9   Key stages of an audit

When considering who will be involved with the audit process the following factors may also prove useful:

➤ The availability of auditors:
  ➤ For the length of time necessary to undertake the audit
  ➤ With the necessary skills
➤ The level of audit experience required
➤ The requirement for fire safety knowledge or technical expertise
➤ Any requirement for audit training
➤ The cost implication of using internal or external auditors
➤ The risk of an internal auditor being overfamiliar or satisfied with the company's arrangements, compared with the approach of an external auditor
➤ The potential for a lack of understanding of unfamiliarity, particularly where complex technical fire safety systems or processes are involved.

11.5 Reviewing performance

As with audits, safety reviews are an essential part of a management system. These reviews are closely linked to the active monitoring process; they assist in providing feedback of the adequacy of the fire safety management system to the management team and the responsible person.

In line with enforcement authority guidance and current standards, it is recommended that reviews should be conducted regularly (at least annually) so that decisions on how and when to rectify substandard conditions and failures in the management system can be taken, prior to there being a loss, failure or potential enforcement action.

11.5.1 Initial status review

According to safety management system guidance, including that of BS 8800, an initial status review should be carried out in organisations that do not have an established fire safety management system. The initial status review should provide information on the scope, adequacy and implementation of the current management system.

Where no formal fire or occupational health and safety management system exists the initial status review should serve as a basis for establishing what arrangements should be made to ensure an effective system is implemented.

Many organisations will have established a system and therefore the initial status review is likely to be based upon appropriate 'benchmarking' standards such as those found in BS 8800, BS 5588 Part 12.

The initial status review should provide the organisation with details of where they currently stand in relation to managing fire risks. The review should be carried out by competent persons in consultation with the workforce or its representatives.

The results of the initial status review should:

➤ Be documented
➤ Become the basis for developing and implementing the fire safety management system and arrangements
➤ Provide a baseline from which continual improvement can be measured.

The initial status review should establish to what extent existing arrangements are in place for:

➤ Identifying and meeting the requirements of relevant legislation and regulations dealing with fire safety management issues
➤ Identifying and implementing best practice and performance in the organisation's employment sector and other appropriate sectors (e.g. from relevant Government guidance industry advisory committees and trade association guidelines)
➤ Obtaining or developing guidance on fire safety management and making it available throughout the organisation
➤ Consulting and disseminating information throughout the organisation
➤ Identifying, anticipating and assessing hazards and risks to safety and health arising out of the work environment and work activities
➤ Developing and implementing effective workplace precautions to eliminate hazards and minimise risk
➤ Developing and implementing effective preventive and protective measures to manage fire risk
➤ Measuring and evaluating fire safety management performance
➤ Ensuring the efficient and effective use of resources devoted to fire safety management.

Many organisations utilise the initial review to prepare policy, organisational arrangements, planning and implementation processes and measuring and monitoring programmes based upon the findings of the report.

11.5.2 Regular reviews

Regular reviews also enable organisations to take into account changes in legislation and guidance, particularly with regard to any standards that affect fire safety installations, etc. together with changes in their own circumstances such as:

➤ Fire safety management procedures including maintenance procedures
➤ Changes of personnel, or in the building or premise, or the usage or activities undertaken
➤ The effectiveness of any automatic fire safety systems to ensure that they remain suitable given any change in usage, activity or layout of the premises
➤ Changes in local fire history – such as the number of arson attacks.

Many organisations undertake continuous reviews of their safety management system. In relation to fire there are a number of key areas that should form part of this process; these are detailed in the following paragraphs.

Figure 11.10 Building works which have resulted in the breaching of a fire compartment wall

Training and fire drills

A review of the fire safety training provided should ideally be undertaken on at least a six monthly basis having completed the fire evacuation exercise. As previously discussed the fire safety training programme will include initial induction training, general fire safety training, specific role oriented fire safety training, such as for fire wardens, marshals and first aid fire fighters which should also be subject to regular reviews to ensure that the programme meets the fire safety demands of the training needs analysis.

The review of the training is likely to revolve around the findings of the evacuation exercises as formal review debriefs should have been completed following the exercise (see Appendix 11.2). The contents of the debrief report should clearly identify effectiveness of the emergency evacuation, including all those involved in the process.

The training programme may also need revision in relation to subsequent risk assessments and the findings of fire safety inspections.

Fire safety systems

A review of the fire safety systems should also be undertaken as part of an ongoing programme to ensure that the fire alarm and detection systems, etc. are kept fully functional and that where third parties are used to undertake periodic testing, inspection, cleaning, etc. that they are providing the required level of service to meet the organisation's statutory responsibilities.

A review of the fire log book and other documentation relating to fire safety systems is likely to indicate specific component failures, which can then assist in identifying an appropriate replacement programme.

Fire safety inspection report reviews

The fire safety manager or nominee will undertake an ongoing review of the findings of the regular fire safety inspections that have been conducted by either themselves, or members of their team. It may be that the fire safety inspection is included in the regular workplace safety inspection regime and that the key fire safety issues will need to be extracted from the report.

Reviews of such inspections should provide an indication of areas such as:

➤ Repeat issues, e.g. wedging open of fire doors or failing to secure areas against arson
➤ Speed of rectification of faults found from previous inspections
➤ Competency of those involved in rectifying faults
➤ Confirm effectiveness of overall fire safety management
➤ The need to review or revise the current fire risk assessment or fire safety training programme.

Ideally, as in the case of reactive monitoring, statistical evidence as to the proactive performance will also be provided in such reviews.

Contractor reviews

As has been discussed within this book the effective management of contractors has a huge bearing upon the overall fire safety arrangements within a premises. It is therefore essential that a review of contractors and their work is conducted regularly.

The degree of monitoring and review required will be dictated by the nature of the work. However, no matter how simple or low risk the work may be there will always be a need for monitoring and review in some form, particularly if the works involve parts of the passive fire safety protection arrangements such as fire compartment walls, fire doors, etc. On-site monitoring may be carried out by the project management team, facilities team or by the local staff in the area of the works.

The staff carrying out the monitoring will need access to information regarding the extent of the work to be carried out and the methods of work to be followed in order to properly judge the actions of the contractor. This may require the staff given the task of overseeing the work being provided with copies of the contractor's method statements particularly in relation to hot work.

The procedure should include reporting channels for non-compliance issues and formal feedback of performance standards.

All staff within an organisation should be encouraged to observe the activities of contractors and report any circumstances they feel to be unsafe to the relevant person within their business area.

On completion of the works, or at predetermined periods during contracts of long duration, the client (for whom the contractor is working) should carry out a formal review of the performance of the contractor against the requirements of the contract and their own safety method statements.

This should include the safety performance of the contractor (such as obstructing fire exits, smoking, etc.) as well as the technical aspects of the works (such as installation of fixed fire fighting systems). The formal review of performance should be used in determining whether the contractor is used on future contracts.

11.5.3 Annual reviews and reports

The process of completing ongoing reviews in order to assess the effectiveness of the fire safety management system will also provide the core information for an annual review and report to be published. In many sectors of industry annual reviews are conducted to ensure the organisation's compliance with, not only the principles of safety managements, but also the requirements of corporate accountability.

Guidance issued by the Institute of Chartered Accountants (known as the Turnbull Report) in September 1999 for companies listed on the London Stock Exchange identified that the board of the organisation should maintain a sound system of internal control in order to safeguard the organisation's shareholders' investment and the organisation's assets.

The guidance goes on to say that the directors should at least annually conduct a review of the effectiveness of control systems and provide a report to shareholders. To meet the requirements of the guidance the review should cover all controls, including financial,

operations and compliance controls, together with risk management, which includes the risks associated with fire, particularly, as already discussed, the consequences of fire having a devastating effect upon a business.

The annual review and its subsequent report are therefore likely to include details of the following:

➤ An outline of the current fire safety policy and changes that may have been made to it to reflect management and operational contingencies
➤ Any changes to management structure from previous reports, e.g. nomination of a 'responsible person' under the RRFSO
➤ An overview of the facilities/premises
➤ The findings of risk assessments and fire safety inspections – particularly those that will require or have required high levels of resources
➤ Details of any fires, false alarms, near miss incidents and the results of investigations and subsequent actions
➤ Details of communication with the enforcing authorities, i.e. the local fire authority and the Health and Safety Executive
➤ Changes to legislation that have or will affect the organisation's fire safety management system
➤ Findings of the insurers when considering insurance premiums or claims
➤ Information from consultation with employees' representative bodies
➤ Effectiveness of controls including emergency evacuation exercises.

The review will take into account the findings of previous review reports and provide an outline of the actions required to ensure continuous improvement in the management of fire safety within the organisation.

The review process, from the initial status review through to a formal annual review report, must be seen as a critical element in the overall fire safety management system.

11.6 Case study

In 2004, a new management company was in the process of taking over the facilities management of a shopping mall in a large town development in the South East. The development included 42 commercial units, a small restaurant facility and large multi-storey car park, including an underground car park and plant room. Shortly after the management company took control of the premises, they were advised by their insurance company of a reasonably substantial increase in the premiums due to a number of fires occurring in similar facilities within the region.

When establishing its risk management system, the management company instigated an 'initial safety management review' (based upon guidance from BS 8800 and BS 5588 Part 12) which focused upon the then current fire safety management systems. This review was undertaken by an external fire safety consultancy organisation. The evidence gathered from the review identified a number of key fire safety management issues, the most significant of which were:

➤ A lack of overall fire safety management, consultation and cooperation with the tenants
➤ No mechanisms in place to update existing fire engineering systems such as sprinklers, detection systems and smoke extract systems
➤ A lack of cohesive management of those undertaking maintenance operations in relation to existing fire safety engineering measures such as alarm systems and detection.

Given the findings contained within the consultant's report, the new management company established an action plan to address the issues. A number of the items in the action plan could be addressed almost immediately such as establishing a maintenance strategy; however, the introduction of new technology to manage smoke and the extension of the existing sprinkler system needed a capital injection which was not available in the short term.

The business case presented by the management company to its board identified a strategy for resolving all of the initial findings and requested capital funds to be made available. Following negotiation with the insurance brokers and insurance company, the increase in premiums and excess was negotiated against the provision of certain fire safety engineering measures.

In this case the insurers reduced their risks and the management company while not substantially reducing its long-term expenditure reduced its short-term expenditure by minimising the increase in premiums and excess.

11.7 Example questions for Chapter 11

1. An employer intends to implement a programme of regular workplace inspections following a workplace fire.
   (a) **Outline** the factors that should be considered when planning such inspections. (6)
   (b) **Outline** THREE additional proactive methods that could be used in the monitoring of health and safety performance. (6)
   (c) **Identify** the possible costs to the organisation as a result of the fire. (8)
2. **Identify** EIGHT measures that can be used to monitor an organisation's health and safety performance. (8)
3. **Outline** FOUR proactive monitoring methods that can be used in assessing the health and safety performance of an organisation. (8)
4. **Outline** the reasons why an organisation should monitor and review its health and safety performance. (8)
5. **Outline** the main features of:
   (a) A health and safety inspection of a workplace. (4)
   (b) A health and safety audit. (4)
6. (a) **Explain** why it is important for an organisation to set targets in terms of its health and safety performance. (2)
   (b) **Outline** SIX types of target that an organization might typically set in relation to health and safety. (6)

Appendix 11.1　Sample of performance indicators

1. Appropriate fire safety policy has been written
2. The fire safety policy has been communicated
3. A director with safety responsibilities including fire has been appointed
4. Fire safety specialist staff have been appointed
5. The extent of influence of fire safety specialists
6. The extent to which fire safety plans have been implemented
7. Staff perceptions of management commitment to fire safety
8. Number of senior managers' safety inspection tours
9. Frequency and effectiveness of safety committee meetings
10. Frequency and effectiveness of staff fire safety briefings
11. Number of staff suggestions for fire and other safety improvements
12. Time to implement action on suggestions
13. Number of personnel trained in fire safety
14. Staff understanding of fire risks and risk controls
15. Number of fire risk assessments completed as a proportion of those required
16. Extent of compliance with risk controls
17. Extent of compliance with statutory fire safety requirements
18. Staff attitudes to fire risks and fire risk controls
19. Housekeeping standards
20. Worker safety representatives and representatives of employee safety have been appointed and are able to exercise their powers
21. False alarms and near misses
22. Fire damage only accidents
23. Reportable dangerous occurrences
24. Fire injuries
25. Fatal accidents
26. Complaints made by the workforce
27. Indicators that demonstrate that the organisation's fire safety objectives have been achieved
28. Criticisms made by regulatory enforcement staff
29. Enforcement actions
30. Complaints made by employees who are not direct employees of the organisation (e.g. self-employed persons, other contractors) or by members of the public.

Appendix 11.2   Sample of fire evacuation review/debrief log

FIRE DRILL RECORD (A fire drill should be held and recorded at least twice per year)

Date of drill:	Time of drill:
Type of drill:	
Number of staff involved:	
Actual evacuation time:	
Expected time of evacuation:	
Time to completion of roll call:	
Person responsible for drill:	
Assessment of drill:	
Remedial action required:	
Remedial action taken:	
Facilities/building supervisor's comments:	
Name: Signature: Date:	

12 Reactive monitoring – reporting, recording and investigation

The investigation of fires and other adverse events is a critical part of a safety management system. This section describes the purpose and process of investigating fires and other adverse events. It also covers the legal and organisational requirements for recording and reporting such events. Reactive systems include reporting, recording, investigation and taking some form of corrective action. The most effective managers are likely to have internal systems that identify, report and investigate all adverse (unplanned, unwanted) events, for example: fires, both accidental and deliberate; injuries and work-related ill health; near misses, false alarms in relation to fire (malicious, good intent, electrical); property damage and other consequential losses; fire and general safety hazards.

12.1 Fires and other adverse events

One of the most significant consequences of failure to adequately report and investigate a near miss safety incident may be that key learning opportunities are lost. It is often the case that a company will have little difficulty gathering data on significant incidents, particularly those that require it to notify the enforcing authority. However, near misses, false alarms and minor injury accidents and other losses often slip through the net of formal reporting and recording, serving to limit the opportunities to address safety systems failures before substantial losses accrue.

This chapter discusses the following key elements:

- ➤ Fires and other adverse events
- ➤ The statutory requirements for reporting fires and other adverse events
- ➤ Civil claims
- ➤ Investigating fire-related events
- ➤ Basic fire investigation procedures
- ➤ Dealing with the aftermath.

Figure 12.1 The consequences of failing to monitor events can lead to serious accidents

12.1.1 Reasons for investigating fires and other adverse events

Investigating the causes of such events highlights failures in health and safety management systems and allows these failures to be corrected.

If an event occurs in the workplace that has not been planned by management there is a risk of loss. The loss may arise from lost production, staff or damage to premises. The potential outcome of any such event is that the organisation and its staff will be liable to criminal prosecution and civil action arising from their negligence.

In addition, any such event may result in physical or physiological harm to employees or other persons. It is also likely that any such unplanned event will result in substantial financial loss including both direct and indirect costs (Chapter 1).

The numbers of work-related fires and other accidents recorded in the United Kingdom for the year 2003/04 are indicated in Table 12.1.

In terms of the economic cost of fire alone in the UK, the direct costs were estimated to be in the region of £8 billion for 2003. The costs of fires that were set deliberately in the same year amounted to £3 billion and the direct costs of the losses associated with false alarms were just under £1 billion.

12.1.2 Role and function of investigation

The purpose of investigating fires and other adverse events is not to apportion blame but to identify:

➣ How/where the management system has failed
➣ Identify additional risk control measures that are required
➣ Implement additional appropriate risk control measures.

Employers that respond effectively and openly to unplanned events not only continually improve the organisation's safety performance, they also demonstrate to the workforce, the industry and enforcers commitment to the safety and health of all those who may be affected by their operations.

Six reasons why organisations should bother to investigate fires and other adverse events:

1. To reduce the human costs
2. To reduce the risk of criminal prosecution and civil action
3. To reduce the economic costs
4. To effectively manage safety
5. To continually improve risk management systems
6. To demonstrate management commitment to safety.

Table 12.1 Safety events reported in the UK during 2003/04

Type of event	Number during 2003/04
Death of employees	235
Death of members of the public	396
Injuries to members of the public	13 575
Major injuries to employees	30 666
Over three-day injuries to employees	129 143
Accidental fires in buildings other than dwellings	31 200
Deliberate fires (total)	355 500
False alarms (total)	364 600

(Source: HSE and ODPM)

Policy requirements – the policy for responding to fires and other incidents should include a statement of intent from the most senior management, the identification of those persons responsible for implementing the policy and detailed organisational arrangements relating to:

➣ The immediate action that should be taken in the event of an incident
➣ Providing the underpinning training required to ensure the continued competence of all those involved in the process
➣ Ensuring that adverse events are recognised by employees
➣ Ensuring that adverse events are reported by employees and others
➣ The appropriate level of investigation is achieved relevant to the event
➣ How the investigation should be conducted
➣ The appropriate third parties involvement as soon as practicable

285

➤ Considering any recommendations made
➤ How recommendations are actioned
➤ How the outcome of the investigation is communicated
➤ The monitoring of the system.

12.1.3 Classifications of adverse events

The first and most important step in investigating an unplanned event is to recognise it when it happens. The HSE describe these events as 'adverse events'. Effective reactive monitoring of systems should allow an organisation to react to such adverse events. The reaction is normally triggered by the outcome of the event, for example:

➤ Injuries and ill health
➤ Short- and long-term sickness absence
➤ Loss of production
➤ Damage to property, tools and vehicles
➤ Near misses/false alarms
➤ Circumstances that have the potential to cause harm.

It is vital to identify those unplanned events that, by luck, have not resulted in any loss. The events that have the potential to cause injury, ill health, loss or damage present a low cost way of identifying and correcting management system failures before they result in loss to the organisation, its employees or the public. These events are often referred to as near misses, although some prefer the more accurate sounding term 'near hit'.

Near misses – research shows that for every major event there are a corresponding number of less serious events. In 1969, Frank Bird carried out a study of accidents and what he termed 'critical incidents' and developed his accident triangle illustration in Figure 12.2.

Eliminating critical incidents (near misses) significantly reduce the risk of an accident that results in a major injury. Near misses provide valuable opportunities to identify and correct failures in management systems that lead to accidents.

Comparisons with more recent figures from the UK tend to indicate that Bird's original findings hold some validity. Table 12.2 shows figures from the HSE for accident rates in 2001/02 for skilled tradesmen and machine operators. The table includes an estimate of the rate of near misses that would have occurred over the same period and gives an indication of the opportunities for reactive monitoring of adverse events that do not result in actual injury or loss.

If there is any doubt as to whether an event should be investigated this simple question should be asked, 'Was this event planned?' If the answer is 'no' then the event was unplanned and as such warrants investigation.

In terms of the resources used for any individual investigation a simple risk assessment will help. If there had been a significant risk of substantial losses through fire or explosion as a result of the 'near miss' then it would seem reasonable to expend resources on investigation at a similar level that would have occurred had the result been more serious.

12.1.4 Categories of injury causation

The HSE categorise adverse events that result in actual injury to people by how the injury occurred. Using these

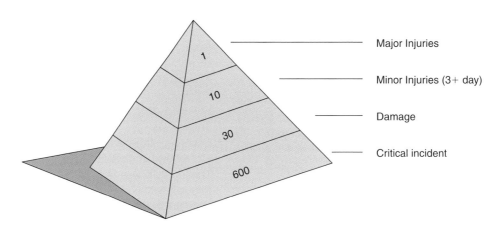

Figure 12.2 The ratio of major injuries to other outcomes

Table 12.2 Accident rates calculated by the HSE in 2001/02 for skilled tradesmen and machine operators

Workgroup	Reported fatal accidents (per 100 000)	Reported major accidents (per 100 000)	Reported minor (3+ day) accidents (per 100 000)	Estimated near misses (per 100 000)
Skilled tradesmen	2	212	2084	2 000 000
Machine operators	3	403	771	800 000

categories the HSE publish information relating to the rates and results of injuries for each category of injury causation. This allows HSE, in partnership with industry, to focus legislation and guidance towards the generic causes of injury resulting from adverse events. The categories of injuries are those that are caused by:

➢ Slips, trips and falls
➢ Falls from height
➢ Falling objects
➢ Collision with objects
➢ Trapping/crushing
➢ Manual handling
➢ Contact with machinery and equipment
➢ Electrocution
➢ Transport
➢ Contact with chemicals
➢ Asphyxiation/drowning
➢ Fire and explosion
➢ Animals
➢ Violence.

Figure 12.3 Unguarded electrical machinery which has the potential to cause injury in many ways including starting a fire

Table 12.3 Example of a matrix to establish the level of investigation required

Likelihood of recurrence	Potential worst consequence of adverse event			
	Minor	Serious	Major	Fatal
Certain				
Likely				
Possible				
Unlikely				
Rare				
Risk	Minimal	Low	Medium	High
Investigation level	Minimal	Low	Medium	High

(Source: HSG 245)

12.1.5 Basic investigation procedures

This section outlines the basic investigation procedures for investigation of all types of adverse events and discusses:

➢ The level of investigation
➢ Involving others
➢ Conducting the investigation
➢ Gathering information
➢ Analysing information
➢ Identifying risk control measures
➢ Agreeing and implementing an action plan.

The level of investigation

As mentioned above the level of resources allocated to the investigation will depend upon the seriousness of the actual or potential outcome of the event.

The decision regarding the level of investigation, particularly for near miss events and undesired circumstances, is quite often complex. To aid this process and achieve a level of consistency some organisations apply a simple matrix which provides an indication of whether the investigation should be conducted at minimal, low, medium or high level, see Table 12.2.

Involving others

Where an adverse event is likely to be investigated by a third party it is best practice to attempt to coordinate the investigation, this minimises disruption to the workplace and employees and demonstrates a willingness by all parties to be open and objective. However, it should be borne in mind that none of the third parties are obliged to take part in a joint investigation of the event, although it is normal that each party will share some information relating to their investigation.

For those investigations that are carried out entirely in-house there is often predetermined levels of investigation. It is important that a clear policy detailing the employer's arrangements for investigating safety events exists to enable managers and other employees at all levels to initiate the necessary investigation with the least delay.

Establishing the cause

Regardless of the level of investigation, its purpose is to establish:

➤ The immediate causes of the loss damage or potential damage
➤ The underlying causes
➤ The root causes.

The immediate causes – the most obvious reason or reasons why the adverse event happened, e.g. cloths left on top of a hot plate.

The underlying causes – there will be a number of unsafe acts and unsafe conditions that have come together to result in the adverse event. For example, a fire has started in a piece of fixed electrical equipment. The underlying cause may be that clothing discarded by an employee has covered the cooling vents on the equipment. The underlying cause in this case is relatively simple, the equipment overheated to the point of ignition caused by blocked ventilation ports. Table 12.4 gives further examples.

The root cause – the root causes of all accidents are failures of management systems, for example:

➤ Lack of adequate fire risk assessment of the workplace
➤ Failure to provide appropriate work equipment
➤ Lack of provision of adequate storage for employees' clothing

Table 12.4 Examples of unsafe acts and unsafe conditions that may lead to damage or injury

Unsafe acts	Unsafe conditions
Unauthorised hot work	Flammable atmosphere cause by flammable liquid leaking from inadequate container
Storing flammable liquid in an unsuitable container	Not having sufficient suitable facilities to store flammable liquids in the workplace
Walking across a slippery floor	Slippery floor caused by leaking oil heater
Using inappropriate equipment to gain access at height	Faulty ladder available in workplace
Opening an electrical supply panel	An unlocked electrical supply panel
Operator's ventilation block, ventilation ports of electrical appliance	No safety device on the equipment which cuts off power in the event of overheating

➤ Inadequate information to/training of employees in the importance of keeping vents clear
➤ Failure to actively monitor the workplace to identify blocked ventilation ports on electrical equipment
➤ Lack of sufficiently competent staff.

In order to establish the immediate, underlying and root causes of an adverse event the HSE suggest a four step approach:

1. Gathering information
2. Analysing information
3. Identifying risk control measures
4. Agreeing and implementing an action plan.

Gathering information

It is important in the beginning of an investigation to gather the information immediately or soon after the adverse event occurs or is discovered because:

➤ The location of the event will be in the same condition, i.e. light levels, temperature, etc.
➤ People's memories will be fresh
➤ There will be limited opportunity for a consensus view to emerge from witnesses.

The amount of time spent on gathering evidence will be proportionate to the outcome/potential outcome of the event; however, consideration should be given early on to the sources of information available for the investigation.

Sources of information

Information relating to the immediate, underlining and root causes of any adverse event should be gathered from as wide a range as possible. Information from different sources will tend to confirm the existence of problems with management systems. For example, it is likely that an unsafe workplace practice has developed as a result of either a lack of policy or a policy that is not supported with adequate training or supervision.

Table 12.5 Examples of the type and sources of information available for an investigation

From people by interview, discussion relating to:	From the location by observation and recording suitability and condition of any:	From the organisation by desk research of:
How the event occurred	Layout	Policy documents
Their observations	Materials	Risk assessments
Background opinions and experiences	Safety equipment	Records of previous or similar events
Relevant safe systems of work	Floors	Training
Normal practices	Work surfaces	Equipment testing
Possible solutions	Environment – lighting noise levels, etc.	Workplace inspections

Analysing information

Once information has been gathered it is necessary to order it in a logical sequence. The most logical approach is to arrange the information chronologically.

Therefore the first question to be asked is exactly what happened. In order to do this a suitable starting point must be identified. In many cases this will be at the point when the work actively started. In some cases the starting point of the investigation may be at a point in time when a significant failure of a management system initiated a sequence of events that led to an adverse event.

After establishing exactly what the sequence of events was the investigation must attempt to understand why these events occurred.

When exploring why a certain sequence of events occurred it will be necessary to consider a number of potential contributory factors. These factors will relate to the job, the people, the organisation, the equipment involved and the environment.

Job factors include:

➤ The nature of the work, e.g. is it routine, boring, or exceptional?
➤ Is there sufficient time available to complete the tasks safely?
➤ Are there any distractions, noise, other jobs, etc.?
➤ Are there adequate safe procedures/systems of work?

Human factors include:

➤ The physical and mental abilities of the individuals involved
➤ The levels of competence
➤ Personal or work-related stress
➤ The effects of fatigue, drugs and alcohol
➤ Human failure, e.g. errors and violations.

Organisational factors include:

➤ Work pressures and long hours worked
➤ The availability of sufficient resources
➤ The availability and quality of supervision
➤ The health and safety culture in the organisation.

Equipment factors include:

➤ The ergonomic design of the controls
➤ The ergonomic layout of the workplace
➤ Built-in safety devices
➤ The condition of the equipment
➤ The history of maintenance and testing.

Environmental factors include:

➤ Temperature
➤ Light levels
➤ Noise levels
➤ Cramped/open working conditions
➤ The provision of adequate welfare arrangements
➤ Cleanliness/housekeeping standards.

Identifying risk control measures

Once the investigation has established how and why the adverse event occurred the next step is to identify any additional risk control measures that should be put in place to reduce or prevent the chances of a recurrence.

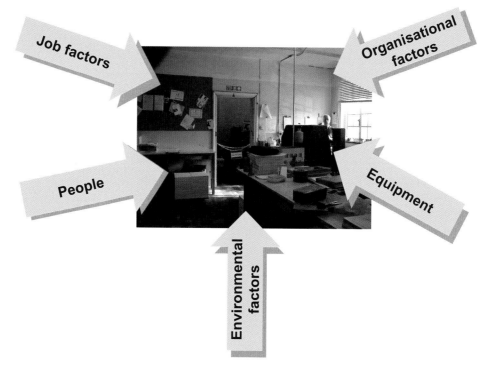

Figure 12.4 Factors that may contribute to adverse events

Figure 12.5 The pressure of work can increase the likelihood of accidents

When attempting to identify additional risk control measures the existing control measures will need to be considered. In particular it is important to identify if:

➤ Any effective control measures do actually exist
➤ If there are any additional control measures available (consider the hierarchy of controls and principles of prevention)

➤ If similar risks exist elsewhere in the organisation where it would be appropriate to extend the additional control measures to.

Agreeing and implementing an action plan

It is important that the experience of the adverse effects and the knowledge developed by the investigation are not lost to the organisation. Therefore it is crucial to develop an action plan for implementing the additional control measures. The action plan should reflect the format developed following a risk assessment (see Chapter 14).

In addition, it is vital that the plan is agreed with those within the organisation who have the authority to expend the necessary resources in terms of engineering solutions or training, for example to sign up to the action plan.

Any action plan should include, as a minimum:

1. The action to be taken
2. The individual who is responsible for taking the action
3. A date for completion
4. Monitoring arrangements to ensure that the action is taken
5. Arrangements for the review of the action to determine its effectiveness in reducing risk.

12.1.6 Internal systems for managing adverse event data

In addition to agreeing and implementing an action plan, it is important to communicate the findings to the workforce. The benefits of this are that management continue to demonstrate their commitment to the management of health and safety and that the lessons learnt through the investigation process are learnt by a wider audience throughout the organisation.

Tracking the response to each event – organisations will want to establish formal systems for recording adverse events to ensure that the enforcing authorities are notified when required, information relating to injuries to staff members is accurately recorded and to monitor the performance of the investigation procedure. As well as learning lessons from each adverse event it is equally important that any longer-term trends that should attract management's attention are identified.

Trend analysis – this can be achieved in a number of ways. As a minimum a paper-based system can be used where the workforce is small and adverse events are few. It is normal for SMEs and larger organisations to employ an electronic system for recording the necessary data. Whichever system is used it must be able to provide a management overview of the event experience and enable an analysis of event trends.

Communication – the lessons learnt from each event and trend analysis including any additional control measures that are to be implemented must be communicated to staff. To be most effective communication will normally be made in a variety of ways including:

> Agenda items for management meetings
> Posters
> Agenda items for health and safety committee meetings
> Intranet websites
> Seminars
> E-mail shots
> Tool box talks
> In-house journals.

12.2 Statutory requirements for recording and reporting adverse events

There are two key pieces of legislation that require employers to record and report adverse events:

> Social Security (Claims and Payments) Regulations and
> The Reporting of Injuries, Diseases and Dangerous Occurrences Regulations 1995 (RIDDOR).

12.2.1 Social Security (Claims and Payments) Regulations (SSCPR)

The SSCPR require that employers keep a record of injuries at premises where more than 10 people work. Under regulation 25, persons who are injured are required to inform their employer and record the details, including how the event occurred, in an accident book (see Fig. 12.6).

The employer is, in turn, obliged to investigate the cause of the event in so far as it establishes the most basic information which should then be recorded in the accident book. The purpose of recording the details in the accident book is to enable the Department for Social Security to have access to basic information in the event of a claim being made as a result of the injury.

In order to aid employers to satisfy the requirements of the SSCP and the Data Protection Act 1998, the HSE has published an accident book BI 510. As with guidance from the HSE, employers are not obliged to use this particular book; however, they are obliged to record the information requested on the form.

Despite the fact that HSE publish a book to record accidents it should be remembered that the requirement for an initial investigation and recording of personal injury accidents is contained within the SSCPR.

To meet the Data Protection Act requirements the personal details that are recorded within the accident book or part of a company's reporting and recording system must be kept securely and only be accessed by authorised persons.

As will be discussed later within this chapter the information recorded within the BI 510 is of limited use when an investigation takes place as it only records personal injury accidents (not near misses or fire, false alarms, etc. if persons are not harmed).

Many organisations therefore provide additional forms for completion, or have decided to dispense with the basic accident book and record in a different manner.

12.2.2 The Reporting of Injuries, Diseases and Dangerous Occurrences Regulations 1995 (RIDDOR)

The Reporting of Injuries, Diseases and Dangerous Occurrences Regulations 1995 (RIDDOR) require that employers notify the relevant enforcing authority of specified injuries, diseases and dangerous occurrences that occur as a result of a work undertaking. The events (specified in RIDDOR) that need to be reported are:

> Death or major injury, where an employee or self-employed person is killed or suffers a major injury

291

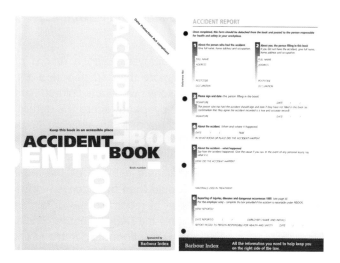

Figure 12.6 Front cover and form from the accident book BI 510

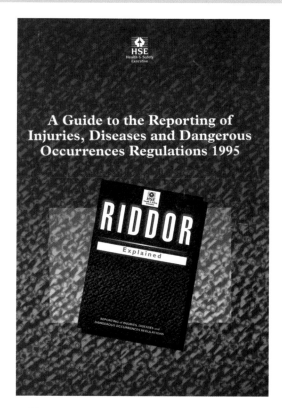

Figure 12.7 The guide to RIDDOR

as specified by the regulations, or when a member of the public is killed or taken to hospital

> Over three-day lost time injury, where an employee or self-employed person suffers an injury and is not available for carrying out their normal range of duties for more than three days. The three days includes weekends and rest days but excludes the day the injury was reported

> Disease, the regulations specify certain industrial diseases. A disease is only reportable to the HSE if it is a disease that is confirmed by a doctor and that is one that HSE have identified as being related to the specific work activity

> Dangerous occurrence – the regulations specify a number of dangerous occurrences, for example a fire that interrupts business for 24 hours or more or an explosive release of a pressurised gas.

The regulations also detail how and when the notifiable events are reported. These reports are routed through the Incident Centre which was set up to assist responsible persons discharge their duty. The Incident Centre will notify the relevant enforcement authority dependent upon the sector the responsible person's organisation operates in. Table 12.6 summarises the requirements of the regulations.

A copy of the HSE form F2508 is provided in Appendix 12.1.

12.2.3 Reporting fire-related events

Personal injuries – any personal injuries, i.e. fatal, major or three-day, that occur as a result of a fire must of course be reported to the Incident Centre under

Table 12.6 Summary of reporting requirements under RIDDOR

Type of event	By e-mail, web, telephone or fax	In writing
Death or specified major injury	ASAP	On form F2508 within 10 days
+3-day injuries	N/A	On form F2508 within 10 days
Certain specified dangerous occurrences	ASAP	On form F2508 within 10 days
Certain specified diseases if related to specified work	As soon as notified by a doctor	On form F2508A within 10 days

RIDDOR in the normal way. Specified injuries that are likely to arise from a fire in a workplace are:

> Chemical or hot metal burns
> Any situation requiring resuscitation
> Loss of consciousness following smoke inhalation or oxygen deficiency
> Admittance to hospital for more than 24 hours.

Dangerous occurrences – any dangerous occurrences that are specified in RIDDOR that occur as a result of a fire must also be reported to the Incident Centre in the normal way. However, specified dangerous occurrences that are likely to arise from a fire in a workplace are:

> The failure of any closed system under pressure that has the potential to cause death
> An electrical short circuit or overload attended by fire which results in:
>> A stoppage for 24 hours or
>> Has the potential to cause death
> Incidents involving explosives
> A failure of a breathing apparatus in use or during test immediately prior to use
> An explosion or fire which is due to the ignition of any material, which results in the stoppage or suspension of normal work for more than 24 hours.

12.3 Civil claims

The most effective way for an organisation to protect itself from the possibility of a successful civil claim is to ensure the safety of its employees and others, to be achieved 'so far as is reasonably practical.'

However, should an injury to an employee or other person be considered as a result of the negligence of the employer, the injured party may wish to pursue compensation for the loss incurred by suing the employer.

It is therefore important for any organisation investigating an adverse event that has resulted in injury to ensure that any documentation or other evidence that is available to defend a claim for negligence is recorded in a credible fashion.

There may be a conflict between the information that is available to the investigation that would harm the defence of a civil claim. However, there is a duty under the civil courts' rules for the employer to disclose all relevant information relating to the event. A claim is more likely to be successfully defended or any compensation minimised if the courts are convinced that the employers have been honest and objective during the course of investigating and recording the event.

12.4 Investigating fire-related events

In the case of a serious fire in the workplace it is inevitable that the Fire Service will conduct its own investigation. Where the Fire Service consider a fire to have been deliberately set, the police will be involved with forensic investigation of the fire in order to detect the persons responsible.

Figure 12.8 The result of an arson attack on an external shed

If the fire has resulted in a death, the Coroners' Court will conduct its own investigation in order to determine the exact cause of the death.

In practice, investigating a fire in the workplace will follow exactly the same principles as described previously in this chapter. The information in this section is provided to allow those with a responsibility for the management of fire safety in the workplace to develop an understanding of the basic principles of fire investigation.

This section explores the following topics:

> The types of fire-related events
> Basic fire-related investigation procedures
> Preserving the scene
> Liaison with other parties
> Identifying root causes and control measures
> Dealing with the aftermath.

12.4.1 Types of fire-related events

There are three distinct types of fire-related incident that should be investigated in order to prevent a recurrence; accidental fires, deliberately set fires and false alarms.

Accidental fires
In the UK in 2003 there were 312 000 accidental fires reported to Fire Services. Analysis of the supposed

causes of these fires indicates that the most common indicators of such fires are:

- Electrical distribution
- Other electrical appliances
- Smokers' materials
- Cigarette lighters
- Matches and candles
- Cooking appliances
- Heating appliances
- Blowlamps, welding and cutting equipment
- Central and water heating devices.

However, when investigating the causes of a fire it is necessary to consider a number of other causes which include chemical reactions, heat from mechanical friction, spontaneous combustion and static electricity.

Deliberately set fires

In the UK in 2003, in contrast to the number of accidental fires, there were 355 500 that Fire Services consider were deliberately set. A great proportion of deliberately set fires involve car fires where vehicles used for criminal purposes, robbery, or more usually joy riding are abandoned in remote areas and set on fire in an attempt to destroy forensic evidence. Arson is the largest single cause of major fires in the United Kingdom, currently costing an estimated £2.2 billion a year, which equates to £44 million every week.

Figure 12.9 Damage to records and stored items following an arson attack

Although the reasons why people will deliberately set a fire may be complex, there are some simple indications that will indicate that a fire has been deliberately set, including:

- The existence of any of the reasons listed above
- The fact that the fire involved accelerants that would not normally be in the vicinity of the fire

The reasons why individuals deliberately set fires are often complex and interrelated but include:

- The excitement/power of seeing a fire develop
- An ongoing dispute with a neighbour
- A grudge held by a disatisfied customer
- A grudge held by a dissatisfied previous employee
- To gain an advantage against a business competitor
- To cover up another crime, for example robbery, murder
- To defraud an insurance company
- To circumvent planning requirements for redevelopment of buildings.

- The fact that the fire may have several 'seats' or places where it started
- Signs of breaking and entering to the premises
- Valuable or sentimental items that have been removed prior to the fire.

False alarms

In the UK in 2003 there was a total of 365 000 false alarms made to local authority Fire Services. About 90% of automatic fire detection and fire alarm systems do not regularly cause false alarms. However, the remaining 10% are involved in most false alarms. Every false alarm causes disruption. The cost of false alarms in the UK is estimated to be about £1 billion a year.

Responding to false alarms diverts the Fire and Rescue Service from their fire prevention duties, or from

Figure 12.10 False alarms degrade the ability of the fire and rescue service to respond to genuine emergencies

Table 12.7 The Fire and Rescue Service classification of false alarms

Type of false alarm	Description
Malicious	Where the operation of a fire alarm system or a call to the Fire Service is made deliberately and maliciously
Good intent	Where a call is made to the Fire Service where the caller believes that there is a fire or a fire alarm system has operated without a fire having started
Good intent – electrical	Where the alarm is initiated by an electrical fault
Good intent – mechanical	Where the alarm system is initiated by a mechanical fault

dealing with real emergencies. They also disrupt work patterns and valuable training programmes. Almost half of the calls to the Fire and Rescue Service are false alarms, and most of these are false alarms from fire detection and fire alarm systems.

According to the Chief Fire Officers Association (CFOA), a well-designed and maintained, fire detection and fire alarm system should produce no more than one false alarm a year for every 50 detectors fitted, and no more than one false alarm in any four-week period. For large fire detection systems, the aim must be to reduce the level of false alarms well below that of one a year for every 50 detectors.

False alarms are classified by the Fire Service by the initial event that caused the alarm to be raised. The classifications are shown in the Table 12.7.

12.5 Basic fire-related investigation procedures

The two types of fire-related events that are covered in these notes are false alarms and fires. As with other adverse events the basic procedure for investigating false alarms and fires involves a number of steps, starting with an assessment of the level of investigation that is required and concluding with the implementation of an agreed action plan:

1. The level of investigation
2. Involving other agencies/parties
3. Conducting the investigation
4. Gathering information
5. Analysing information
6. Identifying risk control measures
7. Agreeing and implementing an action plan.

Investigating the cause of false alarms

The purpose of investigating false alarms is to establish: the immediate cause, for example unauthorised smoking near a smoke detector head; the underlying cause, for example lack of control over contractors smoking in the premises; and the root cause of a failure to have and enforce adequate policies for smoking and the control of contractors.

In order to reduce the numbers of false alarms, a thorough investigation into the circumstances should be undertaken. In some instances managers will want to involve the fire alarm contracting company in order to have the necessary competencies for the investigation. The resultant action plan may involve redesigning part of the fire alarm system and/or improving the levels of inspection, testing and maintenance.

The process

As soon as possible after the false alarm, inspect that area and locate the break glass box, heat detector or smoke detector that set off the alarm in order to ascertain why the alarm was triggered. It should be noted that the detector may be in a duct or above a false ceiling. It will be necessary to call in a competent engineer to assist the process on those occasions when:

➤ It is difficult to locate the detector
➤ The control panel does not show where the relevant detector is
➤ If no detector was triggered.

Whatever the outcome of the investigation it is vital that there is an accurate record all the information about the false alarm in the system log book.

If false alarms continue, it will be necessary to analyse when the false alarms happen and where they come from. This will help identify any pattern that may help identify the cause (for example, cooking before meal times or a boiler switching on early in the morning).

It is very likely that the investigation into the false alarm will indicate that the alarm was caused by equipment faults, malicious acts, human error, or activities near detectors.

Figure 12.11 Investigating equipment faults

Figure 12.12 The result of contractors working on the fire alarm system – removal of detector head

Equipment faults – false alarms will arise from equipment if:

> The equipment is faulty or has not been maintained properly
> Fire detectors or red 'break glass' boxes are in the wrong place and
> The fire detection system is not appropriate for the building.

During the investigation, if it is not possible to identify the cause of the false alarm, or if there seems to be a fault in the system, the alarm should be silenced but the system should not be reset in order to allow an engineer from the alarm company to investigate the fault. In any event if the equipment seems to be faulty, the alarm company that installed or maintains the equipment should be instructed to take the appropriate action.

If the alarm system automatically alerts an alarm receiving centre (an ARC), it should also be contacted immediately and informed of the situation (to avoid needlessly calling out the Fire and Rescue Service) until the problem has been fixed.

Malicious acts – malicious acts can be the most difficult to identify and often need to be investigated carefully.

Examples of malicious acts include:

> Unnecessarily breaking the glass in break glass boxes
> Unauthorised people having and using test keys for break glass boxes
> Deliberately directing smoke (for example, from a cigarette) into a smoke detector.

Human errors – examples of human errors that may result in a false fire alarm include the following:

> Building contractors carrying out hot or dusty work close to smoke detectors or heat detectors
> The fire alarm system not being switched off while its wiring is being altered
> Unsecured control panels being activated, usually as a result of the panel's key or a similar device being left in the panel
> Smoking in unauthorised areas.

Activities near detectors – many false alarms result from activities carried out near fire detectors, particularly smoke detectors.

Common examples of how activities near detectors can be the immediate cause of a false alarm include:

> Smoke from cooking
> Burning toast in a toaster
> Cleaning operations using steam or aerosols
> Building work creating dust
> Hot works.

An example of the record that should be kept of an investigation into a false fire alarm is included at Appendix 12.3.

Investigating the causes of fires

The Fire and Rescue Service (FRS) does not have a statutory duty to investigate the causes of fire; however, the service is obliged to report all fires in buildings to the Department for Communities and Local Government (DCLG) on a form FDR1. Completion of the FDR1 requires FRS to record a 'supposed cause' for every property fire. See Appendix 12.2.

Report of Fire

Date: Day | Month | Year

KEY

Tick the appropriate box ☑ or boxes

Insert code from codelist or enter number ☐

Brigade use ☐

Write in details ☐

1. Brigade Information

1.1 Brigade incident number

1.2 Brigade Area where fire started — Station ground

1.3 Brigade and Home Office Call number — Fire spread box

2. Incident Information

2.1 Address of fire

2.2 Postcode (for buildings) or grid reference (if available)

OS national grid reference

2.3 Risk category

☐ A ☐ B ☐ C ☐ D ☐ R   ☐ Also ✓ if Special risk within area

2.4 Name(s) of occupier(s)/owner(s)

Times

2.5 Estimated interval from

a) Ignition to discovery

☐ Immediately ☐ Under 5 mins ☐ 5 to 30 mins ☐ 30 mins to 2 hours ☐ Over 2 hours ☐ Not known

b) Discovery to first call

☐ Immediately ☐ Under 5 mins ☐ 5 to 30 mins ☐ 30 mins to 2 hours ☐ Over 2 hours ☐ Not known

(use 24 hour clock)

	hour	mins	day*	month*	year*
2.6........ First call to brigade					
2.7 Mobilishing time					
2.8............Arrival of brigade					
2.9.... Under control					
2.10 Last appliance returned					

* Only complete 2.7 to 2.10 if diffrent from 2.6

2.11 Was this a late fire call?

☐ No ☐ Yes

2.12 Discovery and call

a) Discovered by

☐ Person ☐ Automatic system ☐ Other - specify in Section 7

b) Method of call by

☐ Person ☐ Automatic system ☐ Other - specify in Section 7

2.12 Was there an automatic fire alarm system in area affected by fire?

☐ No ☐ Yes

2.14 Alarm activation method

☐ Heat ☐ Smoke ☐ Flame ☐ Other - specify in 2.18 ☐ Not known

2.15 Powered by

☐ Battery ☐ Maths ☐ Mains & battery back up ☐ Other - specify in 2.18 ☐ Not known

2.15 Did it operate?

☐ No ☐ Yes but did not raise alarm ☐ Yes and raised alarm — go to 2.18 or 3.1

2.17 Reason for not operatiing/not raising alarm

2.18 Other details of automatic fire alarm

FDR1 (94)

Figure 12.13 Fire and rescue service fire report form – FDR1

The powers of the Fire and Rescue Service

Fire and Rescue Services in England and Wales are constituted under the Fire and Rescue Service Act 2004 (in Scotland the Fire Scotland Act 2005). Under these Acts, fire service officers have powers to enter and obtain information in two situations, i.e. when they wish to:

- Obtain any information the service may need to assist them in preparing to deal with fires or other emergencies in those premises
- Investigate the cause and spread of a fire.

However, the service may not enter as of right any premises in which there has been a fire if the premises are unoccupied, and it was a private dwelling immediately before the fire, unless 24 hours' written notice is given.

In the case of difficulties, a Fire and Rescue Service officer may apply to a justice of the peace for a warrant authorising the officer to enter the premises by force at any reasonable time.

Supplementary powers

For the purposes of fire investigation, a fire service officer (duly authorised officer) may:

- Take with him any other persons, and any equipment, that he considers necessary
- Inspect and copy any documents or records on the premises or remove them from the premises
- Carry out any inspections, measurements and tests in relation to the premises, or to an article or substance found on the premises, which he considers necessary
- Take samples of an article or substance found on the premises, but not so as to destroy it or damage it unless it is necessary to do so for the purpose of the investigation
- Dismantle an article found on the premises, but not so as to destroy it or damage it unless it is necessary to do so for the purpose of the investigation
- Take possession of an article or substance found on the premises and detain it for as long as is necessary for any of these purposes:
  - To examine it to ensure that it is not tampered with before his examination of it is completed
  - To ensure that it is available for use as evidence in proceedings for an offence relevant to the investigation
  - Require a person present on the premises to provide him with any facilities, information, documents or records, or other assistance, that he may reasonably request.

Figure 12.14 Cash box damaged by fire

It is criminal offence to obstruct a fire service officer who is investigating a fire which is liable on summary conviction to a fine not exceeding level 3 on the standard scale.

The role of the manager in the workplace will be to preserve the scene and assist the Fire Service in their investigation. Local knowledge of the workplace, staff, equipment and normal practices will assist the Fire Service. A basic awareness of factors affecting the initiation and growth of a fire together with an understanding of how materials react to smoke and heat will enable the local manager to identify those issues that will be relevant to the Fire Service.

Late calls – there may be circumstances when a fire occurs that is not immediately evident to those in the workplace and is not detected by the alarm system. In the case where the fire self-extinguishes it may be that it is discovered at some time after the event. In this case the local manager will inevitably conduct the initial investigation to establish that there had in fact been a fire. In these circumstances the Fire Service should be notified by the normal method. Fire Services will treat these events as 'late calls' and will normally respond with a single officer who will take the necessary details to complete the required fire report form.

12.5.1 Health and safety when investigating

As with any investigation, ensuring that those at the scene of the investigation are kept safe must be seen as a priority. The decomposition of materials and potential release of chemicals, asbestos and contaminated water must be guarded against, as should the threat of falling or collapsing materials caused by destabilisation of building structures, etc.

Figure 12.15 Appropriate PPE required for fire investigation

12.5.2 Preserving the scene

After the fire has been extinguished the first action that should be taken is to preserve the scene to facilitate the investigation. This can be achieved in a number of ways and will, of course, be dependent upon a number of factors including:

> The size and extent of the fire
> The level of business disruption
> The time the investigation of the scene is likely to take
> The hazardous nature of any materials involved, including the products of combustion
> The security needs of the site
> The business recovery plan.

Slips and trips together with falls into voids must also be considered and managed to ensure the safety of the investigator.

It is therefore essential that an organisation establishes a management plan for investigation of fires if staff members are to be involved, even if this is only to assist the Fire and Rescue Service in conducting its own investigation.

The following should be considered for those involved with fire investigation:

> Securing the area
> Entry only after confirmation that the building structures are stable (building surveyor, structural engineers or building control officer) and that service supplies are made safe (gas, electrical water, etc.)
> Ensure that sharp tools and equipment are covered/guarded and isolated
> Provision of adequate lighting including handheld torches
> Personal protective equipment such as respiratory protection, head protection, hand protection, foot protection, etc.
> Emergency procedures, particularly if working alone
> Decontamination of PPE together with adequate welfare arrangements.

It is also likely that investigators may well be subject to health surveillance particularly in relation to respiratory functions, blood-borne diseases, etc.

The staff members involved are also likely to have been provided with training not only in relation to investigation but also in relation to the safety aspects required while conducting an investigation and provided with an adequate level of supervision.

One of the most critical actions that is required by the Fire Service for an effective investigation is that the scene of the fire is left undisturbed, so that accurate observations can be made as to the circumstances both before and during the fire.

In the case of a small fire confined to an electrical appliance it may suffice to remove the appliance to a test area for further investigation. In the event the fire is confined to the room of origin, it is likely that the room itself can be isolated during the course of the investigation. For larger fires it may be necessary to cordon off large areas or even the building itself.

It will be important to demonstrate to enforcing authorities, insurance companies and civil courts that the scene was preserved and that no material evidence was added or removed prior to or during the investigation of the scene.

12.5.3 Liaison with other parties

Although there is no direct statutory duty on employers, all fires should be reported to the local Fire Service. Some fires, for example, resulting in a death or a fire that disrupts a workplace for 24 hours or more must (under RIDDOR) be reported to the HSE.

Some fires may have serious implications for the employer/employees but need not to be reported to the HSE though they must be communicated to employees'

Table 12.8 Examples of possible joint investigations

Type of adverse event	Level of investigation	Internal resources	External bodies who may investigate
Explosion and fire in the workplace that results in fatal/ serious injury accident at work	High	Senior manager Employees' representative ROES Section managers Workforce	Coroner Police HSE/local authority Trading standards Insurance company
Accidental fire in the workplace causing stoppage for 24 hours	High	Senior manager Employees' representative ROES Section managers Workforce	Fire authority HSE/local authority Insurance company
Deliberate fire in the workplace	High	Senior manager Employees' representative ROES Section managers Workforce	Fire authority Police HSE/local authority Insurance company
Fire causing a 3-day injury to contractor	Medium	Middle managers Employees' representative ROES	HSE/local authority Contracting company
Fire resulting in minor damage to work equipment	Minimum	Section managers Workforce	Insurance company
Electrical circuit overloaded by the excessive use of adaptors	Medium	Section managers Workforce	HSE/local authority Insurance company

representatives, i.e. a fire in the computer server. Other fires may be investigated by other agencies; Table 12.8 gives further examples of how various types of fire event may be investigated.

12.5.4   The fire investigation process

When examining the site of a fire there are a number of aspects that will enable the investigator to understand the immediate and underlying cause/s of the fire:

Aspect	Comment
The site conditions	An examination of the site will give an indication of whether the circumstances were normal. For example, are all the security systems in place and working? Is there any evidence of unauthorised practices? Is there any evidence of forced entry?
Low points of burn	Under normal conditions fire burns vertically upwards, therefore the identification of the lowest point of burn at the site of a fire will give a good indication of where the fire started. In the case of fires that have been deliberately set it is common to identify a number of different seats of fire by observing a number of low points of burn. Although this is a good indicator of the seat of a fire, low points of burn can also occur when materials or substances melt or fall down from a higher level and create an intense burn at a low level.
Possible heat sources	It is necessary to consider every conceivable heat source (see above) as a possible initiator of the fire. There may be occasions when there are more than one heat source; in these circumstances the Fire Service will attribute a percentage rating to the sources they consider as initiators of the fire.
Evidence of fire spread	When examining a fire scene it is often not clear how the fire has spread, particularly if there are a variety of materials involved. Reference to the mechanisms of fire spread will help the consideration of conduction, convection, radiation and direct burning as methods of how the fire as spread. This will also help to confirm the location of the seat of the fire.
The effects of temperature	Various materials will react differently when exposed to extreme temperatures. The table in (Appendix 6.5) provides an indication of how hot various regions of the fire had become and so will also help to indicate the seat and spread of the fire.

Evidence of fire growth	The rate of fire growth from induction to a fully developed fire will give an indication of the presence or otherwise of fuel and oxygen. In the case of a slow developing fire, glass and plaster tend to remain intact and the affected room becomes heavily smoke logged and therefore sooted.
Evidence of accelerants	The most obvious sign of the presence of some form of fire accelerant is the smell. Petrol and other flammable liquids tend to soak, unburnt, into carpets and other furnishings. Flammable substances will often continue to vaporise after the fire has been extinguished and therefore can be detected by smell or specialist sensing equipment. Another classic sign that a fire has involved flammable liquids is the presence of circular or 'running' burn patterns on floors or walls surfaces.

12.5.5 Identifying causes and risk control measures

It is important to reduce the chances of a recurrence to identify the immediate, underlying and root causes of a fire. The immediate causes of the fire will often involve the initiators of fire. Whereas the underlying causes will be factors that lead to the initiators starting a fire. The root causes, as with all adverse events, will be failures in management systems. Unless the root causes are 'identified', adequate measures to reduce the risk cannot be taken.

Root causes are generally failures of management systems, for example:

➣ Lack of adequate fire risk assessment of the workplace
➣ Failure to provide appropriate work equipment
➣ Inadequate information to/training of employees
➣ Failure to actively monitor the workplace.

It can be seen that for each example in Table 12.9 one of the root causes of all fires is likely to be an inadequate fire risk assessment (FRA). It may be that the FRA failed to identify a particular fire risk or that the fire risk control measures that the FRA identified were not fully effective or in place. It may also be that the FRA had identified

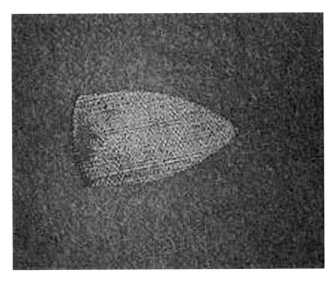

Figure 12.16 Immediate cause of burnt carpet is obvious. But which management systems have failed?

additional risk control measures but there had been a failure to implement them.

Fire risk control strategies, included in the fire risk assessment, will need to reflect the hierarchy of control

Table 12.9 Examples of fires and their possible causes

Event	Immediate cause	Underlying causes	Root causes
Spontaneous combustion of oil soaked cotton rags	Cotton rags soaked in oil and left in the workplace	Lack of employee training Provision and use of inappropriate work materials Poor housekeeping	Lack of management commitment to health and safety Lack of adequate purchasing, training and housekeeping policies Lack of adequate fire risk assessment
Deliberate fire set in premises overnight	Criminal action by person/s unknown	Inadequate security arrangements Lack of effective perimeter security/window locks	Lack of adequate fire risk assessment
Accidental fire in dust extract ducting	Dust ignited by heat from overheated extractor fan bearing	Lack of maintenance of extract system	Failure of planned preventive maintenance programme for the extract system Failure of management to monitor the PPM system Lack of adequate fire risk assessment

measures suggested in both the RRFSO and the MHSW, for example:

➤ If possible eliminate the immediate causes of fire from the workplace
➤ If it is not possible to eliminate causes of fire, substitute the fire hazard for a lesser one (e.g. low voltage equipment, non-flammable substances)
➤ Engineering controls should be used to isolate or insulate heat sources from any combustible material or flammable atmospheres
➤ Hot work should be reduced to a minimum
➤ Maintain good levels of housekeeping
➤ Provide safe systems of work that minimise fire risk
➤ Provide adequate training, instruction and information relating to fire safety
➤ As a last resort provide PPE to protect the individual against the harmful effects of smoke and heat and other harmful products of combustion.

Arson prevention

Due to the high financial and social losses incurred as a result of arson Fire and Rescue Services (FRSs) are investing increasing resources to reduce the numbers of fires that have been deliberately set. Initiatives from the FRSs include educating young persons and identifying and providing life skills mentoring for potential arsonists. If arson is suspected at a fire the FRSs, along with the police, will go to great lengths to identify the guilty parties.

The role of employers and responsible persons in the workplace is to prevent the opportunities and incentives for arson attacks. Such measures should include:

➤ Site security to prevent unauthorised access
➤ Reducing available combustible materials in the premises
➤ Reducing the opportunities for theft in the premises (which may be covered up by arson)
➤ Dealing fairly with employees particularly redundancies or terminating employment contracts.

Further details on arson and arson control can be found in Chapter 8.

12.6 Dealing with the aftermath of fires

Once the investigation of a serious fire is complete the employer or responsible person will need to ensure that there are adequate arrangements to deal with the outcomes of the fire. Very often a fire that has taken just one hour from ignition to extinction will take months and sometimes years to recover from. When dealing with the immediate aftermath of a fire consideration will need to be given to issues relating to the site of the fire, employees, any neighbours and the arrangements for business continuity.

Figure 12.17 Dealing with the aftermath of a fire

12.6.1 The site

The site of the fire will present a number of significant and unusual hazards. The normal substances in the workplace may have undergone a chemical reaction and be more hazardous than before the fire. In addition to the normal products of combustion, including soot, contaminated water used for fire fighting may contain carcinogenic material. Other hazards may be presented by the building materials which will themselves have been affected by the fire. The elements of the building may have been seriously weakened or asbestos may have been disturbed.

Figure 12.18 The effects of fire on hazardous substances in the workplace

In any event, clearing the site of a fire is a hazardous task. It may be necessary to employ specialist contractors to enable the site to be cleared legally and without unnecessary risk to employees.

12.6.2 Employees

Employees will be disconcerted by a serious fire in the workplace and may be concerned for their own future safety and employment prospects.

Management will need to communicate effectively with employees to provide additional reassurance. In addition, management will need to ensure adequate welfare arrangements are made for staff. It may be the case that some employees may be susceptible to post traumatic stress disorder (PTSD) and will therefore benefit from professional counselling to prevent the chronic symptoms developing.

12.6.3 Neighbours

Owners and occupiers of adjacent premises will similarly feel disconcerted by a serious fire. If the fire has disrupted their business they may seek compensation through the civil courts. As a minimum, they will want to be assured that the fire:

- Has not affected any of their building services
- Has not affected the structure of their building
- Has not resulted in the contamination of their premises and that
- Their premises are safe from future fires.

12.6.4 Business continuity

The time to plan for business continuity is prior to the fire. Electronic systems should be backed up and sensitive/important documents should be kept in fire proof storage.

It may be that arrangements can be made to move the business or part of its operation to other premises or to subcontract operations to another company, as part of a disaster management plan. These arrangements should be made prior to the time when they are required and may form part of the conditions for business insurance.

12.7 Case study

Fire in pharmaceutical production site

A multi-national pharmaceutical company occupies an extensive site in the East Midlands region in the UK. The facility comprises a number of buildings with a variety of uses ranging from research to production. The health and safety on the site is a high priority and the company is very well aware of the impact a fire may have on their public image.

In April 2006 a fire was discovered in a small local exhaust ventilation unit in the toilet block of one of the older buildings on the site. Fortunately the fire burnt itself out against the concrete ceiling of the toilet block. The fire was therefore restricted to the compartment of origin and resulted in limited fire spread within the room of origin.

The initial investigation of the site discovered that the ventilation unit had been fitted with the wrong size fuse and overheated. This was thought to be a problem with the supply and fitting of the unit. The unit was replaced, the room redecorated and no further action was taken.

Figure 12.19 Damage to Vent Axia

Figure 12.20 Damage to electrical distribution cupboard

Within a month of this incident, all the power to the ground floor of a research building was completely lost. Upon investigation it was found that there had been a small fire in the external electrical distribution cupboard. The fire had been caused by an internal fuse in the cupboard, having been replaced by a small piece of piping. The direct fire damage was limited to the electrical distribution cupboard; however, the total failure of power to the research laboratory disrupted some significant work and resulted in the loss of time and additional expenses to the research programme.

As a result of the second fire, the company drafted an action plan that included not only dealing with the immediate causes of the fire but also addressed the failures in a number of management systems including those that controlled the work of contractors and the formailsed routine testing and inspection of all electrical equipment on site. As a result of the monitoring systems introduced a number of other instances of overrated fuses were identified and rectified.

12.8 Example questions for Chapter 12

1. A serious fire has occurred and it is necessary to carry out an examination of the fire scene. List the health and safety considerations that must be taken into account before the examination of the scene can be commenced. (8)

2. With reference to the Reporting of Injuries, Diseases and Dangerous Occurrences Regulations 1995:
   (a) **State** the legal requirements for reporting a fatality resulting from an accident at work to an enforcing authority. (2)
   (b) **Outline** THREE further categories of work-related injury (other than fatal injuries) that are reportable. (6)

3. (a) **Outline** the initial actions that should be taken having identified that a fire detector has actuated inadvertently. (4)
   (b) List information that should be recorded in relation to the false alarm. (4)

4. Giving reasons in EACH case, **identify** FOUR people who may be considered useful members of an investigation team. (8)

5. (a) **State** the requirements for reporting an 'over three-day' injury under the Reporting of Injuries, Diseases and Dangerous Occurrences Regulations 1995. (2)
   (b) Giving reasons in EACH case, **identify** THREE categories of persons who may be considered a useful member of an internal accident investigation team. (6)

6. **Identify** FOUR external organisations that may be involved in a fire investigation and indicate their purpose. (8)

Appendix 12.1   F2508 RIDDOR report form

Health and Safety at Work etc Act 1974
The Reporting of Injuries, Diseases and Dangerous Occurrences Regulations 1995

HSE
Health & Safety
Executive

Incident No. [   ]

Report of an injury or dangerous occurrence

[ ] Injury   [ ] Dangerous Occurrence

Part A

About you

1 What is your full name?

[   ]

2 What is your job title?

[   ]

3 What is your telephone number?

[   ]

About your organisation

4 What is the name of your organisation?

[   ]

5 What is its address and postcode?

[   ]

6 What type of work does the organisation do?

Part B   [   ]

About the incident

1 On what date did the incident happen?

[/   /   ]

2 At what time did the incident happen?
   (Please use the 24-hour clock eg 0600)

[   ]

3 Did the incident happen at the above address?

Yes [ ]   Go to question 4

No [ ]   Where did the incident happen?

   [ ] elsewhere in your organisation –
         give the name, address and postcode

   [ ] at someone else's premises –
         give the name, address and postcode

   [ ] in a public place – give details of
         where it happened

[   ]

If you do not know the postcode, what is the name of the local authority?

[   ]

4 In which department, or where on the premises, did the incident happen?

[   ]

Part C

About the injured person

If you are reporting a dangerous occurrence, go to Part F. If more than one person was injured in the same incident, please attach the details asked for in Part C and Part D for each injured person.

1 What is their full name?

[   ]

2 What is their home address and postcode?

[   ]

3 What is their home phone number?

[   ]

4 How old are they?

[   ]

5 Are they?

[ ] male?

[ ] female?

6 What is their job title?

[   ] [ ][ ][ ]

7 Was the injured person (tick only one box)

[ ] one of your employees?

[ ] on a training scheme? Give details:

[   ]

[ ] on work experience?

[ ] employed by someone else? Give details of the employer:

[   ]

305

☐ self-employed and at work?

☐ a member of the public?

Part D

About the injury

1 What was the injury? (eg fracture, laceration)

[   ] ☐☐

2 What part of the body was injured?

[   ] ☐☐

3 Was the injury (tick the one box that applies)

☐ a fatality?

☐ a major injury or condition? (see accompanying notes)

☐ an injury to an employee or self-employed person which prevented them doing their normal work for more than 3 days?

☐ an injury to a member of the public which meant they had to be taken from the scene of the accident to a hospital for treatment?

4 Did the injured person (tick all the boxes that apply)

☐ become unconscious?

☐ need resuscitation?

☐ remain in hospital for more than 24 hours?

☐ none of the above.

Part E

About the kind of accident

Please tick the one box that best describes what happened, then go to Part G.

☐ Contact with moving machinery or material being machined

☐ Hit by a moving, flying or falling object

☐ Hit by a moving vehicle

☐ Hit something fixed or stationary

☐ Injured while handling, lifting or carrying

☐ Slipped, tripped or fell on the same level

☐ Fell from a height

How high was the fall?

[metres]

☐ Trapped by something collapsing

☐ Drowned or asphyxiated

☐ Exposed to, or in contact with, a harmful substance

☐ Exposed to fire

☐ Exposed to an explosion

☐ Contact with electricity or an electrical discharge

☐ Injured by an animal

☐ Physically assaulted by a person

☐ Another kind of accident (describe it in Part G)

Part F

Dangerous occurrences

Enter the number of the dangerous occurrence you are reporting. (The numbers are given in the Regulations and in the notes which accompanying this form)

[   ]

Part G

Describing what happened

Give as much detail as you can. For instance

● the name of any substance involved

● the name and type of any machine involved

● the events that led to the incident

● the part played by any people.

If it was a personal injury, give details of what the person was doing. Describe any action that has since been taken to prevent a similar incident. Use a separate piece of paper if you need to.

☐☐☐☐   [   ]

[   ]

Appendix 12.2   FDR 1 Fire Service fire report form

Report of Fire

Date: Day Month Year

KEY

Tick the appropriate box ☑
(or boxes)

Insert code from codelist ☐
or enter number

Brigade use ☐

Write in details ☐

1. Brigade Information

1.1 Brigade incident number

1.2 Brigade Area where fire started   Station ground

1.3 Brigade and Home Office Call number

Fire spread box ☐

2. Incident Information

2.1 Address of fire

2.2 Postcode (for buildings) or grid reference (if available)

OS national grid reference

2.3 Risk category

☐ ☐ ☐ ☐ ☐
A B C D R   Also ✓ if Special
risk within area

2.4 Name(s) of occupier(s)/owner(s)

Times

2.5 Estimated interval from

a) Ignition to discovery

☐ ☐ ☐ ☐ ☐ ☐
Immediately Under 5 5 to 30 30 mins to Over 2 Not
mins mins 2 hours hours known

b) Discovery to first call

☐ ☐ ☐ ☐ ☐ ☐
Immediately Under 5 5 to 30 30 mins to Over 2 Not
mins mins 2 hours hours known

(use 24 hour clock)

	hour	mins	day*	month*	year*
2.6........ First call to brigade					
2.7.. Mobilising time					
2.8............Arrival of brigade					
2.9 ....Under control					
2.10 Last appliance returned					

* Only complete 2.7 to 2.10 if different from 2.6

2.11 Was this a late fire call?

☐ ☐
No Yes

2.12 Discovery and call

a) Discovered by

☐ ☐ ☐
Person Automatic system Other - specify in Section 7

b) Method of call by

☐ ☐ ☐
Person Automatic system Other - specify in Section 7

2.13 Was there an automatic fire alarm system in area affected by fire?

☐— go to 3.1 ☐
No Yes

2.14 Alarm activation method

☐ ☐ ☐ ☐ ☐
Heat Smoke Flame Other - specify Not
in 2.18 known

2.15 Powered by

☐ ☐ ☐ ☐ ☐
Battery Mains Mains & Other Not
battery specify in known
back up 2.18

2.16 Did it operate?

☐ ☐ ☐— go to 2.18 or 3.1
No Yes but did Yes and
not raise raised
alarm alarm

2.17 Reason for not operating/not raising alarm

2.18 Other details of automatic fire alarm

FDR1 (94)

3. Location of Fire

3.1 a) Type of property where fire started

b) If mobile, give location

3.2 Residential accommodation affected by fire?

☐ No ☐ Yes, where fire started only ☐ Yes, spread to residential ☐ Yes, where both started and spread to residential ☐ Not known

3.3 Main trade or business carried on where fire started

If none ✓ box ☐ eg wholly residential, and go to 3.4

3.4 Multiseated fire

☐ No ☐ Yes

Fires in buildings and ships

If not ✓ box ☐ and go to 3.10

3.5 Occupancy of building where fire started
(leave blank for ship)

☐ Single ☐ Multiple same use ☐ Multiple different use ☐ Under construction ☐ Under demolition

☐ Derelict Unoccupied ☐ Other - specify below eg. under refurbishment ☐ Not known

3.6 Place where fire started

3.7 Use of room, cabin or roof space where fire started

3.8 Floor, deck of origin

Number ☐ if above ground/main deck

✓☐ if ground/main deck

Number ☐ if below ground/main deck

✓☐ if other, specify below

3.9 Total number of floors in building where fire started

☐ (leave blank for ship)

Fires starting in motor vehicles

If not ✓ box ☐ and go to Section 4

3.10 Make/Model

3.11 Fuel of vehicle

☐ Petrol (not fuel injected) ☐ Petrol fuel injected ☐ Petrol (not known) ☐ Diesel/ other oil ☐ Electric ☐ LPG ☐ Other - specify in 3.17

3.12 Was vehicle turbo/supercharged?

☐ No ☐ Yes ☐ Not known

3.13 Registration number (if available)

3.14 Year of manufacture (if available)

3.15 Part of vehicle where fire started

3.16 Was engine running? (immediately before fire)

☐ No ☐ Yes ☐ Not known

3.17 Other information available eg VIN No, Chassis No etc

4. Extinction of fire

Fixed firefighting/venting systems
(in area where fire started)

If none ✓ box ☐ and go to 4.6

	Type 1	Type 2	Type 3
4.1 Type of system (code up to 3) See Code list 4.1	☐	☐	☐
4.2 Manual or automatic M = Manual A = Automatic Z = Not known	☐	☐	☐
4.3 Did it operate? A = Yes and extinguished fire B = Yes and contained (controlled) fire C = Yes but did not contain. (control) fire N = No	☐	☐	☐
4.4 Number of heads actuated	☐	☐	☐

4.5 Reason(s) for not operating/containing/controlling
(Leave blank if answer to question 4.3 is A or B)

Type
1
2
3

Method of fighting the fire

4.6 Before arrival of brigade

4.7 By brigade up to stop

4.8 Number of main jets used ☐

4.9 Number of local authority appliances attending up to time of stop ☐ ☐
Pumping Other

(If further details required by brigade - use Section 7)

5. Supposed cause, damage and other fire details

5.1 Most likely cause

a) ☐ ☐ ☐ ☐ ☐
Accidental   Malicious   Deliberate   Doubtful   Not known

b) caused by

☐ ☐ ☐ ☐ ☐ ☐
Child   Youth   Adult   Animal   Other (not a person or animal)   Not known

Give additional details of person (if known)

[ ]

c) Defect, act, or omission giving rise to ignition

[ ]

5.2 Source of ignition

a) Appliance/installation and other sources

[ ]

b) Powered by

[ ]

c) If source is an appliance, enter the make or model, if known below

[ ]

5.3 Material or item ignited first

a) Description

[ ]

b) Composition

[ ]

5.4 Material or item mainly responsible for development of fire

a) Description

[ ]

b) Composition

[ ]

5.5 Dangerous substances affecting firefighting or development of fire. (Specify up to 2 in order of priority)
If none ✓ box ☐ and go to 5.6

a) Material   b) Circs.

| 1 | |
| 2 | |

Circumstance codes - M = being Made S = in Storage T=in Transit
U = being Used W = combination of circumstances Z = not known

c) Main effect of substance on fire and/or firefighting

[ ]

5.6 Explosion

a) ☐ No go to 5.7   Yes occurred ☐—☐ First fire ☐ During fire ☐ First and during fire ☐ Not known

b) Materials involved in explosion (specify up to 2)

| 1 | |
| 2 | |

c) Containers involved in explosion (specify up to 2)

| 1 | |
| 2 | |

5.7 Abnormal rapid fire development

No ☐   Give additional details (if known)
Yes ☐—[ ]

5.8 Damage caused to:

i)   item ignited first
ii) room, cabin, compartment etc of origin (buildings, ships & vehicles only)
iii) elsewhere on floor, deck, other compartments of origin (buildings, ships & vehicles only)
iv) elsewhere in building, ship, etc of origin
v) outdoors beyond property; beyond building, ship, plant, vehicle etc

a) %: enter percentage of item/room etc damaged eg. 25 = quarter, 50 = half etc

b) Severity: enter code to show severity of damage
I = light,   M = Moderate,   S = Severe

Damage caused by:	to i) a %	b	to ii) a %	b	to iii) a %	b	to iv) a %	b	to v) ✓ box(es) if affected
fire									☐
heat									☐
smoke									☐
other									☐
Total not to exceed 100%									buildings [ ]
% of structure damaged	%		%		%		%		vehicles [ ]

Number of additional:	rooms cabins c'partment etc	floors	other locations [ ]
damaged			
total			

If further description required by brigade use Section 7

5.9 Estimate of horizontal area damaged

(a)   Area - sq m   (b)
	Area - sq m	
☐	under 1 sq m	☐
☐	1-2	☐
☐	3-4	☐
Area damaged by direct burning ☐	5-9	☐ Total area damaged by fire heat smoke etc.
☐	10-19	☐
☐	20-49	☐
☐	50-99	☐
☐	100-199	☐
☐	200 +	☐
[ ]	If over 200 write in to nearest 50 sq m	[ ]

5.10 Animals killed
if none ✓ box ☐ and go to Section 6
if yes record up to 3 main species

	Species	Number
1		
2		
3		

6. Life Risk

Involvement of persons (as known to brigade)

If none ✓ box ☐ and go to Section 7

6.1 Number of non-fatal casualties
(including those who were rescued)

☐

6.2 Number of fatal casualties

☐

6.3 Number of rescues only (exclude those who were casualties)

☐

6.4 Approximate number of persons at discovery of fire in room, cabin, compartment, etc. of origin

☐

6.5 Approximate number of persons at discovery of fire in other parts of building, vehicle, etc.

☐

6.6 Approximate number who left the affected property
(including any who were casualties)

☐

6.7 Fatalities, other casualties and rescues:
Complete one line for each person.
Refer to guidance notes for codes.
Use single code in each column 2 to 7

If no injury leave blank | *If not rescued leave blank*

	Name of person	Age Yrs	Sex	Location	Main circum stance	Status	Nature of injury	Rescued by	Rescue methods up to 2	Brigade use
A										
B										
C										
D										
E										
F										
G										
H										
		1	2	3	4	5	6	7	8	9

7. More detailed description of fire/further information (if applicable)

Section/question

7.1 Further investigation to be carried out

☐ No   Yes by ☐ Fire brigade ☐ Police ☐ Others ☐ Fire of special interest

7.2 Further information to follow

☐ No/Not known   ☐ Yes

Special study boxes

☐ 7.3   ☐ 7.4   ☐ 7.5   ☐ 7.6   ☐ 7.7

Name & rank of person in charge at first attendance (IN CAPITALS) ☐

Name & rank of person in charge of the fire (if different from above) (IN CAPITALS) ☐

Signature

Form completed by (IN CAPITALS) ☐

Rank ☐   Date ☐

Printed in the UK by HMSO Basildon 107194/A02598 400m 8/93 ECRC Supprod.

Appendix 12.3   Sample false alarm report form (adapted from BS 5839-1: 2002)

RECORDS OF FALSE ALARMS/UNWANTED FIRE SIGNALS

1

Date	Time	Zone	Device that triggered alarm signal (identification number and type)

2

Cause of alarm (if known)	Brief circumstances (where cause is unknown record activities in the area)

3

Maintenance visit required?	Findings of maintenance technician (where applicable)
Yes ☐ No ☐	

4

Category of false alarm (e.g. electrical, malicious or good intent)	Further action required (where applicable)	Action completed (date)	Initials

13 Environmental impact of fire

Fire not only poses a risk to life and property it also has a significant impact on the environment. Environmental damage from fires can be both short and long term and, in the case of pollutants from fires affecting groundwater supplies, may persist for decades or even longer. Rivers, sewers, culverts, drains, water distribution systems and other services all present routes for the conveyance of pollutants off-site and the effects of a discharge may be evident some distance away.

In many cases, major pollution incidents can be prevented if appropriate pollution prevention measures are in place or immediately available. Contingency planning is the key to success, therefore preventive and protective measures and incident response strategies should be carefully considered and implemented.

13.1 The sources of pollution in the event of fire

Every combustion process has the ability to cause environmental pollution. In some cases fires will pollute flora and fauna directly and in others the pollution will occur via the air, the earth and/or the ground water. The degree of contamination from fire effluent depends upon how large the fire is, what is burning and the temperature and burning conditions. When considering how a fire from a particular location may pollute the environment, it should be borne in mind that pollution can occur through a number of routes, including:

➤ Via site's surface water drainage system, either directly or via off-site surface water sewers
➤ By direct run-off into nearby watercourses or onto ground, with potential risk to ground waters
➤ Via the foul drainage system, with pollutants either passing unaltered through a sewage treatment works or affecting the performance of the works, resulting in further environmental damage
➤ Through atmospheric deposition, such as vapour plumes.

There are an infinite number of compounds that are produced during a combustion process. Some common building materials and contents are known to give off toxic, corrosive and/or carcinogenic fumes when involved in fire. In addition to fumes, harmful particulates in the form of soot or other fallout present significant risks to the environment.

The most common pollutants that are likely to cause environmental pollution as a result of fires include:

➤ Sulphur dioxide
➤ Carbon monoxide
➤ Benzene
➤ Acetone
➤ Polychlorinated biphenyls
➤ Fire fighting water/foam run-off
➤ Asbestos
➤ Isocyanates – cyanide.

This chapter discusses the following key elements:

➤ The sources of pollution in the event of fire
➤ The legal obligations related to environmental protection in the event of fire
➤ Preplanning to minimise the environmental impact of fire
➤ Containing water run-off from fire.

Figure 13.1 The environmental impact of fire

Sulphur dioxide – sulphur dioxide (SO$_2$) is a colourless gas; it reacts on the surface of a variety of airborne solid particles. It is soluble in water and can be oxidised within airborne water droplets. The major health hazards associated with exposure to high concentrations of sulphur dioxide include affects on breathing, respiratory illness, alterations in pulmonary defences and aggravation of existing cardiovascular disease. In the atmosphere, sulphur dioxide mixes with water vapour producing sulphuric acid. This acidic pollution can be transported by wind over many hundreds of miles, and when present in sufficient quantities can be deposited as acid rain.

Carbon monoxide – carbon monoxide (CO) is an odourless, colourless and toxic gas. It is a narcotic gas that quickly induces sleep and because it is impossible to see, taste or smell the fumes are particular dangerous. Exposure to CO can result in angina, impaired vision and reduced brain function. At higher concentrations, CO exposure can be fatal. While the direct effect on the environment is minimal CO assists in the creation of smog or ground level ozone.

Benzene – is an aromatic hydrocarbon that is produced by the burning of organic products such as wood and other carbonaceous materials – it is also produced in the decomposition of plastics. As an aromatic hydrocarbon benzene contains high quantities of carbon that are rarely fully burnt during combustion. This results in unburnt carbon being released into the atmosphere, which has the distinct appearance of thick black smoke. Benzene is used in the manufacture of plastics, detergents, pesticides and other chemicals. Short-term exposure to high levels of benzene can cause drowsiness, dizziness, unconsciousness and death. Benzene is also known to be a carcinogen causing leukaemia and affecting bone marrow and blood production.

Benzene is more frequently associated with risk to humans than with risk to fish and wildlife. This is partly because only very small amounts are absorbed by plants, fish and birds, and because this volatile compound tends to evaporate into the atmosphere rather than persist in surface waters or soils. However, volatiles such as this compound can pose a drinking water hazard when they accumulate in ground water.

Acetone – is another by-product of the combustion of carbonaceous material. In small amounts the liver breaks acetone down into chemicals that are not harmful and uses them for everyday body functions. Breathing moderate to high levels of acetone for short periods of time (i.e. painting in a poorly ventilated area) can cause nose, throat, lung and eye irritation, headaches, light-headedness, confusion, nausea, vomiting and if the exposure is for a long period, unconsciousness, coma and death. It may damage the kidneys and liver.

It has slight short-term toxicity on aquatic life. Acetone has caused membrane damage, a decrease in size and a decrease in germination of various agricultural and ornamental plants. It may also have slight long-term toxicity to aquatic life. Chronic and acute effects on birds or land animals have not been determined.

Polychlorinated biphenyls – polychlorinated biphenyls (PCBs) is the term for a group of compounds used in the construction of electrical equipment. When PCBs are involved in fires that expose them to temperatures ranging from 500° to 700°C, they form a variety of other compounds. In fires involving electrical PCB-containing electrical equipment it is common for PCBs to form a distribution of black carbonaceous soot. PCBs have been identified in soot following numerous electrical equipment fires and have been detected in soot samples collected following transformer fires. However, PCBs are also released into the environment as a vapour, for example when pressure relief valves of overheated transformers operate. The pressurised release of hot PCB vapours can entrain considerable quantities of liquid PCBs forming a fine aerosol that can be distributed to areas beyond the transformer air currents.

Exposure to PCBs can cause gastrointestinal disturbances and nerve damage. There is suggestive evidence of associations between increased incidences of cancer and exposure to PCBs. However, definite casual relationships between exposure and carcinogenic effects are unclear.

Fire fighting water/foam run-off – obviously the water used for fighting fires is likely to be contaminated to an extent with all of the above contaminants. In addition, if the Fire and Rescue Service use foam to fight a fire, for example in a boiler room, the foam itself is likely to contain perfluorooctane sulfonate (PFOS).

PFOS is an organic compound which does not break down in the environment, and that makes foam spread rapidly at high temperatures, cutting the time it takes to put out a fire. It accumulates in organisms and works its way up the food chain. It has been found to be

Figure 13.2 The use of fire fighting foam

bioaccumulative and toxic to mammals. Studies have linked bladder cancer, although further work is needed to understand this association. It is thought that PFOS interrupts the body's ability to produce cholesterol, thereby affecting almost every system in the body.

Any concentration above three micrograms per litre of water is considered to be detrimental to human health. If PFOS is released into the environment, it can remain present for years. In water courses, it accumulates rapidly in fish. There are moves to make its use illegal in the UK in the future.

Asbestos – asbestos is present in older buildings and was used in hot water systems, ceilings and roof construction. When undisturbed the risks associated with it are minimal; however, in a fire situation is it likely to be disturbed and released into the atmosphere.

To cause disease, asbestos fibres must be inhaled into the lungs. Fibres with a diameter less than three microns (too small to be visible to the naked eye) can enter the lungs and cause disease. The period from exposure to developing symptoms of disease is usually in excess of 10 years.

Several diseases are associated with exposure to asbestos. They are asbestosis, mesothelioma, cancer of the lung and other asbestos-related cancers. These are serious, debilitating diseases that often end in death. In the next 50 years it is estimated that at least 30 people will die each day of an asbestos-related disease.

➤ **Asbestosis** is characterised by a fibrosis (scarring) of the lung tissue, which makes breathing difficult. The most prominent symptom is breathlessness. Early detection of asbestosis is possible by X-ray examination and lung function testing. However, the disease is irreversible and will continue to progress even after exposure is stopped. Rarely a cause

of death itself, asbestosis results in an appreciable reduction in life expectancy due to deaths from related illness.

➤ **Mesothelioma** is a rare cancer arising from the cells of the pleura (lining of the chest cavity and lungs) and the peritoneum is characterised by a long latency period, usually at least 15 years and sometimes more than 40. There is no effective treatment for mesothelioma. A large proportion of mesothelioma patients die within a year of diagnosis; few survive longer than five years. Although asbestos was once thought to be responsible for all mesothelioma, other causes have now been identified. Still, the chance of getting mesothelioma in the absence of asbestos exposure is considered to be extremely remote.

Isocyanates (cyanide) – a large proportion of many day-to-day plastics include isocyanates. These isocyanates, when subject to fire, break down releasing their component parts including among others cyanide (toxic to humans and wildlife). The chemicals released are very likely to become airborne contaminating the air, which when cooled, may be brought to ground with the water used to extinguish the fire resulting in pollution of both the ground and water courses.

13.2 The legal obligations related to environmental protection in the event of fire

In addition to the requirements of the Health and Safety at Work Act, which require employers, the self-employed and those in control of premises to ensure that the safety of persons (in addition to those employed in the work undertaking) are not affected by work activities, there is specific legislation which controls environmental pollution. This legislation is broadly concerned with discharges of effluent into the atmosphere and the ground water systems.

It should be borne in mind that the civil tort of negligence may also apply to incidents that result in loss or harm occasioned by pollution of the environment by fire effluent.

Water Industry Act 1991
The Water Industry Act 1991 controls, among other things, the disposal of waste water and discharges into the surface water drainage.

Waste water disposal – foul drains should carry contaminated water, trade effluent and domestic sewage to a treatment works. Discharges to the public foul sewer require authorisation by the sewerage undertaker and

Figure 13.3 Contaminated water following a fire

may be subject to the terms and conditions of a trade effluent consent. Where there is a disposal to sewer this should always be subject to such approval. In addition to process effluent, trade effluent includes compressor or boiler blow-down, steam condensates, cooling water, pressure testing liquids, air conditioning water, vehicle and plant cleaning effluent, and yard wash-down water. These should all be directed to foul drains.

Surface water drainage – surface water drainage discharges to a watercourse or to groundwater via a soakaway. Surface water drains should therefore carry only uncontaminated rainwater from roofs and clean yard areas. A discharge of waste water to the surface water drain will result in pollution. It is an offence to pollute controlled waters either deliberately or accidentally. In addition, the formal consent of the Environment Agency is required for many discharges to controlled waters, including both direct discharges and discharges to soakaways. Such consents are granted subject to conditions and are not granted automatically. Breaching this legislation can lead to fines, prison sentences and paying remedial costs for damage caused.

The Pollution Prevention and Control Act 1999
The Pollution Prevention and Control Act 1999 introduced a new regime of control for certain industries, who were required to develop Integrated Pollution Prevention and Control (IPPC). Regulations to implement the provisions of the Act were implemented in August 2000.

IPPC is designed to prevent, reduce and eliminate pollution at source through the prudent use of natural resources. Installations are covered where one or more of the following categories of activities are carried out (subject to certain capacity thresholds): energy industries, production and processing of metals, mineral industry, chemical industry, waste management industry and other

activities – paper/board, tanneries, slaughter houses, food/milk processing, animal carcass disposal, intensive pig/poultry units, organic solvents users.

Control of Major Accident Hazards Regulations 1999 (COMAH)
COMAH Regulations came into force on 1 April 1999 and were amended by the Control of Major Accident Hazards (Amendment) Regulations 2005 on 30 June 2005. These regulations require the operators of large industrial sites to take certain precautions against any incident on the site affecting the environment. In order to come within the scope of these regulations industrial sites must store specified quantities of specified hazardous substances (see Appendix 13.1).

COMAH requires that every relevant operator must prepare and keep a 'major accident prevention policy document' setting out the policy with respect to the prevention of major accidents. The policy must be designed to guarantee a high level of protection for persons and the environment by appropriate 'means, structures and safety management systems'.

The major accident prevention policy should be established in writing and should include the operator's overall aims and principles of action with respect to the control of major accident hazards.

The safety management systems should include the general organisational structure, responsibilities, practices, procedures, processes and resources for determining and implementing the major accident prevention policy.

13.3 Preplanning to minimise the environmental impact of fire

Most industrial and commercial sites have the potential to cause significant environmental harm and to threaten both water resources and public health. The Environment Agency has published a series of Pollution Prevention Guidance Notes with the aim to reduce the risk of significant environmental incidents occurring.

However, regardless of the measures that may be taken to prevent environmental pollution, under normal circumstances there is always a residual risk of a spillage or a fire that could cause serious environmental damage. In addition to the obvious threat posed by chemicals and oils, even materials that are non-hazardous to humans, such as foods and beverages, can cause serious environmental harm. The run-off generated in the event of a fire can be very damaging and the discharge of toxic effluent into the atmosphere can have long effects over a wide area.

The environmental impact of a fire may be long term and, in the case of effluent contaminating ground water, may persist for decades. As a result, the legal consequences and clean-up operation can be costly. Rivers, sewers, culverts, drains, surface water soakaways, porous or unmade ground water distribution systems and service ducts all present routes for pollutants to quickly enter the surrounding environment (including surface water and ground water). Thus, the effects of a discharge may not be evident on site, but may become apparent some distance away. Any incident response plan should take into account the vulnerability of ground water both beneath and down-gradient of the site.

In the majority of cases it will be possible to reduce the risk of a fire having a serious impact upon the environment by ensuring that appropriate pollution prevention measures are in place.

13.3.1 Incident response plan (IRP)

The Environment Agency argues that the key to reducing the environmental risk from a fire (or other incident) is to have a contingency or pollution incident response plan in place. The plan need not be expensive to prepare, but could minimise the consequences of an incident.

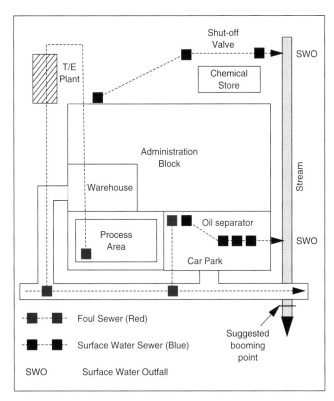

Foul Sewer (Red)

Surface Water Sewer (Blue)

SWO — Surface Water Outfall

Figure 13.4 Incident response plan should include site drainage plan

An example of an incident response plan in shown as Appendix 13.2.

The plan should be drawn up by people with the relevant competencies and may in many cases be developed in consultation with external agencies or specialists in the field. The plan should include:

➤ Company name and full postal address of the site
➤ A brief description of the main business activities on site (specifying those with a high potential for environmental harm)
➤ The date the plan was completed and the date it is due to be reviewed
➤ The signature of the plan by a senior company manager
➤ A list of recipients of the completed plan
➤ A contact list including contact details for:
    ➤ Emergency services
    ➤ Relevant environmental regulators
    ➤ Local water supplier and sewer provider
    ➤ Health and Safety Executive (HSE)
    ➤ Specialist clean-up contractors
    ➤ Site keyholders
➤ A site drainage plan showing:
    ➤ Foul drainage surface water including the direction of flow and any drain covers
    ➤ Discharge points/soakaways for surface water and trade effluent
    ➤ The sewage treatment works to which sewage and trade effluent discharges
    ➤ Any watercourse, spring, borehole are well located within or near the site
➤ General layout of buildings including:
    ➤ Site access routes for emergency services
    ➤ Any on-site treatment facilities for trade effluent or domestic sewage
    ➤ Areas or facilities used for storage of raw materials, products and wastes
    ➤ Any bunded areas together with details of products stored and estimated
    ➤ Retention capacity
    ➤ Location of hydrants, 'fireboxes' and spill kits
    ➤ Inspection points for the detection of pollution
    ➤ Oil separators
    ➤ Retention or balancing tanks
    ➤ Firewater retention ponds.

13.3.2 Emergency procedures

Detailed emergency procedures should be produced which must include details of staff responsibilities and the procedures for dealing with events such as fires, spillages and leaking containers, etc. The level of response will obviously depend on health and safety issues, staff training, the level of personal protective equipment (PPE) available and the nature of any incident. The resultant procedures for dealing with emergencies will therefore need to be site specific. It is important to consider what could happen in the worst case and to take this into account when developing procedures.

The Environment Agency suggests a checklist of actions that may be useful when considering the issues that should influence the development of comprehensive emergency procedures. Any such checklist should include such items as:

➤ The site fire fighting strategy as agreed with the Fire and Rescue Service. If 'controlled burn' is an agreed option, this should be clearly stated
➤ The method of alerting nearby properties, downstream abstractors or environmentally sensitive sites that could be affected by an incident
➤ A quantification of the consequences of an incident at nearby properties
➤ The methods whereby staff on site and, where appropriate, adjacent sites are alerted to an incident
➤ The detailed arrangements for contacting the relevant emergency services, relevant agency, local authority and other organisations, and dealing with the media
➤ Any substances that may present particular risks (these should be recorded in the incident response plan)
➤ The provision and management of any relevant PPE
➤ Arrangements in place for making leaking containers safe
➤ Procedures for containing leaks, spills and fire fighting run-off and for the protection of any on-site effluent treatment plant
➤ The requirement for spill kits, drain blockers and other pollution control equipment and the operation of pollution control devices should be clearly documented
➤ Stocks of pollution control equipment and materials held locally by other organisations should be identified and contact details for clean-up companies should be kept up to date
➤ Procedures for the recovery of spilled product and the safe handling and legal disposal of any wastes arising from an incident.

13.3.3 Training to support the IRP

The effectiveness of any site incident response plan will depend on staff training. All staff and contractors working on site need to be made aware of the plan. They should be aware of their role if an incident occurs. In addition to providing awareness training it is important to provide realistic training, i.e. emergency exercises. Emergency exercises should be carried out periodically

to familiarise staff with the operation of the plan and to test its effectiveness. Records of staff training should be maintained.

The subjects that should be included in any training programme will be, for example:

➤ The potential for harm to both personnel and the environment from the materials held on site
➤ The sensitivity of the environment surrounding the facility
➤ The provision and use of the correct PPE
➤ Arrangements for reporting to relevant agencies if there is a risk of surface, ground water or land contamination
➤ Procedures for reporting to the local sewer provider if a discharge to the foul or combined sewer is involved
➤ Arrangements for clean-up, safe handling and legal disposal of contaminated materials and wastes resulting from an incident (including arrangements for the use of specialist contractors and services)
➤ The appropriate decontamination or legal disposal of contaminated PPE.

Producing an incident response plan

When preparing IRPs organisations are encouraged to liaise with their local agency office for their observations. The finished plan should then be copied to all those parties required to have sight of it. Most importantly a copy must be kept on site in an easily accessible location away from the main building such as a gatehouse or a dedicated 'firebox' to which the emergency services can readily gain access. A notice at the site entrance should be used to indicate the location of the plan.

Finally, in order for the plan to remain effective, it is vital that it is reviewed regularly and that any significant

Figure 13.5 Training to support the incident response plan

changes are reflected in a revised plan. Ensure that revised copies are sent to all plan holders and that old versions are destroyed.

Figure 13.6 Boom used to catch overflow from bunded area

13.4 Containing water run-off

The environmental impact of contaminated water running off from the location of a fire into the surrounding land and water courses can be significant. Water used for fire fighting is always contaminated with the effluent from fire and often contains other contamination as a direct result of a fire. For example, a fire in a cold storage depot may result in quantities of food stuffs degrading and presenting significant biological hazards.

In order to minimise the risks of environmental damage as a result of fire it is necessary to ensure that the site of any potential incident has sufficient emergency containment systems, emergency materials and equipment, arrangements for waste management and consideration has been given to the preplanning of fire fighting strategies and run-off management.

13.4.1 Emergency containment systems

Although permanent containment facilities should be provided at many sites, for example at sites that come under the COMAH Regulations, there may be circumstances where a spillage cannot be dealt with by such facilities, for example if it occurs outside a bunded area.

A bund is a purpose-built dam around an area where a spillage is likely to occur. Typically bunds can be seen around oil storage tanks. In the event of a spill, for whatever reason, the hazardous substance is contained within the bunded area.

In some cases, particularly at smaller sites, containing contaminated firewater run-off will be considered to be impracticable because of cost and space considerations. In such cases, temporary containment systems or pollution control materials are available and should be used to minimise the environmental impact of firewater run-off.

However, if reliance is placed on these secondary measures, consideration of some other form of local containment may be necessary to provide sufficient time to prepare them. Their use and location must be clearly marked in the pollution incident response plan and indicated on site with durable signs explaining their use.

There are a variety of emergency containment measures that may be used including:

➤ Sacrificial areas
➤ Bunding of vehicle parking and other hard standings
➤ Pits and trenches
➤ Portable tanks, overdrums and tankers.

Sacrificial areas – the use of sacrificial areas involves the routing of firewater run-off to a designated, remote area, which is provided to allow infiltration of any contaminant and to prevent run-off from the site. The contaminant is often contained within a layer of permeable soil or other similar material and should be prevented from dispersing into other strata or ground water by an impermeable lining system which should be capable of containing both vertical and horizontal seepage.

The sacrificial area may also be used for other purposes, such as car parking or as a sports ground. Storm water drainage serving the area must be capable of being shut off quickly and effectively in the event of an emergency incident. After use, if the area has been contaminated, the permeable material can be excavated and removed for disposal. Sacrificial areas can also be used for controlling storm water run-off from the site, which helps in the management of flooding and pollution from surface run-off.

Bunding of vehicle parking and other impermeable surfaces – impermeable yards, roads and parking areas can be converted to temporary lagoons using sandbags, suitably excavated soil or sand from emergency stockpiles to form perimeter bunds. Permanently installed bunding, in the form of either a low kerb or roll-over bunds around suitable impermeable areas, the entire site, or just the sensitive area, is a better option. In the event of an incident, all drain inlets, such as gullies, within the area, must be sealed to prevent the escape of the pollutant.

Catch pits and trenches – pits or trenches may be used where other methods have failed or no other method is available. Their use should be considered carefully due to the risk of ground water contamination. If possible, a liner should be employed, particularly in areas of high ground water vulnerability, although the effect of

Figure 13.7 Portable tank mounted on a flatbed lorry

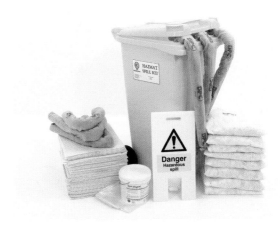

Figure 13.8 Environmental spill kit

the substances being contained on the liner will need to be considered. If no liner is used, the contaminated ground will need to be removed promptly to a disposal site. Pits and trenches may also be used to add reagents for neutralising harmful substances.

Portable tanks, oversize drums and tankers – portable storage tanks made from synthetic rubber, polymers and other materials come in a wide variety of sizes. The portability of the tanks allows them to be moved rapidly to the fire or spillage location, or to where any run-off has been contained. If a portable tank is to be used during an incident, the following measures need to be considered:

➤ Suitable points in the drainage systems must be pre-selected at which the drainage pipe can be blocked and a man-hole chamber used as a pump sump to transfer contaminated waters to the tank. A suitable pump, which may need to be flame-proof, will also be required
➤ Portable tanks must be located in positions, ensuring that there is sufficient space and adequate foundation
➤ Larger portable tanks often need to be supported by a frame
➤ The reuse of the collected water to tackle the fire, taking into account the materials present on site and the risks to equipment through contamination, and to the safety of fire crews
➤ The provision of oversized drums, which are designed primarily to safely store leaking or damaged drums, can also be used as a temporary store for a small quantity of a spilt liquid
➤ Reusable liners are available for oversize drums and portable tanks. These must be resistant to attack by the stored substances

➤ The provision of vacuum or other mobile tankers may also be used for collecting and containing small spills.

13.4.2 Emergency materials and equipment

A wide variety of 'off-the-shelf' products are available to deal with spillages or to contain spills in emergency containment areas. Any materials or equipment used must be located at accessible positions which are clearly marked with durable notices explaining their use. In addition, any equipment must be fit for the purpose and effectively well maintained ready for use. The pollution incident response plan will identify the pollution prevention equipment and materials and their location. The type of materials and equipment that may be provided to mitigate the effects of a fire on the environment include:

➤ Sand and earth
➤ Proprietary absorbents
➤ Sealing substances and devices for containers
➤ Drain seals
➤ Booms.

Sand and earth – these basic containment materials can be used to soak up spillages of oil and chemicals and used in sandbags to block off drains or to direct flows to a predetermined collection point. Sand should be kept dry and sufficient shovels or other means of application must also be readily available. The contaminated sand or earth must be properly disposed of (see below) and, obviously, must not be washed into the drainage system.

Proprietary absorbents – these serve a similar use to sand and earth. They are available commercially in the form of granules, sheets, pillows or a loose powder. Although most absorbents are designed for oil spills, specialised products are available for chemical spills.

Figure 13.9 A boom designed to limit contamination

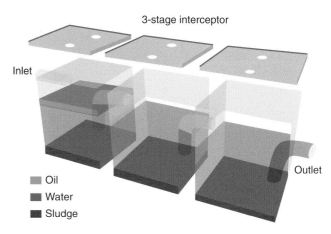

Figure 13.10 Example of a proprietary underground oil interceptor

Sealing devices and substances for damaged containers – these devices and materials are designed for use when a tank, storage drum, valve or pipe has been punctured or damaged by fire. Leak sealing devices may take the form of a pad or clamp which is put over the damaged area like a plaster, or they may be pre-shaped, inserted into the damaged area and then inflated. Leak sealing putties are also available, ready made or supplied in a dry powder form for mixing with water. These are applied over the damaged area to form a temporary seal.

Drain seals – drain seals can be used to seal a drainage grid by covering or blocking the drain and those which fit in a pipe. Again there are several types of drain seal; care should be taken in their installation to avoid exposure to hazardous conditions and to ensure the contained liquid does not overflow from gullies or elsewhere on the drainage system.

Booms – booms designed for use on watercourses may also be used to isolate drains or divert or contain spillages on site. They may be permanently fixed in position or deployed at the time of an incident. There are two distinct types of boom; those that are filled with absorbent material which can be suitable for hydrocarbons, aqueous chemicals or both; and those that are typically plastic and form a physical barrier to limit the spread of the contaminant.

13.4.3 Waste management

Once any spillage, contaminated material or fire fighting water has been adequately contained with a site, it will be necessary to ensure that effective arrangements are in place to dispose of it safely and legally. If it is possible to reuse the spilled material it can be returned to storage on site.

In most cases it will be necessary to arrange disposal off site. In these cases, a registered waste carrier should be used, although if a foul sewer is available it may be possible to discharge to it with the approval of the local sewerage undertaker. It may also be possible to treat water contaminated with hydrocarbons, e.g. petrol or diesel, by using on-site oil separators. Oil interceptors are routinely provided in locations where the likelihood of surface water contamination with oils is high, e.g. service station forecourts (see Fig. 13.10). These interceptors work by allowing the oil to separate from the contaminated surface water via a series of settling tanks. It must be noted that any water that contains fire fighting foam or its constituent parts will adversely affect the efficiency of oil interceptors.

The movement of the waste will need to be documented with a transfer note under the Duty of Care Regulations 1991, or if it is a special waste, with a special waste consignment note under the Special Waste Regulations 1996. There is a statutory responsibility for the producer to keep these notes for a period of two years for waste transfer notes or three years for consignment notes.

In the case of special waste consignments, there is normally a requirement for three days' notice to be given to the Environment Agency prior to its movement. However, in the case of an emergency the Environment Agency may waive this duty, providing that the local Environment Agency office is contacted.

If the fire fighting water contains asbestos it may be necessary to dispose of it in suitably sealed containers that are clearly marked to show that they contain asbestos.

13.4.4 Fire fighting strategies and run-off management

The IRP may consider fire fighting strategies and possible methods of reducing the amount of firewater run-off generated, for example by the use of high pressure

Figure 13.11   Fire fighting run-off

fog, low pressure water sprays rather than main jets, controlled burn and the possible recycling of fire fighting water, where safe and practicable to do so. The fire fighting strategy will need to be discussed with the Fire and Rescue Service and those discussions should be informed by data relating to the environmental impact of each of the options. As a principle of course the best possible strategy to prevent environmental pollution is not to have a fire in the first place.

A main jet used for fire fighting produces large quantities of firewater run-off. In many cases, primary and local containment either by temporary or fixed bunding may prevent an incident from causing pollution.

However, where local containment is not provided, or the fire risk assessment indicates that additional control measures are required, for example to contain firewater run-off, which may amount to thousands of cubic metres, 'remote containment' systems may be appropriate. Remote containment systems may be used by themselves, or in combination with on-site containment arrangements. They may be required to protect both surface and foul water drainage systems.

Quantities of water used for fire fighting

Due of the uncertain nature of an incident and the potential responses of the emergency services, it is difficult to calculate the capacity of containment areas, including bunded enclosures in a way that makes a realistic allowance for fire fighting media that may be used to deal with an incident. However, since many incidents are likely to involve fire, and almost all 'worst case' scenarios involve fire, making adequate provision for retention of fire fighting and cooling water is of critical importance.

When considering the capacity required for containing firewater run-off, a distinction needs to be made between local and remote containment arrangements or combined containment in terms of what may be achievable.

Allowance for fire fighting agents in designing on-site bund capacity

It is impracticable to design a bunded area with sufficient capacity to contain all the potential fire fighting and cooling water that would be used in a major fire. In many cases this would result in very high bund walls. This in turn would not only cause construction, operational and safety problems, but is also likely to adversely affect fire fighting operations. For bunded areas, therefore, it is normal to merely provide sufficient freeboard (above the height of bund required to contain released substances) to retain a blanket of fire fighting foam, and some additional capacity for cooling water. It is normally considered adequate to provide something over 100 mm of freeboard for the containment of foam.

Allowance for fire fighting agents in the design of remote and combined systems

Remote and combined systems should be designed so that they have sufficient capacity to retain such fire fighting and cooling water as could reasonably be expected to be used in a major fire. It is essential to consult fully with the Fire and Rescue Service to agree an estimate of the required capacity. In order to achieve a realistic estimate it will be necessary to consider the following factors:

➤ The size and layout of the plant
➤ The nature of the materials present and the processes carried out
➤ Cooling water from automatic fire fighting systems (e.g. fixed sprinkler installations, on-site fire fighting capability)
➤ The Fire and Rescue Service's own contingency strategy for dealing with an incident
➤ The Fire and Rescue Service's own fire fighting capability.

In the light of this information the designer must then decide, in consultation with the regulators and the plant operators, the capacity of containment required.

On sites where there is the potential for any containment capacity to be significantly overrun, it may be necessary to install additional fixed fire systems, or compartmentalise the plant or site, or gain the agreement of the Fire and Rescue Service, the Environment Agency and the operators to a 'controlled burn' response strategy. For sites with a low hazard or risk rating it will be harder to justify the costs of full containment for fire fighting and cooling water and the designer should seek to strike a reasonable balance between protection and cost, in consultation with the Fire and Rescue Service, the regulators and the site operators.

Appendix 13.1   The quantities of substances that bring an industrial site within the scope of the Control of Major Accident Hazards Regulations (COMAH)

Column 1	Column 2	Column 3
Dangerous substances	Quantity in tonnes	
Ammonium nitrate (as described in Note 1 of this Part)	350	2500
Ammonium nitrate (as described in Note 2 of this Part)	1250	5000
Arsenic pentoxide, arsenic (V) acid and/or salts	1	2
Arsenic trioxide, arsenious (III) acid and/or salts	0.1	0.1
Bromine	20	100
Chlorine	10	25
Nickel compounds in inhalable powder form (nickel monoxide, nickel dioxide, nickel sulphide, trinickel disulphide, dinickel trioxide)	1	1
Ethyleneimine	10	20
Fluorine	10	20
Formaldehyde (concentration >90%)	5	50
Hydrogen	5	50
Hydrogen chloride (liquefied gas)	25	250
Lead alkyls	5	50
Liquefied extremely flammable gases (including LPG) and natural gas (whether liquefied or not)	50	200
Acetylene	5	50
Ethylene oxide	5	50
Propylene oxide	5	50
Methanol	500	5000
4, 4-Methylenebis (2-chloraniline) and/or salts, in powder form	0.01	0.01
Methylisocyanate	0.15	0.15
Oxygen	200	2000
Toluene diisocyanate	10	100
Carbonyl dichloride (phosgene)	0.3	0.75
Arsenic trihydride (arsine)	0.2	1

Phosphorus trihydride (phosphine)	0.2	1
Sulphur dichloride	1	1
Sulphur trioxide	15	75
Polychlorodibenzofurans and polychlorodibenzodioxins (including TCDD), calculated in TCDD equivalent	0.001	0.001
The following CARCINOGENS:		
4-Aminobiphenyl and/or its salts, benzidine and/or salts, bis(chloromethyl) ether, chloromethyl methyl ether, dimethylcarbamoyl chloride, dimethylnitrosomine, hexamethylphosphoric triamide, 2-naphthylamine and/or salts, 1,3 propanesultone and 4-nitrodiphenyl	0.001	0.001
Automotive petrol and other petroleum spirits	5000	50 000

Appendix 13.2   Example of a pollution incident response plan

POLLUTION INCIDENT RESPONSE PLAN

For:

Nature of Business:

Date of Plan:

Approved by:

Environment Agency/SEPA/EHS...............
Fire authority ....
Police ....
Sewer provider..
Water supplier ....
Local authority ....
Other ....

Copies to:

Date sent:

CONTENTS

Page

2. CONTACT DETAILS
3. SITE DRAINAGE PLAN
4. CHEMICAL INVENTORY
5. EMERGENCY PROCEDURES (additional document to pages 1–4)

2. EMERGENCY CONTACT DETAILS

Emergency Services:	999 or 112
Local Police:	
Doctor:	
Environment Hotline:	0800 80 70 60 (24hr Emergency Hotline)
Environment Regulator (Local Office):	

	Office Hours	Out of Hours
Local authority:		
Sewer provider:		
Water supplier:		
Gas supplier:		
Electricity supplier:		
Waste management contractor:		
Specialist advice:		
Specialist clean-up contractors:		

COMPANY CONTACTS: (Out of Hours)

Managing director:	
Site manager:	
Environment manager:	
Foreman:	
Head office contact:	

3. SITE DRAINAGE PLAN

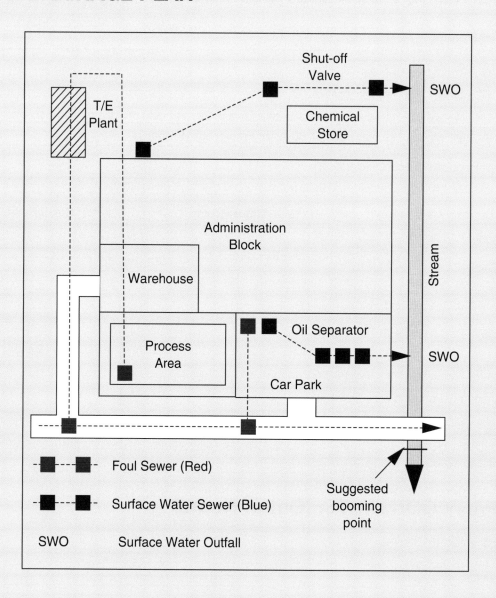

NOTES:

IN THE EVENT OF SPILLAGE OPERATE THE DRAINAGE SHUT-OFF VALVE INDICATED BY TURNING HAND WHEEL FULLY CLOCKWISE.

1. Sewer Treatment Works serving site – Newtown STW
2. Nearest Foul Sewer Pumping Station – Station Road
3. Booming/damming point has been identified

4. CHEMICAL, PRODUCT AND WASTE INVENTORY

Maximum Quantities at Peak Times

Trade name	Substance	Solid/liquid/gas or powder	Container size	Maximum quantity

14 Fire risk assessment

14.1 Introduction

As has been previously discussed in Chapter 5, the purpose of risk assessment in general is to assist an employer and/or a nominated 'responsible person' to identify the preventive and protective measures required to comply with the law. The same may be said for fire risk assessment.

The principles involved with fire risk assessment are very similar to those adopted for task-based and workplace-based risk assessment with a number of discrete differences, in particular many of the protective control measures are built in at the design stage of a building, as has already been discussed in Chapter 9.

It is therefore essential when considering the fire risk assessment process to include not only the task-based or operations-based risks, which may cause a fire, but also the building risks that may prevent persons responding and being able to escape to a place of safety in the event of a fire. The principle of this chapter is to draw together all of the elements previously discussed within the preceding chapters to enable those completing a fire risk assessment to ensure that it is both suitable and sufficient.

This chapter discusses the following key elements:

➤ Definitions relating to fire risk assessment
➤ The process of fire risk assessment
➤ Risk assessment recording and reviewing procedures
➤ The preparation of the emergency plan.

14.2 Definitions relating to fire risk assessment

Chapter 5 of this book included basic definitions relating to general risk assessment. Further definitions specifically related to fire are detailed below.

14.2.1 Fire hazard

A primary fire hazard is something that has the potential to cause harm, by initiating or exacerbating a fire (ignition, fuel or oxygen source).

A secondary fire hazard is something that has the potential to cause harm by preventing an adequate response in the event of a fire (inadequate size of escape route, lack of signage, lack of effective emergency plan).

14.2.2 Fire risk

A combination of the likelihood (chance or probability) of a fire occurring and should it do so, the severity (or consequences) of the outcome.

14.2.3 Fire risk assessment

The process of identifying fire hazards and evaluating the level of risk (including to whom and how many are affected) arising from the hazards, taking into account any existing risk control measures.

14.2.4 Fire risk controls

Workplace precautions, for example sprinkler systems within a building, fire alarm and detection systems, fire emergency plans (procedures), work permit systems and portable fire fighting equipment.

14.2.5 Risk control systems (RCS)

Arrangements that ensure that fire risk controls (workplace precautions) are implemented and maintained. For example, the provision for ensuring that an adequate level of supervision is maintained during hot work or refuelling processes, or a system for planned preventive maintenance for fire safety systems (emergency lighting, fire doors, etc.), establishing a programme of fire safety inspections for buildings, sites and workplaces.

14.3 Risk assessment process

In its guidance documents HM Government (HMG) has used a very similar approach to the HSE's 'Five Steps to Risk Assessment', which is shown below:

1. Look for the hazards
2. Decide who might be harmed and how

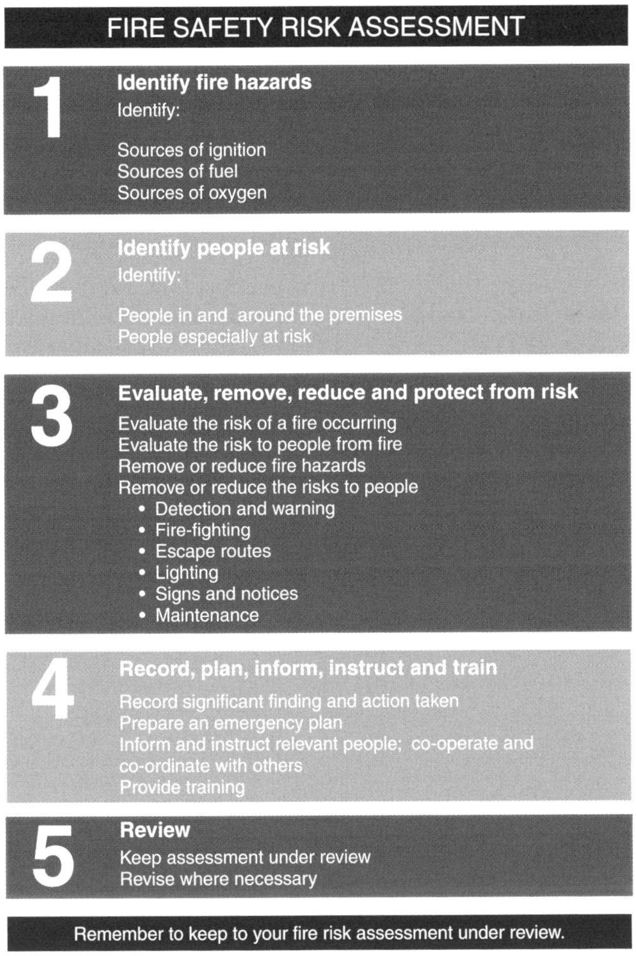

FIRE SAFETY RISK ASSESSMENT

1 Identify fire hazards
Identify:

Sources of ignition
Sources of fuel
Sources of oxygen

2 Identify people at risk
Identify:

People in and around the premises
People especially at risk

3 Evaluate, remove, reduce and protect from risk
Evaluate the risk of a fire occurring
Evaluate the risk to people from fire
Remove or reduce fire hazards
Remove or reduce the risks to people
  • Detection and warning
  • Fire-fighting
  • Escape routes
  • Lighting
  • Signs and notices
  • Maintenance

4 Record, plan, inform, instruct and train
Record significant finding and action taken
Prepare an emergency plan
Inform and instruct relevant people; co-operate and co-ordinate with others
Provide training

5 Review
Keep assessment under review
Revise where necessary

Remember to keep to your fire risk assessment under review.

Figure 14.1 HM Government approach to fire risk assessment

3. Evaluate the risks and decide whether the existing precautions are adequate or whether more should be done
4. Record your findings
5. Review your assessment and revise if necessary.

Essentially Step 3 of HMG's guidance is split into two key areas, that of 'preventive' and 'protective measures', which were discussed in Chapter 6.

14.3.1 Practical steps

Having prepared an inventory of buildings that are to be risk assessed, the next stage in the fire risk assessment process is to determine which buildings are likely to present the most significant risk, in order to establish a prioritised listing for the fire risk assessment process. While it is generally appreciated that the larger, more complex, buildings may present a higher level of risk, this is not always the case.

A smaller workplace that utilises high levels of flammable material, predominantly built of wood, or provides sleeping accommodation, may be considered to present a higher level of risk, particularly if the larger, more complex, building has effective fire safety management systems, is well organised and has workplace controls in place.

Having determined a prioritised order for fire risk assessments, gathering information on each specific facility prior to visiting the site must be seen as the next practical step.

The fire risk assessor, or fire risk assessment team, will benefit from having a range of documentation readily available prior to, during and following the practical risk assessment process (visual observation, verbal information gathering, etc.). The range of documentation that may provide key information for the fire risk assessment process is likely to include:

➤ Fire safety policy document/health and safety policy
➤ Fire certificate (if building was previously certified by Fire Service or HSE)
➤ Enforcement letters/notices
➤ History of any fires in premises or like premises
➤ History of building changes (Building Regulations applications)
➤ Plans of all buildings and plan of site
➤ Construction health and safety file:
  ➤ Schematic diagrams of fire safety systems, e.g. emergency lighting, fire alarm detection systems, etc.
  ➤ Building materials
  ➤ Maintenance, inspection, testing, cleaning arrangements for building services such as fire safety systems, fire doors, etc.

➤ Operations and maintenance (O&M) manual
➤ Records of storage – types/amounts of flammable materials
➤ Details of staff COSHH training (use, handling, storage of flammable substances)
➤ Electrical and gas supply system records
➤ Portable electrical appliance test records
➤ Details of numbers of employees/others who may be on site
➤ Details of employees/others who may have any sensory or physical impairment
➤ Details of staff fire safety training
➤ Details of evacuation exercises (this may be included in the fire log book)
➤ Fire log book:
   ➤ Extinguisher test records
   ➤ Emergency lighting test records
   ➤ Fire alarm and detection test records
   ➤ Sprinkler maintenance records (if fitted to the building)
   ➤ Dry riser/other fixed fire fighting systems records
   ➤ Door holding device test records
   ➤ Smoke extract or pressurised system test records
   ➤ Generator test records
➤ Standard operating procedures for any hot work processes
➤ Hot works permits
➤ Maintenance records (gas units, plant tools and equipment)
➤ Other risk assessments that include fire-related issues.

A large proportion of the above documentation may be available in the form of a fire safety manual that may have been produced as part of compliance with the British Standard BS 5588-12:2004 Managing Fire Safety.

Figure 14.2 Fire safety log book

It is also likely that the fire risk assessor/assessment team may also need to have access to a variety of documents produced as part of the requirements of the Building Regulations (Approved Documents B and M), and a variety of British Standards such as BS 5839 Fire Alarm and Detection Systems and BS 5266 Emergency Lighting.

The risk assessor or assessment team is also likely to require access to key personnel who will have an in-depth knowledge of the facility and activities that are likely to go on within the premises. Such key players may be:

➤ Departmental/employer's representative
➤ Property/building services manager or representative
➤ Facilities manager/supervisor
➤ Health and safety adviser
➤ Fire safety manager/adviser
➤ Landlord's representative (landlord domain areas)
➤ Members of the fire safety team (fire safety coordinator, fire warden/marshal)
➤ Personnel/human resources (for training information)
➤ Members of the janitorial team
➤ Maintenance/electrical engineers (if in-house).

It is therefore prudent to ensure that arrangements are in place to interview or liaise with these key parties and in particular ensure that facilities or areas of facilities such as service risers, plant rooms, roof spaces, etc. are accessible.

Table 14.1 Contents of the introduction section of a fire risk assessment

Section	Example content
Identification	The name, address and type of property
Building/area specific information such as:	The number of storeys, the nature of construction, the means of escape, activities being undertaken Fire safety systems in situ such as fire alarm detection, sprinklers, smoke extraction/ventilation, etc. Landlord's name Number of persons who may resort to the premises
Process specific risks	Activities Flammable materials/processes High noise levels
Specific at risk groups (persons)	Disabled non-ambulant persons Sensory impaired persons Young persons Those with psychological/learning difficulties

14.3.2 Recording the assessment

A recording mechanism must be considered that enables the risk assessor to record the significant findings of the assessment and, as previously discussed in Chapter 5, there is no universal layout for such forms. In principle the majority of forms that are used follow guidance issued by both the HSE and the ODPM; the latter suggests the need for utilising a plan of the building or facilities to assist in identifying locations of specific fire hazards for ease of reference.

Whichever recording layout is adopted it is likely to be broken down into key headings that include the following.

Introduction

The purpose of the introduction is to provide background information that the reader of the fire risk assessment can use to formalise a mental picture of the premises being assessed. It also provides information that will be related to the main contents of the assessment and subsequent action plans.

Management arrangements

This section should give a brief overview of the management arrangements in place to satisfy the requirements of the MHSWR and RRFSO and include, as a minimum:

- Policy and procedures
- Roles and responsibilities
- Training
- Inspection and monitoring systems
- Emergency planning arrangements.

The majority of fire hazards that affect a building or its operations arise out of substandard management arrangements, e.g. when investigating accidents, incidents, fires, false alarms, etc. management failures are seen to have a significant effect upon whether unsafe conditions (e.g. poor management of contractors) or unsafe acts (e.g. wedging open of fire doors) occur. There are very few occasions where both primary and secondary fire hazards arise which are not attributable to management failures and therefore this is a key issue when analysing fire risk.

Hazards

- Primary hazards:
  - Ignition sources
  - Fuel sources
  - Oxygen sources
- Secondary hazards (those that prevent an adequate response in the event of fire):
  - Rapid fire and smoke spread
  - Spreading to adjacent properties

- Inadequate warning arrangements
- Inadequate means of escape
- Persons cannot be accounted for
- Small fire grows rapidly
- Untrained persons at risk
- Fire service unaware of fire
- Fire service unaware of building risks
- Fire service cannot gain access
- Future construction/maintenance works.

The most obvious fire hazards relate to sources of ignition, sources of fuel and sources of oxygen and are known as primary fire hazards (Chapter 7). These will need to be recorded, as will any fire hazards that may affect people being unable to escape to a place of safety in the event of a fire.

It is often the case that a fire safety inspection will not identify issues such as the control of contractors as being a risk, unless contractors and maintenance operations are being undertaken at the time of inspection. This issue must be included in a risk assessment as it is 'reasonably foreseeable' that contractors may either compromise the safety of those on site while undertaking their operations, or compromise fire safety compartments or other control systems.

> While fire fighters are not deemed to be 'relevant persons' under the RRFSO, the assessment must address key areas such as how they will be alerted in the event of fire, the ease of which they will gain access to undertake fire fighting and rescue operations and risks to fire fighter safety from processes, materials, chemicals or other noxious substances. Therefore each of these areas must also be included within the risk assessment.

In relation to secondary hazards associated with fire the following risks may also need to be considered.

- Slips, trips and falls (including those from height) while evacuating
- Handling, lifting or carrying portable fire fighting equipment
- Being trapped by a structural failure such as a wall collapsing
- Being asphyxiated by a gas flood system that operates to extinguish a fire
- Coming into contact with the release of harmful substances such as asbestos

331

➤ Being exposed to fire or explosion while undertaking fire fighting action

➤ Coming into contact with live electrical equipment due to degradation of wiring during a fire

➤ Coming into contact with moving machinery while trying to shut down in the event of an emergency

➤ Being struck by a moving vehicle while evacuating

➤ Being assaulted by a person panicking in the event of a fire.

The above list identifies risks that may also need to be taken into account during the risk assessment process. The health risks, as previously mentioned (Chapter 5), also include the short- and long-term effects of coming into contact with chemicals and biological agents and must also include the potential for harm from psychological effects such as occupational stress, post-traumatic or critical incident stress disorders for those involved in a fire situation.

Many of the above issues may well not find themselves recorded in a building specific fire risk assessment record, neither should they if in the opinion of the risk assessor they present an insignificant risk. However, an assessment of each should be made and where required additional control measures considered and implemented and records of such assessments kept.

It may also be appropriate once the hazards have been identified that a review of the inventories and risk assessment strategy takes place, to ensure that the hazards that pose a significant threat are reprioritised accordingly.

Groups/persons at risk

To ensure that the risk assessment record meets the required standards it must identify the persons or groups of persons who may be at risk, particularly as a number of control measures may be specific to these groups of people, e.g. hearing impaired persons will need to be provided with an alternative to an audible warning device.

A definitive list of those that should be considered can be found in Chapter 5. It is, however, generally the case that any persons who may be on a premises are likely to be at risk in the event of fire but specific groups may be more at risk as indicated in the preceding paragraph. The mechanisms for recording those persons at risk are also wide and varied; however, it is often the case that those groups that are more at risk due to a specific circumstance are clearly identified.

Such groups of people may include the following:

➤ Those directly involved with work within a facility, who may be working in a remote area such as a tank or vessel that could be difficult to evacuate from

➤ Contractors working on plant under noisy conditions such as air conditioning units in a roof plant room, who may not hear the alarm

➤ Visitors including those using meeting rooms who may be unfamiliar with the fire safety management/ escape routes

➤ Members of the public particularly in large places of assembly such as shopping centres who may have no knowledge of where to go and what to do in the event of an emergency

➤ Young persons and children and their parents/ guardians who may attempt to find them if they are separated (e.g. crèche facility in a sports centre)

➤ Those with physical impairments or sensory impairment that may prevent awareness or response, due to their condition. It should also be noted that such conditions may be temporary as in the case of a broken leg

➤ Pregnant/nursing mothers who may be more susceptible to physical/mental stresses that could be affected by the need to undertake an emergency evacuation.

Evaluating existing control measures (workplace precautions and risk control systems)

It is unlikely that a building or premises will have no controls in place for the management of fire, it is therefore essential when evaluating the level of fire risk that any such controls are analysed and included in a risk assessment record. Having identified and recorded the current controls that are in place and any shortcomings they may have, analysis of residual risk can be made.

> **Example.** A fire detection system has been installed to meet current British Standards and guidance by a competent installation company, but is not being subject to regular testing, inspection and maintenance; it may therefore not operate when most needed. The risk in relation to the system being unable to give a warning in the event of fire (to ensure that those within the building can respond and evacuate safely) has not been reduced to the lowest level reasonably practicable and will therefore need to be addressed.

A quantum of the level of risk (likelihood × severity) will reflect the current controls in place, which in the case of the example above may rate a medium risk if a qualitative analysis is applied (see Chapter 5). The risk assessment record should then identify controls that should be implemented to reduce the risk to the lowest level reasonably practicable.

Additional control measures (recommended actions and prioritisation)

Having identified any shortcomings in the preventive and protective arrangements for the management of fire, the next element of the risk assessment record should be completed; that of recording recommended actions based upon a prioritised order. This prioritisation should ensure that those hazards presenting the most significant risk can be identified and addressed quickly and effectively, so that the risk can be reduced.

The recommended actions should not only identify short-term fixes, e.g. removal of obstructions from a fire escape route but also medium- to long-term issues that may need to be addressed to prevent a recurrence.

The prioritisation of any recommendations contained within the action plan will also need to reflect, not only the level of risk, but also the feasibility (practicality) of implementing the prescribed workplace precautions and risk control systems. Risk assessments that merely list long-term recommendations, such as 'review . . . policy' are unlikely to address the immediate requirements for managing fire risks, waiting for a formal policy or procedure, for example a review and revision of contractor selection and management due to poor management of hot work operations, may well take some time to be produced, approved and implemented.

When considering priority ratings it is useful to determine a key which will give a numerical indication of the recommended speed of action. While there are a number of such recording mechanisms available a large proportion will base themselves on similar ratings to those indicated in Table 14.2.

Table 14.2 Example of a table used to identify priority and indicate timescales for action to be taken

Priority rating	Timescale of action required
1	Immediate action – within 24 hours Usually dealt with during the risk assessment
2	Short-term action required – within 1 week
3	Short- to medium-term action – within 1 month
4	Medium-term action – within 3 months or agreed in a formalised plan within 6 months
5	Review as part of annual business plan

Closing the assessment out

Having recorded the recommendations and assigned a priority rating, ideally the action plan should enable a specific named person, or a key role, e.g. human resources manager for training requirements, within an organisation to be nominated to take responsibility for implementing the recommendations contained within the action plan.

As each individual recommendation is implemented it is useful to close out the action by reviewing the risk rating (ideally reducing it to low) and recording the result. This enables the action plan to reflect the ongoing status of the implementation programme until ultimately each element is closed out.

In addition if the organisation is conducting active monitoring as part of its safety management system the scoring or priority rating system of the risk assessment and the ability of the organisation to close out the recommendations in the appropriate timeframes, may prove to be very valuable when measuring the success of the overall fire risk assessment and fire safety management programme.

14.3.3 Worked examples

The following are worked examples that have been drawn from a series of risk assessments.

> **Worked example 1 – Sources of ignition allowing fire to start**
>
> An employer (responsible person) is operating in a small office environment with a large number of pieces of electrical office type equipment. The electrical equipment has been sourced from a reputable supplier in line with the company's procurement policy. Staff undertake pre-user checks on an ad hoc basis as per their induction training guidance. Annual portable appliance testing is carried out by a competent external contractor.
>
> During the physical inspection process of the fire risk assessment it was noted that a cable on a vacuum cleaner within an unlocked cupboard had an unauthorised electrical joint on its cable.

Taking into account the information contained in Worked example 1 the company appears to have a good policy in terms of resourcing its electrical equipment

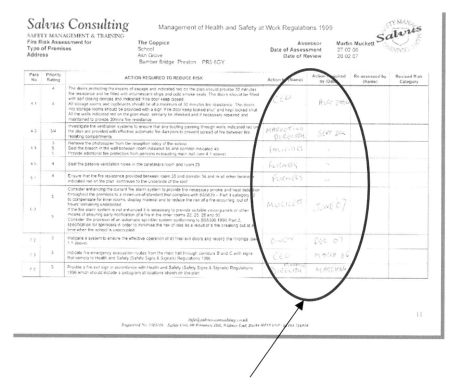

Figure 14.3 Risk assessment with names for actions and anticipated completion dates

and the provision of induction training (training records indicate that pre-use testing is being carried out); this was reinforced by talking to staff members.

Also the electrical equipment was covered under the main health and safety policy arrangements section and that a formalised system for ensuring that contractors PAT test the equipment on an annual basis was in place. Each of these issues is therefore recorded in the existing control measures. Significantly though the vacuum cleaner had had an unauthorised connector to the cable which increased not only the level of risk to the user but may also result in overheating causing a fire – this identifies that there may be inadequate control measures in place.

From a fire risk prospective the risk category with the existing controls is likely to rate a medium/low risk; however, from a health and safety prospective this would rate high. In either case further action is needed to control the risk.

The persons at risk from this hazard are likely to be the user (from the electrical hazard) but as in the majority of cases in relation to fire, any person within the premises, should a fire start from the equipment, could be affected.

The control measures included in the action plan are therefore likely to include, in the immediate phase, the removal of the cleaner from use (Priority 1). Subsequently a competent person would repair/replace the cable prior to its being used again.

It is also likely that a reminder to staff relating to unauthorised alterations to electrical equipment would be issued (this could be by way of e-mail, memo, etc.) (Priority 2). It would be further recommended that electrical appliance leads be included on visual fire safety inspections/tours of the building and records of such inspection tours are kept (Priority 3). This last element may also need the overall fire safety policy to be amended to include formal monthly inspection programmes (Priority 3/4).

The fuel hazard identified by the aerosol containers and the overall lack of any formalised controls could be a significant risk if they are involved in a fire, as they will not only provide a significant fuel source but could precipitate explosions.

There are very limited control measures are already in place, which include commonsense practice by the engineers and that the aerosols appear to be stored within an enclosed area and not left around the workshops (the odd one or two only).

The risk category will depend upon the potential ignition sources available; however, given the volumes of aerosols coupled with the availability of other combustible products it is likely that the risk, with existing controls, will rate a medium/high or high and further action will be required.

As in the case of the previous example all persons may be placed at risk from this hazard. It should also be

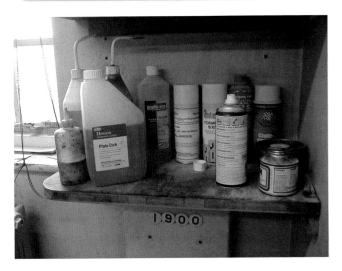

Figure 14.4 Aerosol container

noted that fire fighters not expecting the volume of storage of aerosols may be placed at risk should a fire occur.

The immediate actions that could be taken were to remove a large proportion of the aerosol containers from within the building to a secure metal store outside the building (Priority 1). If no such facility exists it would be recommended that an approved metal storage cabinet be provided (Priority 2). As an interim measure it may be possible to store the aerosols within metal cabinets in another storage area where there is limited ignition and other fuel sources or alternatively remove the ignition and fuel sources from the current storage area (Priority

---

Worked example 2 – Sources of fuel that may assist fire growth

An employer (responsible person) is operating a small service department for mechanical and electrical equipment. They use a variety of different chemicals, a number of which are of an aerosol container type.

They purchase the materials through their authorised dealership and keep the aerosols boxed up in their original containers within a storeroom which is being used to store a range of materials including posters, brochures, etc.

Other than the receipts there is no documentation relating to the management of these highly flammable aerosol units. It was noted during the visual element of the fire risk assessment that some eight to ten boxes each containing a dozen 300 ml aerosols were being stored.

---

2). The recommendations considered are likely to need discussing with the manager responsible to achieve the initial risk reduction measures.

The health and safety/fire safety arrangements section in the policy will need to reflect the storage, use, handling, disposal, etc. of the aerosols and the provision of adequate levels of supervision and training will also need to be addressed.

Ideally, each of these areas will address the need to reduce the quantity of aerosols being stored to an absolute minimum (perhaps making arrangements for 'daily call-off' from the supplier). While it is appreciated that some of the controls will require immediate action the policy and procedures will take a number of months to complete; raising awareness of staff and reducing the stock levels must be seen as a priority in the short term and the recommended risk controls reflect this.

Given that the only existing control measures (wooden doors, some of which were secured and signed, together with a policy on who could gain access) were inadequate the hazard of a fire starting within the service riser would likely present a significant risk of smoke spread throughout the building placing anyone in the building at risk.

The recommendations arising from the findings would therefore relate to the provision of intumescent strips and where appropriate cold smoke seals, to bring the service riser doors/door sets up to British Standard 476 (Priority 3).

Consideration may also be given to the provision of fire stopping (intumescent cushions, pillows, etc.) between each floor within the service risers (Priority 3).

The initial steps (Priority 1) would also be to secure all the doors to the service risers and remind the management team to monitor this issue.

A further recommendation may be to review the policy and arrangements for the management of fire to include a monthly inspection/safety tour programme (Priority 3/4).

On occasions a number of controls may need to be cross-referenced such as the provision of automatic fire detection linked to the existing fire alarm system, which may also be recommended within the service riser units.

The existing control measures identified in the above example have already reduced the risk significantly (low level); people are provided with a warning in the event of fire and they should be able to hear or know when to respond in the event of an emergency.

It is likely that the risk assessor (risk assessment team) will have been provided with a large number of supporting documents, the majority of which are detailed earlier within this chapter. It would appear from the example that the only issue to deal with is the obstruction of an alarm sounder which can be addressed immediately by the removal of the obstruction. To support this

Worked example 3 – Rapid fire and smoke spread in the workplace

A landlord (responsible person) is providing a building for multi-occupant (employer's) use. Service risers (electrical and IT) are located within 'common' parts of the building at each lift/staircase lobby area.

The service risers are fitted with wooden doors on which a sign 'Fire Door Keep Locked Shut' is positioned. Access to the service risers is controlled by the landlord's management team.

Documentation provided indicates that all tenants must refrain from accessing the service risers and that it is the management team's responsibility to manage this area.

During the practical risk assessment process it was noted that a number of service risers were insecure. That the doors to the risers were defective and were not fitted with intumescent strips and that the risers themselves were likely to act as a chimney due to the lack of fire compartmentation between each of the building's floors.

Figure 14.5 Self-closing device

control measure and to ensure that the issue does not recur it may be prudent to consider reminding staff via an e-mail or memo and ensuring such issues are included on the monthly fire safety inspection/tour programme.

The existing control measures in place are that the travel distances calculated for people to escape meet current Building Regulations guidance. The main corridors and staircase enclosures are provided with protection that meets a minimum of FR30 standard and that fire detection systems have been installed to provide advanced warning.

The fire policy currently in place details all the key issues in relation to fire but has not been amended to take into account disabled persons.

Worked example 4 – Persons in premise unaware of fire

An employer (responsible person) occupies an office building which is also provided with a basement storage facility and plant rooms at roof level. A fire alarm system to BS 5839 is installed throughout the premises. The system includes frangible break glass points and audible sounders which are fitted throughout the premises and into all units. Additional visual alarms are fitted within the plant room areas in proximity to equipment which is designated as a 'noise zone'. The entire system was installed by an approved competent company and in line with the policy is tested weekly by the facilities management team.

The policy and documentary records also indicate that the system is tested and inspected on an annual basis by a competent engineer. The audibility of the system is checked at regular intervals when testing the alarm by the facilities management team and employee feedback (a memo is sent to staff requesting information if they are in an area in which they cannot hear the alarm).

The call points are readily accessible and clearly identifiable with appropriate signage.

During the visual element of the risk assessment it was noted that one of the warning devices in the basement storage area had been obstructed by storage materials.

Worked example 5 – Persons cannot safely evacuate in the event of fire

An employer (responsible person) whose main operations involve the use of a four storey office facility has recently employed two physically disabled persons. The hazard that this presents is that the disabled persons may take excessive time to escape to a place of safety.

Specific persons at risk are clearly the disabled persons themselves but also those who may assist them to escape and potentially anyone else whom they may obstruct during the evacuation.

Figure 14.6 HMG Guide to Means of Escape for Disabled People

There are no existing controls such as arrangements for assisting those with disabilities, particularly as they are working on the second floor, access to which is via a lift. There has, however, been discussion by the management team for the provision of two 'evac chairs' to be provided in the staircase areas at either end of the second floor.

The risk category with controls is likely to reflect a medium/high or high risk rating for those who may be affected.

Additional controls will revolve around the production of personal emergency evacuation plans for each of the disabled persons (or a standard emergency evacuation plan which may cover all disabled persons). Controls are likely to include the use of an area within the protected staircases either end (known as a refuge area), the provision of methods for evacuation, e.g. evac chair, trained competent members of staff who may assist persons using the evac chair, the provision of information to all staff in relation to the evacuation procedures for those with disabilities, a review of the fire safety policy arrangements to include persons with disabilities. It may also be possible to consider the upgrading of the existing lift to that of an evacuation lift.

In terms of prioritisation an initial review as to the potential for those with physical disabilities to work on the ground floor (until such time as remedial actions can be implemented) required the provision of evac chairs and training with a Priority 3 rating, together with raising staff awareness, plus the review of the policy with a Priority rating of 3/4 and the provision of an evacuation lift (Priority 5).

Reviewing and revising the fire risk assessment

Having completed the fire risk assessment it must be kept under review to ensure that it remains valid. The fire or health and safety policy must reflect this and establish a programme to ensure that assessments are reviewed as part of an organisation's management system and also to comply with the legal duties.

There are a number of circumstances that may affect the validity of the fire risk assessment which include changes to the following.

The internal or external layout of the premises, which may affect:

➤ Means of escape (numbers, time taken to reach, accessibility, etc.)
➤ Compartmentation (changes to the lines of fire resistance)
➤ Emergency lighting layout (numbers of units, positioning, illumination)
➤ Fire warning/detection systems (positioning, zoning, staged system).

The people or groups of people who may be affected:

➤ Larger numbers may affect the ability of the escape routes to cope in an emergency
➤ Ages of the persons who may resort to the premises such as children or elderly persons who may require additional assistance for evacuation
➤ Those with disabilities who may not be able to evacuate on their own, hear the fire alarm or otherwise respond.

The plant affecting the primary hazards or one or more control measures such as:

➤ Heating systems (oil to electrical)
➤ Air handling circulatory units affecting the possible movement of smoke and smoke control
➤ Introduction/removal of equipment that may produce static electricity.

The procedures in place:

➤ Refuelling, plant isolations, etc.
➤ Emergency response arrangements
➤ Fire plan.

The work processes:

➤ Introduction of or removal of hot work operations
➤ Introduction of powder coating processes or those that may generate dust
➤ Addition or reduction in the use of flammable substances or LPG.

In addition changes to legislation, the results of investigations into fires or false alarms, or enforcement action may also establish a need to review and revise the fire risk assessment.

Where intermediate control measures are introduced such as the provision of additional fire wardens/security teams to cover for a lack of fire warning/detection systems, the fire risk assessment should be reviewed regularly to ensure that any such intermediate control measures remain effective.

> The enforcement agencies also consider that regular reviews of assessments are undertaken as a matter of course; such reviews are likely to be included in an overall annual review into the effectiveness of the safety management system, which should be reported to the board or senior management team or an organisation.

14.3.4 The emergency plan

The completed fire risk assessment will form the basis for producing the emergency plan (see Chapter 5). The plan must be made available to all employees, their representatives (where appointed), other employers (in a multi-occupied building) and the enforcing authority.

The purpose of producing an emergency plan is to establish a formal management system that will ensure that the premises can be safely evacuated and to communicate its contents so that the people who may come into the building know what to do if there is a fire.

As in the case of written risk assessments, if an employer employs five or more persons, details of the emergency plan must be recorded. In addition, if the premises are licensed or an alterations notice requiring an emergency plan is in force, then details of the emergency plan must also be recorded.

Figure 14.7 Emergency fire plan

> In multi-occupied, larger and more complex premises, the emergency plan will need to be more detailed to reflect the findings of the fire risk assessment. Where the building is multi-occupied it should also have been compiled following consultation with other occupiers, employers and other responsible persons, e.g. owners who have control over the building, who have themselves completed a fire risk assessment.
>
> In the majority of cases a single emergency plan covering the whole of a multi-occupied building will be necessary if the emergency plan is to be effective.
>
> It is therefore recommended that one responsible person is nominated to coordinate the task.

Note: The provision of fire action notices and a basic fire safety training programme should meet the emergency planning needs of a small premise with normal or low risk rating.

The contents of an emergency plan have been discussed in Chapter 5.

As a completed fire risk assessment is more than likely to incorporate a plan of the premises, it is also strongly recommended for larger more complex buildings that the emergency plan also incorporates a similar plan of the premises, which will provide information on key fire safety elements and systems within the building.

When produced this plan may also form a key component of the information that may be provided to the emergency services on arrival at a fire incident.

14.4 Example questions for Chapter 14

1. (a) **Explain**, using an example, the meaning of the term 'fire risk'. (2)
   (b) **Outline** the key stages of an assessment of fire risk. (6)

2. A workplace fire safety risk assessment must consider persons who could be placed 'especially at risk' in the event of fire. **Explain** what is meant by the term 'especially at risk' and identify the risk groups who should be considered in your assessment. (8)

3. As part of your workplace risk assessment, internal self-closing fire doors should be inspected to ensure their adequacy. **Outline** what should be examined. (8)

4. Outline four reasons that a completed risk assessment may become invalid. (8)

5. (a) List four groups of persons that should be considered when conducting a fire risk assessment. (2)
   (b) Outline ways in which an emergency plan may be communicated with employees. (6)

Appendix 14.1 Example of a fire risk assessment record and action plan

FIRE RISK ASSESSMENT RECORD

	Acme Company Ltd	Number	
Risk assessment for	Assessor's name	Date of assessment	Review date
	Joe Bloggs		

Overall assessment of risk

HIGH   NORMAL   LOW

Circle as appropriate

No.	Primary hazards	Persons at risk	Risk rating (H,M,L)	Control measures in place *(Control measures required)*	Action yes/no
1.	Sources of ignition allowing fire to start:				
2.	Sources of fuel that may assist fire growth:				
3.	Sources of oxygen that may assist fire growth:				

No.	Secondary hazards	Persons at risk	Risk rating (H,M,L)	Control measures in place (Control measures required)	Action yes/no
4.	Rapid fire and smoke spread in workplace:				
5.	Fire spread to adjacent properties:				
6.	Persons in premises unaware of fire:				
7.	Persons cannot safely evacuate in the event of fire:				

No.	Secondary hazards	Persons at risk	Risk rating (H,M,L)	Control measures in place (Control measures required)	Action yes/no
8.	Persons cannot be accounted for:				
9.	Small fire grows rapidly/untrained persons at risk:				
10.	No emergency plan – Fire Service unaware of fire:				
11.	Fire Service unaware of building risks:				

No.	Secondary hazards	Persons at risk	Risk rating (H,M,L)	Control measures in place (*Control measures required*)	Action yes/no
12.	Fire Service cannot gain access:				
13.	Future construction/ maintenance alterations:				

	Major	Serious	Slight
High	High risk	Medium risk	Low risk
Medium	Medium risk	Medium risk	Low risk
Low	Low risk	Low risk	Insignificant risk

FIRE RISK ASSESSMENT ACTION PLAN

Acme Company Ltd

Number

No.	Controls required	Priority or target date	Person responsible	Reassessment by (name)	Revised risk rating

1	Immediate action required – within 24 hours
2	Short-term action required within 1 week
3	Undertake action within 1 month
4	Action within 3 months or agree plan within 6 months
5	Review as part of business plan

The relevant paragraph numbers are indicated on the attached plan to enable easy identification of the significant uncontrolled material hazards identified in this assessment.

Summary of key legal requirements

15

This chapter contains a summary of the following legislation:

15.1 Acts of Parliament

- The Health and Safety at Work etc. Act 1974 (HSW Act) (pg. 346)
- The Environmental Protection Act 1990 (pg. 350)
- The Fire and Rescue Services Act 2004 (pg. 352)
- The Occupiers' Liability Acts 1957 and 1984 (brief summaries only) (pg. 355)
- The Employers' Liability (Compulsory Insurance) Act 1969 (and supporting regulations) (pg. 355)
- The Disability Discrimination Act 1995 (pg. 349)
- The Water Resources Act 1991 (pg. 359)

15.2 Regulations (listed alphabetically)

- The Building Regulations 2000 (SI 2000/2531)
- The Chemicals (Hazardous Information and Packaging for Supply) Regulations 2002 (SI 1689)
- The Confined Spaces Regulations 1997 (SI 1713)
- The Construction (Design and Management) Regulations 2007 (SI 320)
- The Control of Major Accident Hazards Regulations 1999 (SI 734)
- The Control of Substances Hazardous to Health Regulations 2002 (SI 2677)
- The Dangerous Substances and Explosive Atmospheres Regulations 2002 (SI 2776)
- The Electricity at Work Regulations 1989 (SI 0635)

- The Gas Appliances (Safety) Regulations 1992 (SI 0711) (brief summary only)
- The Gas Safety (Installation and Use) Regulations 1998 (SI 2451) (brief summary only)
- The Health and Safety (Consultation with Employees) Regulations 1996 (SI 1513)
- The Health and Safety (First Aid) Regulations 1981 (SI 0917)
- The Health and Safety (Information for Employees) Regulations 1989 (SI 0682)
- The Health and Safety (Safety Signs and Safety Signals) Regulations 1996 (SI 0341) (brief summary only)
- The Management of Health and Safety at Work Regulations 1999 (SI 3242)
- The Personal Protective Equipment Regulations 1992 (SI 2966)
- The Provision and Use of Work Equipment Regulations 1998 (SI 2306)
- The Regulatory Reform (Fire Safety) Order 2005 (SI 1541)
- The Reporting of Injuries, Diseases and Dangerous Occurrences Regulations 1995 (SI 3163)
- The Safety Representatives and Safety Committees Regulations 1977 (SI 0500)
- The Supply of Machinery (Safety) Regulations 1992 (SI 3073) (brief summary only)
- The Workplace (Health, Safety and Welfare) Regulations 1992 (SI 3004)

Note: The summaries provided in this chapter should not be regarded as direct quotes of either statutory requirements or ACoP standards. For full details of the legal text you should refer to the specific legislation and any supporting ACoP or guidance.

Recent amendments are also included in the summaries where appropriate.

The summaries provided focus upon the key areas relevant to this publication as they relate to fire safety and risk management.

15.3 The Health and Safety at Work etc. Act 1974 (HSW Act)

Often referred to as the Primary or Umbrella Act, the HSW Act details the principal statutory duties in relation to occupational health and safety.

Supporting regulations made under section 15 of the HSW Act expand these duties and are referred to as 'subordinate' or 'delegated' legislation.

The HSW Act is divided into four parts:

1. Part 1 covers health, safety and welfare in connection with work and the control of dangerous substances and emissions into the atmosphere
2. Part 2 covers the Employment Medical Advisory Service (EMAS)
3. Part 3 covers amendments to Building Regulations
4. Part 4 covers miscellaneous and general provisions.

Parts 2, 3 and 4 are not included in this summary.

15.3.1 General duties

The HSW Act places general duties on all those involved with work activities including employers, the self-employed, persons in control of premises, employees and manufacturers and suppliers. These general duties are detailed in sections 2 to 9 and the penalties for breaching the requirements are detailed in section 33.

15.3.2 Section 2 – Duty of the employer to their employees

2 (1) General duty
The employer is required to ensure, so far as is reasonably practicable, the health, safety and welfare at work of their employees.
2 (2) Specific duties
Employers are required to:

a) Provide and maintain safe plant and systems of work
b) Ensure safety in the use, handling, storage and transport of articles and substances for use at work
c) Provide information, instruction, training and supervision as necessary to ensure employee safety
d) Provide and maintain a safe place of work with safe access and egress

e) Provide and maintain a safe and healthy working environment with adequate welfare facilities.

2 (3) Requirement to have a written policy
Where the employer normally employs five or more employees, they must prepare a written statement covering their Policy for health and safety at work including the organisation and arrangements for Policy implementation. The Policy must be kept up to date, revised as necessary and brought to the attention of the employees.

In addition to the above duties, section 2 also covers the appointment of trade union safety representatives, consultation with these appointees and the establishment of a safety committee to review the employer's measures for ensuring the health and safety of their employees. The Safety Representatives and Safety Committees Regulations 1977 (SRSC Regs) expand these particular duties and the Health and Safety (Consultation with Employees) Regulations 1996 extend the duty to consult to non-union represented employees (see later summaries).

15.3.3 Section 3 – General duty of the employer and the self-employed to other persons

Employers and the self-employed are required to carry out their undertaking in such a way that, so far as is reasonably practicable, they do not expose other persons to risks to their health and safety.

15.3.4 Section 4 – General duty on persons in control of premises

This section applies to all persons who have control of non-domestic premises and requires that, so far as is reasonably practicable, the premises, the access, egress and any plant or substances provided for use there are safe and without risks to the health of anyone using the premises, plant or substances provided there whether for work or not.

It should be noted that common parts of residential premises are deemed to be non-domestic premises and fall within this section.

15.3.5 Section 5 – Emissions

This section was repealed by the Environmental Protection Act 1990.

15.3.6 Section 6 – General duty on manufacturers and suppliers

Section 6 places duties on persons in relation to both articles and substances for use at work.

Anyone who designs, manufactures, imports or supplies any article for use at work must:

> ensure, so far as is reasonably practicable, that the article is designed and constructed so that it will be safe and without risk to health while it is being used (including setting, cleaning and maintenance)
> carry out tests or examinations as necessary to comply with these duties
> provide adequate information (including any revisions) to those supplied with the article on the safe use of the article (including dismantling and disposal).

Anyone who erects or installs any article for use at work must ensure, so far as is reasonably practicable, that the article is safely erected or installed.

Anyone who manufactures, imports or supplies substances must:

> ensure, so far as is reasonably practicable, that the substance will be safe and without risks to health at all times while it is being used, handled, processed, stored or transported by a person at work or in a premises covered by section 4 (above)
> carry out tests or examinations as necessary to comply with these duties
> provide adequate information (including any revisions) to those supplied with the substance on any risks to health or safety posed by the substance, the results of any testing and information about conditions necessary to ensure the safe use of the substance (including disposal).

15.3.7   Section 7 – General duty on employees

Employees are required to:

> take reasonable care for the health and safety of themselves and other persons who may be affected by their acts or omissions at work
> cooperate with their employer and others to enable the employer to fulfil their duties.

15.3.8   Section 8 – Duty on all persons

No person must intentionally or recklessly interfere with or misuse anything provided in the interest of health, safety or welfare.

15.3.9   Section 9 – Charges

The employer is required not to charge employees for anything provided to them in compliance with the employers duty, e.g. first aid facilities, welfare provisions, personal protective equipment, training, etc.

15.3.10   Enforcing authorities

Sections 10 to 28 of the HSW Act relate to the establishment, functions and powers of the enforcing authorities and the introduction of subordinate regulations and Approved Codes of Practice.

15.3.11   Sections 10 to 14

These sections cover the establishment of the Health and Safety Commission (HSC) and the Health and Safety Executive (HSE). The HSC is responsible for both administering the HSW Act and associated regulations and for establishing the enforcement policy of their enforcing arm, the HSE.

15.3.12   Section 15

This section enables the Secretary of State to make new regulations for any of the general purposes covered by Part 1 of the HSW Act.

15.3.13   Sections 16 and 17

These sections cover the provision of practical guidance by the HSC on health and safety regulations in the form of Approved Codes of Practice (ACoPs) and the use of ACoPs in criminal proceedings.

15.3.14   Sections 18 and 19

These sections cover the responsibilities for enforcement of the HSW Act and supporting regulations by the HSE and the local authorities and are expanded by the Health and Safety (Enforcing Authority) Regulations 1989.

15.3.15   Sections 20 to 25

These sections cover the appointment and powers of inspectors responsible for the enforcement of the HSW Act and supporting regulations.

Appointed inspectors have the following general powers:

> enter any premises at any reasonable time
> take a constable with them if necessary
> take another authorised person with them and to take any equipment or material needed for the purposes of their entry
> make any necessary investigations or examinations
> direct that premises or anything in them are left undisturbed
> take measurements, photographs and recordings as necessary

- take samples of articles or substances
- require articles or substances to be dismantled and tested as necessary
- take possession of articles or substances as necessary
- require any person to give them information, to answer questions as necessary and to sign a declaration of the truth of their answers
- inspect and take copies of documents
- require facilities and assistance as necessary to exercise their powers
- do anything else necessary to exercise their powers.

In addition to the above, appointed inspectors have the following specific powers:

- the power to serve enforcement notices
- the power to seize, destroy or render harmless any article or substance which the inspector believes is a cause of imminent danger or serious personal injury.

15.3.16 Enforcement notices

Where an inspector is of the opinion that a person is contravening a statutory provision, or has contravened a statutory provision in circumstances that make it likely that the contravention will continue or be repeated, they may serve an improvement notice requiring remedial action to be taken by a specified date. The inspector may give written confirmation of the remedial action required in the form of a schedule attached to the notice but they are not required to do so.

Where an inspector is of the opinion that an activity being carried out or likely to be carried out involves, or will involve, a risk of serious personal injury, they may serve a prohibition notice requiring that the activity is suspended. The activity will be suspended either with immediate effect or after a specified time. Again, the inspector may give written confirmation of the remedial action required in the form of a schedule attached to the notice but they are not required to do so.

A person on whom a notice has been served may appeal to the Employment Tribunal within 21 days of the service date. Where an appeal is lodged against an improvement notice, the notice is suspended pending the outcome of the appeal. In the case of a prohibition notice, the notice will stay in force pending the appeal unless otherwise directed by the Employment Tribunal.

15.3.17 Sections 26 to 28

These sections cover the power of the enforcing authority to indemnify their inspectors and provide detail regarding the obtaining and disclosure of information by the HSC, HSE and enforcing authorities.

15.3.18 Sections 29 to 32

These sections were repealed.

15.3.19 Section 33

This section covers offences and the penalties that may be imposed in criminal proceedings for breaches of the HSW Act and supporting regulations. Examples of offences and penalties for both summary conviction (magistrates' court) and conviction on indictment (cases taken in the Crown Courts) are summarised in Table 1.

The enforcing authorities can initiate criminal proceedings regardless of whether any other action has been taken. There is no obligation placed upon the inspector to warn an offender of their intention to prosecute although in practice prosecution is often only used when warnings and persuasion have been ignored. An inspector may prosecute either an individual person or the employer organisation, or both.

15.3.20 Section 34

This section extends the time for bringing summary proceedings, e.g. following delays for relevant inquiries or inquests.

15.3.21 Section 35

This section enables the location of plant or substances involving a breach of relevant statutory provisions to be regarded as the location of the breach for the purposes of enforcement.

15.3.22 Section 36

Where an offence committed by one person is due to the act or default of another person, this section enables the enforcing authority to initiate criminal proceedings against the other person for the offence regardless of whether any proceedings are taken against the first person.

15.3.23 Section 37

Where an offence committed by a body corporate is committed with the consent or connivance or is attributable to the neglect of any director, manager or similar officer of the body corporate, this section enables the enforcing authority to initiate criminal proceedings against both the corporate body and the individual person for the offence.

15.3.24 Sections 38 and 39

Under these sections the power to initiate criminal proceedings for summary trial in England and Wales is

Table 15.1 Offences and penalties

Offence	Summary conviction	Conviction on indictment
HSW Act s 33(1A): • breach of HSW Act s 2, 3, 4 or 6.	£20 000 maximum fine	Unspecified fine
HSW Act s 33(2): • preventing someone from appearing before an inspector or from answering any questions from an inspector • obstructing an inspector • impersonating an inspector.	£5000 maximum fine	Summary offence only
HSW Act s 33(2A): • failing to comply with an enforcement notice or court order.	£20 000 maximum fine and/or up to 6 months' imprisonment	Unspecified fine and/or up to 2 years' imprisonment
HSW Act s 33(3): • breach of HSW Act s 7, 8 or 9 • breach of regulations made under HSW s 15 • making false statements or entries • using or possessing a document with intent to deceive.	£5000 maximum fine	Unspecified fine
HSW Act s 33(4): • operating without or contravening the terms or conditions of a licence when one is required • acquiring or attempting to acquire, possessing or using an explosive article or substance in contravention of relevant statutory provisions.	£5000 maximum fine	Unspecified fine and/or up to 2 years' imprisonment

restricted to appointed inspectors, the Director of Pubic Prosecutions (DPP) or other persons with the consent of the DPP.

In Scotland, cases are initiated by the Procurator Fiscal.

15.3.25 Section 40

This section places the onus of proving that the standard of reasonable practicability was achieved on the accused. In discharging this onus of proof, the accused must prove their case on the balance of probabilities.

15.4 The Disability Discrimination Act 1995

The Disability Discrimination Act 1995 (DDA) introduced new laws aimed at ending the discrimination that many disabled people face.

The DDA defines disability, and identifies who is protected under it. The definition is broad: 'a physical or mental impairment which has a substantial and long-term adverse effect on a person's ability to carry out normal day-to-day activities'.

The DDA makes it unlawful to treat a disabled person less favourably than others for a disability-related reason in relation to the disposal or management of residential, commercial and other premises, unless that treatment can be justified under the Act.

Part III of the Act introduced specific duties on organisations that provide services; these duties were introduced in three phases:

➢ from December 1996, it became unlawful for service providers to refuse to serve a disabled person, offer a lower standard of service or provide a service on worse terms to a disabled person for a reason related to his/her disability
➢ from October 1999, service providers have had to make reasonable adjustments for disabled people in the way they provide their services
➢ from October 2004, service providers have to make reasonable adjustments in relation to the physical features of their premises to overcome physical barriers to access.

From October 2004, where a physical feature (i.e. anything on the premises arising from a building's design or construction or the approach to, exit from or access to such a building; fixtures, fittings, furnishings, equipment or materials and any other physical element or quality of land in the premises) makes it impossible or difficult for disabled customers to make use of a service offered to the public, service providers will have to take measures, where reasonable, to:

➤ remove the feature, or
➤ alter it so that it no longer has that effect, or
➤ provide a reasonable means of avoiding the feature, or
➤ provide a reasonable alternative method of making the service available to disabled people.

15.5 The Disability Discrimination Act 2005

This Act builds on and extends the Disability Discrimination Act 1995.

15.5.1 New provisions

The DDA 2005 addresses the limitations of the initial legislation by extending disabled people's rights in respect of premises that are let or to be let, and commonhold premises.

From 4 December 2006, landlords and managers of let premises and premises that are to let will be required to make reasonable adjustments for disabled people.

Under the duties, provided certain conditions are met, landlords and managers of premises which are to let, or of premises which have already been let, must make reasonable adjustments, and a failure to do so will be unlawful unless it can be justified under the Act.

These new duties of reasonable adjustment do not apply to:

➤ prospective lettings where landlords let their only or principal home and do not use the services of an estate agent to arrange the letting; or
➤ a letting where the landlord lets their only or principal home and does not use a professional management agent to manage the letting; or
➤ certain small dwellings, for example where a landlord or manager lives on the premises and there is not normally residential accommodation on the premises for more than six persons.

15.5.2 Commonhold

The duty of reasonable adjustment also applies to commonhold. This is a system of freehold ownership for blocks of flats, shops, offices and other multiple occupation premises in England and Wales. A commonhold is made up of individual freehold properties which are known as commonhold units.

15.6 The Environmental Protection Act 1990

The central theme of the Environmental Protection Act 1990 (EPA) is to integrate all forms of pollution control.

The EPA is divided into nine parts:

1. Integrated Pollution Control and Air Pollution Control by Local Authorities
2. Waste on Land
3. Statutory Nuisance and Clean Air
4. Litter Control
5. Amendment to the Radioactive Substances Act 1960
6. Genetically Modified Organisms
7. Nature Conservation in Great Britain
8. Miscellaneous Provisions
9. General Provisions.

This summary concentrates on key parts of Parts 1 and 2 of the EPA.

15.6.1 Definitions

The EPA defines the following:

➤ The environment – consists of all, or any, of the following media, namely, the air, water and land; and the medium of air includes the air within buildings and the air within other natural or man-made structures above or below ground.
➤ Pollution of the environment – means pollution of the environment due to the release (into any environmental medium) from any process of substances which are capable of causing harm to man or any other living organisms supported by the environment.
➤ Harm – means harm to the health of living organisms or other interference with the ecological systems of which they form part and, in the case of man, includes an offence caused to any of his senses or harm to his property; and 'harmless' has a corresponding meaning.
➤ Controlled waste – means household, industrial and commercial waste or any such waste.
➤ Special waste – means controlled waste that is so hazardous that it can only be disposed of using special procedures.

15.6.2 Part 1: Integrated pollution control (IPC)

Processes and substances covered by the EPA are listed in the Environmental Protection (Prescribed Processes and Substances) Regulations 1991, as amended. Those classified as Part A processes are regulated by the Environment Agency under a system of integrated pollution control (IPC), and those listed as Part B are regulated by the relevant local authority (LA).

This system is slowly being replaced by one of integrated pollution prevention and control (IPPC). IPPC regulation will cover installations currently regulated under the existing IPC and LA systems but will extend integrated control to many more industrial companies and will regulate the installation rather than the process.

Operators of the most potentially environmentally polluting processes (Part A processes) have to apply for authorisation from the Environment Agency prior to operating the process and under the system of integrated pollution control (IPC) they must consider the total environmental impact of all releases to air, water and land in their application. The operator must advertise their application and the details are held in a public register which is available for the public to inspect. Exclusions from this public register may only be granted on grounds of commercial confidentiality or national security.

In granting a licence to operate Part A processes, the Environment Agency must include conditions to ensure that:

➤ The 'best available techniques not entailing excessive costs' (BATNEEC) are used
➤ If a process involves release into more than one medium (e.g. air and water), the operator uses the 'best practicable environmental option' (BPEO) to achieve the best overall environmental solution
➤ The operator complies with any directions given by the Secretary of State for the Environment or other applicable requirements or standards.

BATNEEC – best environmental technique not entailing excessive cost

BATNEEC strikes a balance between the best available technology and management techniques with what the industry sector can generally afford. In reducing emissions to the lowest practicable level, account will be taken of local conditions and circumstances, both of the process and the environment.

The 'BAT' refers to:

➤ Best – the most effective in 'preventing, minimising or rendering harmless polluting emissions'
➤ Available – does not necessarily imply that the technology is in general use but that it is generally accessible
➤ Techniques – refers both to technology or the process and how it is operated.

The concept of 'BAT' consists not just of the technology but the whole process and includes matters such as staff numbers, working methods, training, supervision and the manner of operating the process.

BATNEEC is usually expressed as emission limits for the particular substances released by a process.

Best practicable environmental option

The Royal Commission on Environmental Pollution states that the aim of 'BPEO' is:

> To find the optimum combination of available methods of disposal so as to limit damage to the environment to the greatest extent achievable for a reasonable and acceptable total combined cost to industry and the public purse.

Operators are required to monitor their emissions and report these to their relevant enforcing authority (EA) on an annual basis. If the EA believes that the operator is breaching their conditions of authorisation, they may serve enforcement notices to:

➤ Revoke the authorisation
➤ Specify steps that must be taken with a time limit for compliance
➤ Prohibit the process.

Ultimately, the EA may initiate criminal proceedings for a breach of the EPA with a maximum fine of £20 000 for summary cases and an unspecified fine and/or up to two years' imprisonment for indictable cases.

In a similar way to the HSW Act, the onus of proving that BATNEEC and BPEO are achieved rests with the accused.

15.6.3 Part 2: Waste on land

The EPA replaced previous controls in the Control of Pollution Act 1974. It introduced changes in both the waste management licensing system and the bodies responsible for waste regulation and disposal. Local authorities remain the key enforcement bodies, but with extra demands on the waste disposal industry and all producers or handlers of waste.

It is not possible for a producer of waste to rid themselves of it simply by handing it over to another. Anybody who carries, keeps, treats, or disposes of waste, or who acts as a third party and arranges matters such as imports or disposal must satisfy a 'duty of care'.

This requires the duty holder to take all reasonable steps to keep waste safe. If waste is given to someone else, the duty holder must ensure that they are authorised to take it and can transport, recycle or dispose of it safely.

The duty of care does not apply to domestic householders unless the waste is from somewhere else, e.g. the householder's workplace or someone else's premises.

Duty holders are required to:

➤ Stop waste escaping from their control and store it safely and securely to prevent pollution or harm
➤ Keep waste in a suitable container – loose waste on a lorry or in a skip should be covered
➤ If waste is given to someone else, the duty holder must check that this person is authorised to take the waste – see below
➤ Where waste is transferred to someone else, the duty holder must describe the waste in writing and complete, sign and retain a copy of a transfer note for the waste.

Authorised persons to whom waste is transferred must:

➤ Ensure they are authorised to take the waste
➤ Ensure they obtain a written description of the waste
➤ Complete and retain a copy of a transfer note for the waste.

Authority for handling waste

The following are example of persons who may be authorised to handle waste:

➤ Council waste collectors – the duty holder does not have to do any checks but if they are not a householder, they will still need to complete a transfer note
➤ Registered waste carriers – carriers who are registered with the Environment Agency. Details of their authorisation will be on their carrier's certificate of registration
➤ Exempt waste carriers – charities and voluntary organisations may be exempted by the Environment Agency
➤ Holders of waste management licences – the licence will specify what waste the holder is licensed to handle
➤ Exempted businesses – the Environment Agency may grant exemptions to certain businesses, e.g. those recycling scrap metal
➤ Authorised waste transporters – persons authorised to receive waste for transport only.

Authorised waste handlers must be licensed by the Environment Agency and their licence will specify the conditions that they must take to ensure that their activities do not cause pollution of the environment, harm to health or serious detriment to local amenities.

In submitting an application for a licence to handle waste, the applicant must submit a formal application and prepare a working plan describing how the licensee intends to prepare, develop, operate and restore (where relevant) the site or plant. The submission of a comprehensive working plan may help to prevent delay in granting the licence and could also lead to less restrictive conditions being imposed.

During consideration of licence applications, the application form will be put on the public register although this may be restricted on grounds of commercial confidentiality. The Environment Agency must also be satisfied that the applicant is a 'fit and proper person' in that:

➤ They have made adequate financial provision to cover licence obligations
➤ A technically competent person will be managing the licensed activity.

The Environment Agency will also consider whether there have been any previous convictions for relevant offences. Once a licence is issued, the licensee must comply with the licence conditions at all times and the Environment Agency will make visits to check this.

15.7 The Fire and Rescue Service Act 2004 (FRSA)

The FRSA applies to England and Wales only, with the exception of the provisions relating to pensions. For the main part, the Act does not extend to Northern Ireland. Both Scotland and Ireland have their own similar Acts.

The FRSA repeals the previous legislation, which had been in force since 1947. The purpose of the Act was to modernise the structure and responsibilities of local authority Fire and Rescue Services. The Act is divided into the following parts:

Part 1 – Fire and rescue authorities (sections 1 to 5): determines which body is the fire and rescue authority for an area, and provides for the combination of two or more fire and rescue authorities by order. Fire and rescue authorities may be metropolitan administrations or regional or county local governments.

Part 2 – Functions of fire and rescue authorities (sections 6 to 20): sets out the duties and powers of fire and rescue authorities.

Part 3 – Administration (sections 21 to 31): provides for the preparation of a Fire and Rescue National Framework setting out the strategic priorities of the Fire and Rescue Service, and for the supervision of fire and rescue authorities. It makes supplementary provision for the Secretary of State to provide equipment and training centres for fire and rescue authorities.

Part 4 – Employment (sections 32 to 37): covers employment by fire and rescue authorities, including the creation of negotiating bodies to determine the terms and conditions of employees, and pension schemes.

Part 5 – Water supply (sections 38 to 43): places duties on fire and rescue authorities and water undertakers to ensure an adequate supply of water for fire fighting activities.

Part 6 – Supplementary (sections 44 to 54 and Schedules 1 and 2): details the powers of fire and rescue authority employees to undertake rescue work and investigations, as well as a number of consequential provisions and repeals, including the abolition of the Central Fire Brigades Advisory Council.

Part 7 – General (sections 55 to 64): makes general provision in relation to pre-commencement consultation, interpretation, statutory instruments, territorial extent, etc.

Core functions of Fire and Rescue Services – Sections 6–12 detail the functions of the Fire and Rescue Service authorities.

15.7.1   Section 6: Fire safety

For the first time in the history of the fire service in the UK this section imposes a duty onto fire and rescue authorities to conduct fire safety activities in the community.

15.7.2   Sections 7 and 8: Fires and road traffic accidents

Section 7 imposes a duty on fire and rescue authorities to provide arrangements for fighting fires and protecting life and property from fires within its area. Authorities are required to:

➤ Provide sufficient personnel, services and equipment to deal with all normal circumstances
➤ Ensure adequate training
➤ Have arrangements to receive and respond to calls for help
➤ Obtain information which it needs to carry out its functions.

Section 8 places a duty on fire and rescue authorities to make provision for rescuing persons from road traffic accidents and for dealing with the aftermath of such accidents.

Under sections 7 and 8 fire and rescue authorities must seek to mitigate the damage, or potential damage, to property in exercising their statutory functions. As a consequence, the actions a fire and rescue authority must take in responding to an incident which could damage property must be proportionate to the incident and the risk to life.

15.7.3   Section 9: Emergencies

An 'emergency' is defined as an event or situation that causes or is likely to cause:

➤ One or more individuals to die, be seriously injured or become seriously ill or

➤ Serious harm to the environment (including the life and health of plants and animals).

This section empowers the Secretary of State, by order following consultation, to place a duty on fire and rescue authorities to respond to other, specific types of emergency, as defined by order, such as flooding and terrorist incidents. The order to respond can be outside the geographic area which is the responsibility of any particular authority.

Under this section the Secretary of State may also direct fire and rescue authorities as to how they should plan, equip for and respond to such emergencies.

15.7.4   Sections 11 and 12: Power to respond to other eventualities; and other services

Section 11 allows fire and rescue authorities the discretion to equip and respond to events beyond its core functions provided for elsewhere in the Act. A fire and rescue authority will be free to act where it believes there is a risk to life or the environment. This would allow, for example, a fire and rescue authority to engage in specialist activities such as rope rescue.

Section 12 allows fire and rescue authorities the power to use its equipment or personnel for any purpose it believes appropriate and wherever it so chooses. For example, a fire and rescue authority may agree to help pump out a swimming pool as a service to its community.

15.7.5   Section 19: Charging

Section 19 allows the Secretary of State to issue an order allowing authorities to charge for their services (for example, pumping out a swimming pool or attending a premises as a result of a false fire alarm). Fire and rescue authorities are not able to charge for extinguishing fires or protecting life and property in the event of fires, except in respect of incidents at sea or under the sea. There is also an exemption for charging for emergency medical assistance.

15.7.6   Section 38: Duty to secure water supply, etc.

Fire and rescue authorities have a duty to take all reasonable measures to ensure the adequate supply of water for use in the event of fire.

This section allows an authority to use any suitable supply of water, whether provided by a water undertaker or any other person. Any charges that may be made by the owner of the water are limited by the Water Industry

Act 1991, which expressly forbids charging by a water undertaker in respect of: water taken for the purpose of extinguishing fires or for any other emergency purposes; water taken for testing apparatus used for extinguishing fires; or for fire fighting training.

15.7.7 Section 40: Emergency supply by water undertaker

This section places an obligation on a water undertaker to take all necessary steps to increase supply and pressure of water for the purpose of extinguishing a fire, if requested to do so by a fire and rescue authority. A water undertaker may shut off water from the mains and pipes in any area to enable it to comply with a request to increase supply and water pressure. The fire and rescue authority is excluded from liability for anything done by a water undertaker in complying with its obligations.

15.7.8 Section 42: Fire hydrants

It an offence for any person to use a fire hydrant other than for the purpose of fire fighting or any other purpose of a fire and rescue authority; or other than for any purpose authorised by the water undertaker or other person to whom the hydrant belongs.

It is also an offence for anyone to damage or obstruct a fire hydrant.

15.7.9 Section 43: Notice of works affecting water supply and fire hydrants

This section places a duty on any person proposing to carry out any work relating to the provision of water to any part of the area of a fire and rescue authority to give at least six weeks' written notice to the authority. A person proposing to carry out any works affecting a fire hydrant is required to give at least seven days' notice in writing.

15.7.10 Section 44: Powers of fire fighters, etc. in an emergency, etc.

This section of the Act gives certain powers to an employee of the fire and rescue authority (provided that they are authorised in writing) to deal with emergencies; specifically an authorised employee may do anything he reasonably believes to be necessary:

➤ If he believes a fire may have broken out (or to be about to break out), for the purpose of extinguishing or preventing the fire or protecting life or property and preventing or limiting damage to property
➤ If he believes a road traffic accident to have occurred, for the purpose of rescuing people or

protecting them from serious harm and preventing or limiting damage to property
➤ If he believes an emergency of another kind to have occurred, for the purpose of discharging any function conferred on the fire and rescue authority in relation to the emergency and preventing or limiting damage to property.

In addition to the general power to 'do anything' an authorised employee has got specific powers to:

➤ Enter premises or a place, by force if necessary, without the consent of the owner or occupier of the premises or place
➤ Move or break into a vehicle without the consent of its owner
➤ Close a highway
➤ Stop and regulate traffic
➤ Restrict the access of persons to premises or a place.

It is an offence to obstruct or interfere with an authorised officer of a fire and rescue authority without reasonable cause. A person guilty of an offence is liable on summary conviction to a fine not exceeding level 3 on the standard scale.

15.7.11 Section 45: Obtaining information and investigating fires

An authorised officer may at any reasonable time enter premises:

➤ For the purpose of obtaining information
➤ If there has been a fire in the premises, for the purpose of investigating what caused the fire or why it progressed as it did.

An authorised officer may not enter premises by force, or demand admission to premises occupied as a private dwelling unless 24 hours' notice has been given to the occupier of the dwelling.

If an authorised officer considers it necessary to enter premises and is unable to do so, without using force he may apply to a justice of the peace for a warrant authorising entry to the premises by force at any reasonable time.

15.7.12 Section 46: Supplementary powers of an authorised officer

In addition to the powers of entry, an authorised officer has the power to:

➤ Take with him any other persons, and any equipment, that he considers necessary

- Require any person present on the premises to provide him with any facilities, information, documents or records, or other assistance, that he may reasonably request
- Inspect and copy any documents or records on the premises or remove them from the premises
- Carry out any inspections, measurements and tests in relation to the premises, or to an article or substance found on the premises, that he considers necessary
- Take samples of an article or substance found on the premises, but not so as to destroy it or damage it unless it is necessary to do so for the purpose of the investigation
- Dismantle an article found on the premises, but not so as to destroy it or damage it unless it is necessary to do so for the purpose of the investigation
- Take possession of an article or substance found on the premises and detain it for as long as is necessary for any of these purposes:
  - To examine it and do anything he has power to do under paragraph (c) or (e)
  - To ensure that it is not tampered with before his examination of it is completed
  - To ensure that it is available for use as evidence in proceedings for an offence relevant to the investigation
- Require a person present on the premises to provide him with any facilities, information, documents or records, or other assistance, that he may reasonably request.

It is an offence to obstruct or fail to comply with any requirement made by an authorised officer while exercising his powers.

15.7.13 Section 49: False alarms of fire

This section provides that a person who knowingly gives or causes someone else to give a false alarm of fire to a person acting on behalf of a fire and rescue authority is liable to a maximum level 4 fine, prison sentence of 51 weeks, or both.

15.8 The Occupiers' Liability Acts 1957 and 1984

An occupier of premises (the person who has control over the premises, venue or site) owes a duty of care to all their lawful visitors to ensure that they will be reasonably safe using the premises for the purpose for which they are invited or permitted by the occupier.

This duty does not impose an obligation in respect of risks willingly accepted by the visitor if adequate steps are taken by the occupier to enable the visitor to be reasonably safe and they warn the visitor of any remaining risks.

The occupier must be prepared for the fact that children are less careful than adults and what may be an adequate warning to an adult may not be so for a child.

The 1984 Act extends the duty of care owed by the occupier to unlawful visitors or trespassers where:

- The occupier knows there is a risk because of the state of the premises and
- They have reasonable grounds to know the unlawful visitor will be on the premises and
- The risk to the unlawful visitor is one that the occupier could reasonably be expected to provide some protection against.

15.9 The Employers' Liability (Compulsory Insurance) Act 1969

The aim of the Employers' Liability (Compulsory Insurance) Act 1969 (ELCI), and the supporting Employers' Liability (Compulsory Insurance) Regulations 1998, is to ensure that employers have at least a minimum level of insurance cover against civil liability claims made by their:

- Employees for injuries sustained at work and
- Employees and former employees who may become ill as a result of their work.

ELCI applies to every employer, other than those listed under schedule 2 who are exempted, for example:

- Most public organisations including government departments, local authorities, police authorities and nationalised industries
- Health service bodies including National Health Service Trusts, Health Authorities, Primary Care Trusts and Scottish Health Boards
- Other public funded organisations such as passenger transport and magistrates' court committees
- Family businesses, i.e. if employees are closely related to the employer (as husband, wife, father, mother, grandparents, children, siblings or half-siblings).

Incorporated limited companies of this nature are not exempt.

15.9.1 Requirements

Employers must have employer liability insurance for all people they employ (including people normally regarded as self-employed) as summarised below:

Employer liability insurance required	Employer liability insurance not required
• The employer deducts National Insurance and income tax from the money paid to the people • The employer has control over where and when the people work • The employer supplies most materials and equipment for the people to use • The employer has a right to profit made by the people • The employer requires the person to deliver the service personally and does not allow them to employ a substitute if they are unable to do the work • The employer treats the person in the same way as other employees.	• The people do not work exclusively for the employer • The people supply most of the equipment and materials for the job themselves • The people are clearly in business for personal benefit • The people can employ a substitute where they are unable to do the work themselves • The employer does not deduct national insurance or income tax from the money paid.

Employer liability insurance will normally be required for volunteers although not in the case of:

➤ Students working unpaid
➤ People taking part in youth or adult training schemes
➤ School students on a work experience programme.

The employer must display a copy of their current certificate of insurance where their employees can easily read it and retain copies of expired certificates for at least 40 years from expiry (applies to policies in force from 31 December 1998 onwards).

Expired certificates may be stored electronically but they must be made available to health and safety inspectors if requested.

Failure to have employer liability insurance may result in:

➤ A fine of £2500 for any day which the employer is without suitable insurance
➤ A fine of £1000 for failing to display the certificate of insurance or for failing to make the certificate available on request.

15.10 Water Resources Act 1991 (as amended by the Environment Act 1995)

Water Resources Act – this legislation is in place to control discharge to watercourses to avoid pollution. Responsibility for discharges lies with the Environment Agency. It is an offence to cause or knowingly permit:

➤ Any poisonous, noxious or polluting matter, or any solid waste matter, to enter any controlled waters i.e. tidal, coastal, lakes, ponds and ground waters.
➤ Any matter other than trade or sewage effluent to be discharged from a sewer in contravention of a relevant prohibition
➤ Any trade or sewage effluent to be discharged into controlled waters or, through a pipe, into the sea
➤ Any trade or sewage effluent to be discharged onto land or into a pond in contravention of a relevant prohibition
➤ Any matter to enter inland waters so as to cause or aggravate pollution by impeding the flow.

Water Resources Act – discharge to watercourses can only be made with the consent of the Environment Agency.

The Environment Agency has the power to issue 'work notices' requiring those responsible for water pollution to carry out works in order to dispose of polluting matter or clean up polluted watercourses. This notice can be issued if pollution is likely to arise.

Enforcement notices may also be served if a company is failing to comply with their consent conditions.

Discharges to surface waters – in order to discharge to watercourses, consent from the Environment Agency is required. Disposal of new discharges by this route is usually subject to tight controls and is therefore not the usually favoured option. It is likely that disposal of effluent by this route will require treatment to strict limits.

15.11 The Building Regulations 2000

The Secretary of State is empowered by the Building Act 1984 to approve and issue documents containing practical guidance with respect to the requirements contained in these regulations. The Building Regulations 2000 (SI 2000/2531) issued under the Building Act came into force on 1 January 2001 and replaced the Building Regulations 1991 (SI 1991/2768) and consolidated a number of revisions and amendments.

15.11.1 Regulation 3 – The meaning of building work

The building regulations define building work as the erection or extension of any building or its services. The definition includes alterations to buildings if the alteration affects the structure, the means of warning and escape, internal fire spread, external fire spread access and facilities for the Fire Service and access and facilities for disabled people.

15.11.2 Regulations 12–17 – Procedural steps

These sections prescribe the procedural steps that are required when building work is being planned; those conducting building work must notify the local planning authority and may do so either by using a brief planning notice or by submitting full plans for approval. Once the building work is completed in accordance with the requirements of the regulations, the local authority must issue a completion certificate.

15.11.3 Schedule to the Building Regulations

The Schedule to the Building Regulations details a number of functional requirements that any building must satisfy. The schedule is divided into a number of parts.

PART A Stability – this part deals with the stability of the building and has three separate functional requirements:

A1 requires that any building must be designed and constructed so that it can withstand its own weight (deadload), the weight of any contents (imposed load) or any loads imposed by the wind, safely and without deflecting.

A2 requires that the stability of any part of the building is not impaired by the swelling, shrinkage or freezing of the subsoil; or land slip or subsidence.

A3 In the event of an accident buildings of five or more storeys will not suffer collapse that is disproportionate to its cause.

PART B Fire safety – this part deals with the fire safety of the building from fire and has five separate functional requirements:

B1 Means of warning and escape – buildings must be designed and constructed so that there are appropriate provisions for the early warning of fire, and appropriate means of escape in case of fire from the building to a place of safety.

B2 Internal fire spread (linings) – buildings must be designed and constructed to inhibit the spread of fire via the internal linings. All partitions, walls, ceilings or other internal structures must adequately resist the spread of flame over their surfaces; and if they are ignited, have a rate of heat release which is reasonable in the circumstances.

B3 Internal fire spread (structure) – the building shall be designed and constructed so that, in the event of fire, its stability will be maintained for a reasonable period. In particular, walls common to two or more buildings must resist the spread of fire.

To inhibit the spread of fire all buildings must be subdivided with fire resisting construction appropriate to the size and intended use of the building.

In addition, all buildings must be protected against unseen spread of fire and smoke within concealed spaces in its structure and fabric is inhibited.

B4 External fire spread – the external walls of buildings must adequately resist the spread of fire over the walls and from one building to another, and roofs of buildings must adequately resist the spread of fire over the roof and from one building to another.

B5 Access and facilities for the fire service – buildings must be designed and constructed so as to provide reasonable facilities to assist fire fighters in the protection of life and to enable fire appliances to gain access to the building. It should be noted that the subsequent amendments that came into effect in April 2007 split the Approved Documents-B into two volumes: one that relates to Dwelling houses (Volume 1) and one that relates to Buildings other than Dwelling houses (Volume 2).

PART C Site preparation and resistance to moisture – this part of the Building Regulations details the functional requirements relating to the ground conditions of the building site. There are four requirements:

C1 Preparation of site – the ground to be covered by the building shall be reasonably free from vegetable matter.

C2 Dangerous and offensive substances – reasonable precautions shall be taken to avoid danger to health and safety caused by substances found on or in the ground to be covered by the building.

C3 Subsoil drainage – adequate subsoil drainage shall be provided to avoid the passage of the ground moisture to the interior of the building and/or any damage to the fabric of the building.

C4 Resistance to weather and ground moisture – the walls, floors and roof of the building shall adequately resist the passage of moisture to the inside of the building.

PART D Toxic substances

D1 Cavity insulation – if insulating material is inserted into a cavity wall reasonable precautions shall be taken to prevent the subsequent permeation of any toxic fumes from that material into any part of the building occupied by people.

PART E Resistance to the passage of sound

E1–E3 Airborne and impact sound – the three requirements for Part E require that walls, floors and stairs that separate a dwelling from another building or part of a building not used for dwelling must have reasonable resistance to the transmission of airborne and impact sound.

PART F Ventilation

F1–F2 Ventilation – require adequate ventilation from buildings and roofs.

PART G Hygiene

G1–G3 – require adequate sanitary conveniences and washing facilities, bathrooms and hot water storage arrangements.

PART H Drainage and waste disposal

H1–H3 – require that buildings are designed and constructed to have adequate rainwater and foul water waste disposal and solid waste storage.

PART J, K and L – deal respectively with heat producing appliances, protection, falling and collision with vehicles and the conversion of fuel and power.

PART M Access for disabled people

M1 Definition – the functional requirements for providing access for disabled people apply to persons who have an impairment which limits their ability to walk or which requires them to use a wheelchair for mobility, or impaired hearing or sight.

M2 Access and use – requires that reasonable access must be made for disabled people to gain access and use a building.

M3 Sanitary conveniences – requires that reasonable provision for sanitary conveniences shall be made in either the entrance storey or principal storey. In addition, if sanitary conveniences are provided in any building which is not a dwelling, reasonable provision shall be made for disabled people.

M4 Audience or spectator seating – if the building contains audience or spectator seating, reasonable provision shall be made to accommodate disabled people.

PART N Glazing – safety in relation to impact, opening and cleaning

N1 Protection against impact – glazing, with which people are likely to come into contact while moving in or about the building, must break in a way which is unlikely to cause injury; or resist impact without breaking; or be shielded or protected from impact.

N2 Manifestation of glazing – glazing, with which people are likely to come into contact while moving in or about the building, must incorporate features which make it apparent.

N3 Safe opening and closing of windows, etc. – windows, skylights and ventilators which can be opened by people in or about the building shall be so constructed or equipped that they may be opened, closed or adjusted safely.

N4 Safe access for cleaning windows, etc. – provision must be made for any windows, skylights, or any other glazing to be safely accessible for cleaning.

15.12 The Chemicals (Hazardous Information and Packaging for Supply) Regulations 2002

The aim of the Chemicals (Hazardous Information and Packaging for Supply) Regulations 2002 (CHIP) is to ensure that recipients who are supplied with chemicals receive information they need to protect themselves, other persons and the environment from risks the chemicals may pose.

CHIP is supported by the Approved Classification and Labelling Guide (APLG) which details the rules for classification and the Approved Supply List (ASL) which lists chemicals that have already been classified.

CHIP applies to most chemicals although there are exceptions detailed in regulation 3. This includes cosmetics and medicines which have their own regulations.

15.12.1 CHIP terminology

The following table below outlines some terms used throughout CHIP:

Category of danger	A description of hazard type from Schedule 1, e.g. explosive, corrosive
Classification	The identification of the chemical hazard by assigning a category of danger from Schedule 1 and risk phrases from the APLG
Risk phrase (R-phrase)	A standard phrase giving simple information about the chemical risks in normal use
Safety phrase (S-phrase)	A standard phrase giving advice on appropriate precautions when using a chemical
Substance	A chemical element or one of its compounds including any impurities
Preparation	A mixture of substances
Chemical	A generic term for substances and preparations
Dangerous substance	A substance listed in the APLG or, if not listed, one which is in one or more of the categories of danger specified in Schedule 1
Tactile warning devices (TWDs)	A small raised triangle applied to a package to alert visually impaired people to the fact that they are handling the container of a dangerous chemical

Child resistant fastenings (CRFs)	A closure which meets standards intended to protect young children from accessing the hazardous contents of a package
Chain of supply	The successive ownership of a chemical as it passes from the manufacturer to the ultimate user

5.12.2 Summary of requirements

Suppliers must:

➤ Identify the chemical hazards, e.g. flammability, toxicity (classification)
➤ Package the chemical safely and
➤ Pass on this information to the recipient together with advice on the safe use, handling, storage, transport and disposal of the chemical. This is usually done by means of labelling on the packaging and the provision of a material safety data sheet (MSDS).

The above are known as the 'supply' requirements where 'supply' is defined as making a chemical available to another person. The term 'supplier' should be construed accordingly and will include manufacturers, importers, wholesalers and retailers of chemicals.

15.12.3 Classification – regulation 4

This regulation prohibits persons from supplying a dangerous substance unless it has been classified in accordance with CHIP. Most commonly used dangerous substances have already been classified and will be listed in the ASL. Where the substance has not been classified, the supplier must decide what kind of hazardous properties the chemical has and explain these by assigning the appropriate 'Risk phrase(s)' as detailed in the ACLG.

15.12.4 Packaging – regulations 7 and 11

Suppliers must not supply dangerous substances unless they are packaged and:

➤ The receptacle containing the substance is designed and constructed so that the contents cannot escape
➤ The package and fastenings are not susceptible to adverse attack by the contents or liable to form dangerous compounds with the contents
➤ The package and fastenings are strong and solid to ensure that they will not loosen and will meet the normal stresses and strains of handling

➤ Any replaceable fastenings are designed so that the receptacle can be repeatedly fastened without the contents escaping.

Where dangerous substances are supplied to the general public, regulation 11 extends the above requirements for listed substances to include the fitting of child resistant fastenings, tactile warning devices (e.g. a raised triangle on the package) and other consumer protection devices.

15.12.5 Information and labelling – regulations 5, 6, 8, 9 and 10

Suppliers are required to provide the recipient of dangerous substances with a safety data sheet containing information detailed under the 16 headings listed below:

Headings for safety data sheets (CHIP Schedule 4)	
1. Identification of the substance and company undertaking	9. Physical and chemical properties
2. Composition/information on ingredients	10. Stability and reactivity
3. Hazard(s) identification	11. Toxicological information
4. First aid measures	12. Ecological information
5. Fire fighting measures	13. Disposal considerations
6. Accidental release measures	14. Transport information
7. Handling and storage	15. Regulatory information
8. Exposure controls and personal protection	16. Other information

The information must be in English and the safety data sheet must be kept up to date. Where any revisions are made, the word 'Revision' must be marked on the data sheet and every person who has been supplied with the substance within the preceding 12 months must be provided with the revised safety data sheet.

Where dangerous substances are supplied to members of the public, e.g. from retail premises, the supplier is not required to provide a safety data sheet although they must still provide sufficient information to enable the user to take the necessary measures as regards health and safety.

The supplier must also ensure that receptacles containing dangerous substances are labelled with key information including the supplier contact details, the

359

substance name and relevant indications of danger with their corresponding symbols, risk and safety phrases. CHIP also specifies the methods, location, colours and sizes of labels for receptacles.

15.12.6 Retention of data – regulation 12

Information used for the classification, labelling, child resistant fastenings and for preparing safety data sheets

for dangerous preparations must be maintained by the first person who was responsible for supplying it for at least 3 years from the date the dangerous preparation was last supplied.

15.12.7 Summary

The flow chart below provides a summary of CHIP:

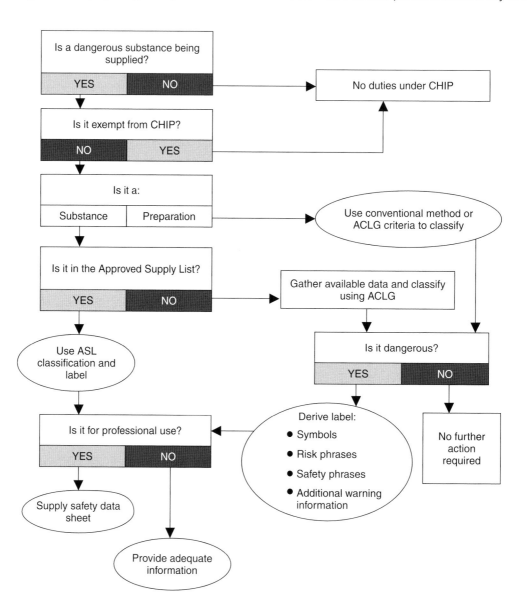

The Confined Spaces Regulations 1997 (SI 1713)

The purpose of the Confined Spaces Regulations 1997 (CS Regs) is to prevent people being exposed to risks to their health and safety from such work activities.

The regulations place an absolute duty to comply with the detailed requirements on both employers for work carried out by their employees and on the self-employed for their own work. In the case of other persons working under the control of the employer or the self-employed person, the duty to comply is qualified by the standard so far as is reasonably practicable.

15.3.1 Definitions

The following definitions are detailed in the CS Regs:

➤ Confined space – any space, including any chamber, tank, vat, silo, pit, trench, pipe, sewer, flue, well or other similar space in which, by virtue of its enclosed nature, there arises a reasonably foreseeable specified risk

➤ Specified risk – any risk of:
  ➤ Serious injury to any person at work arising from a fire or explosion
  ➤ The loss of consciousness of any person at work arising from an increase in body temperature
  ➤ The loss of consciousness of any person at work arising from gas, fume, vapour or lack of oxygen
  ➤ The drowning of any person at work arising from an increase in the level of liquid
  ➤ The asphyxiation of any person at work arising from a free flowing solid or the inability to reach a respirable environment due to entrapment by a free flowing solid.

15.13.2 Preventing the need for entry – regulation 4

This regulation requires that entry to a confined space to carry out work for any purpose is prevented unless it is not reasonably practicable to achieve that purpose without such entry.

Where the work cannot be carried out other than by entry, so far as is reasonably practicable, it must be carried out in accordance with a safe system of work which renders the work safe and without risks to health.

The supporting Approved Code of Practice refers to the need to carry out a risk assessment (see the Management of Health and Safety at Work Regulations 1999) for work in confined spaces, the findings of which would form the basis of a safe system for such work. In designing a safe system, the ACoP further states that priority be given to eliminating the sources of danger before deciding what precautions are needed for entry.

In assessing the risks that may arise, the ACoP states that the factors detailed below should be assessed.

Factors to be assessed	
Factor	Example
The general conditions of the confined spaces	Previous contents, residues, contamination, oxygen deficiency/enrichment, physical dimensions
Hazards arising directly from the work to be undertaken in the confined space	Cleaning chemicals that may be used, sources of ignition, e.g. hot work or sparks from tools to be used
The need to isolate the confined space to prevent dangers from outside	Entry of substances from outside by inadvertent operation of machinery, automatic delivery lines or external processes generating fumes, etc.
The requirement for emergency rescue arrangements	Possible emergencies should be anticipated and the equipment/facilities needed for a rescue identified

15.13.3 Safe system of work

The precise precautions required in a safe system of work will depend on the nature of the confined space and the risk assessment findings although the main elements to consider in this are detailed below:

Elements to consider in a safe system of work			
Supervision	Ventilation	PPE and RPE	Lighting
Competence for the work	Removal of residues	Portable gas cylinders and internal combustion engines	Static electricity
Communications	Isolation from gases, liquids and other free flowing materials	Gas supplied by pipes and hoses	Smoking
Testing/ monitoring the atmosphere	Isolation from mechanical and electrical equipment	Access and egress	Emergencies and rescue
Gas purging	Selection and use of suitable equipment	Fire prevention	Limited working time

15.13.4 Training

The ACoP states that in complying with the requirements of the HSW Act to provide training, the following areas would need to be covered:

➤ An awareness of the CS Regs, especially the need to avoid entry

➤ An understanding of the work to be undertaken, hazards and necessary precautions

➤ An understanding of safe systems of work, in particular, permits to work

➤ How emergencies arise, the need to follow prepared emergency arrangements and the dangers of not doing so.

15.13.5   Emergency arrangements – regulation 5

This regulation requires that entry to a confined space to carry out work is prevented unless suitable and sufficient arrangements have been prepared for the rescue of persons in the event of an emergency.

These arrangements must:

➤ Reduce, so far as is reasonably practicable, the risks to the health and safety of any person required to put the arrangements into operation and

➤ Include the provision and maintenance of necessary equipment for resuscitation procedures to be carried out

➤ It is not reasonably practicable to achieve that purpose without such entry.

The rescue arrangements should include consideration of the factors detailed below:

Factors to be considered in the emergency arrangements	
Rescue and resuscitation equipment	Control of plant
Raising the alarm and rescue	First aid
Safeguarding the rescuers	Public emergency services
Fire safety	Training

15.14   The Construction (Design and Management) Regulations 2007 (SI 320)

The Construction (Design and Management) Regulations 2007 (CDM) are aimed at the design and management aspects of a construction project together with securing the safety of those undertaking or affected by construction operations. The regulations impose duties on the 'key players' involved:

➤ The client: any person for whom a project is carried out

➤ The designer: any person who prepares a design

➤ Coordinator: a competent person appointed by the Client

➤ Principal contractor: a competent person appointed by the Client

➤ Other contractors: all other contractors involved with construction work.

There is nothing to stop the same person filling all of the key roles as long as they are competent and adequately resourced to do so. For example, clients could carry out their own design work, appoint themselves as coordinator and principal contractor and carry out the work themselves.

The regulations are broken down into three parts with an additional five Schedules the last of which are revocations, etc. from previous legislation.

15.14.1   Part 1 – Introduction which includes the application

CDM application:

➤ Applies to all construction work

➤ Notifiable projects to (HSE) involving more than 30 days work on site or more than 500 person days have a greater level of control due to the complexity and risks involved.

15.14.2   Part 2 – Duties

The legislation clearly states that only competent persons can work or be involved with work under the CDM Regulations, with the exception of those being subject to competent supervision.

Under the regulations there is the duty for all involved with the work to cooperate with all other parties and provide information regarding health and safety to other parties involved.

Client duties

The client is responsible for:

➤ Making arrangements to ensure that adequate resources are available to ensure health and safety is implemented and maintained during the project

➤ Ensuring that reviews and revisions of all health and safety arrangements are undertaken throughout the project

➤ Appointing a competent and adequately resourced coordinator (Notifiable projects)

➤ Appointing a competent and adequately resourced principal contractor (Notifiable projects)

➤ Acting as the coordinator and principal contractor until such time as each has been appointed

➤ Providing relevant health and safety information about existing structures and/or the site which is to be developed to the coordinator for inclusion in the pre-tender health and safety plan

> Allowing sufficient time within the project for design and construction work to be carried out safely and properly
> Ensuring that construction work on a project only starts when there is a suitably developed construction phase health and safety plan
> Making the health and safety file available for any future construction work and ensuring that it is kept up to date and handed over to the new owner when a sale takes place.

The client may appoint another person to act as their agent, provided that the person is competent to carry out the duties of the client on their behalf. Where an agent is appointed, the client must ensure that HSE is notified of the appointment and the notification must be supported by a written declaration with the following information:

> Name of declared client and statement signed by or on behalf of the declared client that they will act as client for the duration of the project
> The address where documents may be served on that person
> The address of the construction site to which the appointment relates.

Coordinator functions

The coordinator is assigned functions and not responsibilities (as these rest with the client), the key proportion of which is to coordinate health and safety arrangements prior to the project starting, where changes are required and post-construction works by preparing the health and safety file.

The coordinator functions include:

> Advising and assisting the client in undertaking the measures needed to comply with the regulations including determining the adequacy of the resources
> Notifying the HSE of the construction project where applicable
> Advising on the suitability and compatibility of designs and on any need for modification
> Coordinating design work, planning and other preparation
> Liaising with the principal contractor in relation to any design or change to a design requiring a review of the construction phase plan, during the construction phase
> Providing relevant health and safety information to every person designing the structure, the principal contractor; and every contractor who has been or is likely to be appointed by the client
> Preparing, where none exists, and otherwise reviewing and updating the health and safety file.

Designer duties

Designers play a key role in construction projects – although contractors have to manage risks on site, designers can often eliminate or reduce them at source. The specific duties of the designer are to:

> Advise clients of their duties
> Take positive account of health and safety hazards during design considerations
> Apply the principles of prevention and protection during the design phase to eliminate, reduce or control hazards
> Consider measures that will protect all workers if neither avoidance nor reduction to a safe level is possible
> Make health and safety information available regarding risks that cannot be designed out
> Cooperate with the coordinator and any other designers involved.

Principal contractor duties

The key duties of the principal contractor are to:

> Plan, manage and monitor health and safety during the construction phase of the project
> Ensure that Schedule 2 is complied with (see below)
> Coordinate the activities of all contractors in terms of legal requirements
> Ensure that every contractor complies with their duties under other health and safety legislation
> Draw up health and safety site rules
> Ensure cooperation between contractors and their employees to enable them to comply with their statutory duties
> Ensure that every contractor and employee complies with the construction phase plan
> Ensure that only authorised people are allowed on the construction site during the project
> Ensure, so far as is reasonably practicable, that all persons concerned with the project are suitably trained and kept informed of the health and safety requirements
> Ensure a copy of the notification of the project is displayed on site
> Provide information to the coordinator, as necessary, for inclusion in the construction phase file
> Provide suitable induction training and further information and training as required
> Establish a consultation process with contractors working on site.

15.14.3   Health and safety file

The coordinator is responsible for ensuring that a health and safety file is compiled and handed to the client on completion of the project. The client must ensure that the file is kept safe and is available for any interested party in the future.

The purpose of the health and safety file is to provide all of the relevant details regarding the structure to anyone in the future who may be carrying out further construction work, repairs, maintenance or refurbishment to the structure, and should include:

> A brief description of the work carried out
> Residual hazards and how they have been dealt with (e.g. surveys or other information concerning asbestos, contaminated land, water bearing strata, buried services, etc.)
> Key structural principles incorporated in the design of the structure (e.g. bracing and sources of substantial stored energy) and safe working loads of floors and roofs, particularly where these may preclude placing scaffolding or heavy machinery
> Any hazards associated with materials used (e.g. hazardous substances, lead-based paint and special coatings which should not be burnt off)
> Information regarding the removal or dismantling of installed plant and equipment
> Health and safety information about equipment provided for cleaning or maintaining the structure
> The nature, location and marking of significant services including fire fighting services
> Information and as-built drawings of the structure, its plant and equipment (e.g. the means of safe access to and from service voids, fire doors and compartmentation).

15.14.4   Schedules

The key duty holders with responsibility for complying with the schedules are:

1. Employers whose employees are carrying out construction work
2. Self-employed persons carrying out construction work
3. Persons who control the way in which construction work is carried out by a person at work
4. Employees carrying out construction work and other persons at work.

In addition, every person at work is required to cooperate with the duty holders to enable them to comply with the requirements of CDM. Where persons at work are under the control of an identified duty holder, they are required to report any defects that may endanger people's health and safety to the duty holder.

Schedule 1 – Particulars to be notified to the Executive

Schedule 1 lays down the information that must be provided to the HSE as part of the notification of a project.

1. Date of notification
2. Address of site
3. Site Local Authority location
4. Description of Project
5–7. Contact details
8. Start Date
9–10. Duration of planning and construction
11–12. No. of people on site
13–14. Details of contractors and designers
15. Signed declaration that the Client is aware of his duties under CDM

Schedule 2 – Welfare facilities

Welfare facilities must be provided on site or made available at readily accessible places. This includes:

1. Sanitary conveniences
2. Washing facilities, including showers as necessary
3. An adequate supply of drinking water
4. Accommodation for clothing
5. Changing facilities where necessary
6. Rest facilities.

Schedule 3 – Safe place of work

Duty holders must ensure, so far as is reasonably practicable, that:

1. Every place of work has a suitable and sufficient safe access and egress which must be both without risks to health and safety and properly maintained
2. Every place of work is safe and without risks to health for persons at work
3. Suitable and sufficient steps are taken to prevent persons gaining access to places that do not comply with CDM (other than for the purpose of making the place safe and provided that all practicable steps are taken during this work to ensure their safety)
4. Every place of work has sufficient working space for those working or likely to work there.

15.14.5   Good order

Construction sites must be kept in good order, so far as is reasonably practicable, and places of work must be kept reasonably clean. The perimeter of sites should be identified with signs where necessary and the extent of the site should be readily identifiable.

Timber or other materials with projecting nails must not be either used in any work or allowed to remain in any place where they may be a danger to any person.

15.14.6 Stability of structures

All practicable steps must be taken to prevent danger to any person from the accidental collapse of any new or existing structure which may either become unstable or be in a temporary state of weakness or instability due to construction work.

15.14.7 Demolition or dismantling

Suitable and sufficient steps must be taken to ensure that demolition or dismantling of a structure or part of any structure which gives rise to a risk of danger to any person is planned and carried out under the supervision of a competent person and in such a manner as to prevent, so far as is practicable, such danger.

15.14.8 Explosives

Explosive charges must only be used or fired if suitable and sufficient steps have been taken to ensure that no person is exposed to a risk of injury from either the explosion or by projected/flying material caused by the explosion.

15.14.9 Excavations

This part of the schedule requires that:

1. All practicable steps are taken to prevent danger to persons from accidental collapse of excavations
2. Suitable and sufficient steps are taken to prevent, so far as is reasonably practicable, any person from being trapped or buried by falling or dislodged materials
3. Suitable and sufficient equipment is provided to support excavations and is installed, altered and dismantled under the supervision of a competent person
4. Suitable and sufficient steps are taken to prevent any person, vehicle, plant and equipment, earth or other materials from falling into any excavation
5. No materials, vehicles, plant and equipment are placed or moved near to any excavation where this would be likely to cause the excavation to collapse
6. No excavation work is carried out unless suitable and sufficient steps have been taken to prevent any risk of injury from underground cables or services.

15.14.10 Cofferdams and caissons

These must be suitably designed, constructed and maintained and be of sufficient strength and capacity for their purpose. The construction, installation, alteration and dismantling must be under the supervision of a competent person.

15.14.11 Reports of inspections

Places of work including scaffolds, excavations, cofferdams and caissons must be inspected by a competent person before persons first use the place of work and at scheduled intervals to determine that they are safe.

Reports of the inspections required under the schedules must be prepared by the competent person who carried out the inspection before the end of the working period and a copy provided to the duty holder within 24 hours. A copy of the report must also be retained on site.

15.14.12 Energy distribution installations

Where there is a risk of contact with power supply cables the power supply should be in the first instance directed away from the area or isolated; where this cannot be undertaken, barriers, signs and rerouting of vehicle movements should be considered.

15.14.13 Prevention of drowning

Where any person is liable to fall into water or other liquid with a risk of drowning, suitable and sufficient steps must be taken to prevent, so far as is reasonably practicable, a person from falling; and to minimise risk of drowning if people fall. Suitable rescue equipment must also be provided and maintained.

Where persons are conveyed to work by water, suitable and sufficient steps must be taken to ensure their safe transport and the vessel used to convey them must be suitable, properly constructed and maintained. The vessel must be under the control of a competent person and not be overcrowded or overloaded.

15.14.14 Traffic routes

Construction sites must be organised to ensure that, so far as is reasonably practicable, pedestrians and vehicles can move safely and without risks to health.

In organising traffic routes, the following requirements must be satisfied:

1. Traffic routes must be suitable for those using them, sufficient in number, in suitable positions and of sufficient size
2. Pedestrians or vehicles must be able to use the traffic route without causing danger to others near it
3. Any doors or gates used by pedestrians which lead onto a vehicle traffic route must enable the pedestrians to see any approaching vehicles from a safe position
4. There must be sufficient separation between vehicles and pedestrians where reasonably practicable

or there must be other means to protect pedestrians and effective warning arrangements

5. Loading bays must have at least one exit for the exclusive use of pedestrians
6. Where it is unsafe for pedestrians to use a gate intended for vehicles, a clearly marked pedestrian gate must be provided in the immediate vicinity and it must be kept clear of obstructions
7. Vehicles must not use a traffic route unless it is, so far as is reasonably practicable, free from obstruction and permits sufficient clearance
8. Traffic routes are indicated with suitable signage.

15.14.15 Vehicles

Suitable and sufficient steps must be taken to:

1. Prevent or control the unintended movement of any vehicle
2. Ensure that the person in control of the vehicle gives a warning to any person who may be endangered by the vehicle movement, e.g. while reversing.

In addition, this regulation requires that:

1. Vehicles must be operated or towed in a safe manner and be loaded to enable this
2. No person rides in an unsafe place on any vehicle
3. No person remains on a vehicle during loading or unloading of loose materials (other than in a safe place)
4. Where vehicles are used for excavating or handling materials, suitable and sufficient precautions must be taken to prevent the vehicle from falling into any excavations or pits or water or from overrunning embankments.

15.14.16 Prevention of risks from fire, etc.

Suitable and sufficient precautions shall be taken to prevent, so far as is reasonably practicable, the risk of injury to any persons from:

1. Fire or explosion
2. Flooding
3. Any substance liable to cause asphyxiation.

15.14.17 Emergency procedures

Where necessary in the interests of the health and safety of any person on site, suitable arrangements must be prepared and implemented to deal with any foreseeable emergency. These arrangements should include procedures for any necessary evacuation of the site and

should take account of the factors listed under regulation 19 above.

Suitable and sufficient steps should be taken to ensure that every person covered by these arrangements is familiar with them and the arrangements must be tested by putting them into practice at suitable intervals.

15.14.18 Emergency routes and exits

A sufficient number of suitable emergency routes and exits must be provided to enable any person to reach a place of safety quickly in the event of danger.

The emergency route or exit should:

1. Lead as directly as possible to an identified safe area
2. Be kept clear and free from obstruction
3. Be provided with emergency lighting where necessary so that it may be used at any time
4. Be indicated by suitable signs.

In determining the number of emergency routes and exits, the following factors should be considered:

1. The type of work being carried out on site
2. The characteristics and size of the site
3. The number and locations of places of work on the site
4. The plant and equipment being used
5. The number of persons likely to be present on the site at any one time
6. The physical and chemical properties of any substances or materials likely to be on site.

15.14.19 Fire detection and fire fighting

Suitable and sufficient fire fighting equipment, fire detectors and alarm systems must be provided on site and in suitable locations. In determining this, the duty holder must take account of the factors listed in emergency routes and exits above.

Any fire fighting equipment, fire detector or alarm system must be properly maintained and subject to examination and testing to ensure that it remains effective. Where fire fighting equipment is not designed to come into use automatically, it must also be readily accessible.

Every person at work on site must, so far as is reasonably practicable, be instructed in the correct use of any fire fighting equipment which they may be required to use.

Where a work activity on site may give rise to a particular risk of fire, the work must only be carried out if the person is suitably trained to prevent that risk, so far as is reasonably practicable. Fire fighting equipment must also be indicated by suitable signs.

15.14.20 Fresh air

Suitable and sufficient steps must be taken to ensure, so far as is reasonably practicable, that every workplace has a sufficient supply of fresh or purified air and any plant used for this purpose must be fitted with a visible or audible device to provide a warning if it fails.

15.14.21 Temperature and weather protection

Suitable and sufficient steps must be taken to ensure, so far as is reasonably practicable, that the temperature in any indoor place of work is reasonable during working hours. Every place of work outdoors should be arranged to ensure, so far as is reasonably practicable, that it provides protection from adverse weather (taking account of any protective clothing or equipment provided for use).

15.14.22 Lighting

Suitable and sufficient lighting should be provided in every place of work and traffic route. Wherever reasonably practicable, this should be by natural light. The colour of artificial lighting should not affect or change the perception of any safety signs or signals and secondary lighting should be provided where there would be a risk to health and safety from the failure of primary artificial lighting.

15.15 The Control of Major Accident Hazards Regulations 1999 (SI 743)

These regulations shall apply to an establishment where dangerous substances listed in the following tables are present in the quantities shown. The regulations explicitly do not apply to:

➤ The transport of those substances including temporary storage and vehicles
➤ The transport of those substances in a pipeline or pumping station
➤ An establishment which is under the control of:
   ➤ The Secretary of State for the purposes of the Ministry of Defence
   ➤ The service authorities of a visiting force
➤ Substances that create a hazard from ionizing radiation on a licensed site
➤ Extractive industries concerned with exploration for, and the exploitation of, minerals
➤ Waste landfill sites.
➤ These regulations shall not apply in Northern Ireland.

Table 15.3 shows categories of substances and preparations not specifically named in Table 15.2. Substances and preparations shall be classified for the purposes of this Schedule according to regulation 5 of the Chemicals (Hazard Information and Packaging for Supply) Regulations 2002.

In the case of substances and preparations with properties giving rise to more than one classification, for the purposes of these regulations the lowest thresholds shall apply.

15.15.1 Regulation 4 – General duty

Every operator of a site that comes within the scope of the COMAH must take all measures necessary to prevent major accidents and limit their consequences to persons and the environment.

15.15.2 Regulation 5 – Major accident prevention policy

Operators must prepare a document setting the policy relating to the prevention of major accidents. The policy must:

➤ Guarantee a high level of protection for persons and the environment by appropriate means, structures and management systems
➤ Include sufficient particulars to demonstrate that the operator has established a safety management system
➤ Be kept up to date and revised when establishment or installation is changed
➤ Implemented by the operator.

15.15.3 Regulations 7–8 – Safety report

Prior to the start of construction of an establishment which will come within the scope of COMAH, the operator must produce a report containing information relating to the prevention of major accident hazards and the limitation of their consequences from the site to the competent authority (normally the health and safety executive).

The operator must also provide any further information that the competent authority may request.

Where a safety report has been sent to the competent authority the operator must review it:

➤ At least every 5 years
➤ When there is a change of circumstances or information affecting the plan
➤ Whenever the operator makes a change to the safety management system.

When an operator proposes to modify his site or installation he must, before making any changes, review, and where necessary revise, the safety report prepared in

Table 15.2 Criteria for sites that come within the scope of the Act

Column 1 Dangerous substances	Column 2 Quantity in tonnes	Column 3
Ammonium nitrate (as described in Note 1 of this Part)	350	2500
Ammonium nitrate (as described in Note 2 of this Part)	1250	5000
Arsenic pentoxide, arsenic (V) acid and/or salts	1	2
Arsenic trioxide, arsenious (III) acid and/or salts	0.1	0.1
Bromine	20	100
Chlorine	10	25
Nickel compounds in inhalable powder form (nickel monoxide, nickel dioxide, nickel sulphide, trinickel disulphide, dinickel trioxide)	1	1
Ethyleneimine	10	20
Fluorine	10	20
Formaldehyde (concentration >90%)	5	50
Hydrogen	5	50
Hydrogen chloride (liquefied gas)	25	250
Lead alkyls	5	50
Liquefied extremely flammable gases (including LPG) and natural gas (whether liquefied or not)	50	200
Acetylene	5	50
Ethylene oxide	5	50
Propylene oxide	5	50
Methanol	500	5000
4,4-Methylenebis (2-chloraniline) and/or salts, in powder form	0.01	0.01
Methylisocyanate	0.15	0.15
Oxygen	200	2000
Toluene diisocyanate	10	100
Carbonyl dichloride (phosgene)	0.3	0.75
Arsenic trihydride (arsine)	0.2	1
Phosphorus trihydride (phosphine)	0.2	1
Sulphur dichloride	1	1
Sulphur trioxide	15	75
Polychlorodibenzofurans and polychlorodibenzodioxins (including TCDD), calculated in TCDD equivalent	0.001	0.001
The following carcinogens:		
4-Aminobiphenyl and/or its salts, benzidine and/or salts, bis(chloromethyl) ether, chloromethyl methyl ether, dimethylcarbamoyl chloride, dimethylnitrosomine, hexamethylphosphoric triamide, 2-naphthylamine and/or salts, 1,3-propanesultone and 4-nitrodiphenyl	0.001	0.001
Automotive petrol and other petroleum spirits	5000	50 000

Table 15.3 General criteria for sites to come within the scope of the regulations

Column 1	Column 2	Column 3
Categories of dangerous substances	Quantity in tonnes	
1. VERY TOXIC	5	20
2. TOXIC	50	200
3. OXIDISING	50	200
4. EXPLOSIVE (where the substance or preparation falls within the definition given in Note 2(a))	50	200
5. EXPLOSIVE (where the substance or preparation falls within the definition given in Note 2(b))	10	50
6. FLAMMABLE (where the substance or preparation falls within the definition given in Note 3(a))	5000	50 000
7a. HIGHLY FLAMMABLE (where the substance or preparation falls within the definition given in Note 3(b)(i))	50	200
7b. HIGHLY FLAMMABLE liquids (where the substance or preparation falls within the definition given in Note 3(b)(ii))	5000	50 000
8. EXTREMELY FLAMMABLE (where the substance or preparation falls within the definition given in Note 3(c))	10	50
9. DANGEROUS FOR THE ENVIRONMENT in combination with risk phrases:		
(i) R50: 'Very toxic to aquatic organisms'	200	500
(ii) R51: 'Toxic to aquatic organisms'; and	500	2000
(iii) R53: 'May cause long-term adverse effects in the aquatic environment'		
10. ANY CLASSIFICATION not covered by those given above in combination with risk phrases:	100	500
(i) R14: 'Reacts violently with water' (including R14/15)		
(ii) R29: 'In contact with water, liberates toxic gas'	50	200

respect of the establishment, installation, process or dangerous substances as the case may be; and inform the competent authority of the details of the revision.

15.15.4 Regulation 9 – On-site emergency plan

All operators of COMAH sites must prepare an emergency plan in consultation with:

> Employees
> The competent authority
> The emergency services
> The local health authority
> The local authority.

The plan must be written and should reflect the following principles. It must include:

> The organisational structure, responsibilities, practices, procedures, processes and resources for

determining and implementing the major accident prevention policy

> The roles and responsibilities of personnel involved in the management of major hazards at all levels in the organisation
> Training needs of such personnel and the provision of the training so identified
> Identification and evaluation of major hazards
> Procedures and instructions for safe operation, including maintenance of plant, processes, equipment and temporary stoppages
> Arrangements for managing change
> Procedures to identify foreseeable emergencies by systematic analysis and to prepare, test and review emergency plans to respond to such emergencies
> Procedures for the ongoing assessment of compliance with the objectives set by the operator's major accident prevention policy and safety management system

➤ Procedures for reporting major accidents or near misses, particularly those involving failure of protective measures, and their investigation and follow-up on the basis of lessons learnt

➤ Procedures for periodic systematic assessment of the major accident prevention policy and the effectiveness and suitability of the safety management system; the documented review of performance of the policy and safety management system and its updating by senior management.

15.15.5 Regulations 10–12 – Off-site emergency plan

The local authority, in whose area there is an establishment, shall prepare an emergency plan in respect of that establishment. The off-site emergency plan shall be prepared no later than 6 months (or such longer period, not exceeding 9 months, as the competent authority may agree in writing) after:

➤ The receipt by the local authority of a notice from the competent authority informing the local authority of the need to prepare an off-site emergency plan in respect of the establishment

➤ The time an on-site emergency plan is required to be prepared for the establishment pursuant to regulation 9.

Once the plan has been prepared it must be tested and reviewed every three years.

A person who has prepared an emergency plan must immediately implement the plan when either a major accident occurs or when an uncontrolled event occurs which could reasonably be expected to lead to a major accident.

A local authority may charge the operator a fee for performing its functions under the Act.

15.15.6 Regulation 14 – Provision of information to the public

The operator of an establishment that comes within the scope of COMAH must provide any persons mentioned in the emergency plans with information relating to the safety measures provided on site and the role they may play in the event of an emergency.

The operator must also make salient information available to the public.

The operator must consult the local authority in whose area the COMAH site is situated but remains responsible for the accuracy, completeness and form of the information he supplies.

The information that is provided by the operator must be reviewed if there is a relevant change of circumstances or new information or in any case within three years.

15.15.7 Regulation 15 – Provision of information to competent authority

When requested to, site operators must provide sufficient information to demonstrate that they have taken all measures necessary to comply with these regulations, and the information shall be so provided within such period as the competent authority specifies in the request.

15.15.8 Regulation 16 – Provision of information to other establishments

The competent authority shall, using the information received from operators of sites where the likelihood or consequences of a major accident may be increased because of the location and proximity of other COMAH sites, pass appropriate information to other sites including information relating to their major accident prevention policy documents, safety reports and on-site emergency plans.

15.16 The Control of Substances Hazardous to Health Regulations 2002 (SI 2677)

This section summarises the key requirements of the Control of Substances Hazardous to Health Regulations 2002 (COSHH) as amended by the COSHH (Amendment) Regulations 2004.

COSHH defines substances hazardous to health as a substance:

➤ Which is listed in Part 1 of the ASL (see previous section on CHIP)

➤ For which the HSC has approved a workplace exposure limit

➤ Which is a biological agent

➤ Which is a dust of any kind within specified concentrations in air

➤ Which because of its chemical or toxicological properties and the way it is used or is present in the workplace creates a risk to health and safety.

The term 'workplace exposure limit' (WEL) means the exposure limit approved by the HSC for that substance in relation to the specified reference period when calculated by a method approved by the HSC as contained in the HSE publication 'EH40 – Workplace Exposure Limits' as updated from time to time (currently EH40 2006).

COSHH does not apply:

➤ To lead or asbestos which are covered by other specific regulations

➤ Where the substance is hazardous to health solely by virtue of its radioactive, explosive or flammable properties or because it is at high or low temperature or a high pressure

➤ Where the substance posing the risk to health is administered to a person in the course of their medical treatment.

15.16.1 Duties under COSHH

With the exception of health surveillance, monitoring, information and training, the duties on the employer extend to other persons who may be affected by the work activity, whether they are at work or not. This could include visitors to the premises, suppliers/contractors and their employees, self-employed persons and members of the public.

15.16.2 Assessment of the risks to health – regulation 6

Employers must not carry out work which is liable to expose employees to any substance hazardous to health unless the employer has:

➤ Made a suitable and sufficient assessment of the risks to the employees' health and safety created by that work and the steps that need to be taken to comply with COSHH and

➤ Implemented the above steps.

The assessment must be reviewed regularly:

➤ If there is reason to suspect it is no longer valid

➤ If there has been a significant change in the work to which it relates

➤ If the results of any monitoring carried out in accordance with regulation 10 show it to be necessary.

Where the review identifies that changes to the risk assessment are required, the employer must make these changes.

Employers with five or more employees must record the significant findings of the assessment including the preventive and control measures as detailed under regulation 7.

In carrying out a suitable and sufficient assessment, the employer must consider the points detailed below:

Points to consider in a suitable and sufficient COSHH assessment	
The hazardous properties of the substance	The effect of preventive and control measures which have been or will be taken
Information from the supplier on health effects – see the SDS	The results of relevant health surveillance
The level, type and duration of exposure	The results of monitoring of exposure
The circumstances of the work including the amount of the substance involved	The risk presented by substances in combination where the work will involve exposure to more than one substance
Activities such as maintenance where there is the potential for a high level of exposure	The approved classification of any biological agent
Any relevant workplace exposure limits	Any additional information the employer may need to complete the assessment

15.16.3 Prevention and control of exposure – regulation 7

The employer is required to ensure that exposure of their employees to hazardous substances is either prevented or, where this is not reasonably practicable, adequately controlled.

In determining adequate control, the employer should give preference to avoiding the use of hazardous substances by substituting them with a safer alternative which either eliminates or reduces the risk to health and safety, so far as is reasonably practicable.

Where prevention of exposure is not reasonably practicable, the employer is required to apply the following protection measures in order of priority:

➤ The design and use of appropriate work processes, systems and engineering controls and the provision and use of suitable work equipment and materials

➤ The control of exposure at source, including ventilation systems and appropriate organisational measures

➤ The provision of suitable personal protective equipment (PPE), where adequate control cannot be achieved. PPE should be provided in addition to the above measures.

The protection measures should include:

➤ Arrangements for safe handling, storage and transport

➤ The adoption of suitable maintenance procedures

➤ Reducing to the minimum required for the work:
  ➤ The number of employees subject to exposure
  ➤ The level and duration of exposure
  ➤ The quantity of substances present at the workplace

➤ The control of the working environment including general ventilation

➤ Appropriate hygiene measures including adequate washing facilities.

Control of exposure should only be treated as adequate if:

➤ The principles of good practice as detailed in Schedule 2A are applied

➤ Any workplace exposure limit approved for that substance is not exceeded

➤ For substances listed in Schedule 1 or with R45, R46 or R49 risk phrases (carcinogens) and those listed as asthmagens or with R42 or R42/43 risk phrases, exposure is reduced to as low a level as is reasonably practicable.

Schedule 2A – Principles of good practice

Design and operate processes and activities to minimise emission, release and spread of substances hazardous to health

Take into account all relevant routes of exposure when developing control measures – inhalation, skin absorption and ingestion

Control exposure by measures that are proportionate to the health risk

Choose the most effective and reliable control options which minimise the escape and spread of substances hazardous to health

Where adequate control of exposure cannot be achieved by other means, provide, in combination with other control measures, suitable PPE

Check and review regularly all elements of control measures for their continuing effectiveness

Inform and train all employees on the hazards and risks from the substances with which they work and the use of control measures developed to minimise the risks

Ensure that the introduction of control measures does not increase the overall risk to health and safety

---

Note: Additional precautions are detailed for carcinogens and biological agents

Where PPE is provided, it must be suitable for the purpose and:

➤ Comply with the Personal Protective Equipment Regulations 1992 as amended or

➤ In the case of respiratory protective equipment (RPE), be of a type approved by or conform to a standard approved by HSE.

15.16.4 Use of control measures – regulation 8

Employers must take all reasonable steps to ensure that all preventive and control measures, including PPE, are applied or used properly.

Employees are required to:

➤ Make full and proper use of any preventive or control measure and

➤ Take all reasonable steps to return the measures to any accommodation after use and report any defects to their employer immediately.

15.16.5 Maintenance, examination and testing of control measures – regulation 9

Employers are required to ensure that any control measures they provide:

➤ In the case of plant and equipment, including engineering controls and PPE, are maintained in an efficient state, in efficient working order, in good repair and in a clean condition

➤ In the case of safe systems of work and supervision of any other measure, it is reviewed at suitable intervals and revised as necessary

➤ In the case of engineering controls, thorough examination and testing of these, are carried out:
  ➤ At least once every 14 months for local exhaust ventilation
  ➤ As specified in Schedule 4
  ➤ In other cases, at suitable intervals

➤ Ensure the thorough examination and appropriate testing of RPE at suitable intervals (with the exception of disposable RPE)

➤ Keep a record of all examinations and tests for at least 5 years from the date of the test

➤ Ensure that PPE is:
  ➤ Properly stored in well-defined spaces
  ➤ Checked at suitable intervals
  ➤ Repaired or replaced before further use when it is discovered to be defective

➤ Ensure that any PPE that is contaminated by a substance hazardous to health is either decontaminated and cleaned or destroyed if necessary.

Any PPE which may be contaminated must be removed on leaving the working area and kept apart from uncontaminated clothing and equipment.

15.16.6 Monitoring exposure at the workplace – regulation 10

Where the risk assessment reveals that it is necessary either for the maintenance of adequate control of employee exposure to substances hazardous to health, or for protecting their employees' health, the employer must ensure that employee exposure to hazardous substances is monitored.

Where the employer can demonstrate compliance with regulation 7 by another evaluation method, this requirement does not apply.

The monitoring must take place:

➤ At regular intervals
➤ When any changes occur which may affect that exposure
➤ As specified in Schedule 5.

The employer must make and maintain a suitable record of any monitoring for:

➤ At least 40 years where the monitoring relates to the personal exposure of identifiable employees
➤ At least 5 years in any other case.

15.16.7 Health surveillance – regulation 11

Where it is appropriate for the protection of their employees who are liable to be exposed to a substance hazardous to health, the employer must ensure employees are under health surveillance.

The circumstances where health surveillance would be deemed appropriate are:

➤ The employee is exposed to a substance as specified in Schedule 6 and there is a reasonable likelihood that an identifiable disease or adverse health effect will result from that exposure
➤ The employee exposure is such that:
  ➤ An identifiable disease or adverse health effect may be related to the exposure or
  ➤ There is a reasonable likelihood that the disease or effect may occur under the particular conditions of work and
  ➤ There are valid techniques for detecting indications of the disease or effect (these must be of low risk to the employee).

The employer must:

➤ Keep a health record for the surveillance containing particular details approved by HSE for at least 40 years from the date of the last entry
➤ Allow the employee access to their personal health records
➤ Provide HSE with copies as required
➤ Notify HSE if they cease to trade and make the records available to them.

Where the employee is exposed to a substance as specified in Schedule 6, the health surveillance should include medical surveillance under the supervision of a relevant doctor. This should be at intervals of not more than 12 months. Where a relevant doctor certifies that an employee should either not be engaged in work which exposes them to a particular substance, or that medical surveillance should continue after exposure has ceased, the employer must ensure this.

Employees for whom health surveillance is necessary are required to present themselves for this and provide the relevant doctor with any necessary health information.

Where health surveillance reveals that an employee has an identifiable disease or adverse health effect which a relevant doctor considers to be the result of exposure to a substance hazardous to health, the employer is required to:

➤ Ensure that the employee is informed by a suitably qualified person who provides them with information and advice regarding health surveillance
➤ Review their risk assessment
➤ Review any controls and preventive measures, taking account of any advice by a relevant doctor
➤ Consider assigning the employee to alternative work
➤ Provide for a review of any other employee's health who has been similarly exposed where recommended.

A relevant doctor may, for the purposes of carrying out their function, inspect any workplace or relevant record and the employer must permit this.

15.16.8 Information, instruction and training – regulation 12

Where their employees are likely to be exposed to substances hazardous to health, the employer is required to provide them with suitable and sufficient information, instruction and training including:

➤ Details of the substances hazardous to health to which the employee is liable to be exposed:
  ➤ The names of those substances and the risk they present to health
  ➤ Any relevant workplace exposure limit
  ➤ Access to any relevant safety data sheet
  ➤ Other legal relevant requirements
➤ The significant findings of the risk assessment
➤ The appropriate precautions and actions to be taken by the employee to safeguard themselves and others
➤ The results of any monitoring (where a WEL is approved, the employee should also be informed immediately if the results show that this has been exceeded)
➤ The collective results of any health surveillance (these results must not be identifiable to an individual).

373

15.16.9 *Arrangements to deal with accidents, incidents and emergencies – regulation 13*

Employers are required to prepare procedures including first aid facilities and relevant drills to protect the health of their employees from an accident, incident or emergency related to substances hazardous to health.

Information on the procedures must be made available and displayed at the workplace as appropriate. This information should also be made available to the relevant emergency services to enable them to prepare their own response.

In the event of such an accident, incident or emergency, the employer must:

➤ Take immediate steps to mitigate the effects, restore the situation to normal and inform those who may be affected
➤ Ensure that only persons who are essential for carrying out repairs are permitted in the affected area and then only if they are provided with appropriate PPE and any other necessary specialised equipment.

Note: The above requirements are without prejudice to the requirements for serious and imminent danger procedures under the Management of Health and Safety at Work Regulations 1999 (see later section).

15.17 The Dangerous Substances and Explosive Atmospheres Regulations 2002 (SI 2776)

These regulations impose requirements for the purpose of eliminating or reducing risks to safety from fire, explosion or other events arising from the hazardous properties of a 'dangerous substance' in connection with work.

Dangerous substances are defined as:

(a) substances or preparations which are explosive, oxidising, extremely flammable, highly flammable or flammable, whether or not that substance or preparation is classified under the Chemicals (Hazard Information and Packaging for Supply) Regulations 2002
(b) substances or preparations which due to their physico-chemical or chemical properties and the way they are used or are present at the workplace create a risk or
(c) any dust, whether in the form of solid particles or fibrous materials or otherwise, which can form an explosive mixture with air or an explosive atmosphere.

Dangerous substances include: petrol; liquefied petroleum gas (LPG); paints; varnishes; solvents; and dusts

which when mixed with air could cause an explosive atmosphere, for example dusts from milling and sanding operations. Dangerous substances can be found, in varying quantities, in most workplaces.

An explosive atmosphere is an accumulation of gas, mist, dust or vapour, mixed with air, which has the potential to catch fire or explode. An explosive atmosphere does not always result in an explosion, but if it caught fire the flames would quickly travel through it and if this happened in a confined space (e.g. in plant or equipment) the rapid spread of the flames or rise in pressure could also cause an explosion.

15.17.1 *Employers general duties – regulation 4*

The employer must:

➤ Carry out a risk assessment of any work activities involving dangerous substances
➤ Provide measures to eliminate or reduce risks as far as is reasonably practicable
➤ Provide equipment and procedures to deal with accidents and emergencies
➤ Provide information and training to employees
➤ Classify places where explosive atmospheres may occur into zones and mark the zones where necessary.

15.17.2 *Employers' specific duties – regulation 5*

Employers are required to carry out suitable and sufficient assessment of the risks to their employees where a dangerous substance is or may be present at the workplace. This is also a requirement under the Management of Health and Safety at Work Regulations (regulation 3):

> 'Risk' is defined as meaning 'the likelihood of a person's safety being affected by harmful physical effects being caused to him from fire, explosion or other events arising from the hazardous properties of a dangerous substance in connection with work and also the extent of that harm'.

Employers are required by these regulations to eliminate or reduce risk so far as is reasonably practicable. Where risk is not eliminated, employers are required, so far as is reasonably practicable and consistent with the risk assessment, to apply measures to control risks and mitigate any detrimental effects (regulation 6(3)).

The risk assessment shall include consideration as to:

(a) the hazardous properties of the substance;
(b) information on safety provided by the supplier, including information contained in any relevant safety data sheet;

(c) the circumstances of the work including:
   (i) the work processes and substances used and their possible interactions;
   (ii) the amount of the substance involved;
   (iii) where the work will involve more than one dangerous substance, the risk presented by such substances in combination; and
   (iv) the arrangements for the safe handling, storage and transport of dangerous substances and of waste containing dangerous substances;

(d) activities, such as maintenance, where there is the potential for a high level of risk;

(e) the effect of measures which have been or will be taken pursuant to the regulations;

(f) the likelihood that an explosive atmosphere will occur and its persistence;

(g) the likelihood that ignition sources, including electrostatic discharges, will be present and become active and effective;

(h) the scale of the anticipated effects of a fire or an explosion;

(i) any places which are or can be connected via openings to places in which explosive atmospheres may occur; and

(j) such additional safety information as the employer may need in order to complete the risk assessment.

Where two or more employers share a workplace where an explosive atmosphere may occur, the employer responsible for the workplace is to coordinate the implementation of the measures required by these regulations (regulation 11).

15.17.3 Recording the assessment

When employers have five or more employees a record must be made of the significant findings of the assessment as soon as is practicable after that assessment is made, including:

> Measures taken to eliminate or reduce risk
> Sufficient information to show that the workplace and work equipment will be safe from risk of fire and explosion during operation and maintenance
> Details of any areas zoned as hazardous due to the likely presence of explosive atmospheres
> Where employers share a workplace, any special measures to ensure coordination of safety requirements to protect workers from explosive atmospheres.

15.17.4 Control measures

Employers are required to ensure that the safety risks from dangerous substances are eliminated or, where this

is not reasonably practicable, to take rneasures to control risks and to reduce the harmful effects of any fire, explosion or similar event, so far as is reasonably practicable (mitigation).

15.17.5 Mitigation measures

An employer should apply mitigation measures, which are consistent with the risk assessment and appropriate to the nature of the activity or operation. These can include:

> Preventing fires and explosions from spreading to other plant and equipment or to other parts of the workplace
> Reducing the numbers of employees exposed to a minimum
> In the case of process plant, providing plant and equipment that can safely contain or suppress an explosion, or vent it to a safe place.

Measures taken to achieve the elimination or the reduction of risk should take into account the design, construction and maintenance of the workplace and work processes, including all relevant plant, equipment, control and protection systems.

15.17.6 Workplace classification and marking – regulation 7

Places at the workplace where explosive atmospheres may occur must be classified as hazardous or non-hazardous and hazardous places must be classified into zones on the basis of the frequency and duration of the occurrence of an explosive atmosphere (regulation 7(1) and Schedule 2).

Equipment and protective systems in hazardous places must comply with the requirements of Schedule 3 (regulation 7(2)) and, where necessary, hazardous places must be marked with signs at their points of entry in accordance with Schedule 4 (regulation 7(3)).

15.17.7 Regulation 10

Containers and pipes used at work for dangerous substances must, where not already marked, clearly identify their contents:

> Where necessary, areas classified into zones are marked with a specified 'EX' sign at their points of entry
> Where employees work in zoned areas they are provided with appropriate clothing that does not create a risk of an electrostatic discharge igniting the explosive atmosphere

➤ Before coming into operation for the first time, areas where hazardous explosive atmospheres may be present are confirmed as being safe (verified) by a person (or organisation) competent in the field of explosion protection. The person carrying out the verification must be competent to consider the particular risks at the workplace and the adequacy of control and other measures put in place.

15.17.8 Emergency arrangements – regulation 8

Employers are required to make arrangements for dealing with accidents, incidents and emergencies.

The requirements of DSEAR build on existing requirements under the Management Regulations. An employer will need to supplement those existing arrangements if it is assessed that an accident, incident or emergency could arise, for example a fire, or a significant spillage, because of the quantity of dangerous substances at the workplace. In these circumstances an employer must arrange:

➤ Suitable warning (including visual and audible alarms) and communication systems
➤ Escape facilities, if required by the risk assessment
➤ Emergency procedures to be followed in the event of an emergency
➤ Equipment and clothing for essential personnel dealing with the incident and
➤ Practice drills.

The scale and nature of the emergency arrangements should be proportionate to the level of risk.

Information on emergency procedures should be made available to employees and emergency services should be contacted to advise them that the information is available (and provide the emergency services with any information they consider necessary).

15.17.9 Information and training – regulation 9

Employers will also need to provide employees with precautionary information, instruction and training where a dangerous substance is present at the workplace.

Employers are required to provide employees (and their representatives), and other people at the workplace who may be at risk, with suitable information, instruction and training on precautions and actions they need to take to safeguard themselves and others, including:

➤ Name of the substances in use and risks they present
➤ Access to any relevant safety data sheet
➤ Details of legislation that applies to the hazardous properties of those substances and
➤ The significant findings of the risk assessment.

Employers need only provide information, instruction and training to non-employees where it is required to ensure their safety. Where it is provided, it should be in proportion to the level and type of risk.

15.18 The Electricity at Work Regulations 1989 (SI 0635)

The purpose of the Electricity at Work Regulations 1989 (EAWR) is to require precautions to be taken against the risk of death or personal injury from electricity in work activities and are supported by guidance.

Definitions in EAWR	
Term	**Meaning**
Conductor	A conductor or electrical energy
Danger	Risk of injury
Electrical equipment	Anything used, intended to be used or installed for use to generate, provide, transmit, transform, rectify, convert, conduct, distribute, control, store, measure or use electrical energy
Injury	Death or personal injury from electric shock, electric burn, electrical explosion or arcing or from fire or explosion initiated by electrical energy where such death or injury is associated with the 'use' of electrical energy
System	Any electrical system in which all the electrical equipment is or may be electrically connected to a common source of electrical energy and includes such source and such equipment

15.18.1 Duties – regulation 3

The key duty holders under EAWR are employers and the self-employed, both of whom are required to comply with the requirements for matters within their control.

Employees at work also have duties to:

➤ Cooperate with their employer to enable them to comply with their requirements and
➤ Comply with the EAWR for matters under their control (this duty is equivalent to the employer duty above and reflects the level of responsibility of employees in electrical trades and professions).

15.18.2 Systems, work activities and protective equipment – regulation 4

This general duty requires that:

➤ All systems are at all times of such construction as to prevent, so far as is reasonably practicable, danger
➤ All systems are maintained so as to prevent, so far as is reasonably practicable, danger
➤ Every work activity including operation, use and maintenance of and near a system is carried out so as not to give rise, so far as is reasonably practicable, to danger
➤ Any equipment provided for the purposes of protecting people at work on or near electrical equipment is suitable for use, suitably maintained and properly used.

The remainder of EAWR provides further detail on these general duties.

15.18.3 Strength and capability of equipment – regulation 5

Electrical equipment must not be put to use where its strength and capability may be exceeded in such a way as may give rise to danger. In satisfying this requirement, consideration should be given to the expected current flow including load and transient currents, fault currents and pulses of current.

15.18.4 Adverse or hazardous environments – regulation 6

Where it is reasonably foreseeable that electrical equipment may be exposed to adverse or hazardous environments, it must be of such construction or as necessary protected to, so far as is reasonably practicable, prevent danger.
This includes exposure to:

➤ Mechanical damage
➤ The effects of the weather, natural hazards, temperature or pressure
➤ The effects of wet, dirty, dusty or corrosive conditions
➤ Any flammable or explosive substance, including dusts, vapours or gases.

In selecting suitable equipment for such hazardous environments, reference should be made to the Index of Protection Ratings (IP Ratings) which is an internationally recognised system for classifying the degree of protection provided.

15.18.5 Insulation, protection and placing of conductors – regulation 7

All conductors in a system which may give rise to danger must be either:

➤ Suitably covered with insulating material and protected to prevent, so far as is reasonably practicable, danger or
➤ Have such precautions taken as will prevent, so far as is reasonably practicable, danger.

Note: The precautions that may be required for this range from the safe placement of cables as overhead cables to formal safe systems of work for high voltage switching where the risk of danger cannot be avoided.

15.18.6 Earthing or other suitable precautions – regulation 8

Precautions should be taken, either by earthing or by other suitable means, to prevent danger arising when a conductor other than a circuit conductor which may reasonably foreseeably become charged as a result of either the use of or fault in the system, becomes so charged.
A conductor will be regarded as earthed when it is connected to a general mass of earth by conductors of sufficient strength and capability to discharge electrical energy to earth.

15.18.7 Integrity of referenced conductors – regulation 9

If a circuit conductor is connected to earth or to any other reference point, nothing which could give rise to danger by breaking the electrical continuity or introducing high impedance is allowed in the conductor unless suitable precautions are taken against that danger.

15.18.8 Connections – regulation 10

All joints and connections in a system must be both mechanically and electrically suitable for use. This includes all plugs, sockets, joints and terminals and must also take account of both the environmental conditions and the location of the connections.

15.18.9 Means for protecting from excess current – regulation 11

Efficient means shall be provided and suitable located to protect every part of a system from excess current as necessary to prevent danger. This covers short circuits and overloads and the usual means of protection is by fuses or circuit breakers.

15.18.10 Means for cutting off the supply and isolation – regulation 12

Suitable means should be available for cutting off the supply to any electrical equipment and the isolation of any electrical equipment where necessary to prevent danger. In this context, isolation means the disconnection and separation of the electrical equipment from every source of electrical energy in such a way that this is secure and inadvertent reconnection is prevented.

Electrical equipment that is in itself a source of electrical energy (e.g. the windings of generators) is exempt from this requirement although precautions must still be taken to prevent danger.

15.18.11 Precautions of equipment made dead – regulation 13

Adequate precautions must be taken to prevent electrical equipment, which has been made dead to prevent danger while work is carried out, from becoming electrically charged during that work if this would result in danger.

This is the system of switch-off, isolate, lock-off and prove dead, all of which should be included in a formal safe system of work or a permit to work.

15.18.12 Work on or near live conductors – regulation 14

The EAWR are very clear on the issue of live work – it should be avoided wherever possible.

This regulation illustrates this principle clearly and requires that no person should work on or so near any live conductor where danger may arise unless:

➤ It is unreasonable in all circumstances for it to be dead and
➤ It is reasonable in all circumstances for the person to be at work on or near it while it is live and
➤ Suitable precautions are taken to prevent injury, including the provision of protective equipment.

15.18.13 Working space, access and lighting – regulation 15

Adequate working space, means of access and lighting should be provided at all electrical equipment on which or near which work is being done which may give rise to danger for the purpose of preventing injury.

Where this work involves work on or near live conductors, this would include having sufficient space for persons to pass one another safely and to pull back from the conductors.

15.18.14 Persons to be competent to prevent danger – regulation 16

This regulation requires that no person is engaged in any work activity where technical knowledge or experience is necessary to prevent danger or injury unless the person has such knowledge or experience or is under appropriate supervision for the activity.

The supporting guidance to this regulation lists the scope of this 'technical knowledge or experience', including the ability to recognise at all times whether it is safe for work to continue.

15.19 The Gas Appliances (Safety) Regulations 1992 (SI 0711)

These regulations require that no person should supply any gas appliance or fitting unless the appliance conforms to EC type examination and satisfies essential safety requirements listed in Schedule 3 of the regulations. Examples of some of these requirements are listed below:

➤ Appliances are designed and built as to operate safely and present no danger to persons, domestic animals or property
➤ When placed on the market, they are:
　➤ Accompanied by technical instructions intended for the installer
　➤ Accompanied by instructions for use and servicing, intended for the user
　➤ Bear appropriate warning notices, which must also appear on the packaging
　➤ The instructions and warning notices must be in English or in the languages of the Member States of destination
➤ The instructions for use and servicing intended for the user must contain all the information required for safe use, and must in particular draw the user's attention to any restrictions on use
➤ The warning notices on the appliance and its packaging must clearly state the type of gas used, the gas supply pressure and any restrictions on use, in particular the restriction whereby the appliance must be installed only in areas where there is sufficient ventilation
➤ Materials must be appropriate for their intended purpose and must withstand the technical, chemical and thermal conditions to which they will foreseeably be subjected
➤ The properties of materials that are important for safety must be guaranteed by the manufacturer or the supplier of the appliance

- Appliances must be so designed and constructed as to minimise the risk of explosion in the event of a fire of external origin
- Appliances must be so designed and constructed that failure of a safety, controlling or regulating device may not lead to an unsafe situation
- Appliances must be so constructed that gas release during ignition and reignition and after flame extinction is limited in order to avoid a dangerous accumulation of unburned gas in the appliance
- Appliances intended to be used in indoor spaces and rooms must be fitted with a special device which avoids a dangerous accumulation of unburned gas in such spaces or rooms
- Appliances must be so constructed that, when used normally, there will be no accidental release of combustion products
- Parts of the appliance which are intended to be placed in close proximity to the floor or other surfaces must not reach temperatures which present a danger in the surrounding area.

15.20 The Gas Safety (Installation and Use) Regulations 1998 (SI 2451)

These regulations focus on the safe installation and use of gas appliances. Some of the key requirements are summarised below:

- Installation is only carried out by competent persons (they must be registered with CORGI – the Council for Registered Gas Installers)
- General safety precautions are met during installation including:
  - Ensuring the gas fitting is of good construction and sound material
  - Working in such a way as to prevent a dangerous release of gas
  - Not searching for gas leaks with a source of ignition
  - Not installing a gas storage vessel in a basement or cellar
- Gas fittings are properly supported and protected
- Alterations to existing fittings do not adversely affect the safety of the fitting
- An emergency control (isolation valve) is provided before gas is first supplied in any premises
- Gas appliances are not installed unless they can be used without danger to people
- Flued domestic gas appliances must be connected by fixed rigid pipes
- Installers of gas appliances must not leave the appliance connected unless it can be used or is sealed off from the gas supply

- Flues are suitable and in a safe position
- Only room sealed appliances must be installed in bath/shower rooms
- Gas fire or heaters intended to be used in sleeping accommodation must be room sealed appliances or incorporate a safety device to shut the device down
- Appliances are only used if they are safe
- Landlords must ensure that gas appliances are tested at 12 monthly intervals.

15.21 The Health and Safety (Consultation with Employees) Regulations 1996 (SI 1513)

The Health and Safety (Consultation with Employees) Regulations 1996 (HSCWE Regs) only apply where employees are not represented by union appointed safety representatives (see the Safety Representatives and Safety Committees Regulations 1977).

15.21.1   Duty to consult – regulation 3

The employer is required to consult with their employees in good time on matters relating to their health and safety and, in particular:

- The introduction of any measure which may affect their health and safety
- The employer's arrangements for appointing or nominating competent persons under the Management of Health and Safety at Work Regulations 1999 (see later)
- Any information the employer is required to provide under other relevant legislation
- The planning and organisation of any training the employer is required to provide under other relevant legislation
- The health and safety consequences of the introduction of any new technologies into the workplace.

Note: The supporting guidance makes it clear that consultation involves both listening to the views of employees and taking account of these before any decision is made.

15.21.2   Persons to be consulted – regulation 4

The employer must either:

- Consult with their employers directly or
- Consult with one or more persons from a group of employees where the group has elected that person or persons to represent them. (Persons elected

in this way are termed representatives of employee safety (ROES).)

Where the employer consults with ROES, they must inform their employees of both the name of the ROES and the group of employees the ROES represents.

The employer should not consult with ROES where:

➤ The ROES notifies the employer that they do not intend to represent the group of employees
➤ The ROES ceases to be employed in the group of employees they represented
➤ The election period has expired
➤ The ROES becomes incapacitated from carrying out their functions.

Where the employer discontinues consultation with ROES, they must inform their employees in the group concerned. Where the employer has previously consulted with ROES and then decides to consult with their employees directly, they must inform both the ROES and their employees of their decision.

15.21.3 Duty to provide information – regulation 5

Where the employer consults employees directly, they must also make information available to the employees as necessary to enable the employees to participate fully and effectively in the consultation.

The guidance to the HSCWE Regs indicates that this would include information on:

➤ The hazards and associated risks to health and safety that may arise as a result of their activities
➤ The preventive and protective measures required for these
➤ The action employees should take if they encounter hazards and associated risks.

In the case of consultation with ROES, the employer must also provide the ROES with any records required under the Reporting of Injuries, Diseases and Dangerous Occurrences Regulations 1995 (RIDDOR – see later).

The employer is not required to provide information which:

➤ Would be against the interests of national security
➤ Relates specifically to an individual without their consent
➤ Could cause substantial injury to the employers' undertaking
➤ Is obtained by the employer for bringing, prosecuting or defending any legal proceedings
➤ Is not related to health and safety.

15.21.4 Functions of ROES – regulation 6

Where the employer consults with ROES, they have the following functions:

➤ To make representations to the employer on potential hazards and dangerous occurrences at the workplace which either affect or could affect the group of employees the ROES represents
➤ To make representations to the employer on general matters affecting the health and safety of the group of employees the ROES represents, and in particular on the matters detailed in regulation 3 above
➤ To represent the group of employees in consultation at the workplace with appointed enforcing authority inspectors (either HSE or EHOs).

Note: The above are not legal duties on ROES and they do not absolve the ROES of their duties as employees under relevant legislation. It is becoming increasingly common for employers to involve ROES in safety committees, critical incident investigation and workplace inspections in the same way that union appointed safety representatives are and such good practice should be encouraged wherever possible.

15.21.5 Training, time off and facilities for ROES – regulation 7

Where the employer consults with ROES, they must:

➤ Ensure that the ROES is provided with training in respect of their functions and meet any reasonable costs associated with this training
➤ Permit the ROES sufficient time off with pay during the ROES working hours as necessary for the ROES to perform their functions or attend training
➤ Provide the ROES with other facilities and assistance to enable the ROES to perform their functions.

The employer must also permit a candidate standing for election as an ROES to take reasonable time off with pay during their working hours to fulfil their functions as a candidate.

15.21.6 Civil liability – regulation 9

Civil action for a breach of the HSCWE Regs is precluded although the ROES may complain to the employment tribunal where their employer either fails to permit them to take time off as detailed under regulation 7 or has failed to pay them while the ROES was either performing or attending training in their functions.

15.22 The Health and Safety (First-Aid) Regulations 1981 (SI 0917)

These regulations place a general duty on the employer to make adequate provision of their employees if they are injured or become ill at work and are supported by an Approved Code of Practice.

15.22.1 Definition of first aid

The regulations define 'first aid' as:

> Treatment for the purposes of preserving life and minimising the consequence of injury and ill health until help from a medical practitioner or nurse is obtained.

and

> Treatment of minor injuries which would otherwise receive no treatment or which do not require treatment by a medical practitioner or nurse.

15.22.2 Provision for first aid – regulation 3

The employer is required to provide, or ensure that there are provided:

➢ Adequate and appropriate equipment and facilities to enable first aid to be rendered to their employees if they are injured or become ill at work
➢ Adequate and appropriate numbers of suitable persons for rendering first aid.

The ACoP provides extensive guidance on how to determine the adequacy of first aid equipment and facilities which should largely be determined by the nature of the activities performed as opposed to the numbers employed.

As a minimum, there should be:

➢ At least one first aid kit which:
  ➢ Is placed in a clearly identifiable and readily accessible location
  ➢ Contains a sufficient quantity of suitable first aid materials and nothing else (guidance on suitable contents is detailed in the ACoP)
  ➢ Is checked regularly so that the contents are not used past their expiry date and are replenished as soon as possible after use
  ➢ Is stored in a clearly identifiable and suitable box designed to protect the contents from damp and dust and labelled with a white cross on a green background (see later section on the Safety Signs and Safety Signals Regulations 1996)
➢ Soap and water and disposable drying materials for first aid purposes (individually wrapped moist cleansing wipes which are not impregnated with alcohol may be used instead)
➢ Disposable plastic gloves, aprons and suitable protective equipment should also be provided as should plastic disposable bags for soiled or used first aid dressings. These should be disposed of in sealed bags as necessary.

In deciding on the provision of 'suitable persons', the employer should again concentrate on the nature of the work as opposed to the numbers employed. As a minimum there must be an 'appointed person' at all times when employees are at work and in low risk situations, e.g. offices, an employer will need one 'suitable person' for every 50 employees. Supporting guidance from the HSE provides a simple chart to assist in determining the level of provision required.

A 'suitable person' is defined as a first-aider who holds a current first aid certificate issued by an HSE approved organisation and any other person who has undergone training and obtained qualifications approved by the HSE. Certificates of qualification are currently valid for 3 years after which a refresher course of a minimum of 12 contact hours (normally over 2 days) followed by an examination is required for recertification.

An 'appointed person' is someone who is provided by the employer to take charge of the situation if a serious injury or illness occurs in the absence of a first-aider. They are not required to be first aid trained although it is good practice for them to be trained in emergency first aid (minimum 4 hours' contact time repeated every 3 years).

15.22.3 Information for employees – regulation 4

The employer must inform their employees of the first aid arrangements including the location of equipment, facilities and personnel. This should be provided at induction and reinforced should they move to another part of the establishment.

15.22.4 Self-employed – regulation 5

The self-employed are required to provide such first aid equipment as is adequate for themselves. This may be as simple as a small travel kit. Often, the self-employed will share facilities with other employers although this should be agreed and not assumed.

15.23 The Health and Safety (Information for Employees) Regulations 1989 (SI 682)

These regulations require the employer to either display an approved poster (Health and Safety – What you should know) or to issue their employees with an approved leaflet covering the same information. These approved documents contain information on the general requirements of health and safety law and have additional spaces for the employer to include the following:

➤ The address of their local enforcing authority and the address of the Employment Medical Advisory Service (EMAS)
➤ Details of either trade union or other safety representatives
➤ Details of the competent persons appointed to assist the employer to comply with their duties under health and safety legislation.

15.23.1 Regulation 4 – Provision of poster or leaflet

All employers must provide for each of his employees an approved poster which is displayed in a readable condition:

➤ At a place which is reasonably accessible to the employee while he is at work and
➤ In such a position in that place as to be easily seen and read by that employee or
➤ Give to the employee the approved leaflet.

Every employer must be treated as having complied with this paragraph if he gives to the employee the approved leaflet as soon as is reasonably practicable.

Where the form of poster or leaflet is revised on or before the date the revision takes effect an employer shall provide his employees with revised posters of leaflets.

15.23.2 Regulation 5 – Provision of further information

The information contained within the poster or leaflet must include:

➤ The name of the enforcing authority for the premises where the poster is displayed and the address of the office of that authority for the area in which those premises are situated and
➤ The address of the office of the employment medical advisory service for the area in which those premises are situated.

15.23.3 Regulation 6 – Exemption certificates

The Health and Safety Executive may, by a certificate in writing, exempt any person from all or any of the requirements imposed by these regulations. However, the Executive cannot grant any exemption unless it is satisfied that the health, safety and welfare of persons who are likely to be affected by the exemption will not be prejudiced in consequence of it.

15.23.4 Regulation 7 – Defence

A defence will be successful if an employer demonstrates that he took all reasonable precautions and exercised all due diligence to avoid the commission of that offence.

15.24 The Management of Health and Safety at Work Regulations 1999 (SI 3242)

The Management of Health and Safety at Work Regulations 1999 (MHSWR) originally came into force in 1993 and following significant amendment, most recently in 2003, the original MHSWR have been revised and republished with a new ACoP.

15.24.1 Risk assessment – regulation 3

This requires that the employer carry out a 'suitable and sufficient' assessment of the risks to the health and safety of both their employees at work and other persons who may be affected by their undertaking. The same duty is placed on the self-employed in relation to risks to their own health and safety and to the health and safety of other persons who may be affected by their undertaking.

The ACoP and supporting text provides detailed guidance on 'suitable and sufficient' which is summarised below:

Suitable and sufficient

The assessment should identify the risks arising from or in connection with work and should be proportionate to these risks. Once the risks have been assessed, insignificant risks can usually be ignored as can the risks arising from routine activities of life in general.

The duty holder must take reasonable steps to help identify the risks, e.g. by looking at appropriate information sources including legislation, trade manuals, seeking competent advice and following industry best practice.

The assessment should be appropriate to the nature of the work and should identify the period of time for which it is likely to remain valid.

The assessment must take account of the particular risks to the health and safety of young persons and to new or expectant mothers or their babies.

The assessment must be reviewed if:

➤ There is reason to suspect that it is no longer valid or
➤ There has been a significant change in the matters to which it relates.

Where the review identifies that changes are required, these must be made.

Where the employer has five or more employees, they are required to record the significant findings or their assessment including any group of employees who are especially at risk.

Note: The purpose of the risk assessment is to enable the duty holder to identify the measures they need to take to comply with other relevant legislation.

15.24.2 Principles of prevention – regulation 4

In implementing any preventive and protective measures, the employer is required to follow the principles of prevention specified in Schedule 1:

Schedule 1: General principles of prevention
Avoid risks to health and safety
Evaluate risks which cannot be avoided
Combat the risks at source
Adapt the work to the individual
Adapt to technological processes
Replace the dangerous with the non- or less dangerous
Develop a coherent overall prevention policy
Give priority to collective protective measures over individual protective measures
Give appropriate instructions to employees

15.24.3 Health and safety arrangements – regulation 5

Employers are required to have arrangements in place to cover the effective planning, organisation, control, monitoring and review of the preventive and protective measures. Where the employer has five or more employees they must record these arrangements.

15.24.4 Health surveillance – regulation 6

Employers should provide their employees with such health surveillance as is appropriate having regard to the risks to their health and safety.

The risk assessment should identify the need for health surveillance under specific legislation, e.g. COSHH (see earlier section). The employer should also provide health surveillance where:

➤ There is an identifiable disease or adverse health condition related to the work concerned
➤ Valid techniques are available to detect indications of the disease or condition
➤ There is a reasonable likelihood that the disease or condition may occur under the particular work circumstances
➤ Surveillance is likely to further protect the health and safety of the employees concerned.

15.24.5 Health and safety assistance – regulation 7

Every employer is required to appoint one or more competent persons to assist them to comply with their legal obligations under both health and safety and fire safety legislation.

The employer must also ensure that:

➤ The number of persons they appoint and
➤ The time available for them and
➤ The means at their disposal

are adequate.

In judging their competence, the employer must take into account the training and experience or knowledge and other qualities the appointee has to properly assist the employer to comply.

Note: Industry best practice gives preference to the appointment of in-house competent persons as opposed to external consultants although they are commonly used to support appointees and for specialist advice.

15.24.6 Procedures for serious and imminent danger – regulation 8

The employer must:

➤ Establish procedures to be followed in the event of serious and imminent danger to persons at work in their undertaking
➤ Nominate a sufficient number of competent persons to implement evacuation procedures of those persons
➤ Ensure that none of their employees have access to an area the employer occupies where it is necessary to restrict access on health and safety grounds unless the employee has received adequate health and safety training.

The serious and imminent danger procedures should:

➤ Require any persons at work who are exposed to serious and imminent danger to be informed of the nature of the hazard and the steps taken or to be taken to protect them from it

➤ Enable the persons concerned to stop work and immediately proceed to a place of safety in the event of their exposure to serious, imminent and unavoidable danger

➤ Require the persons to be prevented from resuming work in any situation where there is still a serious and imminent danger.

15.24.7 Contact with emergency services – regulation 9

Employers are required to ensure that any necessary contacts with emergency services are arranged, particularly as regards first aid, emergency medical care and rescue work and the ACoP provides clear instructions to the employer on establishing procedures for this and what the procedure should cover.

15.24.8 Information for employees – regulation 10

Employers are required to provide their employees with comprehensible and relevant information on:

➤ The risks to their health and safety identified by the assessment

➤ The preventive and protective measures required

➤ The procedures for serious and imminent danger

➤ The identity of nominated competent persons responsible for implementing evacuation procedures

➤ The risks arising out of the undertaking of other employers sharing the workplace.

The information should be pitched appropriately taking account of the level of training, knowledge and experience of the employee and any language difficulties or disabilities.

Before employing a child (a person who is not over compulsory school age) the employer must provide a parent of the child with comprehensible and relevant information on:

➤ The risks to their health and safety identified by the assessment

➤ The preventive and protective measures required

➤ The risks arising out of the undertaking of other employers sharing the workplace.

The information can be provided verbally or through other organizations, e.g. the school in the case of work experience projects.

15.24.9 Cooperation and coordination – regulation 11

Where two or more employers share a workplace, each of them should:

➤ Cooperate with the other to enable them to comply with relevant fire and health and safety legislation

➤ Take all reasonable steps to coordinate the preventive and protective measures required by relevant fire and health and safety legislation

➤ Take all reasonable steps to inform the other employer of the risks to their employees' health and safety arising out of their undertaking.

15.24.10 Persons working in host employer's undertakings – regulation 12

Where employees from outside are working in a host employer's undertaking, the host employer must provide their employer with information on:

➤ The risks to their health and safety from the host employer's undertaking

➤ The preventive and protective measures taken in so far as they relate to that employee

➤ The identity of nominated competent persons responsible for implementing evacuation procedures.

The host employer must also ensure that any person working in their undertaking is provided with appropriate instructions and comprehensible information on the risks to their health and safety from the host employer's undertaking and the identity of nominated competent persons responsible for implementing evacuation procedures.

15.24.11 Capabilities and training – regulation 13

Employers are required to take account of the capabilities of their employees when entrusting tasks to them and should ensure that they are provided with adequate health and safety training on their being recruited and on their being exposed to new or increased risks because of:

➤ Their being transferred or given a change in responsibilities

➤ The introduction of new work equipment or a change in existing equipment

➤ The introduction of new technology

➤ The introduction of a new system or a change in a system of work.

This training must be repeated periodically where appropriate, adapted to take account of new or changed risks and take place during working hours.

15.24.12 Employees' duties – regulation 14

Employees are required to:

➤ Use any machinery, plant, equipment, dangerous substances, etc. in accordance with their training and instruction
➤ Inform their employer of any work situation which would represent a serious and imminent danger to health and safety and any matter representing a shortcoming in the employer's protection arrangements for health and safety.

15.24.13 Temporary workers – regulation 15

Employers are required to provide any person they employ on a fixed-term contract with comprehensible information on:

➤ Any special occupational qualification or skills required to carry out the work safely
➤ Any health surveillance required by relevant health and safety legislation.

This information must be provided before the employee commences their duties.

Employers must also provide this information to any person employed in an employment business who is to carry out work in the employer's undertaking. In addition, the employer must provide every person carrying out an employment business whose employees are to carry out work in the employer's undertaking with comprehensible information on any special occupational qualification or skills required to carry out the work safely and specific features of the job in so far as those features are likely to affect their health and safety.

15.24.14 Risk assessment of new or expectant mothers – regulations 16, 17 and 18

Where persons working in the undertaking include women of child bearing age and the work could involve risk by reason of her condition, risks to the health and safety of the new or expectant mother or to her baby, the employer must take account of this in their risk assessment (see regulation 3).

This requirement will only apply where the employee has notified the employer in writing that she is pregnant, has given birth within the last 6 months or is breastfeeding.

Where a new or expectant mother works at night and a certificate from a registered medical practitioner or midwife shows that it is necessary for reasons of her health and safety that she should not be at work, the employer should suspend her from work for as long as necessary for her health and safety.

15.24.15 Protection of young persons – regulation 19

Where employers employ young persons, they must ensure that they are protected from risks to their health and safety arising out of their lack of experience or the fact that they have not yet fully matured.

Young persons must not be employed for work:

➤ Which is beyond their physical or psychological capacity
➤ Involving harmful exposure to toxic or carcinogenic agents or those which cause heritable genetic damage or harm
➤ Involving harmful exposure to radiation
➤ Involving the risk of accidents which may not be recognised or avoided by young persons owing to their insufficient attention to safety or lack of experience or training
➤ Where there is a risk to health from extreme heat or cold, noise or vibration.

The above should be taken into account in the risk assessment which should be reviewed with the above in mind.

15.24.16 Provision as to liability – regulation 21

Employers are not able to use an act or default by an employee or by a competent person as defence in any criminal proceedings for a breach of MHSWR. This does not affect the established duty on the employee to take reasonable care and to cooperate with their employer.

15.24.17 Exclusion of civil liability – regulation 22

A breach of these regulations shall not confer a right of action in any civil proceedings except where the duty is owed to:

➤ Women of child bearing age and their work could involve risk to the new or expectant mother or to her baby
➤ Young persons.

15.25 The Personal Protective Equipment Regulations 1992 (SI 2966)

The Personal Protective Equipment Regulations 1992 (PPE Regs) define the term personal protective equipment (PPE) as:

all equipment (including clothing affording protection against the weather) which is intended to be worn or held by a person at work and which protects them against one or more risks to their health and safety, and any addition or accessory designed to meet that objective.

The following items are not included in this definition:

> Ordinary working clothing and uniform which do not specifically protect the health and safety of the user
> An offensive weapon used as self-defence or as a deterrent weapon
> Portable devices for detecting and signalling risks and nuisances
> PPE used for protection while travelling on the road
> Equipment used during the playing of competitive sports.

Where existing regulations (covering lead, ionising radiations, asbestos, hazardous substances, noise and head protection) apply, the PPE Regs do not apply with the exception of regulation 5 (compatibility) although the general principles covering selection and maintenance will still be of relevance.

In summary, the PPE Regs require:

> Employers to ensure that suitable PPE is provided to their employees who may be exposed to risks to their health and safety except where the risks have been adequately controlled by other means which are equally or more effective (the same duty applies to the self-employed in relation to risks to their own health and safety). In selecting suitable PPE, it must:
>> Be appropriate for the risks involved, the conditions where it is used and the period for which it is worn
>> Take account of ergonomic requirements, the health of the user and the characteristics of the user workstation
>> Be capable of fitting the wearer correctly
>> So far as is practicable, be effective at preventing or adequately controlling the risks involved without increasing the overall risk
>> Comply with UK legislation covering design and manufacture (CE marking)
>> Be for personal use only where necessary to ensure it is hygienic
> Where there is more than one risk to health and safety for which PPE is used, the employer (and self-employed person) must ensure that the PPE required is compatible and remains effective
> Before selecting PPE, the employer (and self-employed person) must ensure that an assessment is made to determine if the PPE they intend to provide is suitable. This assessment should follow on from the risk assessment carried out in accordance with the MHSW Regs and should include:

> An assessment of the health and safety risks which have not been avoided by other means
> The characteristics the PPE must have to be effective against both these risks and any created by the use of the PPE itself
> A comparison of the characteristics of the PPE available with the above characteristics
> An assessment of compatibility with other PPE in use
> A review of the assessment if there is reason to suspect it may no longer be valid or if there has been a significant change
> The employer (and self-employed person) must ensure that any PPE they provide is maintained (including replaced or cleaned as appropriate) in an efficient working state, in efficient working order and in good repair
> Where PPE is provided, the employer (and self-employed person) must ensure that appropriate accommodation is also provided for the PPE when it is not in use
> The employer must provide their employees with such information, instruction and training as is adequate and appropriate (including demonstrations in wearing the PPE) to enable the employees to know:
>> The risks which the PPE will avoid or limit
>> The purpose for and manner in which the PPE should be used
>> Any action the employee needs to take to ensure the PPE remains efficient and effective
> Employers must take reasonable steps to ensure that PPE they provide to their employees is properly used (the self-employed have the same duty in relation to PPE they provide for themselves)
> Employees are required to:
>> Use any PPE they are provided with as per their training and instruction
>> Return it to the accommodation provided for it after use
>> Report any loss or obvious defect in the PPE to their employer.

15.26 Provision and Use of Work Equipment Regulations 1998 (SI 2306)

The aim of the Provision and Use of Work Equipment Regulations 1998 (PUWER) is to ensure that equipment provided for use at work can be used safely and without risks to health.

PUWER also has specific requirements covering mobile work equipment and power presses. These are not covered in this summary.

The regulations may be divided into two groups:

- Management duties – covering selection, maintenance, inspection, conformity, information, instruction and training
- Physical aspects of equipment – covering guarding, control systems, emergency controls, stability, lighting and warnings.

Definitions in PUWER

Term	Meaning
Work equipment	Any machinery, appliance, apparatus, tool or installation for use at work
Use	Any activity involving work equipment including starting, stopping, programming, setting, transporting, repairing, modifying, maintaining, servicing and cleaning

PUWER applies to the following duty holders:

- Employers in respect of work equipment provided for use or used by their employees at work
- The self-employed for work equipment they use
- People who have control of work equipment.

15.26.1 Management duties – regulations 4 to 10

PUWER requires that:

- Work equipment is constructed or adapted so that it is suitable for the purpose for which it is provided
- Work equipment is selected having regard to the working conditions, the risks to health and safety in the premises or undertaking where the work equipment is used and any additional risks posed by the use of the work equipment
- Work equipment is used only for operations and under conditions for which it is suitable
- Work equipment is maintained in an efficient state, in efficient working order and in good repair (and that where any machinery has a log, it is kept up to date)
- Work equipment is inspected after installation and before first use (or after assembly at a new site or location) where the safety of the work equipment depends on the installation conditions (a record must be made of this inspection)
- Work equipment is inspected at suitable intervals where it is exposed to conditions causing deterioration liable to result in dangerous situations (a record must be made of this inspection)

- Work equipment is inspected each time that exceptional circumstances liable to jeopardise the safety of the work equipment have occurred (a record must be made of this inspection)
- Where the use of work equipment is likely to involve specific risk to health and safety, the use of the equipment must be restricted to those given the task of using it and any repairs, modifications, maintenance or servicing restricted to people specifically designated to perform those operations. These designated persons must also be given adequate training related to these operations
- People who use work equipment or who supervise or manage the use of work equipment have adequate health and safety information and, where appropriate, written instructions pertaining to the use of the work equipment
- In providing adequate information and instruction, it must be readily comprehensible to those concerned and the following should be included:
  - The conditions in which and the methods by which the work equipment may be used
  - Foreseeable abnormal situations and action to be taken should these occur
  - Any conclusions drawn from experience in suing the work equipment
- Persons who use, supervise or manage the use of work equipment receive adequate training for the purpose of health and safety including training in:
  - The methods which may be adopted when using the work equipment
  - Any risks which such use may entail
  - The precautions to be taken against these risks
- Work equipment is designed and constructed in compliance with any essential requirements detailed in EU Directives (see later section on the Supply of Machinery (Safety) Regulations 1992).

15.26.2 Physical requirements – regulations 11 to 24

PUWER requires that:

- Measures are taken to either:
  - Prevent access to any dangerous parts of machinery or to rotating stock bars or
  - To stop movement of dangerous parts or rotating stock bars before a person enters a danger zone
- The measures taken should consist of the provision of:
  - Fixed guards enclosing every dangerous part or rotating stock bar, where practicable or

387

- ➤ Other guards or protection devices, where practicable or
- ➤ Jigs, holders, push-sticks or similar protection appliances, where practicable or
- ➤ Information, instruction, training and supervision
- ➤ All guards and protection devices must:
  - ➤ Be suitable for the purpose
  - ➤ Be of good construction, sound material and adequate strength
  - ➤ Be maintained in an efficient state, in efficient working order and in good repair
  - ➤ Not give rise to any increased risk to health and safety
  - ➤ Not be easily bypassed or disabled
  - ➤ Be situated at sufficient distance from the danger zone
  - ➤ Not unduly restrict the view of the operating cycle of the machinery
  - ➤ Be constructed or adapted to allow operations necessary to fit or replace parts and for maintenance without having to dismantle the guard or protection device
- ➤ Measures (other than the provision of PPE or information, instruction, training or supervision, so far as is reasonably practicable) are taken to ensure that exposure of persons to the following specified hazards is either prevented or adequately controlled:
  - ➤ Articles or substances falling or being ejected from the work equipment
  - ➤ Rupture or disintegration of parts of work equipment
  - ➤ Work equipment catching fire or overheating
  - ➤ Unintended or premature discharge of any article or of any gas, dust, vapour or other substance produced, used or stored in the work equipment
  - ➤ Unintended or premature explosion of the work equipment or any materials or substance produced, used or stored in the work equipment
- ➤ Where work equipment, parts of work equipment and any article or substance produced, used or stored in the work equipment is at a very high or low temperature, it should have protection where appropriate to prevent injury to any person by burn, scald or sear
- ➤ Where appropriate, work equipment is provided with one or more controls (which require deliberate action to operate them) for:
  - ➤ Starting, including restarting after a stoppage
  - ➤ Controlling any change in speed or pressure or other operating condition where such a change in conditions results in a different or greater risk to health and safety
- ➤ Where appropriate, work equipment is provided with one or more readily accessible controls to bring

the work equipment to a safe condition in a safe manner. Where necessary for health or safety, the control should:

- ➤ Bring the work equipment to a complete stop
- ➤ Switch off all energy sources after stopping the functioning of the work equipment
- ➤ Operate in priority to any control which starts or changes the operating conditions of the work equipment
- ➤ Where appropriate, work equipment is provided with one or more readily accessible emergency stops which operate in priority to any of the controls mentioned above
- ➤ All controls must be clearly visible and identifiable (by appropriate marking where necessary) and be positioned so that no person operating the control is exposed to a risk to their health or safety
- ➤ Where appropriate, systems of work are effective in preventing any person being exposed to a risk to their health or safety when equipment is about to start
- ➤ Where appropriate, audible, visible or other suitable warnings are given whenever work equipment is about to start
- ➤ Control systems are safe and chosen taking account of expected failures, faults and constraints in the planned use of the work equipment
- ➤ Control systems do not create increased risk to health or safety and that any fault in or damage to any part of the control system or loss of supply of energy used by the work equipment cannot result in additional or increased risk to health or safety or impede the operation of stop and emergency stop controls
- ➤ Where appropriate, work equipment is provided with suitable means (which are clearly identifiable and readily accessible) to isolate it from all its sources of energy
- ➤ Appropriate measures are taken to ensure that reconnection to any energy source does not expose people to any risks to their health or safety
- ➤ Work equipment or any part of the work equipment is stabilised by clamping or otherwise for the purposes of health or safety
- ➤ Suitable and sufficient lighting is provided at any place where people use work equipment, taking account of the operations to be carried out
- ➤ Appropriate measures are taken to ensure that work equipment is:
  - ➤ Constructed or adapted to ensure that maintenance operations which involve a risk to health or safety can be carried out while the work equipment is shut down or
  - ➤ Can be carried out without exposing people to risk to their health or safety or

> Appropriate protection measures can be taken for the person carrying out the maintenance operation

> Where appropriate, work equipment is marked in a clearly visible manner with any marking appropriate for health and safety

> Where appropriate, work equipment incorporates any warnings or warning devices for reasons of health or safety. These must be unambiguous, easily perceived and easily understood.

Note: Any warnings or markings must comply with the Health and Safety (Safety Signs and Signals) Regulations 1996 (see later section).

PUWER also has specific requirements covering mobile work equipment and power presses. These are not covered in this summary.

15.27 The Regulatory Reform (Fire Safety) Order 2005 (SI 1541)

The Regulatory Reform (Fire Safety) Order (RRFSO) replaces or amends 118 pieces of fire-related legislation, the most significant being the repeal of the Fire Precautions Act 1971 and the revocation of Fire Precautions (Workplace) Regulations 1997(as amended). It develops and extends many of the concepts from the Fire Precautions (Workplace) Regulations 1997 and the Management of Health and Safety at Work Regulations 1999.

The RRFSO applies to the majority of premises and workplaces in the UK. Broadly it does not apply to dwellings, the underground parts of mines, offshore installations, anything that floats, flies or runs on wheels.

15.27.1 Definitions and meanings

Responsible person – this is the person who more than likely owns the premises or business or the person with control over the premises, business or activity. Where two or more responsible persons share responsibility (e.g. landlord/tenant, multiple occupied building or adjacent premises) the responsible persons must cooperate, share information and collaborate to provide fire safety measures.

Relevant person – this is anyone who is legally on the premises or anyone who is not on the premises but who may be affected, such as in nearby property.

The RRFSO specifically excludes fire fighters from this group of people.

Employee – this has a broad definition and may include subcontractors, self-employed and casual workers.

Competent person – anyone appointed by the responsible person and could be a company employee, fire safety manager, fire warden, etc. or an external consultant. But must be someone who has enough training and experience or knowledge and other qualities to undertake their role.

Inspector – person appointed by the enforcing authority (normally a fire officer from the local Fire and Rescue Service).

Enforcing authority – normally the Fire and Rescue Service but may be the HSE, MOD or local authority.

Enforcement – failing to comply with articles 8 through 21 and 38 may result in the serving of a variety of notices or on successful prosecution a fine or up to two years in jail.

Articles 8 through 21 cover all the main provisions of the RRFSO, from carrying out risk assessments to maintenance and training.

Fire risk assessment – the cornerstone of the RRFSO is the risk assessment. This must be reviewed regularly and if necessary amended. The risk assessment must be formally recorded if the responsible person employs five or more people, the premises are licensed or an alterations notice requires it.

Maintenance – all equipment provided for the purpose of fire safety or for the use or protection of fire fighters must be maintained and kept in good order.

Note: Reference to the 'Articles' of the RRFSO are made in the following paragraphs by the use of brackets, e.g. [4.1(b)].

15.27.2 Application

The RRFSO applies to the vast majority of workplaces and premises in the UK. There are notably very few exceptions. Workplaces and premises vary considerably in size and risk and as a consequence the RRFSO contains fewer concise words such as 'adequate', 'appropriate' and 'reasonable'. The following summarises the RRFSO in the context of a typical organisation for which you are responsible, or own the premises, business or operation, i.e. the responsible person.

15.27.3 The responsible person

As the responsible person you must ensure the safety from fire of your employees and any person who may legally come onto your premises, or any person not on the premises who may be affected (relevant persons) [8(a) and (b)]. Where you share responsibility with other responsible persons (e.g. adjacent premises, employers, tenant/landlord or multiple tenancy building) you must cooperate by sharing information and collaborating in providing measures. This also extends to premises in which an explosive atmosphere could occur [22].

The RRFSO places a duty on the responsible person (RP) and outlines all the measures that must be taken to ensure the safety of all the RP is directly or indirectly responsible for. At the same time it allows the enforcing authority to make sure that it is enacted and sets penalties if it is not.

It requires the RP to:

➢ Carry out a fire risk assessment
➢ Develop and produce a policy
➢ Develop procedures (particularly with regard to evacuation)
➢ Provide staff with information and training and carry out fire drills.

The responsible person must also provide and maintain:

➢ Clear means of escape
➢ Signs
➢ Notices
➢ Emergency lighting
➢ Fire detection and alarm and
➢ Fire fighting equipment.

The RRFSO contains the phrase '…The responsible person must…appoint one or more Competent Persons to assist him…'. In addition, the competent person must have '…sufficient training, experience and knowledge…'. Where the competent person is directly employed the responsible person must ensure that he or she is properly trained.

When there is no competent person within the responsible person's employment to assist in the preparation of a fire risk assessment an external consultant can be engaged, they must be provided with information to ensure that they can undertake their tasks.

15.27.4   Means of escape

The RRFSO requires that the responsible person provides a means of escape [4.(1)(b)] and ensures that they are available at all times [4.(1)(c)]. In particular the RRSFO requires that:

Escape routes are established and always available and lead as directly as possible to a place of safety. Doors must open easily and in the direction of escape, no sliding or revolving doors. They should be adequate in size and number and provided with emergency lighting and signs [14(1) and (2)].

Signs and notices must be provided that are appropriate and that:

➢ Give instruction to employees and other relevant persons [Schedule 1 Part 3(h)] of emergency procedures including fire action notices [19(1)(c) 4.(1)(f)]

➢ Indicate the position of extinguishers [13(1)(b)]
➢ Indicate emergency routes and exits. [14(2)(g)]

Fire detection and alarm system must be provided [13(1)(a), 4.(1)(e) and 15(2)(a) and (b)]. The type and coverage of the fire alarm would be subject to the requirements of the risk assessment.

Emergency lighting must be provided in escape routes [14(2)(h)].

Compartments and doors. The RP must take measures to ensure all fire resisting walls and doors are kept in good order, walls, ceilings and floors are not breached and fire doors have appropriate intumescent strips, cold smoke seals and closing devices [4.(1)(a)].

15.27.5   Fire fighting

The RP must provide appropriate fire fighting equipment [4.(1)(d), 13(1)(a), 13(1)(b) and 13(3)(a)]. In general this means portable extinguishers but may include hose reels, sprinklers and other types of fixed installations where appropriate.

15.27.6   Maintenance

All equipment installed for the purposes of fire safety must be maintained in good working order by either the responsible person or occupier [17(1) and (2)]. The person who undertakes the maintenance must be competent [18(1) through (8)]. You may need to provide evidence that they are third party accredited, e.g. ISO 9001 certified, LPCB approved, etc.

15.27.7   Training

All employees must be given adequate training, including 'action to be taken', when they commence employment and receive refresher training as appropriate. Further training would be required if there were any change that may affect fire safety, e.g. change of work, new systems or alteration to the building.

The training should be provided during normal working hours. [4.(1)(f)(i), 15(2)(a), 21(1) and (2) and Schedule 1 Part 3(h)]. The employer must carry out regular fire drills as part of the training [15(1)(a) and (b)].

15.27.8   Article 13/15 – Competent persons

A responsible person must appoint one or more competent persons to:

➢ Carry out fire fighting duties [13(3)(b)]
➢ Make contact with emergency services [13(3)(c)]
➢ Assist in evacuations [15(1)(b)].

A competent person can only be regarded as competent if they have appropriate training, experience, knowledge or other qualities [13(4) and 15(3)].

15.27.9   Articles 40–41 – Employee rights and responsibilities

Employees also have rights and responsibilities under the RRFSO broadly similar to those found in current health and safety legislation. The responsible person must consult employees on fire safety matters and provide information to them and ensure that they have been suitably trained to fulfil their role.

An employee must not act in a way that endangers themselves or others, they must also bring to the attention of the employer any shortcomings in the fire prevention and protection arrangements and cooperate with the employer. The responsible person cannot charge an employee for providing any fire safety measures that may be required under the RRFSO. An employee is entitled to recover his losses via civil proceedings if the responsible person fails to comply with the RRFSO.

Responsible persons must ensure that any person they employ (directly or indirectly) is provided with all information related to fire safety [19 and 20]. Employers must consult with employees with regard to fire safety issues [41(1) and (2)].

Employees cannot be charged for anything provided for, e.g. training or safety equipment [40]. However, if an employee suffers loss or damage as result of a responsible person failing to comply with the RRFSO the employee can recover their losses from the employer [39(2)].

Employees must take reasonable care of themselves and others and cooperate with their employer. They must also report any serious dangers and shortcomings in the fire safety arrangements [23(1)(a–c)].

15.27.10   Young persons

When considering employing a young person the responsible person must include particular aspects in his risk assessment [9(5)] such as their inexperience and lack of maturity, the type and extent of training, degree of exposure to risk, the activities and premises that they may be undertaking their work in [Schedule 1 Part].

As such it is therefore likely that you will include specific reference to these issues in your risk assessment.

15.27.11   Article 25 – Enforcement

The enforcing authority is usually the local Fire and Rescue Service [25(a)]; however, there are other enforcing authorities:

> HSE (Health and Safety Executive) – nuclear installations, ships under construction and building sites [25(b)]

> MOD (Ministry of Defence) Fire Service – military and defence establishments [25(c)]
> Local authority – sports grounds and stadia [25(d)].

15.27.12   Article 27 – Inspectors

In most cases inspectors are officers from the local Fire and Rescue Service. An inspector has the right to:

> Enter into your premises or workplace if he is inspecting it but cannot force an entry [27(1)(a)]
> Make enquiries to establish the limits of the premises and who the responsible persons are [27(1)(b)]
> Inspect or take copies of any records [27(1)(c)]
> Take samples [27(1)(e)].

15.27.13   Article 29–31 – Notices and penalties

An 'alterations notice' may be served by an enforcing authority when in their opinion any changes may increase the level of risk significantly (high risk premises) [29(1)].

An 'enforcement notice' may be served when the RRFSO has not been complied with, such as providing adequate fire fighting equipment or emergency lighting [30(1)].

A 'prohibition notice' may be served when the enforcing authority is of the opinion that the premises pose a serious threat to life [31(1)].

Failing to comply with Articles 8 through 21 and 38 of the RRFSO can result in a fine or a term of imprisonment not exceeding 2 years. Articles 8 through 21 cover all the main provisions referred to above.

15.27.14   Schedule 1 Part 3 – Documents and records

Policy

A responsible person must:

> Develop a coherent overall policy minimising risk [Schedule 1 Part 3(f)]
> Reduce the risk of the outbreak of fire [4.(1)(a)]
> Reduce the risk of the spread of fire [4.(1)(a)]
> Provide means of escape [4.(1)(b)]
> Demonstrate preventive action [10 and Schedule 1 Part 3].

Procedures

A responsible person must establish a procedure for dealing with a fire including regular fire response exercises and drills. The procedure should identify the circumstances that trigger the emergency. Who and how the evacuation should take place and when people should be allowed to return to the workplace [15(2)(a), (b) and (c)].

In addition procedures should be established to:

- Ensure that the means of escape are available at all times [4.(1)(c)]
- Provide adequate numbers of fire extinguishers [4.(1)(d)]
- Give appropriate instruction to employees [Schedule 1 Part 3(h)]
- Ensure people from outside organisations are properly controlled and informed [20], e.g. permits to work, signing in and out, induction training.

Fire risk assessment

A responsible person must carry out a risk assessment [9.(1)]. The risk assessment must be recorded if:

i.   You employ five or more people [9(6)(a)]
ii. There is a licence in force [9(6)(b)]
iii. An alterations notice requires you to do so [9(6)(c)]
iv. The risk assessment must record the significant findings and the measures already in place or that are required [(7)].

You must review the risk assessment regularly and where a change may affect your fire safety arrangements, such as a change of work activity, a change in its use or an alteration to the building. The risk assessment must be amended accordingly [9(3)] and kept up to date.

Fire safety arrangements

Employers must have a system for managing fire safety [11], which includes measures for **planning, organisation, control and monitoring** of your fire preventive and protective measures. This must be recorded if:

i.   You employ five or more people [11.(2)(a)]
ii. There is a licence in force [11.(2)(b)]
iii. You are subject to an alterations notice [11.(2) (c)].

Records

Records should include reviews of the fire risk assessment [9(3)], fire safety policy, procedures or arrangements [11.(1)(c)], training records, drills, certificates for the installation and maintenance of any fire safety systems or equipment. The inspector can demand to see the records.

15.28   The Reporting of Injuries, Diseases and Dangerous Occurrences Regulations 1995 (SI 3163)

The Reporting of Injuries, Diseases and Dangerous Occurrences Regulations 1995 (RIDDOR) require that specified injuries and dangerous occurrences resulting from accidents arising out of or in connection with work and specified occupational diseases are reported to the enforcing authority.

15.28.1   Interpretation – regulation 2

The following terms are defined in RIDDOR:

- Accident – includes an act of non-consensual violence done to a person at work and an act of suicide which occurs on or in the course of the operation of a relevant transport system (a railway, tramway, trolley vehicle system or guided transport system)
- Dangerous occurrence – a specified class of occurrence arising out of or in connection with work including:
  - Electrical short circuit or overload attended by fire or explosion which results in the stoppage of the plant involved for more than 24 hours or which has the potential to cause the death of any person
  - Unintentional explosion or ignition of explosives
  - The unintentional ignition of anything in a pipeline
  - Any incident involving a road tanker or tank container or vehicle used for the carriage of dangerous goods in which there is a fire which involves the dangerous goods being carried
  - Any explosion or fire occurring in any large vehicle or mobile plant which results in the stoppage of the vehicle or plant for more than 24 hours and which affects any place where people normally work or the egress from such a place
- Disease – includes a medical condition
- Major injury – means a specified injury or condition including:
  - Any fracture other than to fingers, thumbs or toes
  - Any amputation
  - A chemical or hot metal burn to the eye
  - Any injury resulting from an electric shock or electrical burn leading to unconsciousness or requiring resuscitation or hospital admittance for more than 24 hours
- Responsible person – includes the employer and the person having control of the premises at which the reportable event happened.

15.28.2   Notification and reporting of injuries and dangerous occurrences – regulation 3

RIDDOR requires the responsible person to notify the relevant enforcing authority by the quickest practicable means of any accident arising out of or in connection with work which results in:

- The death of any person at work
- Any person at work suffering a major injury

- Any person not at work suffering an injury and being taken to hospital for treatment
- Any person not at work suffering a major injury at a hospital
- A dangerous occurrence.

Within 10 days of the date of the accident, the responsible person must follow up the notification with a report on the approved form (F2508).

Where a person at work is incapacitated for work for more than three consecutive days because of an injury resulting from an accident arising out of or in connection with work, the responsible person must send a report on the approved form (F2508) within 10 days.

15.28.3 Reporting of the death of an employee – regulation 4

Where an employee suffers a reportable injury resulting from an accident arising out of or in connection with work which is a cause of death within one year of the date of the accident, the employer must inform the relevant enforcing authority in writing as soon as it comes to the employer's knowledge.

15.28.4 Reporting of cases of disease – regulation 5

Where a person at work suffers from a specified occupational disease and their work involves a specified activity the responsible person must (as soon as they receive a written diagnosis by a registered medical practitioner) report the case to the relevant enforcing authority on the approved form (F2508A).

Examples of specified diseased and associated activities include:

- Traumatic inflammation of the tendons or associated tendon sheaths of the hand or forearm associated with physically demanding work, frequent or repeated movements or constrained postures
- Hand-arm vibration syndrome associated with work involving the use of chain saws in forestry or woodworking
- Hepatitis from work involving contact with human blood or blood products
- Skin cancer from work involving exposure to mineral oil.

15.28.5 Reporting of gas incidents – regulation 6

Where a conveyor of flammable gas through a fixed pipeline or a filler, importer or supplier of LPG in a refillable container receives notification of the death or major injury arising out of or in connection with the gas, they must notify the HSE forthwith and send a written report within 14 days (F2508G1 or G2).

Where a CORGI registered employer or employee suspects that a gas fitting, flue or ventilation used in connection with the fitting is or has been likely to cause death or major injury by way of accidental leakage or inadequate combustion, they must send a written report to the HSE within 14 days (F2508G1 or G2).

15.28.6 Records – regulation 7

The responsible person must keep a record of reportable events for at least three years from the date on which it was made.

15.29 The Safety Representatives and Safety Committees Regulations 1977 (SI 0500)

The Safety Representatives and Safety Committees Regulations 1977 (SRSC Regs) prescribe:

- The cases in which recognised trade unions may appoint safety representatives
- The functions of appointed safety representatives and
- The obligation of the employer to appointed safety representatives.

15.29.1 Appointment – regulation 3

A recognised trade union may appoint safety representatives in all cases where one or more employees are employed by the employer by whom the union is recognised. When the employer is notified in writing of the names of those appointed and the group of employees they represent, the safety representative have the functions detailed under regulation 4 (see inset).

To be appointed, the person must either have been employed by the employer throughout the preceding two years or have had at least two years' experience in similar employment. (Regulation 8 extends this in relation to safety representatives appointed to represent members of Equity or the Musicians Union).

A person will cease to be a safety representative when:

- The trade union who appointed them notifies the employer that the appointment has been terminated
- They cease to be employed at the workplace
- They resign.

Regulation 4 – Functions of appointed safety representatives
To investigate potential hazards and dangerous occurrences at the workplace and to examine the causes of accidents at the workplace
To investigate complaints by any employee they represent relating to that employee's health, safety or welfare at work
To make representations to the employer on matters arising out of the above
To make representations to the employer on general matters affecting health, safety or welfare at work of the employees at the workplace
To carry out inspections in accordance with regulations 5, 6 and 7 (see below)
To represent employees they were appointed to represent in consultation at the workplace with enforcing authority inspectors
To receive information from enforcing authority inspectors
To attend meetings of safety committees in their capacity as a safety representative

Note: The above functions do not impose duties on safety representatives nor do they absolve them of their duties as employees.

15.29.2   Employer duties – regulations 4(2) and 4A

The employer is required to permit the safety representative to take time off with pay during working hours to fulfil their functions and to undergo training in their functions.

Where the employer fails to permit safety representatives time off or fails to pay them for time off, the safety representative may complain to the employment tribunal (regulation 11).

The employer is also required to provide the safety representative with facilities and assistance to enable the representative to fulfil their functions and to consult with them in good time on the following matters:

➤ The introduction of any measure which may affect the health and safety of the employees they represent
➤ The employer's arrangements for appointing or nominating competent persons under the Management of Health and Safety at Work Regulations 1999 (see later)
➤ Any information the employer is required to provide under other relevant legislation
➤ The planning and organisation of any training the employer is required to provide under other relevant legislation

➤ The health and safety consequences of the introduction of any new technologies into the workplace.

15.29.3   Inspections – regulations 5 and 6

Safety representatives are entitled to inspect the workplace every three months (or more often with the employer's agreement) if they have given the employer reasonable notice in writing of their intention to do so. Further inspections may be carried out if there has been a substantial change in work conditions or because of the introduction of new machinery or otherwise.

The employer must provide facilities and assistance to enable the safety representative to carry out inspections and the employer may be present in the workplace during the inspection.

Where there has been a reportable injury, dangerous occurrence or a case of disease has been contracted in the workplace, the safety representative may inspect the workplace, if it is safe to do so, and the employer must provide facilities and assistance to enable this including facilities for independent investigation and private discussion with the employees.

15.29.4   Information – regulation 7

If they have given reasonable notice to the employer, safety representatives are entitled to inspect and take copies of relevant documents which the employer is required to keep under relevant statutory provisions, except documents consisting of or relating to any health record of an identifiable individual.

The following documentation is excluded from this requirement:

➤ Any information, the disclosure of which would be against the interests of national security
➤ Any information which the employer could not disclose without contravening a prohibition under any enactment
➤ Any information relating specifically to an individual unless they have given their consent to its disclosure
➤ Any information, the disclosure of which could cause substantial injury to the employer's undertaking
➤ Any information obtained by the employer for the purpose of bringing, prosecuting or defending any legal proceedings.

15.29.5   Safety committees – regulation 9

Where two or more safety representatives make a written request to the employer to set up a safety committee, the employer must:

➤ Consult with the safety representatives in respect of the function of the committee

➤ Post a notice stating the composition of the committee and the workplace covered

➤ Establish a safety committee within three months of receiving the written request.

Note: See previous section covering the Health and Safety (Consultation with Employees) Regulations 1996.

15.30 The Health and Safety (Safety Signs and Safety Signals) Regulations 1996 (SI 0341)

These regulations apply to all workplaces and to all activities where people are employed but exclude signs used in connection with transport or supply of dangerous substances (see previous section covering CHIP).

The regulations require employers to use safety signs where there is a significant risk to health and safety that has not been avoided or controlled by other methods required in relevant law, provided that use of the sign can help reduce the risk.

Safety signs are not a substitute for other risk control methods such as engineering controls and safe systems of work but should be used to warn of any remaining significant risk or to instruct employees of the measures they must take in relation to these risks.

15.30.1 Provision and maintenance of safety signs

Where the employer's risk assessment indicates that, having adopted all appropriate techniques, the risks cannot be adequately reduced except by the provision of appropriate safety signs to warn or instruct of the nature of the risks and of the protective measures required, the employer must:

➤ Provide and maintain any appropriate safety sign and

➤ Ensure, so far as is reasonably practicable, that any appropriate hand signal or verbal communication is used.

15.30.2 Information, instruction and training – regulation 5

The employer must ensure that:

➤ They provide comprehensible and relevant information to each of their employees on the measures to be taken in connection with safety signs

➤ Each of their employees receives suitable and sufficient training in the meaning of safety signs and the measures to be taken in connection with the safety signs.

15.30.3 Minimum requirements for safety signs and signals

Schedule 1 to the Regulations specifies the minimum requirements covering colour, shapes and symbols to be used in safety signs as illustrated below:

Example use of safety signs

Significant risk	Safety sign	Example
Flammable materials catching fire	Prohibition sign– no smoking	
Contact with live electrical systems or equipment	Warning sign – danger electricity	**DANGER** ⚡ **ELECTRICITY**
Outbreak of fire	Acoustic signal – fire alarm	

Example safety signs

Type	Example	Meaning
Prohibition sign		No smoking
Warning sign	FLAMMABLE SOLID	Flammable material

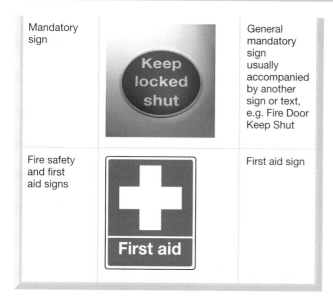

Mandatory sign		General mandatory sign usually accompanied by another sign or text, e.g. Fire Door Keep Shut
Fire safety and first aid signs		First aid sign

Acoustic signs need to be set at a considerably higher level than the ambient noise (e.g. 10 dB above) although this must not be excessive or painful. They must also be easily recognisable.

15.30.4   Fire safety signs

These are signs which:

➤ Provide information on escape routes and emergency exits in case of fire

Example fire safety signs

Example	Meaning
	Emergency escape route sign
	Fire extinguisher
	Fire point sign

➤ Provide information on the identification or location of fire fighting equipment
➤ Give warnings in case of fire.

Note: Where employers use fire safety signs that comply with BS 5499, or other safety signs that comply with BS 5378, the regulations do not require any changes to those signs.

15.31   The Supply of Machinery (Safety) Regulations 1992 (SI 3073)

These regulations require that no person should supply machinery unless:

➤ It satisfies relevant essential health and safety requirements
➤ It has been subject to the appropriate conformity assessment or a declaration of incorporation has been issued
➤ A declaration of conformity has been issued (unless a declaration of incorporation has been issued)
➤ It carries CE marking and other information (unless a declaration of incorporation has been issued)
➤ It is in fact safe.

The manufacturer of the machinery must also carry out necessary research or tests on components, fittings or the whole machine to determine whether it is capable of being erected and put into use safely.

In most cases (other than machinery listed in Schedule 4 of the regulations), the manufacturer must also draw up a technical file which comprises:

➤ An overall drawing of the machinery and the control circuits
➤ Detailed drawings (including calculation notes, test results and similar data) to check conformity with the essential health and safety requirements
➤ A list of the following which were used when the machinery was designed:
  ➤ The essential health and safety requirements
  ➤ Transposed harmonised standards
  ➤ Standards
  ➤ Other technical specifications
➤ A description of methods to eliminate hazards presented by the machinery
➤ Any technical report or certificate obtained from a competent body
➤ The technical report and results of tests carried out (where the manufacturer declares conformity with a transposed harmonised standard)
➤ A copy of the machinery instructions.

A copy of the technical file must be held for at least 10 years.

15.31.1 Essential health and safety requirements

The essential health and safety requirements are detailed in Schedule 3 to the regulations and cover a wide range of issues including:

- Materials and products used to construct machinery
- Integral lighting
- Design of machinery to facilitate handling
- Controls, control systems and control devices
- Failure of power supplies and control circuits
- Software
- Protection against mechanical hazards
- Characteristics of guards and protection devices
- Protection against other hazards
- Maintenance
- Indicators (warning devices and instructions).

Certain categories of machines (agi-foodstuffs machinery, portable machinery, etc.) have additional essential health and safety requirements that must also be satisfied.

Where machinery is manufactured in conformity with published EU standards which have also been published as identically worded national standards (transposed harmonised standards) it is presumed that the machinery will comply with the listed essential health and safety requirements.

15.31.2 Declaration of conformity and CE marking

The declaration of conformity is the procedure whereby the manufacturer declares that the item of machinery complies with the essential health and safety requirements and CE marking is used to indicate this on the machinery.

The declaration should state:

- The business name and address of the manufacturer
- A description of the machinery (including make, type and serial number)
- All relevant provisions that the machinery complies with
- The name and address of the approved body and the number of the certificate of approval where applicable
- The name and address of the approved body to which the technical file was sent or has drawn up a certificate of adequacy
- The transposed harmonised standards used

- The national standards and any technical specifications used
- The person authorised to sign the declaration.

CE marking should be fixed to machinery in a distinct, legible and indelible manner and no other marking which may be confused with the CE mark should be used on machinery.

15.32 The Workplace (Health, Safety and Welfare) Regulations 1992 (SI 3004)

These regulations define a workplace as:

> *any premises or part of premises which are not domestic premises and are made available to any person as a place of work, and includes any place within the premises to which such a person has access while at work and any room, lobby, corridor, staircase, road or other means of access to or egress from that place of work.*

They apply to all workplaces (including temporary workplaces) with the exception of docks, construction sites, mines and quarries.

The regulations place duties on both employers and persons who have control of workplaces (with the exception of self-employed persons in respect of their own work), to comply with the requirements and to ensure that the workplace and any equipment, devices and systems to which the regulations apply are maintained in an efficient state, in efficient working order and in good repair.

The requirements are summarised below under the broad headings of health, safety and welfare.

Note: These headings are not used within the regulations although they do provide a useful way of summarising the principal requirements.

15.32.1 Health requirements – regulations 6 to 11

- Enclosed workplaces should have effective and suitable provision for ventilation by a sufficient quantity of fresh or purified air, including a visible or audible warning device in the event of failure of the system where appropriate for health or safety reasons
- The temperature in all workplaces during working hours should be reasonable (at least 16°C), and heating or cooling methods which result in the escape of injurious or offensive fumes, gas or vapour should not be used

- Thermometers should be provided to enable persons at work to determine the temperature inside a workplace and excessive effects of sunlight on temperature should be avoided
- Workplaces should have suitable and sufficient lighting (by natural means where reasonably practicable), including emergency lighting where failure of artificial lighting will expose people at work to danger
- Workplaces, furniture, furnishings and fittings should be kept sufficiently clean and all surfaces of floors, walls and ceilings inside buildings should be capable of being kept sufficiently clean
- Waste materials should not be allowed to accumulate in a workplace except in suitable receptacles
- Rooms where people work should be of sufficient dimensions for health, safety and welfare purposes
- Workstations should be arranged so that they are suitable for both the person at work and the work activity that will take place and should:
  - Provide protection from adverse weather
  - Enable any person to leave swiftly in the event of an emergency
  - Ensure that the person is not likely to slip or fall
- A suitable seat (and footrest as necessary) should be provided where the work is of a kind that can or must be done seated.

15.32.2 Safety requirements – regulations 12 to 19

- Floors and traffic routes should be:
  - Suitably constructed (free from holes, slopes and uneven or slippery surfaces posing a risk to health or safety and have effective drainage where necessary)
  - Kept free from obstructions and an article or substance which may cause a person to slip, trip or fall
- Traffic routes which are staircases should be provided with handrails and guards where necessary
- Where there is a risk of a person falling into a dangerous substance in a tank, pit or structure, these must be securely fenced or covered
- Windows or other transparent or translucent surfaces in a wall, partition, door or gate should be made of safety material or protected against breakage and be appropriately marked to make it apparent
- Windows, skylights or ventilators which can be opened must be capable of being opened, closed or adjusted without exposing people to a risk to their health or safety
- When open, windows, skylights or ventilators must not expose any person to risks to their health or safety

- Windows and skylights must be designed or constructed so that they may be cleaned safely
- Workplaces should be organised in such a way that pedestrians and vehicles can circulate safely
- Traffic routes should be suitably indicated and must be of sufficient number, size and position and suitable to ensure that:
  - Pedestrians or vehicles can use it without causing danger to persons at work near it
  - There is sufficient separation of any vehicle traffic route from doors, gates or pedestrian routes which lead onto it
  - Where pedestrians and vehicles use the same route, there is sufficient separation between them
- Doors and gates should be of suitable construction (including being fitted with any necessary safety devices) and:
  - Sliding doors or gates must have a device to prevent them coming off their track during use
  - Any upward opening door or gate has a device to prevent it falling back
  - Powered doors or gates must have suitable and effective measures to prevent injury by trapping people
  - Where necessary, powered doors or gates should open manually unless they open automatically if the power fails
  - Doors or gates which can be opened by either being pushed or pulled must have a clear view of both sides when closed
- Escalators and moving walkways must function safely, have any necessary safety devices and be fitted with one or more emergency stop controls.

15.32.3 Safety requirements – regulations 20 to 25

- Suitable and sufficient sanitary conveniences should be provided which are:
  - Adequately ventilated and lit
  - Kept in a clean and orderly condition
  - Separate for men and women (except where they are in a separate room and can be secured from the inside)
- Washing facilities should be provided (including showers where required), which are:
  - In the immediate vicinity of every sanitary convenience
  - In the vicinity of changing rooms where required
  - Available with a supply of clean hot and cold (or warm) water, soap and means of drying
  - Sufficiently lit and ventilated
  - Kept in a clean and orderly condition

➤ Separate for men and women (except where they are in a separate room, can be secured from the inside and are intended to be used by one person at a time)

➤ An adequate supply of wholesome drinking water should be provided for all persons in the workplace and this should:
  ➤ Be readily accessible
  ➤ Be conspicuously marked
  ➤ Include a suitable sufficient number of cups or drinking vessels unless the supply is a drinking jet

➤ Suitable and sufficient accommodation should be provided for persons to store both clothing not worn during work and special clothing required for work

➤ The accommodation should be in a suitable location, easily accessible, of sufficient capacity and include:
  ➤ Facilities to change clothing, including separate facilities where necessary for propriety
  ➤ Suitable security for personal clothing not worn at work
  ➤ Separate accommodation for work and non-work clothing where necessary
  ➤ Facilities to dry clothing
  ➤ Seating

➤ Suitable and sufficient rest facilities should be provided, including facilities to eat meals where food would otherwise become contaminated

➤ Rest rooms should include:
  ➤ Suitable arrangements to protect non-smokers from discomfort caused by tobacco smoke
  ➤ Be equipped with an adequate number of tables and seating (with backs)
  ➤ Suitable facilities for pregnant women or nursing mothers to rest
  ➤ Suitable facilities to eat meals where these are regularly eaten in the workplace.

Abbreviations

AFFF	Aqueous Film Forming Foam
APAU	Accident Prevention Advisory Unit
ARC	Alarm Receiving Centre
ASET	Available Safe Escape Time
BCF	Bromochlorodifluoromethane
BS	British Standard
BSI	British Standards Institution
BTM	Bromotrifluoromethane
CCTV	Closed Circuit Television
CDM	The Construction (Design and Management) Regulations 2007
CFOA	Chief Fire Officers Association
CHIP	Chemicals (Hazards Information and Packaging for Supply) Regulations 2002
CO_2	Carbon Dioxide
COMAH	Control of Major Accident Hazards Regulations 1999
CORGI	Council Registered Gas Installers
COSHH	Control of Substances Hazardous to Health Regulations 2004
DCLG	Department of Communities and Local Government
DDA	Disability Discrimination Act 1995
DSEAR	Dangerous Substances and Explosives Atmospheres Regulations 2002
EMAS	Employment Medical Advisory Service
EU	European Union
EVC	Emergency Voice Communications
FFS	Fire Fighting Systems
FIC	Fire Incident Controllers
FMEA	Failure Modes and Effects Analysis
FMT	Facilities Management Team
FPA	Fire Protection Association
FRA	Fire Risk Assessment
FRRS	Fire Resisting Roller Shutters
FRS	Fire and Rescue Services
FTA	Fault Tree Analysis

HAZOP	Hazard and Operability Study
HSC	Health and Safety Commission
HP	Hydraulic Platform
HSCER	The Health and Safety (Consultation with Employees) Regulations 1996
HSE	Health and Safety Executive
HSG65	HSE's Successful Health and Safety Management
HSWA	Health and Safety at Work etc. Act 1974
HTM65	Health Technical Memorandum 65
IBCs	Intermediate Bulk Containers
IFE	Institute of Fire Engineers
ILO-OSH	International Labour Organisation – Occupational Safety and Health
IOSH	The Institution of Occupational Safety and Health
IPPC	Integrated Pollution Prevention and Control
IRP	Incident Response Plan
JSA	Job Safety Analysis
KMFRA	Kent and Medway Fire and Rescue Authority
LEL	Lower Explosion Limit
LEV	Local Exhaust Ventilation
LFL	Lower Flammable Limit
LHDC	Linear Heat Detecting Cable
LPG	Liquefied Petroleum Gas
Lx	Lux
MCCBs	Moulded Case Circuit Breakers
MHSW	Management of Health and Safety at Work Regulations 1999
MoE	Means of Escape
MSDS	Materials Safety Data Sheets
MW/m^2	Megawatts per square metre
O&M	Operations and Maintenance
ODPM	Office of the Deputy Prime Minister
OHSAS	Occupational Health and Safety Assessment Series

OPSI	The Offices of Public Sector Information
PAT	Portable Appliance Testing
PCBs	Polychlorinated Biphenyls
PEEP	Personal Emergency Evacuation Plan
PFOS	Perfluoroctane Sulfonate
PFPF	Passive Fire Protection Federation
PIR	Passive Infrared
PPE	Personal Protective Equipment
PPM	Planned Preventive Maintenance
PTSD	Post Traumatic Stress Disorder
PTW	Permit To Work
PUWER	Provision and Use of Work Equipment Regulations 1998
RCDs	Residual Current Devices
RCS	Risk Control Systems
RF	Radio Frequency
RIDDOR	Reporting of Injuries Diseases and Dangerous Occurrences Regulations 1995
ROES	Representatives of Employee Safety
RORO	Roll on Roll off
RRFSO	Regulatory Reform (Fire Safety) Order 2005
RSET	Required Safe Escape Time
SIESO	Sharing Information and Experience for Safer Operations
SME	Small to Medium Enterprises
SO$_2$	Sulphur Dioxide
SRSC	Safety Representatives and Safety Committees Regulations 1977
SSCPR	Social Security (Claims and Payments) Regulations
SSOW	Safe Systems of Work
TL	Turntable Ladder
UEL	Upper Explosion Limit
UFL	Upper Flammable Limit
UK	United Kingdom
WEL	Workplace Exposure Limited

Index